LITHOGRAPHED PLATES.

SMOOTHBORE CAST IRON ORDNANCE *faces Table III.*

AT END OF VOLUME.

RIFLED ORDNANCE.

7-prs.	Plate I.
9-prs.	II.
16 and 25-prs.	III.
40-prs	IV.
64-prs	V.
Converted Guns	VI.
6·3-inch Howitzer	VII.
8-inch Howitzer	VIII.
7-inch 7 tons	IX.
do. 6½ "	X.
7-inch { 90 cwt. / 7 tons, Mark IV. }	XI.
8-inch 9 tons	XII.
9-inch 12 "	XIII.
10-inch 18 "	XIV.
11-inch 25 "	XV.
12-inch 25 "	XVI.
12-inch 35 "	XVII.
12·5-inch 38 "	XVIII.
16-inch 80 "	XIX.
17·72-inch 100 tons.	XX.

WOODCUTS.

(These accompany the respective descriptions of Stores, &c., &c., and can thus be readily ascertained by the Index.)

EXPLANATION OF ABBREVIATIONS USED IN THIS WORK.

R.M.L.	Rifled Muzzle-Loading.
R.B.L.	Rifled Breech-Loading.
M.L.	Muzzle-Loading.
B.L.	Breech-Loading.
S.B.	Smooth-Bore.
W.D.	War Department.
W.O.	War Office.
D. of A. and S. ..	Director of Artillery and Stores.
L.S.	Land Service.
S.S.	Sea Service.
R.L.G.	Rifle Large Grain (powder).
R.L.G.^{2}	Rifle Large Grain (3 to 6 in mesh).
P.	Pebble (powder).
P.^{2}	Cubical (powder of $1\frac{1}{2}''$ size).
R.G.F.	Royal Gun Factories.
R.C.D.	Royal Carriage Department.
R.L.	Royal Laboratory.
R.A.I.	Royal Artillery Institution.
O.S.C.	Ordnance Select Committee.
E.O.C.	Elswick Ordnance Company.
Expl.	Experimental.
§	This refers to the paragraph in List of Changes of War Stores.
Extracts	From the Proceedings of the Department of Director of Artillery.
M.V.	Muzzle Velocity.
f.s.	Feet-seconds (as to velocity).
f.t.	Foot-tons (as to energy).

TABLES.

			Page
Table	I.	Elastic limit and tenacity of metals used in R.G.F.	18
"	II.	Position of Vents in R.M.L. guns.	49
"	III.	Cast iron and bronze S.B. ordnance to be retained	65
"	IV.	Cast iron and bronze S.B. ordnance to be abolished	66
"	V.	Comparison between S.B. and R.B.L. service field guns, 1860	90
"	VI.	Comparison between English and Foreign field guns, 1860–61	91
"	VII.	Comparison between English and Foreign field guns, 1872	92
"	VIII.	Comparison between English and Foreign Field guns, 1878	93
"	IX.	Comparison between S.B. cast iron, and R.M.L. converted guns	94
"	X.	Service and experimental heavy guns of Foreign Powers, 1878	95
"	XI.	Power of steam hammer used in R.G.F. ..	99
"	XII.	Dimensions of R.G.F. cylinder gauges and R.L shot gauges	121
"	XIII.	Service tangent sights, R.B.L. guns.. ..	143
"	XIV.	Number in service of ditto .. .	143
"	XV.	Weights of vent-pieces and breech-screws ..	144
"	XVI.	Fittings and stores, R.B.L. guns	145
"	XVII.	Tangent sights, R.B.L.	146
"	XVIII.	Dimensions, &c. &c., R.B.L. guns. .. .	160
"	XIX.	Service tangent sights, R.M.L. guns.. ..	212
"	XX.	Tangent sights, heavy R.M.L.	214
"	XXI.	" medium "	217
"	XXII.	" light "	219
"	XXIII.	Centre hind sights, R.M.L.	225
"	XXIV.	Wood scales, R.M.L.	227
"	XXV.	Small stores and fittings, R.M.L	231
"	XXVa.	Sights and fittings required for guns Mounted in different methods	231a

			Page
TABLE	XXVI.	Construction of built up guns	245
"	XXVII.	Dimensions, &c. R.M.L. guns	291
"	XXVIII.	Tools for cleaning and examining cast iron and bronze S.B. ordnance	317
"	XXIX.	Tools for cleaning and examining rifled ordnance	318
"	XXX.	Sighting tools for S.B. ordnance	320
"	XXXI.	Facing implements, R.B.L. guns	321
"	XXXII.	Venting tools for S.B. and R. pieces	322
"	XXXIII.	Venting tools, distribution of, (list B.) ..	326
"	XXXIV.	Velocities of projectiles, heavy service guns ..	354
"	XXXV.	Energy " " "	356
"	XXXVI.	Comparison of results with varying charges of powder and weights of projectiles, from an experimental gun	358
"	XXXVII.	Pressures in terms of density	359
"	XXXVIIA.	Densities and Volumes of charges referred to cubic inches of space per. lb. of powder ..	359
"	XXXVIII.	Work of fired gunpowder	361
"	XXXIX.	" per lb. of powder realised, service guns ..	362
"	XL.	" Values of $\left(\frac{d^2}{w}s\right)$ for ogival-headed shot	366
"	XLI.	" $\left(\frac{d^2}{w}t\right)$ " "	368
"	XLII.	Service Ordnance	428
		Ranges for Rifled pieces .. Appendix V.	429

CONTENTS.

	Page
PREFACE	ix
CHAPTER I.—Materials used in the Construction of Ordnance	1
CHAPTER II.—Theory of Construction, Venting, Rifling, and Sighting, generally	19
CHAPTER III.—Smoothbore Ordnance and Stores	56
CHAPTER IV.—Short History of our Rifled Ordnance from their introduction up to the present date	67
CHAPTER V.—Manufacturing Operations in Royal Gun Factories	
CHAPTER VI.—R.B.L. Ordnance and Stores	122
CHAPTER VII.—Different natures of R.B.L. guns in the Service	147
CHAPTER VIII.—Manufacture of R.M.L. (built up) Woolwich Guns...	161
CHAPTER IX.—Sights, Fittings, and Stores for R.M.L. (built up) Guns	183
CHAPTER X.—Manufacture of R.M.L. Converted Guns, Their Sights and Stores	233
CHAPTER XI.—Recapitulation of all Natures of R.M.L. Guns in the Service.	243
CHAPTER XII.—Examination, Preservation, and Repairs of Ordnance and Stores	293
CHAPTER XIII.—Power of Guns, how ascertained, compared, &c.	332

	Page
APPENDIX:—	
I. Rules for furnishing Annual Returns	371
II. Blueing, Browning, and Bronzing, Deepening centre hind sight sockets. Preparing guns for derricks, &c.	377
III. Gatling Guns	381
IV. Mode of obtaining Range Tables and Trajectories List of Service Ordnance	385
V. Tables of Ranges for Service Guns..	386
INDEX	429

PREFACE.

In the preparation of the following work I have very largely profited by the valuable labours of my predecessors, Major Stoney and Captain Jones, Royal Artillery, whose "Text Book of Rifled Ordnance" constitutes the backbone as it were of this Treatise, some chapters being taken in great part from the Text Book.

I have, however, found it necessary to introduce a great deal of new matter, to leave out much information that has now lost its interest, and to make an entirely new arrangement of the contents.

The question of the materials used for manufacture and the theoretical considerations of construction are rather more fully entered into,* while a separate chapter has been devoted to the question of sights and fittings for R.M.L. guns, which have so increased of late years in number, that further explanation was required of details which necessarily become more complicated as our ordnance increase in size and accuracy of shooting.

The various natures of sights used in the service are given in a tabular form, which I hope will be found of much convenience in Royal Artillery District Offices, by Firemasters and Inspectors of Warlike Stores, and generally by all officers having charge of ordnance.

Tables showing the comparative value of field guns, &c. at various typical epochs are also added; these may serve as landmarks to show how great our progress has been of late years, and how we are still going forward in search of perfection.

All the necessary information regarding S.B. ordnance† has also

---

* A mere outline of theory, however, is given. Officers wishing to study that part of the subject of gun construction may well refer to such works as Mallett's "Construction of Artillery," Holley's "Ordnance, and Armour," the works of Major (now Col.) Sir W. Palliser, and others of more recent date.

† Epitomized from Notes on S.B. Ordnance, 1874, which were prepared by me from former Notes by Captain Molony (late R.A.) and Major Stoney, R.A.

been embodied in this Treatise, so that an artillery officer need only have a single book to refer to for information regarding any gun whatever with which he may have to do.

Lastly, in order to keep pace in some degree with that increase in knowledge, and interest in the study, of artillery which each year developes itself still more, a chapter has been added at the end of the Treatise which will enable any officer to work out for himself ordinary questions regarding the power of our own and foreign ordnance. In this chapter advantage has been taken of the valuable additions to artillery science rendered by Captain A. Noble, F.R.S., (late R.A.,) Professor Abel, F.R.S., and Professor Bashforth, B.D., from whose works the tables given at pp. 366–9, have been taken. With regard to the formulæ and mathematical terminology here and elsewhere in the book, I am much indebted to my friend Mr. Greenhill, Professor of Mathematics to the Advanced Class of artillery officers, who has kindly corrected the same where necessary.

I cannot close this short preface without expressing the obligation I am under to my assistant instructors, Sergeant Major White, R.A., and Quarter Master Sergeant Ashworth, R.A., in the preparation of this Treatise for the press; to Sergeant Major White I am in particular indebted for the ability with which he has completed the tables of sights, &c., which tables I am confident will be found a very useful addition to the book.

JOHN F. OWEN, Captain R.A.,
Captain Instructor, Royal Gun Factories.

Instruction Rooms, R.G.F.,
April, 1877.

A reprint of this Treatise being necessary we have brought the work up to date as far as possible. So rapid has been the progress of new and experimental pieces, and so varied and successful the attempts at increasing the power of existing pieces during the short period which has elapsed since this Treatise was commenced, in 1876, that it has been necessary to introduce much new matter. Owing to the same causes, however, it has been possible to leave out, or reduce in bulk a great deal which appeared in the first edition both in the form of letterpress and tables.

We trust therefore that instead of increasing the size and cost of the book, we have succeeded in somewhat reducing both, without decreasing its utility as a Treatise for Instruction and a work of reference for Artillery and other Officers.

JOHN F. OWEN, Major R.A.,
Assistant Superintendent, Royal Gun Factories.

MORTON PORTER, Captain R.A.,
Captain Instructor Royal Gun Factories.

Royal Gun Factories, October, 1878.

CHAPTER I.

METALS USED IN GUN MANUFACTURE.

Late advances in metallurgy.—**Physical properties of metals generally.**—Malleability.—Brittleness.—Ductility.—Softness.—Soft steel.—Hard steel.—Toughness.—Tenacity.—Tensile strength.—Elasticity.—Elastic limit.—Modes of measuring tenacity and elasticity.—**Metals used for construction of ordnance.**—**Bronze.**—Its advantages and disadvantages.—Tin spots.—Bronze no longer used for gun construction in our service, but still employed by some foreign nations.—Attempts to improve bronze.—Phosphor bronze.—Italian experiments.—Russian experiments.—Uchatius bronze steel.—Austrian experiments.—Iron.—Different forms of cast iron.—Wrought iron and steel.—Impurities in cast iron.—Distinctive qualities.—Advantages and disadvantages.—Suitable for S.B. guns.—Its employment by foreign nations.—Wrought iron.—How obtained.—Fibrous structure.—Impurities.—Physical properties.—Welding.—Malleability.—Ductility.—Tenacity.—Elastic limit.—Summary of properties.—Why used for the exterior of our guns.—Test applied to wrought-iron for guns.—Steel.—Definition.—Chemical and structural difference between steel and wrought iron.—Puddled or cement steel.—Cast steel.—Crucible steel.—Siemens steel.—Siemens-Martin steel.—Bessemer steel.—Flaws in wrought iron.—Steel free from flaws.—Uncertainty of steel.—Remedies tried to overcome it.—Tests applied to steel for gun purposes.—Summary as to the gun metals.

Necessity for understanding the physical properties of metals.

PRIOR to discussing the theoretical and practical considerations which govern the construction of ordnance, and before describing the processes of actual manufacture, it is necessary to know what metals are employed in such manufacture, and what are the properties they possess which make them suitable for the purposes required. With regard to these properties, we must have a clear and definite knowledge of their comparative values, of the methods employed to measure the limits of these values, and of the manner in which they may best be utilised for gun construction.

Rapid strides in metallurgy in recent years.

Metallurgy has of late made such wonderful advances, that in no branch of scientific knowledge has progress been more rapid than in the art of obtaining metals from their ores and of working them subsequently. It is, however, the physical treatment of metals after separation from these ores which more especially concerns us, and in this we find that the greatest advances have been made. It would be

CHAP. I.

out of place here to enter upon this interesting subject in detail. Officers wishing to obtain full information concerning it may study with advantage the works mentioned below,* as well as many other valuable treatises recently published.

Here, however, we need only discuss briefly the qualities of the metals which are employed for the manufacture of guns, and allude in a few words to the methods of production by which such qualities are largely modified.

PHYSICAL PROPERTIES OF METALS.

Qualities of the metals employed.

The qualities with which we are more particularly concerned are the physical properties of malleability, ductility, hardness or softness, toughness, elasticity, and tensile strength, while we must also understand what is meant by tenacity and elastic limit as applied to metals.

Malleability. Brittleness.

Malleability is the property of being permanently extended in all directions without rupture by pressure (as in rolling) or by impact (as in hammering). It is opposed to *brittleness*, which is the tendency to break more or less readily under compression either gradual or sudden.

Ductility.

Ductility is the property of permanently extending, or drawing out, by traction, as in wire drawing.

Softness.

A metal is said to be *soft* when it yields easily to compression without breaking, and does not return to its original form on the removal of the compressing stress.

Only comparative.

These terms are, of course, only comparative; thus we have hard leads and soft leads, while any sort of lead whatever is soft as compared with wrought iron, which latter again is called soft when we compare it with cast iron.

Soft or low steel. Hard or high steel.

Steel is called *soft* or *low* when the proportion of carbon contained in it is small, and *hard* or *high* when the contrary is the case, because when treated in a similar manner one variety is much harder than the other. It should, however, be remembered that a tolerably soft steel may be made very hard by tempering.

Toughness. Dr. Young's definition.

It is easy to understand what *toughness* means, but not so easy to define exactly what it is. Dr. Young (Nat. Phil., i. 142) gives the following explanation of the term as applied to steel. "Steel, whether perfectly hard or of the softest temper, resists flexure with equal force when the deviations from the natural state are small, but at a certain point the steel, if soft, begins to undergo an alteration of form, at another point it breaks if much hardened, but when the hardness is moderate it is capable of a much greater curvature without permanent alteration or fracture, and this quality, which is valuable for the purposes of springs" (and also for gun barrels), "is called toughness, and is opposed to rigidity and brittleness on the one side, and to ductility on the other."

Elasticity.

Elasticity is the property possessed by a metal of resisting permanent deformation when subjected to a stress, and is measured by the ratio of stress to strain, so that the modulus of elasticity is equal to the cotangent of the angle HOJ (in fig p. 8).

Elastic limit.

The "*elastic limit*"† of a metal is the tension which causes permanent elongation, and in the fig. is represented by the ordinate OJ.

---

* Metallurgy, by John Percy, M.D., F.R.S. Metals, their Properties and Treatment, by Professor Bloxam.

† Often termed "limit of elasticity," and very frequently confounded with "elasticity" itself.

Tenacity.

*Tenacity** is the tension required to produce rupture, and is represented above by the ordinate OD.

Tensile strength.

Tensile strength we shall employ to denote the work done upon the metal to produce rupture by traction. It would be measured in the fig. by the area AOB.

Further explanation of terms used.

In order to understand these several terms more clearly, let us take the fig. below, in which the ordinates represent the tensions and the abscissa the extensions of a bar of metal (experimentally determined) corresponding to the tensions.†

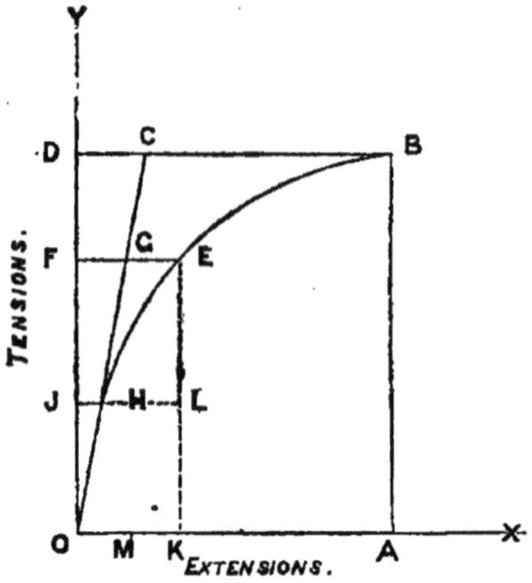

Effects of increasing stress on a bar.

If the bar be subject to a constantly increasing tension the extension is at first in a constant ratio to the tension,‡ increasing after a certain point in a varying ratio. This point, represented in the diagram by the extension HJ and measured by the tension represented by the ordinate OJ, is termed the elastic limit.§

Measure of the limits of fracture.

Total, elastic, and permanent extension.

After this point is reached the extensions increase in a higher ratio for every increment of tension, and the line joining the ordinates becomes a curved line, as shown by HB in the figure. As we continue to increase the tension we arrive at a point when the bar will fracture. Suppose the total extension of the bar at that point to be represented by BD, and the breaking tension by the ordinate AB, which is the measure of the tenacity or limit of fracture, we have then, as will be seen by the figure, three extensions of the bar, the total, elastic, and permanent, the former being in all cases the sum of the two latter; while, until the elastic limit is reached, the total extension is synonymous with the elastic extension.

---

* Or "limit of fracture," or "breaking tension." This term is sometimes employed to express tensile strength.

† It should be remembered that work done = force × distance through which the force acts, and also that a strain is the effect of a stress.

In the figures given the elasticity of the metal $= \frac{OJ}{HJ} =$ Cot. angle HOJ.

‡ This extension increases very slightly in proportion of the duration of the tension.

§ This, however, according to Mallet, p. 57, will not always be the case if we reimpose the weight for any time. With wrought iron, for instance, should the tension exceed one-fourth the tenacity, the extension will slowly increase with time.

CHAP. I.

The abscissa of the curve (a straight line as far as H) OB represent the total extensions, and the abscissa of the straight line OC the elastic extensions of the bar, while the work required to produce rupture is measured by the area AOB, which thus measures the tensile strength.

Similarly the work necessary to produce a total extension EF is measured by the area KOE.

If we remove the tension represented by OF, after the bar has been extended by EF, and then re-impose it, the greatest extension of the bar caused by this re-imposition of the tension will not exceed FG;* because it has been permanently extended by the amount GE, and if we once more remove the tension the bar will revert to its former length, *i.e.*, its original length + the permanent extension EG:

Work done upon a bar of metal.

Here, of the total work done on the bar, represented by the area KOE, that portion corresponding to difference of areas EHOK, and triangle OGM has been absorbed by it and applied to the re-arrangement of its molecules, and so to the permanent extension of the bar by the amount EG, being the measure of the loss sustained in the tensile strength of the bar. Its "tenacity" † may, however, be increased, and we see that its elastic limit is so, for any ductile metal increases (within certain limits) in elastic limit and ultimate strength (as represented by the tenacity), though not in absolute or tensile strength (as shown by total work required to produce rupture) when subjected to drawing, hammering, or rolling. In fact, a material strained beyond its elastic limit will exhibit the same characteristics as an originally harder metal.

To recapitulate, then, we must remember that increase in the tenacity (or breaking tension) and limit of elasticity do not necessarily imply greater working strength in a given bar of metal.‡

We do, however, gain very much, as we all know, by subjecting metals to the operations of rolling, hammering, &c., for we obtain a higher limit of elasticity and tenacity in smaller bulk by making the mass more homogeneous.

How tensile strength and tenacity are measured.

It will be seen that the tensile strength of a metal is by no means the same as the tenacity, which latter is often erroneously termed tensile strength, and which is measured here by the weight in tons that a bar of a square inch in sectional area will just support without breaking. The former is proportional to an area and the latter to a straight line in the figure above.

In order to fracture steel of great tenacity, less work may in fact be done than is required to fracture a similar bar of soft wrought iron.

Measure of elasticity.

Again, the elasticity of the iron may equal that of the steel, but the limit of elasticity might be very different in the two cases. The elasticity is measured by the cotangent of the angle COD in the figure, while the elastic limit is represented by the tension measured by the line OJ.

METALS USED FOR CONSTRUCTION OF ORDNANCE.

Metals employed in constructing ordnance.

The metals employed in constructing ordnance are bronze, cast or wrought iron, and steel.§

---

* This extension increases very slightly in proportion of the duration of the tension.

† By some termed the "limit of cohesion."

‡ The total work necessary to break a bar, depends principally upon the expansion at the limit of fracture (vide Fig. p. 3), and the tougher the gun metal the safer the gun.

§ As explained further on, the three latter are really only alloys or mixtures of the metal iron with other substances.

Bronze.

Bronze is a mixture or alloy of copper and tin. That particular sort of bronze formerly used for our guns is often called gun metal, and consists of about 90 parts of copper and 10 of tin. **Bronze. Its composition.**

Bronze is a tough and tenacious metal, but when cast or founded in the ordinary way it is comparatively soft and is easily indented and damaged by the projectile. When heated, as for instance by rapid firing, this metal becomes still softer and so more readily damaged. **Advantages. Disadvantages.**

For the small S.B. guns formerly used in the field, bronze answered tolerably well, as the weight of shot was comparatively small; but with rifled guns using much heavier charges bronze is not found to be a sufficiently good material. **Suitable for S.B. guns. Unsuited for rifle guns.**

Besides the faults above mentioned, which are inherent to bronze cast in the ordinary way, even when the casting is sound this alloy has the serious defect of never being quite homogeneous.* The tin has a much lower melting point than copper (442° F. as compared to 1,800° F.), while its specific gravity is also very different (and although definite alloys can be found, they are not represented by the above proportion nor such as would answer for gun metals). **Tin spots and their origin.**

While cooling, the two metals forming the alloy seem to separate more or less from one another, the tin liquating or sweating out in parts and causing white spots or blotches called "tin spots,"† which are readily acted upon by the powder gas and eaten away, leaving flaws or holes in the bore of the gun. In rifled guns this defect is much more serious than in S.B. pieces, for the grooves cut in the bore lay open a further surface and expose more tin spots, while the powder gas acts with greater force (on account of the larger charges and heavier projectiles used), and eats away the spots more quickly.

With S.B. bronze guns much inconvenience was occasioned by the softness of this alloy, especially when heated by rapid firing,‡ and as experience was gained concerning rifled ordnance in the field, it was found that the defects inherent to ordinary bronze were, as mentioned above, still more serious in such pieces. Their accuracy was affected by much firing, and the greater pressure in the powder chamber quickly developed flaws by burning out the tin spots. The cutting of the grooves also laid bare many of these spots, which otherwise would not have been apparent. **Softness of bronze.**

To reap the full advantages of rifling, it became evident that some better material than bronze would have to be employed in the con- **Bronze no longer used by us as a gun material.**

---

* These defects existed very largely in the S.B. bronze guns, for guns cast at various dates as far back as 1790 have been cut open, and tin spots are found to a greater or less degree in all of them.

† "Tin spots," so called, being patches or veins of white alloy, rich in tin, always found in ordinary bronze. Such patches are burnt away with a comparatively small amount of heat; they are harder than bronze.

‡ In his "Employment of Artillery," Sir John May tells us that in the first siege of Badajoz by our forces in 1811, we had to borrow from the Portuguese a siege train of bronze ordnance. The rapid firing of these pieces soon disabled 18 out of 40 guns, and the siege was consequently raised. Shortly afterwards, in 1812, we again attacked that fortress, but with an English siege train of iron ordnance, when it soon fell into our hands.

CHAP. I.

struction of field guns. We therefore no longer construct guns of bronze, and the only ordnance remaining is our service of that material, besides a number of old S.B. pieces in list A. p. 65, are a few 7-pr. mountain guns which were made in 1866–70.

Attempts at converting S.B. bronze into rifled guns.

Most nations have attempted to utilise their stock of old S.B. bronze guns by melting up the material and re-making from it rifled guns.* In 1869 we ourselves made some 8 cwt. 9-pr. guns of bronze (vide p. 78), but the uncertain nature of the material was soon shown by a number of them becoming unserviceable.

Improvements attempted in bronze by mixture of foreign substances.

Various attempts have been made to discover a modification of bronze, or some analogous alloy, sufficiently hard, elastic, and strong for the purpose required, but as yet, unless the bronze steel mentioned below be a success, all these attempts have failed to satisfy completely the conditions required.

It has also been attempted to improve its quality both in hardness and homogeneity by altering the proportions of the two constituents, and by adding small portions of other metals or non-metals such as manganese, phosphorus.

Phosphor bronze.

Phosphor bronze containing small quantities of phosphorus has been extensively tried, and gives a metal of more uniform character and also stronger than bronze.†

French experiment with above.

In 1872 an exhaustive trial was undertaken at Bourges by the French Government with 4-prs.; four of these pieces were cast, two being of ordinary bronze and two of an alloy of phosphor-bronze. The superiority of the latter over the guns cast from ordinary bronze was so slight, that the committee carrying on the experiments concluded that any advantages were more than neutralized by the necessity of adding the phosphorus in very exact proportions, and so further complicating the manufacture of bronze guns.

Italian experiments as to the same.

Colonel Rosset, of the Italian Artillery, superintendent of the arsenal at Turin, has for some years past been carrying on a series of interesting trials, with regard to bronze and other metals in the arsenal at Turin, where a 7·5c gun of phosphor bronze was tested in comparison with others of ordinary bronze. This gun stood the trial well, and the alloy from which it was cast (in an iron mould) showed a tenacity of about 25 tons per square inch. Notwithstanding this, however, Colonel Rosset concluded that it was not advisable to employ such an alloy in gun manufacture on account of the unstable character of phosphorus, and the great difficulty of securing uniformity of result in the mixture of this element with bronze.

Russian experiments.

Russia, too, has been making much advance in the working of bronze, and is manufacturing some powerful experimental field guns (Lavroff guns)—shown in Table VIII—of this alloy treated in the same way as described with regard to the Austrian "Uchatius" field pieces.

Austrian experiments.

Uchatius bronze steel.

In Austria great attention has always been paid to bronze and analogous alloys, and General Von Uchatius, the director of the gun foundry at Vienna, has for years studied the subject. In a lecture lately delivered by him, he tells us that about three years ago his attention

---

* The present French field guns are of bronze, pending the introduction of steel pieces, now being rapidly manufactured; as are also the new Austrian field guns, the Italian light field guns, and a large proportion of the Russian field pieces.

† This alloy is now considerably used for bearings, sheaves of pullies, &c. It seems doubtful, however, whether this mixture, as well as the so-called manganese-bronze, &c., do not owe most of their good qualities in great part to the fact of being cast in chill rather than to their chemical constitution.

was particularly called to a fragment of bronze, cast under pressure, which the Archduke William had brought from Russia. He found the properties of this metal so far superior to those of bronze cast in the ordinary way, that he was led to researches which resulted in his casting bronze in an iron mould or chill casting, and at the same time chilling the interior of the mass by means of a core of solid copper or otherwise.

How manufactured.

This bronze, which is an ordinary alloy containing 8 p.c. of tin, can be forged cold and possesses many of the properties of steel, and has consequently been termed *bronze steel;* it possesses, however, apparently the advantage over steel of being a safer material when employed alone.

Result of experiments with bronze steel.

The results obtained with these experimental guns bore out the expectations of General Von Uchatius; and the new Austrian field guns with which the active army is now completely equipped are constructed of this material. Our own experience, however, of the uncertain nature of bronze,* so far as we know of that alloy, and of the delusive effects of experiments made with a few pieces, make us pause before we can accept the success, so far obtained, as proof sufficient, that bronze steel is in all respects adapted for the construction of field guns, even if the theoretical principles are correct upon which its subsequent treatment is based.†

It seems unlikely, indeed, that such a metal as bronze, however improved, can long hold its own against the great advances certain to take place in steel manufacture, and in the construction on sound theoretical principles of field guns from that material, either alone or combined with wrought iron.

Iron.

Iron.

Although there exists but one elementary metal which can properly be termed iron, yet when this is mixed or alloyed with comparatively small quantities of other elements, we have in these alloys or mixtures virtually distinct metals, which differ from one another in their external characters more than many chemically distinct metals.

Three varieties.

Cast iron, wrought iron, and steel.

These different varieties of iron may be divided into groups termed respectively cast iron, wrought iron or piled metal, steel or ingot metal. Regarding the two latter we shall also see that, according to the latest nomenclature, they differ from each other more in the mode of mechanical treatment they undergo during their production than in chemical constitution.

How iron is usually obtained from its ores.

Iron is usually obtained from its ores by melting these in large blast furnaces with coke or coal, various fluxes being added, according to the

* A number of these Uchatius guns are rejected in manufacture owing to the presence of tin spots, while we have had to withdraw our rifled bronze guns, although the experimental pieces fired at Shoeburyness gave us most excellent results.

† In trials made prior to the introduction of 9-pr. R.M.L. bronze guns into our service in 1870 (vide p. 73, as to subsequent failure of these pieces), some experimental 9-prs. were subjected to very severe trials. They were fired at the rate of 50 rounds in 7 minutes for rapidity; 5 rounds in 13 minutes at a 9-ft. square target, at 1,000 yards off, for rapidity and accuracy, giving 27 hits: and on one occasion 140 rounds were fired from one gun, without stopping, at the rate of 3 rounds per minute, the metal becoming so hot as to boil water placed in the bore. In all cases the results were satisfactory, and one of the guns in question actually fired 2,732 rounds without destructive injury.

15-cm. (6-inch) siege guns are now however being made in Austria of this metal, which are said to give excellent results, while the Germans are carrying out extensive experiments with the same material for siege guns.

CHAP. I.

nature of the ore, to carry off the earthy matters; the metal so obtained is run into sand moulds in the shape of the well known rough looking bars technically termed "pigs" or "pig iron."

Pig iron.

The metal in this state, as run from the blast furnace, is termed "cast iron," and contains many foreign elements, principally carbon and silicon, the former being mostly derived from the fuel with which the ore was smelted. Besides these impurities, small quantities of sulphur, phosphorus, and manganese are commonly found in cast iron when first run out.

Cast Iron.

Cast iron.

Effects of silicon, sulphur, and phosphorus.

By refining, &c. a portion of the carbon and other impurities may be removed, but so long as the proportion of carbon* is not less than 2°/$_o$ the metal will possess the chacteristic properties of cast iron mentioned below. The presence of silicon, sulphur, and phosphorus modify the strength, brittleness, &c. of cast iron very much, that of sulphur in particular increasing its tenacity, which is always, however, comparatively low.

Per centage of carbon.

We may say that cast iron contains from 2°/$_o$ to 5°/$_o$ by weight of carbon, which exists in two states, either chemically combined with the iron or mechanically mixed with it.

Various qualities of cast iron.

In the trade cast iron is distinguished by numbers, from one to eight, the lower numbers being given to those descriptions in which the surface when broken presents a "grey" or "mottled" appearance, and in which the larger part of the carbon is in the state of graphite, that is, uncombined with the iron. The higher numbers represent "white" or "bright" iron, and in these the carbon is almost entirely in the combined state.

Cast iron easily fused.

Brittle.

Cast iron is easily fuzed,† and can be readily cast into a homogeneous mass of any size or shape we choose, but it is brittle and cannot be worked under the hammer either hot or cold. If, indeed, we heat a mass of cast iron to a red heat and hammer it, it will crumble to pieces, a fact taken advantage of in the breaking up of our obsolete smooth bore cast iron guns.

Cheap and easily manufactured.

Hard.

Uncertainty.

Liable to fly to pieces.

Cast iron has the advantage of being cheap and easily worked, by melting and casting, and also by cutting tools, &c. It is also comparatively hard, and is homogeneous, so that we can obtain a smooth and uniform surface. It is, however, a brittle and uncertain metal, and although its elastic limit is not very low, it elongates but little between that limit and the limit of fracture, and when strained beyond its stretching power it is therefore apt to fly to pieces without warning, a serious defect in gun manufacture.

Tenacity.

The tenacity of cast iron is low, that of good average cast iron being about the same as that of ordinary bronze (gun metal), viz., 12 to 14 tons (vide Table, p. 18).

Suitable for S.B. ordnance.

Unsuited for rifled guns.

For smooth bore guns, where the pressure in the bore was comparatively small, cast iron was a good material, but for rifled guns it is much too weak, and as it cannot be worked under the hammer we are not able to build up a gun in different pieces from this metal, but must employ it in the shape of a homogeneous mass (vide p. 20); it is there-

---

* This limit, like others of a similar nature, has reference only to varieties of iron nearly free from other hardening elements except carbon; for some irons, containing much less carbon than the above, but a sensible portion of silicon and phosphorus, are directly produced from the blast furnace, and must be classed as cast iron.

† The melting point of ordinary cast iron being about 1580 c.

fore at an additional disadvantage as a material for rifled guns. We ourselves employ cast iron in no case as the sole material for such guns, though for the sake of economy we utilise as second class guns, old smooth bore cast iron pieces, strengthened by means of wrought iron barrels (vide p. 80), which latter gives us a sufficient margin of safety.

Cast iron only used by us for S.B. guns converted on the Palliser principle.

For economical reasons, cast iron has been employed for rifled guns in America and elsewhere, but has, when unstrengthened by other materials, always failed signally when subjected to a heavy strain (vide p. 17). In Sweden and Denmark, where exceptionally good cast iron is produced, field guns are still made of this metal. These pieces are very cheap, but the velocity obtained from them is low. In France and Italy heavy rifled guns are made of this material and strengthened with steel,* but such guns would appear, though economical, to be unsafe.

Its employment in America.

Sweden and Denmark.

France and Italy.

Wrought Iron.

If we remove the carbon still further from cast iron, so that the amount becomes less than 2°/₀, we obtain either *steel* or *wrought iron*, according to the amount removed, the subsequent treatment of the metal, or to both combined.

Wrought iron.

Wrought iron approaches to, theoretically, pure iron, and according to the old nomenclature is iron containing from 0·1 °/₀ to 0·3°/₀ of carbon, though we shall find that certain characteristic steels now made have no larger proportion of this element.

Wrought iron theoretically pure iron.

Wrought iron is obtained by removing the carbon from cast iron by puddling† or otherwise, and then working it up by hammering or rolling into a useful or marketable form, whence the term *wrought* or *piled* is applied to this particular nature of iron.

How obtained. Puddling, hammering, and rolling.

If the cast iron is very impure, some of the impurities will be retained by the wrought iron, and will affect it seriously if present in any quantity, thus a small proportion of phosphorus (0·25°/₀) will make a bar of wrought iron "cold short" or brittle when cold, while a little sulphur present, makes wrought iron "red short" or brittle when heated.

Impurities.

Cold shortness. Red shortness.

While this more or less pure metal obtained from cast iron by removal of the carbon, through puddling or otherwise, is worked under the hammer or rolls, it is drawn out, and so given a fibrous structure, the fibres running lengthways in the direction in which it is drawn. The fibre runs along the length, in fact, as the fibre runs in the branch of a tree.

Fibrous structure.

This structural arrangement of the material can be readily demonstrated by subjecting a piece cut off a bar of wrought iron to the action of acid, when certain portions of the mass are eaten away, and the fibres stand out clearly, presenting the appearance of a bundle of fine wires. This fibrous quality of bar iron renders it much stronger in one direction than in the other, for—just as in the case of wood—it requires about twice the force to break a piece of wrought iron across its fibre that it does to tear it asunder along the fibre. In the latter case the fibres need only be separated, not broken, and the cohesion which binds them together is not much greater than that of the crystals which compose good cast iron.

Shown by action of acids.

Unfortunately the process of working up wrought iron into a proper condition for use (vide p. 102), does not remove all the foreign matter,

Difficulties in removing impurities.

---

* A 100-ton gun is now being made in Italy of cast iron strengthened by superimposed steel hoops.

† The operation of puddling is described at p. 102.

CHAP. I.

Not thoroughly homogeneous, and why.

such as minute portions of slag, &c., which have been entangled amongst the particles and fibres, so that owing to its mode of treatment a mass of wrought iron is never thoroughly homogeneous, and if laid open always exhibits flaws of some description, whence it is difficult to obtain a uniform and smooth surface of wrought iron perfectly free from defects.

Practically infusible by ordinary means.

Valuable property of welding.

Wrought iron may be said to be practically infusible in any ordinary furnace, but it has the property of welding (which cast iron has not);* that is, if two clean surfaces of wrought iron, heated to a white heat (about 3,000°) be brought into contact and pressed together, either by rolling or hammering, they will unite so perfectly, that the mass when broken, will part as readily at any other place as at the point of union; this is a most valuable property, and is largely taken advantage of in the construction of our service ordnance.

Malleability. Ductility. Tenacity. Tensile strength. Elastic limit.

Wrought iron is very malleable and ductile, and is also of great tensile strength, although its tenacity is much below that of most natures of steel. The tenacity is about 25 tons per square inch for good average wrought iron.

The elastic limit of wrought iron is not very high, about 12 tons per square inch; but after that limit is exceeded much work must be done in permanently stretching this very ductile metal before the limit of fracture is reached, giving us a large margin of safety.

Principal properties of wrought iron summed up.

To repeat the principal physical properties of wrought iron which make it suitable or otherwise as a material for ordnance.

Wrought iron is very malleable and ductile, with a low elastic limit and a tenacity of about 25 tons or double that of cast iron. Its ductility, however, necessitates much work being performed in stretching the metal between the limits of elasticity and fracture.

Wrought iron barrels discontinued except for converted guns.

In our service ordnance no heavy guns are now made with wrought iron barrels, as this material is too soft, and very liable to flaws. Most of our rifled B.L. guns are, however, entire built up of wrought iron, and also the earlier natures of of M.L. guns. We still use this material for the inner barrel of our converted cast iron guns, in order to give sufficient safety to those pieces, which in the outer casing have so brittle a material as cast iron.

Why used for the outer coil.

For the exterior portions of our R.M.L. ordnance we always employ wrought iron, which is not only cheap, easily worked and welded, but from its ductility gives, as before mentioned, a large margin of safety.

Testing Wrought Iron in R.G.F.

Testing of wrought iron for guns.

Appearance of fracture.

In the Royal Gun Factories, the quality of the iron used for bars is tested previous to using it for the manufacture of guns, the amount of drawing out or extension of the specimens previous to rupture being taken into consideration, as well as the weight which the metal will bear without breaking, and the appearance of the broken surface. In good tough wrought iron, the fracture presents an irregular silky appearance, light grey in colour, and shows distinctly the fibrous structure and the extension of the fibres; should the surface be largely crystalline, and the specimen have broken off short without drawing out, it shows that the wrought iron is deficient in fibre, and consequently not well suited to resist the continued stresses to which it would be subjected in a heavy gun.

---

* And steel in a very inferior degree.

Method of ascertaining the steely property of wrought iron.

The "steely" property of the wrought iron, that is, the amount of carbon it contains, is ascertained by an empirical test. The specimen is raised to a red heat and cooled suddenly in water; the effect of this treatment is to increase the tenacity, but the appearance of the fracture (even in good fibrous iron) often becomes crystalline or granular instead of fibrous, and the wrought iron has moreover lost to a great degree its property of extensibility, so that it breaks off short. If the increase in the tenacity when hardened is excessive (more than 2 tons per square inch) and the appearance of the fracture is such as to indicate the presence of a considerable amount of carbon, the iron is rejected, for the iron required for gun building must not only be tenacious, but should also possess in a marked degree, the properties of extensibility and toughness under sudden pressures, so that should the gun be at any time strained beyond its powers of endurance, it may fail gradually and not suddenly, without giving any indication to the gun detachment of its approaching rupture.

Steel.

To give a short definition of steel is a difficult task.

Definition of steel according to chemical constitution.

Until lately steel has been usually defined to be iron containing a small amount of carbon, an amount less than that present in cast iron, but greater than the maximum quantity to be found in characteristic wrought iron, *i.e.*, iron containing between 0·3 per cent. and 2·0 per cent. of carbon was termed "steel."

This proportion of carbon is however only approximate, and Dr. Percy gives from 0·5 to 0·65 per cent. of carbon as the limit at which, when free from other foreign matter, iron may be considered as passing into steel, so that when hardened by quenching in water it will strike fire readily with flint. According to this definition when carbon is present "in certain proportions, the limits of which cannot be exactly prescribed, we have the various kinds of steel, which are highly elastic, malleable, ductile, forgable, weldable, capable of receiving very different degrees of hardness by tempering, and fusible in furnaces."

Percentage of carbon in steel.

Owing, however, to the gradual development of new modes of manufacture, and to the enormous increase in recent years in the production of cast steel of all kinds, the arbitrary definition of steel above given, leads to much confusion and serious mistakes,* and though it may be called a mere question of words, the above definition has no doubt exercised a deleterious effect upon the introduction of steel, in place of worked or wrought iron, for many purposes for which it is well suited.

Definition of steel according to its physical condition and mode of manufacture.

A simpler definition of steel seems likely to be adopted, a definition which possesses the advantages of precision, and is in harmony with the current modes of manufacture. According to this, steel is an *alloy of iron, cast while in a fluid state into a malleable ingot.*

It is held, according to this nomenclature, that steel and wrought iron cannot be always distinguished by chemical analysis (for the same proportions of carbon, manganese, silicon, &c. may exist in any malleable alloy of iron), and that the fundamental and essential difference between steel and compounds of iron, merely worked or wrought, is a structural difference easily determined.

Structural difference between steel and wrought iron. Wrought iron is piled metal.

All malleable products of iron industry, that is to say, all varieties of iron, except cast iron, may be divided into piled metal (wrought iron)

* Thus, if steel be defined as an alloy, "containing carbon enough to harden it, when it is heated and plunged into water, then puddled iron, though laminated and heterogeneous in structure, may be steel, and the finest product of the crucible, though crystallised and homogeneous in structure, may not be steel."

CHAP. I.

and ingot metal (steel); the former embracing all malleable iron or alloys of iron produced without fusion of the metals while in a malleable state, and the latter applying to all irons, however produced, which are cast into a malleable ingot.

Steel is ingot metal.

Differences in mode of treatment, &c.

"These two classes differ more widely in mode of manufacture, appearance, and in many important properties than the varieties of each class among themselves, and form two parallel and continuous series, the corresponding members of which are chemically identical, differing only in mode of production, and in mechanical structure, and rising in each series from the purest and softest iron, to the hardest and most highly carburetted varieties."*

"M. Adolph Grenier of Seraing, adopting this definition of steel, classifies the two parallel series of products, the irons and the steels, as follows:—

Parallel series of steels and irons.

Percentage of Carbon.

0· to 0·15	0·15 to 0·45	0·45 to 0·55	0·55 to 1·50 or more.

Series of the Irons.

Ordinary irons.	Granular irons.	Steely irons or puddled steels.†	Cemented steels. Styrian steel.

Series of the Steels.

Extrasoft steels.	Soft Steels.	Half soft steels.	Hard steels.

Puddled or cement steel.

The production of the more highly carburetted varieties of the iron group, such as puddled steel or cement steel, except with the view to casting the metal subsequently‡ is now carried on upon a comparatively limited scale, while that of the steels proper, or cast steel, is increasing enormously each year.

Cast steel used by us.

Definition of such steel.

For the construction of ordnance we may say that cast steel is at present, and will in the future be as a rule employed, so we may safely take the new definition of steel, and look upon this metal as a *melted malleable alloy of iron*, produced in *any way whatever, and containing a smaller proportion of carbon or other hardening element than is contained in cast iron.*

Though the mode of production of wrought iron is not given in this chapter, but in Chapter V., under the head of "Manufacturing Opera-

* Journal of the Iron and Steel Institute, 1873, p. 499.

† This classification, placing all varieties of "puddled steel" in the group of irons containing between 0·45 and 0·55 of carbon, is not correct, as many specimens of puddled iron contain a much higher percentage. Thus Percy ("Iron and Steel," p. 799) gives the following analysis, by W. Brauns, of puddled steel from Königshütton,—

Carbon, combined	1·380
Carbon, graphitic	trace
Silicon	0·006
Sulphur	—
Phosphorus	trace
Manganese	0·012

‡ The manufacture of such steels is generally speaking carried on to a further stage, so that they become steel according to our definition, for they are very often subsequently cast. Thus the steel used by us for the barrels of Woolwich guns is obtained from bars of good wrought iron, which are converted by cementation into steel. These bars are afterwards broken up, and the pieces whose fracture indicate a mild nature are melted in crucibles, the contents of which are poured into a metal mould of a size sufficient to give the required ingot.

tions," it is necessary here to explain generally the processes of steel-making, as the steel forgings for our gun barrels are not manufactured in the Arsenal.

Manufacture of steel.

Steel may be produced in a variety of ways, by more or less decarburetting cast iron under such conditions as to obtain a melted product.

By dissolving wrought iron or steel scrap or spongy reduced iron in melted cast iron.

By the direct melting of puddled iron or other variety of iron of the requisite degree of hardness (the hardness referring to the amount of carbon in the iron).

By melting a mixture of soft wrought iron or iron sponge with carbon or with cast iron, or of malleable iron, which is too hard for the variety of steel required, with oxidising agents, or by directly melting together a mixture of iron ore and carbon as a single operation.

Crucible steel. Siemens or Siemens-Martin steel. Bessemer steel.

The processes by which steelmaking is practically carried out are by melting in pots or crucibles, giving "pot" or "crucible" steel; on the open hearth or bed of a reverberatory furnace, giving Siemens or Siemens-Martin steel; or by blowing air through molten cast iron, producing Bessemer steel. In whatever way made, the material is essentially the same, depending on its chemical composition and physical structure for its properties. It differs much from wrought iron, even when chemically the same metal, because a mass of wrought iron is made up from a number of bars or blooms, heated and welded together, each of these again being composed of separate granules with impurities interposed, such as slag, &c., entangled so that the mass is unavoidably full of flaws and imperfect welds. Steel, on the other hand, has been brought into a state of perfect fusion, and cast while liquid into a malleable ingot, which is homogeneous* throughout, and free from flaws or intermixed impurities.

Flaws likely to occur in wrought iron. Steel free from such flaws.

We now can understand fully what steel means in accordance with either of the definitions given of the term, whether depending merely upon its chemical constitution as to the amount of carbon present or upon the physical treatment of the metal, which must be melted and cast into a malleable mass.

Varying properties of steel.

In either case it is evident that the properties of the metal must vary very much, but more especially if we adopt the second definition. The soft or low steels approximate, like ordinary wrought iron does, to pure iron, and the hard or high natures of steel approach cast iron in their properties.

Advantage of steel over wrought iron. Uncertainty in the quality of steel.

Over wrought iron, steel has always the advantage of being homogeneous, but it is unfortunately uncertain in quality, a surprising fact when we think how it is produced; still it is so, and we find that ingots of this metal of the same chemical constitution, produced in the same way and from identical materials, often differ much from one another in the properties of elasticity and tenacity.

Remedies tried.

Many nostrums have been proposed for overcoming this defect, which appears to be due to the molecular condition of the metal as affected during the melting and the subsequent treatment of the ingot.

Casting under pressure.

It is in modification and improvements of the latter that the solution appears to lie, and already many plans are being attempted for the insurance of greater certainty in the quality of steel, such, *e.g.*, as casting it under pressure, to get rid of the bubbles or blow holes.

Steel will not, like wrought iron, stand a high welding heat, especially if it be of a "hard" or "high" nature. If hammered at too great a heat the ingot will fall to pieces.†

---

* Homogeneous, that is as to surface and apparent structure, though not necessarily so in tensile strength, tenacity, &c.

† Vide interesting examples given in "Remarks on the Manufacture of Steel," by D. Chernoff, London: Clowes and Son.

CHAP. I.

Hammering and rolling makes steel fibrous.

After being cast, steel is hammered or rolled, and it acquires a fibrous structure during this process, as may be easily shown if a bar of this metal be acted on by a strong acid.

Hardening steel.

When first cast, steel, excepting the higher varieties, is comparatively soft and inelastic, and can thus be treated in a similar manner to wrought iron; but with the exception of the very low or soft varieties,* it may be subsequently hardened by what is called "tempering," when the metal is heated† and plunged into mercury, water, oil, or some other liquid, in order to cool it more or less quickly. The effect of such tempering depends very much upon the amount of carbon present in the steel, as well as upon the degree of heat it is previously raised to, and has a marvellous effect in increasing both tenacity and elastic limit in the case of good steel.

Steel inferior in property of welding.

Besides the great uncertainty of steel, we must remember that it is very inferior to wrought iron in another important point, that of being easily weldable. The importance of this property with regard to facility and economy in working large masses of metal can hardly be exaggerated.

To sum up the qualities of steel—

Tenacity and elastic limit.

In tenacity it varies, roughly speaking, from 30 to 60 tons per square inch, and its limit of elasticity lies between 13 and 30 tons.

Effects of rapidly cooled steel.

The softer the steel the more ductile it is, the less readily can it be tempered, the lower is its limit of elasticity and its tenacity. Other things being equal, the more rapidly it is cooled the harder it becomes, but at the same time the higher natures are rendered more brittle.

Cooling in a bad conductor of heat.

When cooled in a comparatively bad conductor of heat, such as oil, the mass parts with its heat slowly, and is not only made more elastic and harder but also toughened,‡ while in any case the limits of elasticity and tenacity approach one another more nearly.

Tempering.

The effect of tempering upon the metal is, of course, comparative; thus experiments have shown that certain soft steels containing from 0·15 per cent. to 0·22 per cent. or 0·24 per cent. of carbon, were decidedly toughened by heating to a red heat, and then being quenched in water. The steel used for our Woolwich gun-barrels, containing about 0·3 per cent. of carbon, would be made very hard but also very brittle if so heated; therefore it is tempered in oil, as being a worse conductor of heat than water, and is thus toughened.

Change produced.

The change in the character of steel produced by tempering depends not only upon the constitution of the steel itself, and upon the nature of the medium into which the steel is plunged, but also upon the degree of heat imparted to the mass beforehand.

This is shown well by the following table given by Mr. Kirkaldy:—

Nature of Specimen.	Tenacity (tons per square inch).	Elongation per cent.	Character of Fracture.
1. Highly heated, and cooled *in water*	30	0	Entirely granular.
2. Highly heated and *cooled slowly*	36¾	22	Entirely fibrous.
3. Moderately heated, and cooled *in oil*	53	14½	⅓rd granular. ⅔rds fibrous.
4. Highly heated, and cooled *in oil*	58	2½	Almost entirely granular.

* Thus, "some varieties of soft steel used for boiler-plates. Steel having less carbon than many wrought irons, may be heated to redness, quenched in water, and subsequently bent double without cracking, or may be twisted into any form, like as much lead." Vide Proceedings Institute C.E., vol. xlii., Part IV.

† This heating, however, must reach a certain minimum of temperature in each case, the minimum depending upon the nature of the steel.

‡ Vide p. 163 as to toughening tubes for service guns.

Should the steel be too highly carburetted, or be in too large a mass, it is liable to fracture during the process of toughening.

Effects on highly carburetted steel.

With the higher natures we can obtain very great strength and elasticity, though we lose in ductility and have a more brittle material. We have also the great advantage of hardness and homogeneity, giving us, when required, a very smooth surface quite free from flaws.

Uncertainty and expense of steel.

Until further improvements are made, these advantages are counterbalanced by great uncertainty* as to the behaviour of the material, especially in the higher or harder steels. The greater cost of steel of sufficiently good quality is also a disadvantage when we compare it with wrought iron.†

Why well suited for gun barrels.

The brittleness and uncertainty of this material prevent us employing steel as a sole material for any gun save the small 7-pr. for boat and mountain service (vide p. 173). Although, however, steel is not such a material as we should employ for the sole manufacture of a gun, yet it is admirably suited for the inner barrel, as it is very strong and gives us a hard clean surface, while its limit of elasticity is high, so that even a heavy strain does not stretch it permanently and deform the bore. It may indeed split if subject to too great a pressure when not properly supported by the exterior layers, but by putting a wrought iron jacket outside we prevent any danger from that cause; for should the steel tube burst, the wrought iron exterior will prevent any explosive rupture.

Tests applied to Steel in the R.G.F.

Quality of steel required for gun barrels.

For our own gun barrels we require steel of a certain quality which is not too high or brittle when toughened, and which will yet give us after that operation sufficiently hard, tough, and yet elastic material.

On account of the uncertainty of steel, each ingot, when received from the contractor, is tested in the R.G.F. (as described in Chapter V, p. 99).

Summary of the Gun Metals.

Bronze.

Summary of the gun metals.

To sum up as to gun metals generally,—

Bronze, an alloy of copper and tin, is an expensive material, too soft for the bore of a rifled gun, and further liable to flaws, due to the want of proper mixture of its constituents. It is tough, but wanting in elasticity and tenacity.

---

* If this uncertainty be overcome, steel, according to our definition of an alloy of iron, cast into a malleable ingot, may, perhaps, prove a satisfactory gun metal. The several parts being made of steel of different qualities, that portion for the inner barrel being hard and elastic, and that for the outer portions becoming softer and more ductile as we approach the exterior.

† Numbers of steel guns are made, both in England and by foreign gunmakers, the outer portion being of a much milder or less carburetted nature of steel than the barrel, but as yet it is doubtful whether such guns can be trusted, on account of the uncertainty of their material. Many such steel guns have burst under service conditions.

CHAP. I.

Bronze.

Much advance has recently been made in the treatment of this alloy, but the defect of want of homogeneity does not appear to be overcome, while any such advance will probably be easily distanced by improvements in the much stronger metal, steel.

In our service ordnance we no longer employ bronze, though we retain some smooth-bore bronze ordnance and a few rifled 7-prs. made of that metal.

Cast iron.

Cast iron.

Cast iron is cheap, easily worked, but of low tenacity as compared with other iron, being on an average about half that of wrought iron. It can be fused and cast without difficulty, and is comparatively hard. It is not malleable, and therefore incapable of welding. It is brittle and uncertain in character, on account of which and of its comparative low tenacity, we no longer employ this metal as a material for ordnance, and only retain it in the case of old smooth bore guns, and for economy's sake in rifled guns converted from such pieces by strengthening them with wrought iron barrels.

Wrought iron.

Wrought iron.

Wrought iron or malleable iron obtained from cast iron without being melted subsequently, when it contains less than 0·3 per cent. of carbon (as in the case of the wrought iron used by us in gun manufacture), has the following qualities:—

It is almost infusible, but readily welded. It is soft, having about one half of the hardness of cast iron; it is further, from its mode of production, liable to flaws, which make it difficult to obtain with this metal a fair and clean surface. Its elastic limit is also low, and these qualities prevent our employing it any longer as a material for gun-barrels, except in the case of our converted guns, when we employ it in conjunction with the brittle cast iron for the sake of safety. Most of our rifled breech-loading guns, however, and the earlier R.M.L. guns are made entirely of this metal.

The comparative softness of wrought iron causes the bore and grooves to be worn and indented when firing the enormous charges now used, and the very property of extensibility which renders it so valuable for the exterior is to a certain extent a disadvantage in the inner tube, which becomes more and more enlarged with repeated firing, until at last the material is stretched so far that it must yield, and the barrel splits, while the exterior of the gun remains perfectly sound.

Wrought iron has a large amount of tensile strength between its limits of elasticity and fracture due to its ductility, and it is therefore a very safe material, stretching considerably before it ruptures. This quality combined with its certainty, cheapness, facility of working, and the capability it possesses in a high degree of being welded, make wrought iron a very good material for the outer portion of a gun where small flaws are of no consequence, but where we require above all a material to be depended on in case of rupture of the inner barrel; we consequently employ wrought iron for the exterior of all the built up rifled guns now manufactured.

Steel.

Steel. Steel is a fusible, malleable alloy of iron with carbon and other impurities. When it contains above 0·2 per cent. of carbon, as that employed by us in gun manufacture, it possesses some of the qualities both of cast iron and wrought iron. It can be easily fused, and on the other hand is malleable and more or less weldable,* according as it is comparatively soft or hard.

It is tough and elastic, with a much higher tenacity, but not necessarily of greater tensile strength than wrought iron. By tempering this material we can increase very much its limit of elasticity and tenacity. We can also by the same method make it extremely hard.

As it has been melted when in a malleable condition, it is free from the foreign matters entangled between its particles such as we find in wrought iron; consequently it is homogeneous and allows us to obtain a clean surface.

The qualities of hardness, homogeneity, high elasticity, and great tenacity, make steel a good material for the inner barrel of a gun.

Cast steel as at present made has, however, the great defect of uncertainty, while such steels, when suitable for an inner barrel, from being hard and of high tenacity, have little margin between their limits of elasticity and rupture, and are consequently liable to fly to pieces suddenly without warning. We therefore use steel alone for the manufacture of but one nature of gun, the small rifled 7-pr.† but employ this metal for the barrels of all our R.M.L. guns, except in the earlier natures, before mentioned, which have wrought iron tubes.

Materials at present used in Service Ordnance.

We manufacture our present ordnance from mild steel toughened in oil for the inner barrel, and from good ductile wrought iron for the remaining portions.

Not a single rifled gun of our own manufacture burst on service.

So admirably do these materials answer as we now employ them, that we may boast of having thousands of rifled guns thus built up, from the 9-pr. weighing but 6 cwt., to the 16-inch gun of 80 tons weight, not one of which has burst explosively on service, nor has the sacrifice of a single life been due to their breaking up. Under service conditions in fact, no built up gun of wrought iron and steel of our present manufacture has failed from any defect due to the materials of which it was made. With this safety and certainty in our material we may well be satisfied, though our authorities are by no means blind to the advances in metallurgy, and will be always ready to adopt new materials if proved really superior for our purposes to those we now employ.

* Though always far inferior in this property to wrought iron.

† That further advances in steel manufacture, especially in its treatment after casting, may at a future time enable us to make guns wholly of steel, and still stronger than our service pieces, is perhaps possible. It will, however, in that case, be little more than a difference of terms, as the very soft steel which would be used for the exterior portions would in most of its characteristics, in durability, margin of safety, &c., resemble the wrought iron we now use. Before there is any chance of our thus using steel, however, we must have much greater certainty in quality ensured to it than we have at present.

TABLE I.

TABLE showing the ELASTIC LIMIT and TENACITY of Average Specimens of the Metals used in the R.G.F.

Materials.	Tons per Square Inch at		Elongation per Inch at Breaking.
	Yielding.	Breaking.	
Bronze	6·8	14·9	0·29″
Cast iron { from	} about 4*	9·0	*
Cast iron { to		14·0	*
Wrought iron along its fibre	11·0	22·0	0·3″
Steel { soft	13·0	31·0	0·21″
Steel { tempered in oil..	31·0	47·0	0·11″

These numbers show of course only a rough average approximation with reference to the particular natures of the metals mentioned which are used in the R.G.F.

Roughly speaking, wrought iron is twice as hard as bronze, cast iron is twice as hard as wrought iron, and hardened steel is twice as hard as cast iron.

* The specimens of cast iron tested in the R.G.F. are too short to enable the yielding point and elongation to be very accurately determined.

CHAPTER II.

CHAP. II.

GUN CONSTRUCTION GENERALLY.*

Gun construction generally.—Casting.—Building up.—Bore.—Tangential and longitudinal stresses.—Strength of a cylinder limited.—Barlow's theory.—American Rodman guns.—Austrian Uchatius guns.—Systems of initial tension.—Varying elasticities.—Armstrong principles.—Wrought iron coils.—Arrangement of fibre tangentially and longitudinally.—Palliser guns.—Shrinkage.—Compression.—Tension.—Disposition of metal in service guns.—**Construction of our R.B.L. guns.**—**Fraser construction.**—First trial of 9-inch guns on the cheap plan.—Trial of 64-prs.—Adoption of **cheap construction.**—Bursting of a 9-inch gun at proof.—Final trial of 9-inch guns.—Why steel is used for our gun barrels.—**Converted guns,** why wrought iron barrels used.—General question of B.L. *v.* M.L.—Some advantages and disadvantages of both systems.—Chambering.—Enlarged chambers and results produced.—**Air-spacing** and its object.—**Rifling.**—Angular velocity.—Velocity of translation.—Twist.—Angle of rifling.—Uniform or increasing twist.—Centring of projectile.—Loading and driving side of groove.—Pressures upon studs, groove, and bore.—With uniform and increasing twist.—Advantages of the increasing twist.—The grooves.—Stud system.—Grooves of B.L. guns and M.L. guns.—Shunt groove.—Plain.—Woolwich.—French modified.—French.—Modern polygroove.—**Stops** in bore for projectile.—**Venting** and vents.—Copper bush.—Steel bush.—Wrought iron bush.—Position of vent.—R.B.L., S.B., and R.M.L. guns.—In turret ships.—Effects consequent on position.—Axial vent.—Position of vents in small guns.—Forward vents.—Vents in converted guns.—Howitzers and 40-pr. downwards.—Cone and through vents.—For converted guns.—7-pr. steel gun exceptional.—80-ton gun.—Wear of vent.—Plans for prevention of wear.—**Sighting** generally.—Fore sights.—Tangent sights.—Sights for S.B. ordnance.—Sights for rifled ordnance.—Angle of deflection.—Determination of permanent angle of deflection.

How should a gun be constructed? This question, so often asked, is not so readily answered; when we begin to think over the matter, we discover that the problem set before us is so complicated, and depends upon so many variables that its solution is no easy matter.

Gun construction.

A very complex problem.

It has indeed been solved by many an inventor to his own complete satisfaction, but otherwise there remain points on which gunmakers so far must agree to differ.

To take seriatim those which we have principally to consider in constructing a piece of ordnance.

* This chapter does not pretend to deal in any way exhaustively, nor to lay down the law as to the complex question of gun construction, but rather to draw attention to the main points to be considered with the view of inducing thought on the subject.

CHAP. II.

Principal considerations.

In the first place there is (1) the absolute casting or building up of the gun; then follows (2) its completion as a B.L. or M.L., and also the question of what should be (3) the length of bore, calibre, &c., and the dimensions of chambers, if any; subsequently (4) the adoption of a suitable system of rifling, and of a mode of stopping the projectile always at a definite point in the bore and lastly the minor considerations of (5) venting and (6) sighting.

1. *Casting or building up the Gun.*

Setting on one side then, for the present, the vexed question of B. L. versus M. L., and the equally disputed point as to the system of rifling which should be employed, let us look at a gun merely as a tube destined to withstand a given pressure from within when throwing a projectile which shall produce certain effects at the required distance or distances.

Gun tube should be safe as to strength.

In constructing such a tube, we must first consider what pressure it will have to withstand at the various points in its length, and then make it strong enough to ensure perfect safety. The bore also should be of such a material as to stand the wear and tear of firing a large number of rounds without being so damaged by expansion or abrasion as to interfere with the shooting.

Every part should perform its work.

Not only must the gun be sufficiently strong, but it must not be too costly, nor too heavy, so it is important that the material should be arranged in such a manner that there should be no waste of its strength, arranged in fact so that every part should perform its own share in withstanding the pressure from within.

Pressure greater in rifled than S.B. ordnance.

When the shot begins to move, or more usually shortly afterwards, the pressure inside the gun gets less (vide p. 348), and continues to decrease as the projectile approaches the muzzle; for this reason we have to make the gun stronger about the powder chamber than towards the muzzle end, more especially in rifled guns where the pressure is much greater than in S.B. ordnance, and where the shot does not move so soon.*

Circumferential tension.

Longitudinal tension.

Looking simply to the construction of a gun cylinder, we find that the two principal stresses† to which such a cylinder is subjected upon the explosion of the charge are: first and most important, (a), a circumferential tension tending to split the gun open longitudinally; secondly, (b), a longitudinal tension tending to pull the gun apart in the direction of its length.

(a) *Circumferential Tension.*

The simplest way of making guns.

Now the least complicated method of making a gun would seem to be that of casting it as a homogeneous hollow cylinder, but if we take such a tube, made of one material throughout, we find that its tangential strength to resist a pressure from within does not increase uniformly with its thickness.

We find practically that with a given material we soon reach a limit

---

* The greater the weight of a projectile, the greater is the opposition from inertia which it offers in the bore to the expansion of the ignited charge, and this opposition is also slightly increased if the projectile is obliged to travel through the bore in a spiral course, so it is easy to understand why a rifled gun must be of stronger, tougher, and more elastic material than a smooth-bore gun, in which the round shot yields more promptly to the first pressure of the powder gas (to which it presents half its surface), while the stress on the gun is much relieved by the comparatively great windage.

† Besides these, we have (c) the stress of compression or crushing stress, tending to diminish the thickness of the metal, and (d) a transverse stress, tending to stretch the gun transversely.

CHAP. II.

beyond which any additional thickness of wall aids but little in enabling the cylinder to withstand such a pressure; supposing the metal to be incompressible, this limit is taken at about half a calibre, so that—for example, in the cylinder of an hydraulic press—if the thickness of its walls be equal to one half the diameter of the piston which works inside, then the cylinder will be nearly as strong as if it were ten times as thick.

Limits of useful thickness of metal.

The hydraulic press.

To give an instance from our own ordnance; the sides of the 13-inch S.S. mortar are twice as thick as those of the 13-inch L.S., but the former piece is not by any means twice as strong as the latter.

Thickness of 13-inch S.S. mortar.

In fact, however thick we make the sides of such a cylinder, we can only obtain a certain strength from it, depending upon the nature of the material used.

It is generally agreed that no possible thickness can enable a cylinder to bear a continual pressure from within, greater on each square inch than the tenacity of a square inch bar of the material; that is to say, if the tenacity of cast iron be 11 tons per inch no cast iron gun however thick could bear a charge which would strain it beyond that point, for on the first round the interior layer would be ruptured before the outer portion could come into play and every succeeding round would tend to make matters worse.*

Strength of a homogeneous cylinder limited by the strength of its material.

To begin with, supposing the metal incompressible, Professor Barlow, F.R.S. (whose theory is generally adopted, though known to be but approximately true), arrived at the conclusion that the circumferential tension of every concentric layer of metal of a gun is inversely as the square of its distance from the centre of the bore. The following being his method of proof:—

Professor Barlow's theory as to the tensions of the several layers of metal.

Suppose the metal of which a hollow cylinder is made to be incompressible, which would be the most favourable case, then the tension of any cylinder due to a pressure acting from within is inversely as the square of the distance from the centre.

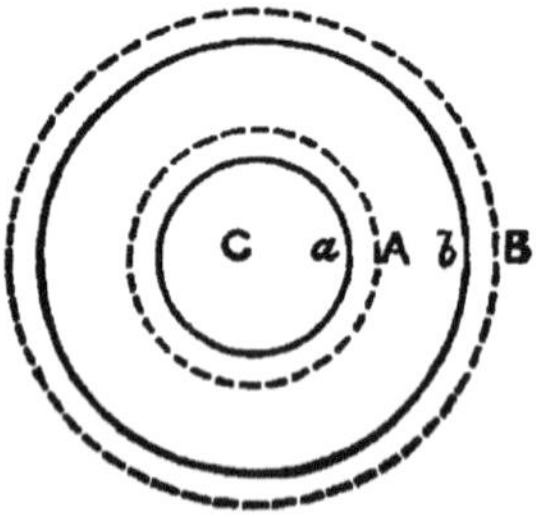

Let $Ca = r = Cb = R$

$CA = r' = CB = R'$

and let the solid lines in the Fig. represent the inner and outer surfaces of a hollow cylinder before extension, and the dotted lines represent the same after extension.

Then in this case we have supposed the area between the dotted lines is equal to the area between the solid lines or

$$\pi(R^2 - r^2) = \pi(R'^2 - r'^2) \text{ or } A = A'.$$

But as A′ is further removed from the centre its length is greater,

---

* This assumption is founded on the supposition that the metal is incompressible, and that in consequence the tension on each layer is inversely as the square of its distance from the centre; no account is taken of the compressibility of the metal, which would allow of a greater tension of the exterior to that given, while again it supposes that the tension is transmitted instantaneously throughout the thickness which gives an error in the opposite direction.

Taking extensibility and compressibility into consideration, the stress on the exterior layers is greater than that shown by Barlow's formula; the actual tension of each layer probably lies somewhere between the results given by the undermentioned formula of Dr. Hart, and that of Professor Barlow.

The latter, however, is the simpler, and its results have been shown to be fairly trustworthy in practice.

Dr. Hart, Fellow of Trinity College, Dublin, having taken into account the extensibility and compressibility of the metal, the latter of which Professor Barlow

CHAP. II.

Proof of Professor Barlow's theory.

and therefore its breadth less, so that the absolute extension of A' compared to that of A will be inversely as their radii.

In other words, if $R' - R = dR$

$$r' - r = dr$$

$$\text{Then } dR : dr :: r : R$$

$$\text{or } dR . R = dr . r$$

$$\therefore \frac{dR}{R} . R^2 = \frac{dr}{r} . r^2.$$

$$\text{or } \frac{dR}{R} : \frac{dr}{r} :: r^2 : R^2$$

Now the strain is the increase in length produced, divided by the original length, so that if $\frac{dR}{R}$ represents strain on A, then $\frac{dr}{r}$ represents strain of A', so that the strain upon any concentric layer diminishes as the square of its distance from the centre, or the strains of such layers are to one another inversely as the square of their radii.

Iow to find ie total resistnce of any ollow homoeneous cynder.

We can now find what the total resistance of any hollow cylinder would be to a pressure from within according to the formula. In the figure below, showing section of a homogeneous cylinder, let $\begin{matrix} oa = r \\ ob = R \end{matrix}$ and o centre of the bore, being the origin of the axes of x and y, the abscissæ will represent the distance from the centre, and the ordinates $a p$, $b q$ the stress or tension of the different concentric layers.

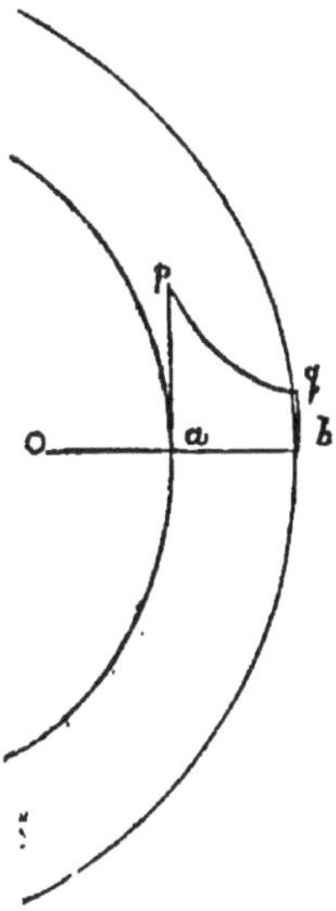

For the curve pq we have the equation $y = \frac{c}{x^2}$, for the tension y varies inversely with x^2 and directly with c, which is a constant depending upon the pressure exerted, and the radius of bore (r).

Now the total useful effect (T) to be got out of such a cylinder is equal to the sum of the tensions on the different layers multiplied by the thickness of each layer, and will be represented by area $a\,p\,q\,b$.

$$\text{Then } T = \int_R^r \frac{c}{x^2} . dx = c\left(\frac{1}{r} - \frac{1}{R}\right)$$

$$\text{or } T = \frac{c}{rR} (R - r)$$

where c is a constant depending on the pressure, *etc.*

---

neglects, calculated that there is a greater tension of the exterior. His formula may be written thus.

$$\frac{\sigma}{s} = \frac{r^2}{\rho^2} \frac{R^2 + \rho^2}{R^2 + r^2}.$$

R and r being the external and internal radii, ρ the radius of an intermediate lamina of which σ is the tension, and s the tension of the inside.

If we want to compare the strain on the inside with that on the outside, $\rho = R$, and we have by inversion,—

$$\frac{s}{\sigma} = \frac{R^2 + r^2}{2r^2}.$$

In the case of the 10-inch gun before referred to, the tension of the interior compared with that of the exterior, would, according to Barlow, be as 100 to 25, or four times as great, but according to Hart, only as 125 to 50, or $2\frac{1}{2}$ times as great.

CHAP.

If the tension were uniformly the same as at r, we should have each layer of metal offering the same amount of resistance, *i.e.*, we could utilize the whole strength of the metal in the thickness of the cylinder. Therefore, were the tension the same throughout, the total strength (S) would plainly be $\left(\frac{c}{r^2}\right)$ multiplied by thickness of metal, or

$$(S) = \frac{c}{r^2}(R - r)$$

$$\frac{T}{S} = \frac{\text{Total useful effect}}{\text{Total strength of metal in wall.}} = \frac{\frac{c}{rR}(R-r)}{\frac{c}{r^2}(R-r)} = \frac{r}{R}$$

or the value of such a cylinder may be represented by the ratio of its external to its internal diameter multiplied by the strength of the wall, which clearly varies directly as the thickness of the wall of the cylinder. It is clear also that the thinner the wall of the cylinder the greater proportional amount of work we get out of the metal composing it, as represented by the ratio of (T) the total useful effect to (S), the total strength of the metal in the wall of the gun, while the greater the calibre the more advantageously can we increase the thickness of our gun cylinder.

How to find greatest possible resistance.

To find the greatest possible resistance (Z) for a cylinder of infinite thickness, we will integrate $y = \frac{c}{x^2}$ between the limits $\begin{cases} x = r \\ x = \infty \end{cases}$

$$\text{now } c\int_{\infty}^{r}\frac{dx}{x^2} = c\left\{\frac{1}{r} - \frac{1}{\infty}\right\}$$

$$\therefore Z = \frac{c}{r}.$$

But to find the value of c we know that when y is a maximum then $x = r$. And if the maximum value of y be expressed by Y,

$$\text{Then } c = Yr^2.$$

$$\text{And } Z = \frac{Yr^2}{r} = Yr.$$

Now the total tension developed by explosion tending to split a gun open longitudinally is pr where p is the pressure per square inch and r is the radius of bore. When an explosion takes place and p is greater than Y the cylinder must burst; or, in other words, no thickness of metal, however great, in an homogeneous cylinder can withstand a pressure greater than the tenacity of a bar of the same metal.

Initial tensions and varying elasticities.

Initial tensions and varying elasticities.

From the above it is clear that guns made, as S.B. pieces used to be manufactured, by casting iron or bronze into homogeneous cylinders, cannot be made strong enough to stand more than a certain pressure from within, for the inner portions receive the brunt of the explosion, while the outer portions are hardly affected by it at all, and consequently there is a certain amount of dead weight about every homogeneous gun. What we require, then, in manufacturing a gun cylinder is not to make it a homogeneous hollow cylinder, but so to arrange that the inner portions shall be stronger than the outer parts, and to obtain from each concentric layer of metal as much useful effect as possible in withstanding a pressure from the interior. How to carry

CHAP. II. this out in the best way is the problem which all gunmakers try to solve.

American Rodman guns.

In America Rodman S.B. guns for L.S. were made by casting the piece hollow and cooling it from the interior, so that the inner portion is compressed and supported by the contraction of the outer more liquid portions upon it while cooling.

This was a great step in advance theoretically,* for the layers of metal become stronger and stronger as we approach the surface of the bore, where the strain is greatest. But however well cast iron may be disposed, it is naturally too weak and brittle a metal for use with heavy rifled guns, and those nations which employ it by itself for such guns do so because it is cheap and easy to manufacture, and not because they consider it the best material.

Austrian Uchatius guns.

In another case, that of the Austrian Uchatius guns, and also of the Russian Lavroff guns, we have rifled pieces cast from a metal homogeneous at first as to material, but the several layers of which are subsequently to casting in chill placed in a state of tension as regards the exterior, and of compression as regards the interior, by driving through the bore of the gun block a series of steel mandrels gradually increasing in size.

The inner layers are so made harder, and although a portion of their elasticity and tensile strength is overcome, they are better supported by each of the outer concentric layers, and are given higher limits of *ultimate* strength, both as to elasticity and tenacity, as we see from Fig., p. 3 should be the case.

Building-up systems.

Gunmakers at the present day are fortunately no longer limited to manufacturing guns by casting, but can build up a hollow cylinder, the walls of which may be composed of concentric layers of different strength, so that each may arrive at its limit of elasticity at the same time; so that, in fact, when the layer of metal next the bore approaches its elastic limit or its limit of fracture, each successive ring of metal from within outwards, approaches at the same time its elastic limit or its limit of fracture respectively.

This object we may obtain by employing a single metal for the several portions, and so disposing the various layers over each other that the inner layers or tubes are compressed by those outside them, while the exterior tubes are at the same time put into a state of tension, the inner layers being thus strengthened at the expense of the outer portions of the metal. In this case we obtain the whole strength of all the layers except a part of that of the outer and unsupported ring. It must not, however, be for one moment supposed that this theoretical perfection is ever reached; the nearest approach to it is far from perfect, and theoretic advantages have to give way largely to practical considerations of manufacture.

Again, we may arrive at a similar result by employing metals varying in elasticity or in tenacity for the several parts, those possessing the largest amount of strength constituting, of course, the inner portions, so that where the greatest stress is felt it will be borne by the stronger material.

Initial tension. Varying elasticities.

These two methods are sometimes called respectively those of *initial tension* and of *varying elasticities*. They may be, and frequently are, both employed in the manufacture of a gun, as in the case of our so-called Woolwich guns.

Whichever mode we employ, the metal in each ring should so be

* Practically, however, it has long been a disputed point in America as to whether this mode of casting gave in reality the good effects supposed. With the naval Dahlgren gun it was not employed.

disposed as to be in the best position for resisting the circumferential stress it will be exposed to

b. Longitudinal Tension.

Longitudinal stress.

In the above considerations we have confined ourselves to the strength of the gun to resist the circumferential stress. This stress is so much greater than the longitudinal stress to which a piece of ordnance is subjected that we need not discuss the theoretical considerations bearing upon the latter. It has been calculated that the longitudinal tension in a gun of 1 calibre in thickness is only $\frac{1}{3}$ of the circumferential tension.

Much less than circumferential.

As in the case of coils or tubes required to resist a circumferential tension, the fibre should run round the gun; so for the purpose of withstanding a longitudinal tension the fibre should run lengthways, as it does in the steel tubes of Woolwich guns, and in the solid forged breech-pieces of the Armstrong rifled breech-loading guns.*

Armstrong Construction.

Armstrong construction. Wrought iron coils shrunk over each other.

To Sir W. Armstrong is due the credit of first successfully employing the principle of initial tensions for all the parts of a gun.† He employed wrought iron coils shrunk one over another in such a manner that the inner tube is placed in a state of compression, and the outer portions in a state of tension; the amount of tension being so regulated‡ that each coil should perform its maximum amount of useful effect in resisting the pressure from within.

Arrangement of the fibre.

Further, he arranged the fibre in the several portions so as to be in the best position for withstanding the pressures by making it run round the gun in the coils for withstanding the circumferential tension, and by placing a forged breech-piece over the powder chamber, the fibre in which ran lengthways, in order to withstand the longitudinal stress.

Woolwich guns, initial tensions and varying elasticities employed.

In our Woolwich guns, which are an improvement on the original Armstrong construction, we employ both the principles of initial tensions and varying elasticities, using a very strong and elastic steel tube, with coils of wrought iron for the outer portions, which are shrunk on so as to compress the steel tube, while themselves in varying states of tension, the outer coils having the maximum amount.

Palliser guns, varying elasticities.

In our Palliser guns the wrought iron tube is permanently extended by firing proof charges, and the cast iron casing placed slightly in a state of tension.

In fact these two modes of making each concentric layer of metal afford the maximum useful effect are employed, either singly or simultaneously, in all gun construction of the present day.

The most successful constructions then so far employed, and those on which the most powerful guns have as yet been made, are such as have one coil or tube built up over another, in some similar manner to that

---

* The 3-pr. B.L. gun submitted by Sir W. Armstrong in 1855, had a steel barrel, but that material was abandoned, owing to the difficulty of getting it of suitable quality at that time.

† The first Krupp guns were made from solid blocks of steel.

‡ Not that it was attempted rigidly to carry out the tension and compression of the several rings in accordance with actual mathematical calculations, but the layers were put on successively, with sufficient shrinkage to ensure a certain amount of compression of the inner portion.

CHAP. II.

first practically introduced by Sir W. Armstrong, and in such a way that each outer coil compresses those inside it, and so supports them while made weaker itself; this may be done by forcing on the outer tubes by hydraulic pressure, as in the Whitworth guns,* or by expanding them by heat,† dropping them over the inner layers, and then allowing them to cool, when they contract upon the parts inside. Our own guns are built in this way, as are those of Armstrong, Krupp, Vavasseur, and others.

Shrinkage.

In all these cases the inside diameter of the outer tube, when cold, must be rather smaller than the outside diameter of the inner tube; this difference in the diameter is called the "shrinkage."

In the figure below the shrinkage is the difference between AB and CD, AB being a little the larger.

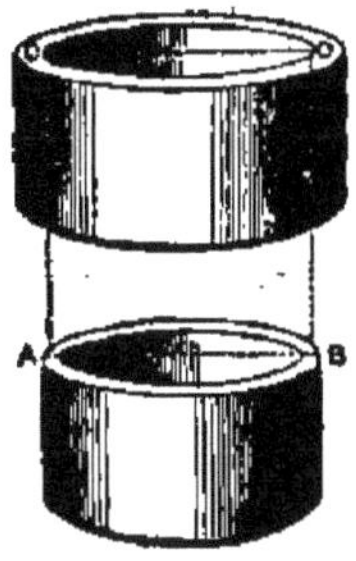

Compression.

While the outer coil is cooling and contracting it compresses the inner one, and makes AB rather smaller than before; the amount by which the diameter AB is decreased is termed the "compression."

Extension.

Again, the outer coil itself is stretched on account of the resistance of the inner one, and its diameter CD is increased; this increase in the diameter of an outer coil is called "extension."

The shrinkage is equal to compression *plus* the extension, and the amount must be regulated by the known extension and compression under certain stresses and given circumstances.

The compression varies inversely as the density and rigidity of the interior mass, the first layer of coils will therefore undergo more compression than the second, and the second more than the third, and so on. Accordingly, in the Armstrong or original construction, a greater proportion of shrinkage was given to the inner layers than the outer, because so much of it was absorbed by compression. The shrinkage, however, never exceeded ·002 of the diameter. Much too will depend on the thickness and strength of the coil to be shrunk on; for the heavier it is, the tighter will be its grip, and the more will the inner parts be compressed and supported, whereas a thin weak coil, if shrunk on the same mass, would probably suffer from over extension. In the guns of present construction the heavy breech coil compresses the steel barrel to such an extent that the latter becomes in some instances reduced in diameter by even as much as $\frac{4}{100}$ths of its original dimensions

* It is understood that this mode has been abandoned in the heaviest Whitworth guns, and that shrinkage is had recourse to.

† 500 F. is quite sufficient; this will allow a working margin of expansion beyond the amount calculated, so that the iron need never be raised to the high temperature at which scales form.

during the process of shrinking,* whilst one or two cases have occurred of the thin exterior coils in large guns of the original construction splitting during practice, thus indicating that they were strained beyond their strength (see next page).

Circumstances which regulate amount of shrinkage.

The position of the coil must be also considered. The shrinkage over the seat of the charge is greatest of all, as that part of a gun must be the strongest and firmest, whilst the shrinkage over the muzzle is the least for an opposite reason.

Tension of a coil not to exceed elastic limit of the material.

The extension of a coil when shrunk on must in no case exceed that due to the elastic limit of the material. In wrought iron, for example, the elastic limit of bar iron is about 12 tons per square inch and causes an extension of about $\frac{1}{1000}$th of its length,† any extension beyond this would stretch the iron permanently and weaken the fibre. Hence the extension on no coil should exceed $\frac{1}{1000}$th of its diameter.‡

Add to all the foregoing conditions the expediency of shrinking on the several parts so that each shall do its proportion of work on the discharge of the piece (according to a law not exactly known), and it will be seen that the problem is a very difficult one to solve in practice. Indeed Sir William Armstrong, who introduced the system, has admitted that he did not carry out his plan with theoretical precision, but that the coils were simply shrunk together sufficiently to secure the stability of the fabric and that a small variation was immaterial.§ One primary object therefore was to secure cohesion throughout the mass.

Shrinkage not carried out with theoretical precision.

Building up of Service Ordnance in keeping with theory.

We have so far considered generally what we laid down, p. 19, as the first point to be treated of in gun construction, viz., the absolute casting, construction, or building up of the gun, looking upon the latter as a tube destined to withstand a certain strain from within.

---

* TABLE showing CONTRACTION by shrinking and EXPANSION by proof, on the bore of a steel tube for a 10-inch R.M.L. gun, Mark I.

—	Diameter of bore.		Contraction.	After proof.	
	Before Shrinking on Breech Coil.	After Shrinking on Breech Coil.		Diameter of bore.	Expansion.
	inches.	inches.	inch.	inches.	inch.
At 90 inches from muzzle ..	9·998	9·998	—	9·998	—
„ 104 „ „ ..	9·998	9·994	·004	9·994	—
„ 110 „ „ ..	9·998	9·991	·007	9·994	·003
„ 116 „ „ ..	9·998	9·988	·010	9·99	·002
„ 122 „ „ ..	9·999	9·99	·009	9·991	·001
„ 128 „ „ ..	9·999	9·991	.008	9·991	Nil.
„ 134 „ „ ..	9.998	9·994	·004	9·996	·002

† The amount of extension varies not only on the quality of the metal, but with the size and shape of the specimen tested. The specimens tested in the Royal Gun Factories' machine are two inches long and stretch from ·003″ to ·004″ at the limit of elasticity, but from the experiments quoted by Mr. Kirkaldy ("Experiments on Wrought Iron and Steel," p. 66) made from a bar of iron, 10 feet long, and 1 inch square, the elongation was only $\frac{8}{100000}$th of the length, for every ton, per square inch of section up to the elastic limit, after which it rapidly increases up to the breaking tension.

‡ Which is the same thing as saying that the inner layer of metal of the coil should not be stretched more than $\frac{1}{1000}$th of its length.

§ Report on Ordnance, 1863, p. 162. "Construction of Artillery," Inst. Civil Engineers, 1860.

CHAP. II.

Disposition of the metal in our service guns.

Let us see how we carry out in our own service ordnance the correct principles of having a sufficiently hard surface for the bore, and of so disposing of the metal or metals in the several layers, that we obtain as much work as possible out of our material, and have at the same time a tube more than strong enough to withstand the greatest strain which it can be called upon to bear.

Armstrong or Original Construction.

Construction of our R.B.L. guns until 1867.

Early heavy R.M.L. guns.

The Armstrong construction, where the gun was built up of a number of thin coils, and in which solid forged breech-pieces were employed, was that upon which all our R.B.L. guns were made, and also the earliest of the heavy R.M.L. guns, vizt., 64-prs., 7-inch, 8-inch, 9-inch, and 12-inch of 25 tons, all of Mark I.

Introduction of steel tubes.

For R.B.L., 1863.

With this construction steel tubes had been gradually introduced, though their employment was but partial; thus in 1863 it was approved of that the A tubes for all 12-pr. R.B.L. guns should be in future made of toughened steel, and, as will be seen by the table at p. 143, a certain number of our service R.B.L. guns have in consequence steel tubes. After the introduction of R.M.L. pieces it remained for some time an open question which material, steel or coiled iron, should be used for the inner barrel, and at the end of the year 1867 it was proposed by the O.S.C. and approved of that for the 9-inch R.M.L. guns to be manufactured in that year, one half of the barrels should be of steel and the remainder of wrought iron, whether the guns were of the service (original) construction or of the Fraser construction, which was then experimental.

For R.M.L. of original construction in 1867.

For all R.M.L. guns in 1868.

Steel when toughened in oil proved itself, however, a very superior material to wrought iron for the inner barrel of the piece, and in March 1868 it was therefore decided that the inner tubes of all R.M.L. built up guns subsequently manufactured should be made of that metal, so that although some of the early natures of R.M.L. guns had wrought iron tubes, all later patterns have barrels of steel. We shall see further on that the only change since that date has been in decreasing the thickness of the steel tube as compared with those first employed.

Disadvantages of original construction.

Too many pieces.

Costly solid forged breech-pieces.

Liability to split of outer coils.

Although this original construction gave good results, it was expensive, both from the quality of the iron required and from the number of pieces to be made, moved about, and put together. The large forging for the breech-piece was also very costly. Further, the thin outer hoops were not always of sufficient mass properly to compress the parts within them, and although a series of thin coils help us to distribute the circumferential stress between the concentric layers of metal of a gun by shrinking on each coil separately, the method is open to the serious objection that it is practically difficult to calculate the respective proportionate amount of tension, and consequently the greater number of pieces in a gun, within certain limits, the more likely that some weakness will exist in the mass owing to the undue strain on some of the parts; for instance, a 13-inch gun of original construction (Experimental No. 300), split some of its outer coils while the interior ones remained uninjured, thus clearly proving that there was too great strain on the former.

Fraser Construction.

Important modification by Mr. Fraser in 1865.

Mr. Fraser, the Deputy Assistant Superintendent of the R.G.F., therefore proposed in 1865 his modification of the system. This consists of using larger coils made of thicker bars, and so much

CHAP. II.

stronger longitudinally, that the forged breech-piece becomes unnecessary. The greater weight and strength of these outer coils also allows compression to be given more certainly to the steel barrel and inner coils.

Thick coils. No forged breech-piece.

In this construction, moreover, the trunnion ring which was merely shrunk on in the Armstrong guns, and occasionally slipped, is welded to the breech coil.

Trunnion ring welded to breech coil.

This Fraser construction was partly adopted in Mark II. 64-prs., 7-inch, 8-inch, 9-inch, and 12-inch, in which guns the forged breech-piece was retained. In all later marks of these guns introduced into the service since 1869 no forged breech-piece is used.

Partially adopted in 1867. Wholly in 1869.

The following short account of the trials of strength of the "cheap" or "Fraser" construction may be interesting. Colonel, now General, Campbell, who as Superintendent R.G.F., had spent much time and labour in inquiries into the question, reported officially in 1864 that it was quite possible to reduce the expense hitherto incurred in the construction of wrought iron guns, by the employment of steel barrels, with external coils reduced in numbers, and forged from iron of a cheap quality,* and subsequently he submitted tracings of two 9-inch iron guns on this plan, one with a steel barrel and one with a wrought iron barrel; but both had breech-pieces, and the *D* and *C* coils were not hooked together. The tracings were approved in October 1864, and he was directed to manufacture the guns accordingly.

They were tested in 1865, and from the result of the previous trials, the Ordnance Select Committee reported † that the guns of cheap construction were not inferior in strength to those on the more expensive mode; but recommended further trials with 9-inch guns including those on a newer plan without forged breech-pieces, which Mr. Fraser, from the experience recently gained, was led to submit.

First trial of 9-inch guns on this construction in 1865.

In the meantime the naval authorities were anxious to obtain cheap 64-prs. and two "Fraser" guns were tested to destruction, to ascertain whether either pattern would answer. One (Expl. No. 320) was known as the *B* pattern (see Mark II., Plate V.), the other (Expl. No. 317) as the *D* (see Mark III., Plate V.). These experimental guns fired each 2,000 service rounds (charge 8 lbs.), and abnormal charges being subsequently used, the *B* gun burst into 38 pieces, after a total number of 2,270 rounds, and the *D* gun after 2,211 rounds blew out its inner barrel, leaving the breech portion on its carriage, and still sound.‡

64-pr. *B* and *D* guns tested in the meantime.

The great strength of the cheap construction, as displayed by these 64-prs., combined with its apparent success in the 9-inch guns under trial, induced the Ordnance Select Committee to recommend in December 1866 that half the guns for 1867-8 should be on the Fraser method, and it was subsequently approved by War Office and Admiralty, that all the guns estimated for that year should be on this plan.

Adoption of the Fraser as the service construction, 1866.

During the years 1867–8 many of the cheap guns had been proved and issued for service, the majority being on the pattern of Expl. No. 330—*i.e.*, with a steel tube, and reinforced with a triple breech coil (Mark III., Plate XIII.).

Of the 9-inch guns alone about 100 had passed proof, and the soundness of the system of construction was generally recognised, but on the 25th of September, 1868, a 9-inch gun of that pattern burst at the first round of proof. This was no doubt due to a defective tube and

Bursting of a 9-inch gun at proof in 1868.

* Extracts from "Proceedings of Ordnance Select Committee," vol. ii., p. 230.

† Ibid., vol. iv., p. 192.

‡ Ibid., vol. iv, p. 356.

CHAP. II.

to the very heavy proof charges used, 53¾ lbs. R.L.G. powder, which caused an abnormal pressure.*

Final trial of 9-inch guns in 1869.

Further trials with two cheap 9-inch guns were therefore recommended and approved. No. 332, with a steel tube two inches thick, and reinforced with two double coils, survived the ordeal. No. 368, with a steel tube three inches thick, and reinforced with one massive triple coil, did not, it is true, complete the test, but it refused to yield although its tube was split at the 1,008th round.

The result was deemed most satisfactory, not only because the steel tube failed gradually, but because the great strength of the outer fabric—the point at issue—was proved beyond all doubt by the gun actually firing 41 rounds after the tube was split through, and yet remaining sound exteriorly. (For details see Ordnance Select Committee's Proceedings, 1869.)

In fact both guns behaved so exceedingly well under trial that the authorities were left in the dilemma of not knowing which pattern to choose. They, however, decided on constructing 7-inch and 8-inch guns as before, on the No. 368 type, but to make 9-inch guns and upwards on the No. 332 type.†

Thin steel tubes for heavy guns in 1869.

In 1869 an alteration took place in the manufacture for 9-inch guns and upwards, a thinner steel tube being used, and two coils on the breech (see Marks IV. and V., Plate XIII.) instead of one triple coil. The higher natures are made in the same way, but have a "belt" in addition.‡

We thus, step by step, arrived at our present construction of a Woolwich gun where we not only make each portion of the metal do its proper amount of work by shrinking one coil over another, but also use a tube of stronger material, steel, for the inner barrel. We employ this latter not only because we gain by it longitudinal strength, but more especially because steel gives us a hard smooth surface, is much stronger than wrought iron, and free from flaws. Its elastic limit is also much greater than that of wrought iron, especially when toughened in oil (as 30 tons to 12 tons), so that it can stand without deformation a pressure which would permanently stretch a wrought iron barrel.

Steel, why used for our gun barrels.

Woolwich guns.

In our heavy R.M.L. guns, we thus make use of the several methods employed to make every portion of the metal do its proper amount of work. In this way we avoid, as far as possible, extra weight, and succeed in making guns so strong that we may assert without fear of contradiction that in the system of construction now pursued by us, though theoretically it may not be perfect, we have the safest, cheapest, and most simple system of manufacturing heavy ordnance which at present exists.

Converted guns.

Our converted Palliser guns are in theory and practice comparatively weak, for the cast iron exterior is too weak, and the material (wrought iron) used for the inner barrel, though strong to resist rupture, is too soft and inelastic, and also liable to flaws. Still as an economical mode of using up old S.B. guns, and for ordnance not meant to fire large charges, it answers tolerably well. We cannot expect such

Why wrought iron barrels are used with these.

* To prevent, however, a similar occurrence, steel tubes for heavy guns are now subjected, after toughening, to a water pressure on the interior of about 4 tons per square inch, which is sufficient, it is thought, to detect any latent cracks. Any gun that has stood the proof may safely be relied on as free from these dangerous defects.

† This pattern costs about the same as the other, the extra expense of making two breech coils being compensated for by the lighter steel barrel.

‡ See extracts from Proceedings of the Ordnance Select Committee for 1864–9.

pieces to be powerful guns, but they will at any rate be much more formidable than the S.B. guns from which they are made*.

2. *Breech-loading and Muzzle-loading.*

We come now to the second point to be considered by the gun constructor.

The gun tube having been constructed, should the piece be a muzzle loader or a breech-loader,† and if the latter, which of the many systems proposed or in use for closing the breech should be employed.

Question of B.L. v. M.L.

The question of B.L. v. M.L. is one of great complexity, but it hinges broadly upon the value to be assigned to the advantages peculiar to B.L. guns, as against the simplicity in which M.L. guns are necessarily so superior.

Further, the qualities required in the several cases of field guns, siege ordnance, and heavy guns, are in many ways so different that each case should be taken separately if we wish to arrive at satisfactory conclusions. It may indeed be advisable under some circumstances to employ B.L. guns, and under others M.L. guns.

Some advantages claimed for B.L.

The principal advantages usually claimed for B.L. guns are—

1. Reduced size of embrasure.
2. Less exposure of men.
3. Greater rapidity of fire.
4. Increased length of gun.
5. Greater power and accuracy of gun.
6. Superior durability.
7. Facilities for examining the bore.
8. Use of a large chamber, and consequent short cartridge.

Reduced size of embrasure.

As to (1) it applies only to heavy guns, and with the muzzle pivoting system now in use for M.L. guns, much value can hardly be given to it. Thus with our 38-ton guns behind iron shields, the embrasures are not more than 3′ 2″ by 2′ 9″ in some cases.

In siege works of the future it is presumed that embrasures will not be, as a rule, employed.

(2.) With field guns in the open, M.L. or B.L. would be on a par as to exposure of gunners, but when cover, such as hedges, gun pits, &c., is made use of, B.L. might have slight advantages.

Less exposure of gunners.

In the case of siege ordnance, the advantage might in this respect be still more apparent, although we require carefully conducted experiments to make the matter certain. It must not be forgotten that while serving M.L. guns, where no embrasure exists, the men are more immediately under cover of the parapet than when B.L. guns are used, and that both with B.L. and M.L. siege and field pieces an amount of recoil will take place, which must give ample space for loading M.L. guns.

Regarding heavy guns behind embrasures, the B.L. has apparently some advantage, but mechanical methods of loading now being tried with M.L. appear to promise so well, that here also this advantage would be considerably lessened for L.S., while for S.S., when guns are mounted in turrets, where mechanical loading can readily be applied, the M.L. system is undoubtedly the best. In this latter case especially

---

* Vide Table IX, Chapter IV., p. 94.

† It cannot be too often repeated that, save indirectly, the construction of a gun both as to its material and building up, it is quite separate from the question of B.L. or M.L.

CHAP. II.

the disadvantage of the turret being filled with smoke on opening the breech for loading, would be much felt; nor must this disadvantage be lost sight of if guns are mounted in casemates or in a broadside armament.

Facility of loading and rapidity of fire.

(3.) As to facility of loading and rapidity of fire frequent experiments with field, siege, and heavy pieces have shown that if anything, M.L. guns have the advantage (vide p. 75, as to comparison of M.L. and Krupp field guns). An interesting trial was made in 1878, between two 13-pr. field guns built up in precisely a similar manner, and of the same weight and dimensions. One of these pieces was an M.L. gun, and the other a B.L. with B.L. apparatus on the French system with the improvements introduced at Elswick.

The projectiles fired were of the same weight and dimensions, and the M.V.'s obtained were equal, and as to facility and rapidity of loading there was practically no difference, the M.L. having perhaps a slight advantage if any worth mention existed.

As to heavy guns, we may well quote here from Colonel Reilly's valuable report.* This officer had the advantage of seeing German heavy guns tested as to their carriages, &c., in the presence of a select committee of German officers, the guns being 28, 26, 17, and 15 c.m.† The breech arrangements, he says, for all the pieces consisted of Krupp's latest improvements,‡ and goes on to report:—

Length of time to load heavy Krupp B.L. guns.

"The result of this day's practice was to my mind a complete answer to the first argument in favour of B.L. guns, viz., the facility of loading and firing; the guns were worked by well-instructed men, who are continuously employed in the establishment; we therefore saw the breech loading tried under the most favourable circumstances. The operation of loading appeared to be laborious, complicated, and, if not carefully performed, dangerous to the gun, and I should conceive consequently much slower than that of loading a M.L. gun; whilst notwithstanding the care bestowed upon the closing of the breech, an escape of gas was at times observable. But as my experience is limited to one day's fire, it will be well to quote the opinion of an officer‡ given in a report otherwise favourable to Mr. Krupp." He says:

"Krupp's cylindro-prismatical wedge with Broadwell ring, the best breech apparatus known up to the present time, requires for its manufacture uncommon mechanical appliances, and great technical skill.

Numerous precautions necessary.

"Before every part of it is right and works well enough for service, a great deal of time and trouble is required, and very little is enough to disable it, intentionally or involuntarily, for present use. The wedge, the ring, the gas escape plate of one breech-piece and gun, cannot at once be used with another gun, so that only under certain suppositions, would it be of any use to take duplicates of these and other parts of the breech apparatus for the purpose of serving several guns in common.

"Some of the parts, especially the Broadwell ring, are subject to the

---

* Notes of a Visit to Berlin, 1872, by Col. Reilly, C.B., R.A., Assistant-Director of Artillery.

† Corresponding pretty nearly to our 11, 10, 7, and 6·3 inch guns.

‡ "It consists of a single steel block which runs in a slot cut transversely through the breech of the gun; to facilitate the movements of the breech-piece, which is of considerable weight, a roller bar is let into its upper edge, which moves on rollers fixed in the upper part of the slot.

§ Lt.-Col. de Wilheelmi, of the Imperial Royal Austrian Marine Ordnance Corps. Archiv für Seewearn, March and May, 1868.

CHAP. II.

weakness of losing immediately, by rough handling, those properties without which they do not act.

"To keep the breech-piece in order in such a way that it may at all times be used without any delay, renders compulsory many precautions in the daily cleaning, especially as regards polishing and preserving, and demands, at least on the part of the captain of the gun, the knowledge of how to put the apparatus together.

"On board ship, more than anywhere else, on account of the damp, it is difficult to prevent the rusting of the surfaces which are to be kept bright, and uninterrupted attention must be paid to the gun from this cause alone, but where there is a great deal of scouring there is a great deal of wear.

"In the event of the Broadwell ring not closing the breech particularly well, both the ring and the plate must be wiped clean after every shot, which is not easy to do when the wedge is in the position for loading, and can therefore only be done imperfectly. In such case it will be more frequently necessary to bring out the wedge to the end of the groove, but this will involve a loss of time, which time may possibly be invaluable.

* * * * * * *

"Although only three movements are necessary to open and as many to close, the possibility is not excluded of one of these movements being forgotten in the heat of battle.* If this oversight occurred in closing—we mean an omission in pushing to the locking bolt—the breech would not be closed, on discharge the gas would rush violently out, the respective surfaces would be so fouled that they would require a thorough cleaning and, which would be the worst part of the matter, this oversight might possibly lead to the breech-piece being blown out, which would be equivalent to putting the gun *hors de combat.*"

It appears therefore that in facility of loading, and consequent rapidity of fire, the advantage lies with the muzzle-loading system also in the case of heavy guns.

The above extracts refer, it must be remembered, to one of the most improved B.L. apparatus used.

Rapidity of fire.

No data as to heavy B.L. guns.

We have no data to go upon as to rapidity of fire with heavy B.L. guns, but abundant information is at hand to show with what facility our own R.M.L. guns can be loaded and fired. To give a few instances: In 1875 seven detachments of Garrison Artillery competed at Shoeburyness with 10-inch 18-ton guns. They fired 35 rounds with the results shown:—

Rapid fire of our R.M.L. 10-inch gun.

Number of		Hits.			Mean error in range.	Mean time per round.	
Detachments.	Rounds fired.	Direct.	Ricochet.	Total per round.		Loading only.	Loading, laying and firing.
					Yards.	Mins.	Mins.
Seven	35	7	6	0·37	30	1′ 17″	1′ 45″

The targets were 9 ft. × 9 ft. placed at ranges of 1,200 and 1,600 yards alternately.* The day was unfavourable, there being a strong and variable wind across the range, while a heavy rain-storm just before the practice rendered everything wet and slippery.

* The distances were actually measured.

CHAP. II.

It will however be seen that under these disadvantages one shot out of every three hit a small target at the distance of about a mile and ¾ of a mile respectively. The shooting, we are told, was very fair as to line, all the shot being close to the targets.

R.M.L. 38-ton gun practice.

Again in July 1876, experimental practice was carried out, under service conditions, with a 12·5-inch 38-ton gun mounted in a casemate at Shoeburyness, 100 rounds were fired at ranges from 1,000 to 3,000 yards at targets 9 feet square, with charges of 130 lbs. of powder and 800 lbs. projectiles, at the average rate of one round in less than two and a half minutes, the precision of fire being excellent; no two consecutive rounds were fired at the same target.

Number of Rounds fired.	Total number of Hits.	Number per Round.	Average time per Round to load and fire.	
			Mins.	Secs.
100	27	0·27	2	9¼

It must be remembered that the men were unused to the gun, and that we have not yet quite perfected the mechanical means by which the loading and firing of our heavy guns will soon be very much facilitated both on land and sea.

Practice with our heavy guns at sea.

At sea with all the drawbacks due to the rolling of the ship, &c., we find that in 1876 prize firing throughout the fleet gave the following results:—

Nature of Gun.	No. of Guns.	Average time of firing 8 rounds.		Average time per round.		Average points per round.*		Remarks.
		M.	S.	M.	S.			
12 inch, 35 tons	4	14	24	1	48	8	12	Turrets
12 " 25 "	4	15	17	1	55	8	31	Turrets
9 "	67	9	47	1	13	7	76	
8 "	31	8	27	1	8	7	48	
7 "	120	8	21	1	3	6	20	
64 pr.	174	8	8	1	1	6	62	

During the R.A. prize firing in 1876 at Shoeburyness, the best detachment fired five rounds at the rate of 1·2 mins. a round and struck a target 9′ × 9′ at a distance of 1,200 yards, three times or at the rate of 0·6 hits per round.

Increased length of gun.

(4.) As to increased length of gun; with field pieces, M.L. guns can claim this advantage rather than B.L. in which a certain extra length behind the charge is required for the B.L. apparatus. In the matter of siege ordnance meant to be employed under cover, B.L. pieces may have the best of the matter in this respect, except in the case of the short rifled mortar or howitzer now everywhere adopted as a part of a siege train.

With heavy guns, except when mounted in turrets, B.L. pieces have no doubt in many cases an advantage in this important point.

* For system of awarding points, see p. 187 of Gunnery Manual for H.M. Fleet.

CHAP. II.

Comparative power of gun.

(5.) As to power and accuracy of fire, we have lately been told (vide note†, p. 76) that we had in the case of M.L. field guns reached a limit of power beyond which we could not go, but this statement was apparently made under the supposition that it was not possible with M.L. guns to make the powder chamber of different calibre from the rest of the bore, a supposition which experiment and practice have shown to be mistaken.*

No windage with B.L.

It has been further urged that with B.L. guns, windage, and so loss of power could be quite prevented by the forcing of copper rings, lead coating, &c. through a bore smaller than the powder chamber, by which also greater accuracy of fire could be ensured from the better centring of the projectile.

The use of expanding wads or cups attached beforehand or upon discharge to the base of the projectile has proved quite feasible with M.L. guns, and is now largely employed. By their means windage can be completely stopped, and the advantages claimed for B.L. be participated in by M.L. guns.†

In the case of an enlarged chamber, it is more easy to avoid the use of an excessively long cartridge in B.L. guns, so that they have in this respect a theoretical advantage not to be despised.

Accuracy of fire.

As to accuracy of fire, it seems generally supposed that B.L. guns are more accurate than M.L. guns, apparently because it has so often been so asserted. There may formerly have been grounds for this, for in B.L. guns, from the very nature of things, the shot was always stopped at the same point in the bore, while a chamber was left in rear of its base of a constant capacity in which the charge was burnt. This no doubt tended to produce a more regular M.V. than when no such definite chamber existed in M.L. guns, from the shot being rammed not always to exactly the same place. As already explained, however, these advantages no longer exist of necessity in the case of M.L. guns. We have not many recent facts to go upon in the case of heavy B.L. guns, but the following example gives us good reason to suppose that in accuracy of fire our M.L. guns are on a par, if not superior, to any R.B.L. yet manufactured.

In 1872 60 rounds at various elevations were fired at Shoeburyness from a 40-pr. B.L. and a 40-pr. M.L. gun for comparison, service charges being employed. As to accuracy there was very little difference. In reduced deflection they were almost equal, in some cases the M.L. and in others the B.L. being more accurate. In mean difference of range the B.L. had a slight advantage at most ranges, but at two of the elevations used the M.L. proved the better of the two.

The fire of our M.L. field guns as exemplified during the Dartmoor experiments of 1875 was very accurate and proved that up to ranges of 4,000 yards it would be difficult for troops to advance under their fire.

Accuracy of our heavy guns.

The following practice carried on at Shoeburyness in 1878 gives us a good idea of the accuracy of the 12·2 polygrooved 38-ton gun, rotating gas checks being used:—

---

* The increased size of the powder chamber is a source of weakness in the case of B.L. guns, an excessive pressure being given to the most vulnerable part of the gun, but with M.L. this is not the case. Not long ago, indeed, during the trial of a B.L. steel gun fitted with B.L. apparatus similar to that of Krupp, the gun failed through the blowing out of the gas check ring, the breech block being forced back, having previously indented the comparatively soft steel of the rear face of the breech slot.

† Any system of B.L. such as Krupp's, where a soft coated projectile is employed, entails also a serious loss of penetrative power with heavy guns.

CHAP. II.

Range.	Mean Error.		Remarks.
	Range.	Direction.	
Yards. 3399	Yards. 12·4	Yards. 0·6	Mean of 5 rounds.

We have already given other remarkable instances of the accuracy of our heavy R.M.L. guns when fired against time, as in prize firing. Looking at these we can hardly credit the assertion that light or heavy B.L. guns surpass very much in accuracy our own M.L. pieces.

Durability or Endurance.

(6.) Durability.—This point hardly affects the question of siege or field guns, for our smaller field guns can fire a very great number of charges with scarcely any perceptible damage. With heavy guns, however, the case was formerly different, the erosion of the bore, due to windage, and to the rush of heated gas between the projectile and surface of the bore, was an undoubted defect.

By the use of expanding wads or gas checks, as mentioned above, this defect may be considered as done away with.* Our knowledge as to the endurance of very heavy B.L. guns is very small, and it is not improbable that facts may show it to be less than that of our heavy M.L. pieces, now that the latter can be fired with gas checks.

Facilities for examination.

(7.) As to facilities for examining the bore—The complete examination of a M.L. steel lined gun is so simple an operation, that little value can be attached to the slight advantage B.L. guns have over M.L. in this respect.

(8.) As regards (8.), see p. 38.

Advantages of M.L. Guns.

Advantages of M.L. ordnance.

We see then that although the question of M.L. and B.L. generally, is one on which opinions may fairly differ, the advantages claimed on one side or the other are nicely balanced, and that for some uses they incline slightly more towards B.L. than in others. There is, however, one point where M.L. has a most evident and undeniable advantage, an advantage of more especial importance in the case of field guns, and that is in simplicity of gun equipment and stores.

Simplicity.

It was principally on account of this advantage that we abandoned R.B.L. guns, and so far we have not seen such marked superiority in any B.L. system, either proposed or in use, as would lead us to sacrifice this simplicity in the case of field guns, even when the balance seemed for some reasons to incline in favour of B.L.

However excellent a B.L. system may be, B.L. ordnance must have several parts, each and all of which are liable to derangement, and which must be carefully looked to and repaired. This necessitates carrying into the field, spare parts and complicated stores, as well as great care in the handling and service of the gun. The ammunition also becomes more complex, and it is necessary to carry some sort of lubricant to ensure proper working.

Complicated stores and ammunition.

Why simplicity is so requisite in our service.

In the equipment of an artillery like our own, where guns, ammunition, and stores must be interchangeable for and be used in such various climates as those of our widely spread colonies, and in our Indian Empire, as well as for home service, simplicity is of even greater importance than in the case of purely Continental Powers, and unless

* Thus a 10-inch R.M.L. gun lately fired 385 rounds of battering charges with projectiles using gas check; on examination hardly a sign of wear was to be seen. The 80-ton gun has now (March, 1877), fired 94 rounds (since boring up to 16-inch calibre), with projectiles of 1,700 lbs. and 370 lbs. of powder, and shows no perceptible erosion, though the charges used were so enormous.

considerable improvements are made in B.L. apparatus, it does not seem very likely that we shall adopt this system of loading for any nature of ordnance, instead of that which gives us such strength, security and simplicity, combined with accuracy and rapidity of fire in our M.L. gun, which latter are further rapidly improving in all these qualities, as progress is made in the manufacture of our powder, and in that of our projectiles and their adjuncts.

3. *Length of Bore, Calibre, Chamber, &c.,*

Calibre, length of bore and chamber.

In regard to calibre, length of bore, and dimensions of chamber, these points are all comparative, and as their values vary often in an inverse ratio, they have all to be considered, and their relative advantages carefully balanced in each particular case, as to the best mode of consuming the powder charge before and after the vis inertia of the projectile is overcome, and with heavy charges as to the development of abnormal pressures, should the cartridge be too long, then again as to the best weight and length of a projectile. In regard to the distance at which we want the projectile to act, a large calibre will generally give a greater velocity at short, and a lower velocity at long ranges than a small calibre; while the former again is favourable to large shell charges. Here we must also be guided by the purpose for which the gun is intended, whether field, siege, or garrison, and by other extraneous considerations, such as weight, strength of carriage to withstand the recoil, &c. So that both powder, projectile, and carriage, must be thought of as well as the gun itself.

Practically, most constructors have arrived at nearly the same conclusion with regard to these points; and we find that our own service guns, and those of Krupp, though of totally different materials and construction, have about the same calibre and length.*

Length of bore.

As to the length of bore, it is perhaps best considered as giving space for so many expansions of the powder charge; the rate of burning of the powder of course affects the question of length very much, as with a slow burning powder, it is advisable to use a longer bore than with quick burning powder, the tendency in these days is, therefore, to lengthen the gun, and our later pieces have greater length of bore than the earlier guns.

Useful effects of powder.

The diagrams at the end of the book show how rapidly the additional useful effect per pound of powder falls off, for each additional expansion of cartridge, *i.e.*, for each increment of length of bore after a certain length is reached.

The data from which it is made out are taken from a valuable work recently† published, and it refers to the service powders used by us.

Chambers. B.L. chambers.

As to chambers‡ by which are here meant powder chambers of larger diameter than the rest of the bore, B.L. guns have always been furnished with these, and one of the advantages before mentioned claimed for B.L. was the possibility of having such a chamber,§ which allowed of a larger charge of powder being usefully consumed; in

* The latter guns, of more recent date than our own, and made subsequent to the introduction of slower burning powders, are somewhat longer.

† Researches on fired Gunpowder, by Capt. A. Noble, F.R.S., and Professor Abel, F.R.S., L.S.P.C.S.

‡ Enlarged chambers were originally adopted as a necessity for R.B.L. guns, using lead coated projectiles.

§ The conical or cylindrical chambers of less size than a similar length of bore which are used with S.B. pieces, and in some of our R.M.L. guns were for quite other purposes, as explained at p. 174.

CHAP. II.

Advantages and disadvantages of same with B.L.

these guns, however, the chamber was also a source of weakness, for the gastight joint closing the breech end of the bore has always been difficult to secure, and this difficulty naturally increased with the size of the joint, and so, for any given calibre of bore, with the size of the powder chamber, unless an expanding cartridge is employed.

"*Air-spacing.*"

Recent attempts to reduce the pressure in bore.

The tendency at present is to use very large charges of slow burning powder, in order to obtain power without straining the gun too much. However slow burning the powder may be, we soon reach a limit of charge, which gives an excessive pressure, if the charge be rammed tightly up, so as to fill the whole of the chamber in rear of the base of the shot.* By so arranging that the charge of powder does not fill the whole of the chamber, or in other words, by leaving what is technically termed an air space in the chamber, it has been found that the limit of charge first mentioned can be much exceeded, the chamber being of course at the same time enlarged in capacity.

Air space.

Supposing for instance that a charge of 3¼ lbs. in a chamber of 110½ cubic inches capacity gives a M.V. of 1590 f.s. with the maximum pressure allowed in a 13-pr. gun. This gives clearly a space of 34 cubic inches per lb. of powder in the chamber.

Permits increased charge.

If we now increase the charge to 3½ lbs. and at the same time enlarge the chamber, so as to give a greater space per lb. than before, *i.e.* 36 cubic inches instead of 34 cubic inches, we find that we do not exceed the maximum pressure, although using a larger charge. This larger charge, as might be expected, gives us an increase of M.V., though not in the same ratio as the increase of charge itself. This fact is due to the decreased density of the charge in the chamber, as we do not get the same amount of work per lb. of powder, though gaining on the whole by thus increasing the charge, and at the same time the air space per lb. Of course a second limit is soon reached where the increased space occupied by the larger powder charge deducts so much for the total capacity (or number of expansions) of the charge in the bore, would more than make up for the increase of charge.†

Increased M.V. therefrom.

Methods of air spacing.

We may plainly give the increased air space when using a given increased charge in two different ways, either by, (*a*) keeping the length of the powder chamber the same as before, and enlarging its diameter, the given increased charge being of the same diameter as the original smaller one to allow of loading,‡ or (*b*) by simply increasing the length of the chamber. (*a*) When rammed home, the charge will not completely go into the chamber unless some means of expanding the cartridge is available. Such means have been tried, but lead to so much complication in M.L. guns, that it seems likely they will be abandoned. At the same time as explained below, the advantage of getting the powder into a comparatively short length, is of such utility, that the system of a powder chamber of much greater calibre than the bore of the gun cannot be abandoned without regret. (*b*) We can also give the increased air space for a given increased charge, by making the powder chamber longer, so as to be of the length, or nearly so, of the increased charge when loaded, and by increasing the diameter of the chamber only to a slight degree. In this case we require no device for expanding the charge when loaded, but we have the disadvantage of the charge occupying a considerable length in the bore.

Reasons why a large chamber cannot be used for M.L. gun.

* This space it should always be remembered is the powder chamber proper.

† For further information on this theory of the subject, vide "Principles of Rifled Ordnance," by Captain Sladen, R.A.

‡ In a M.L. gun only. In a B.L. gun this need not be the case; and on this point therefore, B.L. guns have an undoubted advantage.

This is a disadvantage when carried to a certain point, for abnormal pressure* are found to be set up when the cartridge to be loaded is of an excessive length.

Hence, from the reasons given (*a*) and (*b*) we now make the enlarged chamber of the new pieces, in which such chambers are employed, slightly larger in diameter than the calibre of the gun.

Thus, in the 80-ton gun the chamber is 18 inches in diameter, and the bore 16 inches. In the experimental 13-pr. gun the chamber is 3″·15 in diameter and the bore 3 inches.

The proper air space to be given in the chamber in relation to the calibre and length, depends very much upon the nature of powder employed, and we have not sufficient data as yet, to lay down the law very definitely upon the subject.

4. *System of Rifling.*

We now come to quite another consideration, viz.:—What system of rifling should be adopted?

Object of any system of rifling. The object of any such system is to give the necessary amount of rotation or spin to the projectile with a minimum pressure on the bore of the piece upon the grooves, and on the studs, ribs, soft coating or otherwise of the projectile itself; the required result should also be obtained by the most simple means, and with the least possible loss of power.

Necessary angular velocity. The angular velocity† imparted must be sufficient to keep the projectile point first and steady, until it arrives at the maximum range required; as yet we have none but experimental data at hand for determining what the amount of this angular velocity should be. It

---

* The tendency to such abnormal pressure, however, seems to decrease as the air space per lb. of powder is increased.

Regarding chambering M.L. guns, interesting experiments were arranged and carried out in 1873, in the Royal Gun Factories, by Lieut.-Colonel Maitland, then Assist. Supt. R. G. F., under General Campbell's direction, with a 10-inch gun of 18 tons and a 9-pr. field gun.

The 10-inch was first chambered to a diameter of 11 inches for a length of 19 inches, and with service charges the velocity as well as the pressure was rather less than in the unchambered gun; the chamber was then enlarged to a calibre of about 12 inches; the service charge of 75 lbs. gave a still less pressure than before, and slightly less velocity than in the unchambered gun; with an increased charge of 85 lbs. of powder, however, a less maximum pressure was given than in the unchambered gun with the service charge, while the velocity was increased by 75 f.s.

It was also observed that no partially unconsumed pebbles were blown out as the case with the service gun, and that a large and brilliant flash was given at the muzzle, showing apparently, that through the employment of the chamber, this larger charge of powder was more completely consumed, than a smaller charge in the unchambered gun.

To continue, however, as to the R. G. F. experiments of 1873. With the 9-pr. of 3″ calibre, chambers of various capacities[1] were employed, and also different lengths of bore: the maximum increase of M.V. by increasing the length by 8 calibres, and using a charge of 2¼ lbs., which was found to be the largest charge that could be advantageously used in such a calibre, was about 100 f.s., but when a chamber of 3″·8 diameter was employed with a charge of 3 lbs. 8 ozs., an of increase of 210 f.s. velocity was gained with a 9 lb. shot, and using the increased length and chamber together, a total increase was observed of 352 f.s. velocity over the service gun.

† The "angular velocity" must not be confused with the "linear velocity of rotation" which is the lineal velocity of any point in a rotating mass, but usually applied to a point on the surface, while the former is the rate at which every particle of the rigid rotating body moves through a given angle.

---

[1] Bronze chambers of various capacities being introduced for the breech end, and secured in position by means of a strong screw in rear.

CHAP. II.

depends much upon the form, weight and distribution of the mass of the projectile itself.*

How given.

The necessary velocity of rotation is given by means of the twist of the rifling, combined with the velocity of translation at the muzzle; this will be evident when we consider what is meant by the twist of rifling.

Twist of rifling.

By *twist* is meant the amount of inclination of the groove, the angular inclination shown when we develop the grooves on a plane, as shown in fig. below.

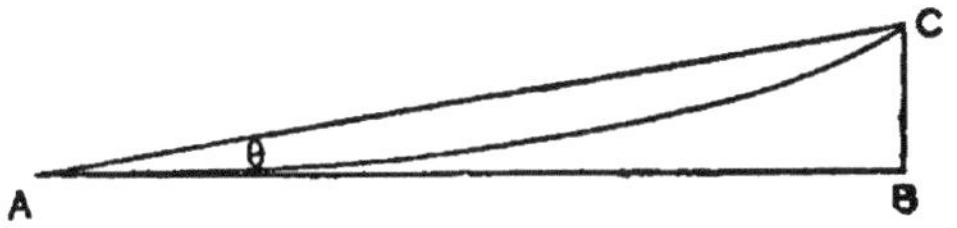

Thus, if AB represents the distance a shot would have to travel along the bore, or in continuation of the same, with a given twist or inclination of groove, while it made one complete revolution round its own axis, and BC at right angles to it, represent the space passed through by any point of the surface of the projectile during the time of its making this complete revolution, then AC will represent the length of spiral due to such twist, and would show the length of a groove in the case of a gun long enough to allow of the projectile making a complete turn in the bore.

BC equals in length the circumference of the projectile $= 2\pi r = \pi d$ if d be the calibre or diameter of projectile, while $AB = n\, d$ if n be the number of calibres in which a complete turn is made.

Angle of rifling.

The angle CAB $= \theta$, is usually termed the angle of rifling.

$$\text{Now, } \tan\theta = \frac{CB}{AB} = \frac{\pi d}{nd} = \frac{\pi}{n}$$

For example, take the 7-inch gun, whose spiral is one turn in 35 calibres, then

$$\tan\theta = \frac{\pi}{35} = \cdot 0897\,; \quad \therefore\ \theta = 5^\circ\ 8'\ \text{nearly.}$$

Did we require the actual amount of twist x in the bore, it is easily obtained from the proportion

$$AB : BC :: l : x,\dagger$$

l being the length of the rifled part of the gun; and we can form a good idea of the course of the groove along the bore, by supposing the position which the line AC would assume were the above figure wrapped round a cylinder equal to AB in length, and BC in circumference.

Uniform or increasing twist.

The twist may be either uniform or increasing: in the former case the inclination of the groove is always the same as shown by the plain line in fig. above; in the latter the inclination increases, and a groove of such rifling, would, when laid out, be some such curved line as shown by the line AC.

The form of this curve may correspond to the locus of any suitable

---

* The best length of projectiles is from about 2½ to 3 calibres, including the ogival head. With this form of projectile it has been ascertained by experiments that a twist of about one turn in 35 calibres ($\frac{1}{35}$) gives the best results, and that any material deviation from this amount in large guns tends to render the projectile more or less unsteady in its flight.

† This is not the case with the increasing twist, where at the muzzle $AB : \frac{BC}{2} :: l : x$, so that the grooves show only half the amount of twist shown in the uniform rifling, which gives the same amount of angular velocity at the muzzle.

equation; we ourselves generally use a parabolic curve where the equation is $x^2 = py$, and in which the increments of inclination are uniform—*Vide fig. below.

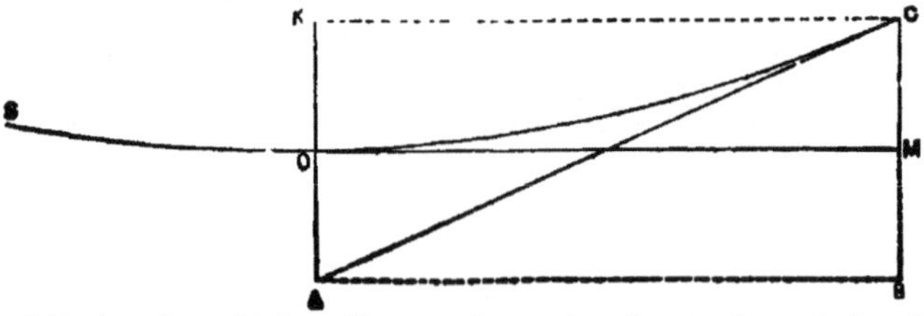

Whether the twist be uniform or increasing the angular velocity of the projectile on leaving the muzzle will depend upon the angle of rifling at that point, and upon the MV of translation.

Let a = the angle which the rifling makes at the muzzle with the axis of the piece.
ϖ' = angular velocity of shot at muzzle.
d = calibre of shot.
V = muzzle velocity.

Then $\frac{d}{2}\varpi'$ in the linear velocity of rotation of a point on the surface of the projectile at the muzzle.

At the muzzle: $\frac{\text{linear velocity of rotation}}{\text{velocity of translation}} = \frac{d\varpi'}{2V} = \frac{\pi d}{n d} = \frac{\pi}{n}$

Angular velocity expressed in terms of V and length of twist.

if we suppose the rifling at the muzzle to be making one turn in n calibres.

$$\therefore \frac{d\varpi'}{2} = \frac{\pi V}{n}$$

(a) $$\text{And } \varpi' = \left(\frac{V}{n\,d}\right) 2\pi\dagger$$

This value of the angular velocity (ϖ') being true, whether the twist is uniform, or increasing according to any law whatever.

We see then, that the angular velocity given depends upon $n\,d$ or the length of twist, and upon V the muzzle velocity of translation, and that by altering either of these variables we can change the rate of rotation of the projectile.

For this reason, with similar projectiles when the muzzle velocity is low, a much sharper twist must be given, *e.g.*—In our 8-inch M.L. guns

---

* In the 80-ton gun, however, the curve used is the semicubical parabola, the equation to which is $x^{\frac{3}{2}} = p.y$.

† Here 2π denotes the ratio of the circumference to the radius of a circle, so that if d is the diameter of a projectile, then πd is the circumference.

Now at the muzzle $\frac{d}{2}\varpi' = \frac{\pi V}{n}$ is the linear velocity of rotation.

Therefore the number of turns per second made by the shot as it leaves the gun,

$$= \frac{\text{linear velocity of rotation}}{\text{circumference of shot}} = \frac{\frac{\pi V}{n}}{\pi d} = \frac{V}{n\,d}$$

Thus with the 10-inch R.M.L., where

$$M\,V = 1364 \text{ f.s.}$$
$$n = 40$$
$$d = \frac{10}{12} \text{ feet.}$$

The shot on leaving the muzzle is making $\frac{1364 \times 12}{40 \times 10} = 40\frac{3}{4}$ turns per second.

CHAP. II.

where the muzzle velocity is 1,400 f.s., the twist found sufficient is 1 in. 40 calibres at the muzzle; while with the 8-inch Howitzer, having a low muzzle velocity, of say 550 f.s., a twist of 1 in 16 calibres is required.

Centring of the projectile. Not only must the system of rifling adopted give sufficient velocity of rotation to ensure stability of flight, but the projectile should also be as far as possible centred as it leaves the gun, *i.e.*, it should be rotating round its longest axis, and that axis should be coincident with the line of flight; as we shall see further on, various modes of obtaining such centring have been tried, certain form of groove, &c. being employed.

Loading and driving side of groove. The two sides of a groove are termed the loading and driving side, meaning respectively the sides against which the projecting parts of the projectile bear, in passing up or in coming out along the bore. In order to centre a shot better, the latter is sometimes inclined at a smaller angle than the loading side, so as to allow the studs, &c. to run up it, and the shot so to be centred.

Pressure upon the studs and grooves. Important points to be considered with regard to a system of rifling are the pressures due to rifling upon the studs or other projecting parts, by means of which the projectile is rotated in the bore, upon the grooves in the gun itself, and upon the walls of the bore.

How ascertained. We can determine the amount of these pressures exactly by a little calculation, if we know the pressure developed upon discharge in the gun, as the former always bears a definite relation to the latter. With a uniform twist, the total pressure on the studs or ribs, and so on the grooves of the gun, is a constant fraction of the pressure on the base of the shot, the value of the fraction depending on the angle of the rifling.*

Pressure on studs with uniform twist. It is evident that with the uniform rifling there is a great pressure at first upon the grooves† or studs, which rapidly decreases towards the

---

* For taking equation (a), p. 41, and differentiating both sides we have—

(b) $\frac{d^2\Theta}{dt^2}=\frac{2\pi}{nd}\cdot\frac{d^2 l}{dt^2}$ where l is the distance travelled in time t, also if ϖ be the angular velocity.

(c) $\frac{d\varpi}{dt}=\frac{d^2\Theta}{dt^2}=\frac{2\pi}{nd}\frac{d l^2}{dt^2}=\frac{2\pi}{nd}\frac{P}{M}\left(\text{because } P=Mf=M\frac{d^2 l}{dt^2}\therefore\frac{P}{M}=\frac{d^2 l}{dt^2}\right)$

Also, looking at the projectile as a cylinder of mass M and radius $\frac{d}{2}$ where p is the pressure producing rotation, and P the pressure on the base of the shot producing translation, which is less than the absolute pressure by about 1¼ per cent. [taking friction at 1 per cent., and the part of p absorbed in direction of bore as proportionate to tan-angle of rifling or $\frac{1}{13}$ of p = ¼ per cent. of P; total 1¼ per cent.], and K being the radius of gyration, we have—

$$(1)\ \frac{d\omega'}{dt}\left(MK^2\right)=p\,\frac{d}{2}$$

$$(2)\ \frac{d\omega'}{dt}\left(M\frac{d^2}{8}\right)=p\,\frac{d}{2}\ \text{in this case.}$$

$$(3)\ \frac{d\omega'}{dt}=\frac{4p}{Md}$$

And from eqnation (3) and (*c*) above—

$$(4)\ \frac{4p}{Md}=\frac{2\pi}{nd}\frac{P}{M}\ \text{or}\ p=\frac{\pi}{2n}P=C^1 P.$$

Where p is a constant fraction of P, the pressure on the base of the projectile producing translation, the value of which depends on n, or what is the same thing upon the angle of rifling.

Supposing $n=40$, that is, that the twist of rifling is one turn in 40 calibres.

Then (5) $p=\frac{1}{25}P$, or the pressure on the studs is 4 per cent. of that on the base of the projectile.

† This is borne out practically by the greater wear which is found to occur in guns rifled with the uniform twist, in that part of the groove near the bottom of the bore. The Russians in their experimental field gun (vide Table, p. 98) are using an increasing twist, as do the French also in their heavy guns, as well as in their new field gun, and for the reasons given in the text.

muzzle; this is owing to the fact that the angle of rifling is the same throughout, while the acceleration, as shown by the increments of velocity is much the greatest to begin with, and falls away very quickly as the shot moves through the bore.

Pressure with an increasing twist.

With an increasing twist, however, the case is different, for here the angle of rifling, which is nothing, or very small when the shot starts, increases rapidly, so that if using the former nomenclature $p = CP$, the value of C increases rapidly, while that of P decreases, and we have a much lower maximum pressure on the studs, but at the same time, a more uniform pressure, so that, for example, in the case of two-10-inch guns, with a twist of 1 in 40 at the muzzle, the one rifled with a uniform, and the other with increasing twist. We see that while the maximum pressure in the one case is 68 tons, coming down to 9 tons at the muzzle, in the other, where the increasing twist is used, the maximum pressure is reduced to one half, being only 36 tons, while the pressure is very uniform throughout.

Advantages of the increasing twist.

It has been found that the substitution of increasing for uniform rifling in the Woolwich guns, decreases the pressure on the studs by about one-half, a matter of much importance.

In order to give rotation to a projectile, the pressure on the bore of the gun must be increased, for not only has the inertia of the shot to be overcome as to translation, but also as to rotation.

With guns rifled like our own this increment of pressure, however, as proved by Captain Noble, late R.A.,* is an exceedingly small one, though small as it is, he says that "it is still less in the parabolic than in the uniform system of rifling."†

His conclusions are borne out by the experiments of the Explosive Committee, who found no sensible differences of pressure in the 10-inch gun fired in the rifled and in the unrifled state.

It is clear that we may obtain a sufficient amount of rotation in the projectile, by the use of any one of the numerous systems of rifling which have been at various times proposed, and then we must choose that particular system which will give the rotation with least damage to bore of gun and projectile.

In 1863, at the time of introducing M.L. guns into the service, exhaustive trials were carried out with various systems of rifling brought forward by inventors, as well as by the official department.‡

On the whole, the Woolwich system suited for studdied projectiles bore away the palm and was adopted.

Modern poly-groove system and introduction of gas checks.

Of late years, the introduction of gas checks, led to the attempt to rotate the projectile by means of a metal base attached to the shot and expanding into a number of shallow grooves. This system § has given such good results, that it has already been adopted in several new natures of ordnance, and bids fair in the future to supersede the system of a few deep grooves and studs (the Woolwich system) in future manufacture.

Form of Grooves.

The grooves.

Regarding the actual grooves in a rifled gun, their size, shape, and number, must depend upon the system of rifling employed, the simpler

---

* Vide paper by Capt. A. Noble (late R.A.), F.R.S., Philosophical Magazine, March, 1873.

† In France this parabolic curve has been adopted for all heavy guns, on account of the smaller maximum pressure on the studs.

‡ For detailed descriptions of trials of these systems, see note at end of chapter, p. 52.

§ By no means new.

CHAP. II.

Shallow grooves preferred to deep ones.

the form and the less sharp the angles, the less will the groove as a rule suffer, and the less liable is the barrel to be split along the angle of the grooves. The strength of the gun tube is less affected by a number of shallow grooves than by a few deep grooves, and the tendency now is to employ the former, which is evidently the best adapted for use with projectile having a soft coating or exterior rings, or an expanding base cup or wad to be cut into by the lands and forced through the bore.

The grooves of our B.L. guns.

With our service B.L. guns, in which lead-coated projectiles are forced through the bore, we have a great number of grooves and narrow lands, so that the lead coating is easily cut through. On account of the number of grooves these guns are said to have polygroove rifling, the form of the groove is very simple as shown below—

SECTION OF RIFLING (full size).

The grooves of our M.L. guns.

With the service M.L. guns using studded projectiles we employ five different forms of grooves as below.

Shunt groove.

(1.) The "*shunt*" which is used with 64-pr. built up guns, Marks I, II, and III which have not been retubed, and which have wrought iron barrels.

PLAN OF MUZZLE SHOWING "SHUNT" GROOVES. Scale 3 ins = 1 foot.

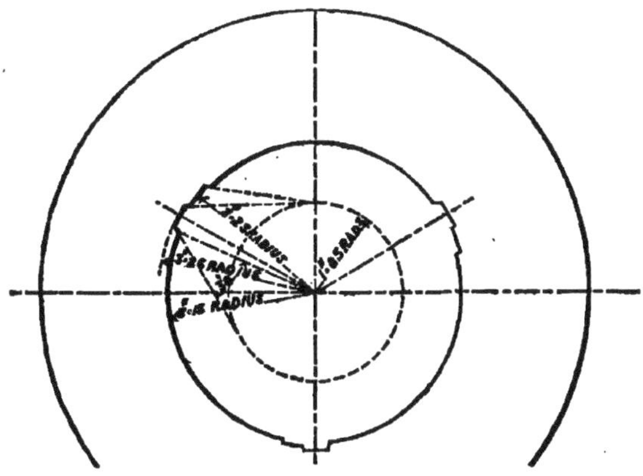

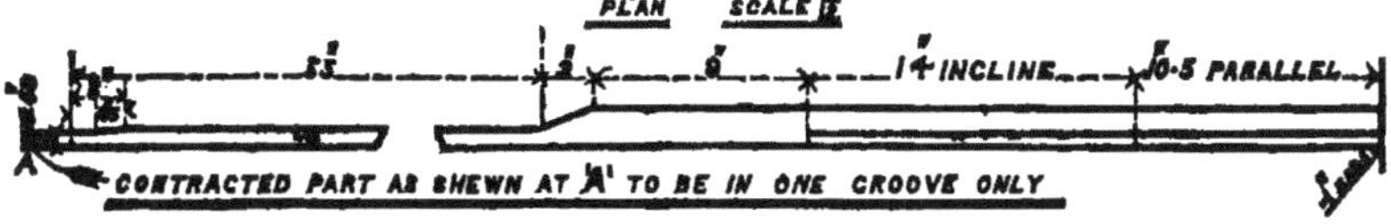

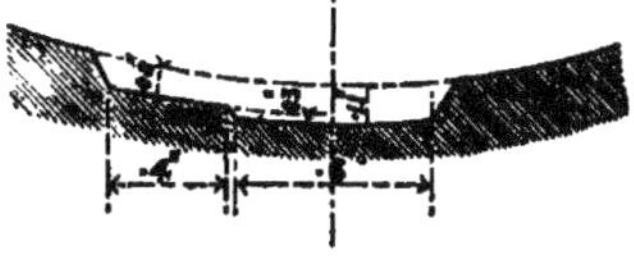

SECTION ON LINE A B
FULL SIZE

CHAP. II.

"Shunt" system.

The peculiarity in this system of rifling is that the depth and width of the grooves varies at different parts, the object aimed at being to provide a deep groove for the studs of the projectile to travel down when the gun is being loaded, and a shallow groove through which they must pass when the gun is fired, so that the projectile may be gripped and perfectly centred on leaving the muzzle. This system was abandoned as being complicated and liable to damage projectile and gun.

Plain groove.

PLAIN GROOVE. Scale $\frac{1}{8}$th.

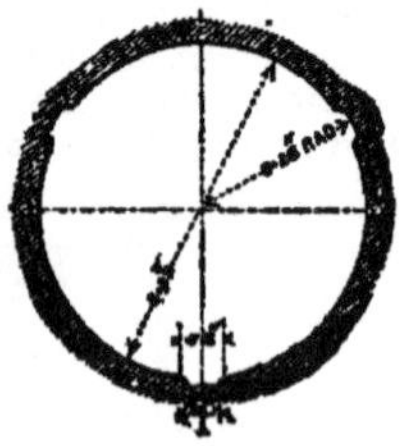

(2.) The "*plain groove.*"*—This is really the narrow deep portion of the shunt groove.

It is employed with all 64-prs., except those of Marks I. II. and III. built up guns, having wrought iron barrels and shunt grooves, and which have not yet been retubed, so that they can all use the ammunition employed with the shunt rifled pieces.

This groove gives very good results as to shooting. It will be seen that the bottom of the groove is concentric with the bottom of the bore.

Woolwich groove.

(3.) The so-called "*Woolwich*" groove is shown in diagram below, and is used with all guns above the 64-pr., as well as with the 80-pr. converted gun, 8-inch howitzer, 40-pr., and 25-pr. guns. The dimensions differ slightly for the different pieces, as will be seen in Table at p. 291.

WOOLWICH GROOVE. Scale, full size.

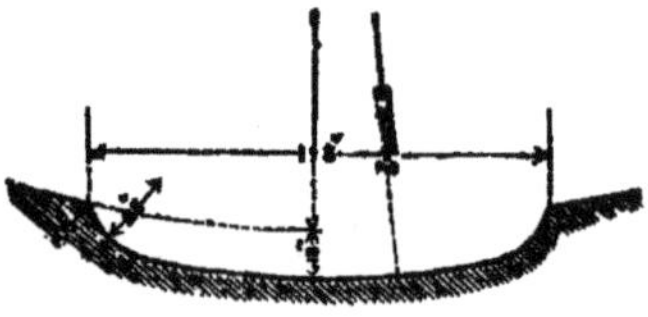

French modified groove.

(4.) The "*French modified,*" with 16-prs. and 9-prs. R.M.L. guns, as shown in diagram below:—

SECTION OF GROOVE. Scale, full size.

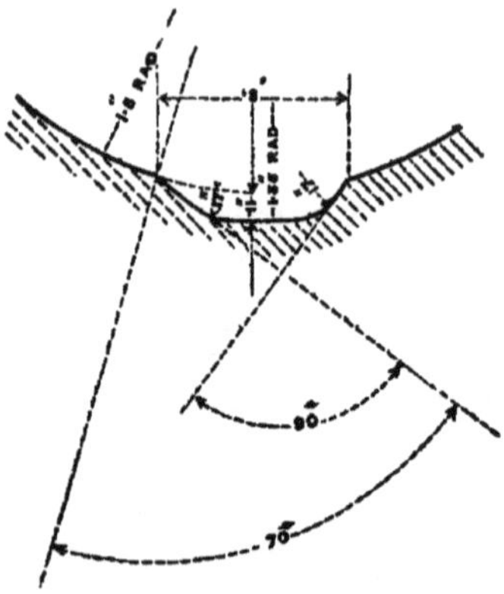

* This groove varies in width and depth according to the nature of gun, vide p. 291.

The driving edge of the groove forms an angle of 70° with the normal to the surface of the bore, and the loading side is at right angles to the driving side.

The width is 0″·8 at top, and the depth 0″·11; the bottom of the groove being eccentric to the bore, and the corners rounded off.

The smaller incline on the driving side is given in order that the studs may run up the incline, and thus be gripped and the projectile centred: the bottom of the grooves are eccentric with the bottom of the bore, so as to assist in this centering action.

French groove.

(5.) The "*French*," with 7-pr. R.M.L. guns, as shown in diagram below:—

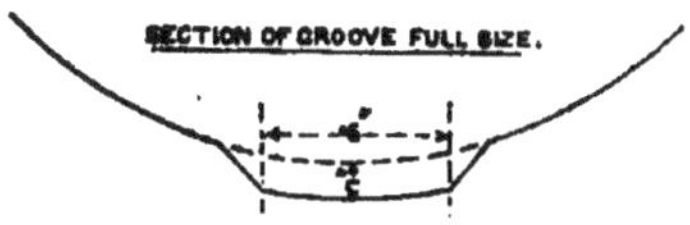

This rifling differs from the modified French system (used in the 9-pr. and 16-pr.) in not having the corners rounded and in the curve of the bottom of the groove being described concentric to the bore. The grooves are 0″·6 wide at the bottom and 0″·1 deep.*

Modern Polygroove.

13-PR. R.M.L. GUN, 8 CWT. SECTION OF GROOVE. Full Size.

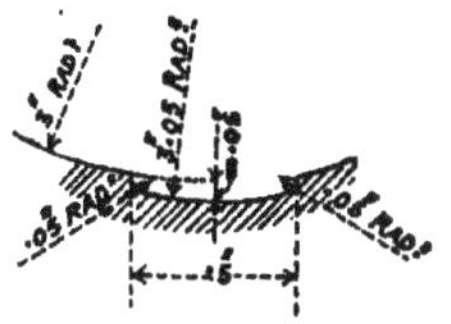

This form of groove is used in conjunction with the driving gas check.

In the 13-pr. it is 0″·5 in width and 0″·05 in depth, the bottom is concentric with the bore and the edges are rounded.

In the 6″·3 Howitzer the groove is 0″·5 in width and 0″·1 in depth.

The lands in these pieces are equal in width to the grooves.

So far this nature of groove has only been introduced for the two pieces given above, and the two experimental Howitzers, 8″ and 6″·6, but its use will be extended to 80-ton guns and other new pieces.

Mode of Stopping the Projectile.

With reference to the explanation given at p. 38, as to the necessity of having a constant space in rear of the projectile, or in other words a powder chamber of constant proportions, it is evident that the projectile must always be stopped at the same point in the bore by some mechanical contrivance, otherwise when different charges are employed, or when the projectile is rammed home with varying strength we should have the charge burnt with varying air space per pound of powder.

---

* With this nature of groove, and also with the plain groove, the width is measured at the bottom, the angles being rounded off in all other grooves the width at bottom cannot be accurately measured.

Two modes are employed:—

1. Making use of the termination of the grooves themselves in conjunction with the studs on the shot, or the ribs on the gas check.

This mode is employed in the case of the 40-pr. siege gun, the 8-inch and 6·3 inch service Howitzers, and the two new Experimental Rifled Howitzers. With these latter pieces, the termination of the grooves is made very abrupt by doing away with the slope at the end of the grooves as far as possible.

No doubt the use of this expedient will be further extended, and it will in all probability be employed in the 80-ton gun.

6·3-INCH R.M.L. HOWITZER 18 CWT. STOP FOR PROJECTILE. Full size.

2. By reducing the diameter of the bore immediately in front of the powder chamber, or in other words putting a choke in the bore which stops the gas check attached to the shot when it is not furnished with ribs or projections.

This method has so far been adopted only with the new 13-pr. field guns. It cannot be applied to guns already made, while in many cases the mode first mentioned can be so utilized for such guns.

13-PR. R.M.L. GUN OF 8 CWT. STOP FOR PROJECTILE. Full size.

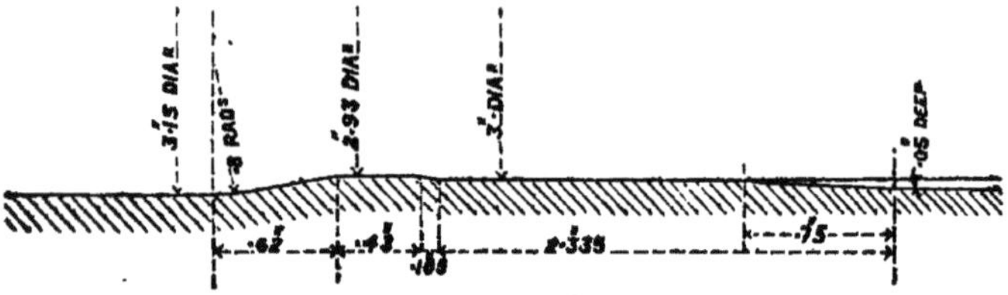

Systems of Rifling, and Construction generally.

It should be always borne in mind that, so far as at present known, nearly the same results can be obtained from almost any system of rifling, if the advantages and disadvantages which each one possessed are properly balanced. At times, as progress is made, each in its turn may seem superior to all others, until some new improvement brings another system more to the front, to be later eclipsed by yet another, and so on.

What is above all necessary to success in this particular detail, as in the whole of gun construction, is that no minor points be overlooked, and that as progress continues in the manufacture of powder, projectiles, gun carriages, &c., equal advances should be made to meet the improved requirements as regards the gun itself, so that the benefits reaped from progress in one direction may not be lost by want of progress in another.

The problem of how to construct a piece of ordnance is indeed a most complicated and vexed question into which so many variables enter that to lay down rigid and arbitrary rules is not permissible; all we can do is to utilize theory so far as known and to avail ourselves of those experimental data, the store of which each year increases.

CHAP. II.

5. *Venting and Vents.*

Venting.

Having constructed our gun, and chosen the system of rifling to be adopted, we have to determine what means should be employed to fire the charge; this is done as regards all existing ordnance by means of a percussion tube or otherwise, the flash from which passes down a fire-hole bored through the gun itself, through the breech, stopper, or through both together, as in the German field guns.

Vent bush.

As the rush of heated powder gas quickly wears away the metal round the fire hole,* it is usual to bore a hole in the metal of the gun or breech stopper, into which is fitted a screw plug or vent bush,† through which the fire channel or vent proper is made; when the vent becomes much damaged by wear or otherwise, the vent bush can be removed and replaced by a new one.

Copper. Steel.

As copper withstands the action of the powder gas very well, the bush is usually made of this metal, though steel is sometimes employed; wrought iron was formerly used for this purpose, and vents of that material are still to be found in some of our S.B. cast iron† pieces.

Wrought iron. Position. In R.B.L. guns. In S.B. and R.M.L.

The position of the vent in the gun may either be what is usually called axial or central, *i.e.*, when the vent channel runs through the axis of the breech and strikes the cartridge in the centre, at the bottom of the bore, as with our R.B.L. guns; or it may be such as to strike the chamber at an angle, or perpendicularly to the axis near the bottom or elsewhere, as is done in our S.B. and R.M.L. ordnance.

In heavy guns.

For convenience the vent generally opens at the top of the gun; but in 10-inch guns and upwards, where the size of the gun renders this position awkward, the bush is placed at an angle of 45° with the perpendicular, and the vent hole will therefore be at the top right side in such guns for broadside and garrison service, while in turret guns it is sometimes placed on the left side and sometimes on the right as convenient.

Effects on position of vent.

When powders of comparatively small grain are employed, the rate of ignition of the charge, and so the maximum pressure on the shot and gun, is influenced considerably by the point at which ignition first takes place, *i.e.*, by the position of the vent.

This is more particularly the case when large charges of considerable length are employed.

Favourable effects obtained. by striking the cartridge $\frac{4}{10}$ths of its length from end of bore.

From experiments made in 1863 with large charges of R.L.G., the powder then employed with heavy guns, it appeared that when the cartridge was ignited at a distance of $\frac{4}{10}$ths of its length from the bottom of the bore that the best results were obtained as to velocity imparted to the projectiles; it was therefore settled that heavy guns were to be vented so that the bush should strike the bore at that distance from the bottom. This rule is still adhered to in all our heavy guns, although the powder grains are much larger than those of R.L.G.

*** We see that the practice of vent-bushing guns is of ancient date, as a scientific soldier of olden time, Captain Hexham, writing in 1642, says, "I have seene in Ostend, upon the west bulwark, that some touch-holes of canon were blowne so great with often and continuall shooting, that I have put my fist into them. Now such a touch-hole," he says, "being blowne some 3 or 4 inches may easely be remedied, for if you bore the hole round and drive in a screw of iyron into it as thick as your finger, and in the midst of the screwe abovesaid bore a touch-hole in yt, you shall find this to last longer than any other way." He also says that new ordnance had such vents screwed in at their first manufacture, and that they lasted longer than brass or copper touch-holes.**

† For description of the actual vent or bush itself, vide p. 50.

CHAP. II.

In the smaller natures of R.M.L. guns, it has, until lately been the rule to place the vent so as to strike the bore near the bottom with a view to ensuring the certain ignition of reduced charges, and further to lessen the supposed danger due to fragments of unconsumed cartridge remaining in the bore in a smouldering condition when firing blank charges. This danger however has disappeared with the introduction of the present service silk instead of serge cartridge. Experiments made with the new 13-pr. field gun, show that a blank cartridge of considerable length can be ignited with a vent 7-inches from the bottom of the bore without danger, and that the unconsumed residue left is not greater than when the same cartridge is ignited by means of a vent near the end of bore.

Velocity gained by forward vent.

Experiments carried out with forward vents in 40-pr. and 25-pr. rifled muzzle-loading guns, firing *service* rifle large grain powder, show that a considerable increase in velocity, without any injurious increase in pressure, may be attained by igniting the cartridge in the centre instead of in rear. The pressures are also more uniform with the forward vent. In the case of the 16-pr. gun, there is no practical difference in velocity with either vent, but the pressure on the base of the shot is reduced by the use of the forward vent, and this reduction might possibly obviate the tendency to premature explosion of shrapnel shell in this gun, and overcome what has hitherto been a serious difficulty.

16-pr. R.M.L.

In consequence of the above, the 40-prs. have now been forward vented (vide Table II.), the old vent being plugged up.

With rifled Howitzers, the rear vent is still adhered to for convenience in igniting the very small charges which would occasionally be used with these pieces, in which the powder chamber is usually also of no great length.

No alteration has as yet been made in the position of the vents in the remaining service guns 64-pr. and downwards which were originally rear vented for the reasons given above.

Converted guns.

With the converted guns, 80-prs. and 64-prs. the vent strikes the conical powder chamber at right angles to the surface, and near the bottom, and is inclined at the same angle as was the bush of the old S.B. guns or nearly so.*

TABLE II.—Table showing position of Vents in R.M.L. Guns.

Nature.	Distance from bottom of bore.	Remarks.	Nature.	Distance from bottom of bore.	Remarks.
17·7 inch of 100 tons	...	...	64-pr., 64 cwt., Mark II & III.	5″·2	Vertical.
16″ of 80 ,,	...	Axial.			
12″·5 of 38 ,,	12″·0	Inclined at 45°, and normal to bore.	40-pr. Mark I or II.	7″·0	,,
12″ of 35 ,,	12″·0	,,			
12″ of 25 ,,	9″·8	,,	25-pr.	1″·0	,,
11″ of 25 ,,	10″·0	,,	16-pr. and 9-pr. ...	0″·6	,,
10″ of 18 ,,	11″·0	,,	9-pr., 6 cwt. Mark I.	1″·0	,,
9″ of 12 ,,	9″·7	Vertical.	13-pr.	7″·0	,,
8″ of 9 ,,	9″·2	,,	7-pr. steel	1″·0	,,
7″ all natures ...	9″·6	,,	6·3 Howitzer... ...	1″·125	
80-pr. of 5 tons ...	1″·85	Normal to cup.	8″ Howitzer of 46 cwt....	1″·75	Normal to coned end.
64-pr. of 71 & 58 cwt.	1″·8	,,	6″·6 Howitzer ...	1″·5	Vertical.
64-pr. 64 cwt. Mark I.	6″·1	Vertical.	8″ Howitzer, 70 cwt.	2″·0	Vertical.

* This is done on account of the conical shape of the cup. Vide p. 174.

CHAP. II.

Vent bushes.

Vent bushes. The service vent bushes, or vents used by us are of copper (except the few iron bushes found with some S.B. pieces, vide p. 60), and for R.M.L. guns, the metal is specially hardened. For vent bushes of R.B.L. guns, vide p. 126.

For S.B. and R.M.L. guns there are two kinds of copper vents or bushes, viz. :—the "through vent" and the "cone vent." *

Through vent. The through vent is a cylinder $1\frac{1}{8}$ inch in diameter, cut with a screw thread $\frac{1}{16}$ inch deep, and having a square head by means of which the bush is screwed into the gun.

Cone vent. The cone vent is of the same shape and size as a through vent, except towards the end, where the screw thread terminates, and the cylinder merges into the frustrum of a cone $1\frac{1}{4}$ inch in length and $\frac{7}{8}$ inch in diameter at the extreme end.

Guns are first vented with a cone vent, the through vent is only used when the wear round the copper is so great that the cone must all be bored out to remove flaws, vide p. .

The copper is 2 inches square in section. It is drawn down square while cold under a light steam hammer to the size required for the screw, the blows being as light and numerous as possible, so that the greatest amount of condensation may be effected. It is afterwards treated in the usual manner, *i.e.*, turned, a seven-thread screw cut on it, and coned at the bottom.

§ 1821. Vent bushes of 18-ton guns. The vents of course vary in length according to the thickness of metal of the gun. In 18-ton guns and upwards the threads are limited to a length of 6 inches above the cone, the upper part being plain. Guns for sea service have the mouth of the vent rimed out to a depth of 1 inch, tapering from ·28 at the top to ·22 at the bottom.

Different threads. Some S.B. bushes have a different thread, vide p. 60, chapter III., while the converted guns are vented with a through vent to begin with, vide p. 236.

Exceptional bushes 7-pr. steel gun. 80-ton gun. The bush of the 7-pr. steel gun is exceptional, having 18 threads to the inch, and being of much smaller dimensions, 0″·625 in diameter, in order not to weaken the gun too much. The vent bush of the 80-ton (experimental) gun is also exceptional, being of steel provided with a mushroom-shaped head. The bar of the vent bush passes through a hole cut through the breech in the prolongation of the axis, and is clamped by a nut on the exterior.

Wear of the vent. It may be remarked that although a copper vent can be readily removed when worn out, as explained at p. 315, yet with large guns firing heavy charges, the wear of the vent caused by the action of the escaping gases is so great that it is very desirable to find some practical method of stopping this rush of gas.

Plans for preventing the wear of the vent. Many plans have been brought forward, some of which are still under trial, for obviating this defect, either by stopping the fire channel immediately the charge is ignited near the top (1); or at the bottom (2); or again by providing the gun with a vent which can be more readily removed than the service vent (3).

The first method seems the easiest to use, though the least efficient, because it allows of a rush of gas along the vent channel; the second would be most complete if it can be attained without too much practical

---

* For R.M.L. guns, except the converted guns, only "cone" vents are issued. When necessary they are used as "through" vents, the cone being entirely removed.

† A simple method of this description is that brought forward by Lieut. Col. Maitland, R.A., where a plate of copper is attached to or loaded in rear of the end of the cartridge, which plate on the explosion of the charge is driven tightly against the aperture of an axial vent. This has the objection that after every fire the plate must

difficulty, as by means of it no wear of the gas channel would take place.† The third method plainly does not solve the difficulty altogether, but partially does away with its ill effects, and is that which will probably be of the simplest and readiest application.

Fraser Removable Vent bush.

This removable vent bush has so far given very good results. It consists of (1) a vent or bush proper of steel with a copper tip, (2) a nut for securing the vent in the gun, (3) a clamp ring with friction tube pin.* The vent proper fits easily into the hole bored for its reception; it is held securely in the gun by the "nut" which fits over it and which is screwed into the metal of the gun, until it presses against a shoulder upon the upper portion of the vent. The clamp ring* is placed in position upon the upper end of the vent, and clamped by tightening a small screw. To remove the vent, the clamp ring is first taken off, the nut screwed out and another nut, which takes the thread made on upper part of vent, inserted instead. This nut is screwed down until its collar rests on the upper surface of the gun, whereas this collar prevents downward movement of the vent; the vent is necessarily drawn upwards as the nut is further screwed round, when both vent and nut can be readily taken out.

6. *Sighting generally.*

Sighting.

Although sighting is not strictly speaking a part of gun construction, yet we may treat it here generally, considering that all ordnance must be provided with some means of laying in order to be of any use.

A gun is usually supplied with one or more† pairs of sights, fixed on so as to be readily removed when necessary.

Front or fore sight. **Tangent or hind sight.**

The front (fore) sight is usually placed on the gun near the muzzle or on the trunnion ring, and the tangent or hind sight near the breech end; the line of sight passing through a notch in the latter and taking in the point of the fore sight and the object aimed at.

Sights for S.B. ordnance perpendicular. **Sights for rifled ordnance placed at an angle.** **Amount of derivation.**

With service S.B. pieces these sights are both perpendicular, or more correctly in the vertical plane passing through the axis of the piece, when the trunnions are horizontal, but with our rifled pieces it is found that the projectile always goes to the right of the object aimed at, which is due to their having right-handed rifling. In order to make up for this drift, or "derivation,"‡ as it is sometimes called, the hind sights are inclined slightly to the left, so that the line of sight is directed somewhat to the left of the target apparently aimed at, the vertical plane passing through the line of sight when the trunnions are horizontal being slightly inclined to the vertical plane passing through the axis of the piece. The amount of derivation will vary with the shape,

---

be removed from the gun. This method, as well as others of the several natures mentioned in the text, is under consideration.

* Required for S.S.

† Should there be three pairs or sets of sights, the middle pair laterally are termed centre sights (hind and fore).

‡ As to the causes which produce this drift in the case of rifled guns, they appear to be complex, and as yet nothing positive can be laid down regarding them. This subject will be found discussed in the books below-mentioned amongst others.

We know as a fact, however, that during their flight through the air, pointed projectiles always deflect to the right with right-handed rifling, and to the left with left-handed rifling. As to the actual drift of flat-headed projectiles, we have not a sufficient amount of data to say that we know to a certainty ~~how they deflect,~~ though it is held by some that they do so in the opposite direction to pointed projectiles. "Motion of Projectiles," p. 65. Professor Bashforth, B.D. Asher and Co., 1873. "Modern Artillery," p. 224. Lieut.-Col. Owen, R.A. Murray & Co., 1871. "De "l'Influence du movement du rotation sur la Trajectione des Power Technologie "Militaire," IV. 1876, p. 18. Helié, Treatise on Ballistics, pp. 436–7.

CHAP. II.

&c., of the projectile, and also with the twist of rifling. As a rule, the sharper the twist the greater is the derivation.

Angle of deflection in B.L. guns.

8-inch howitzer sight not set at an angle.

Regarding the angle at which the hind sights must be inclined in order to make up for derivation, with R.B.L. service guns it is 2° 16′ for all natures, but with R.M.L. it differs considerably in the several guns. The only R.M.L. service pieces of ordnance where the sights are not set at an angle are the howitzers. With the 8 and 6·3 inch howitzers a perpendicular hind sight is used, and the drift at the several ranges is made up for by means of a long deflection leaf.

Method of ascertaining th angle of deflection.

In order to determine the "permanent angle of drift" when a new nature of gun is about to be introduced into the service, actual practice is carried on at Shoeburyness from the specimen gun sent there for trial of range and accuracy; the gun is either sighted perpendicularly or in some cases it is not sighted at all, elevation being given by temporary means.

A still day being chosen,* a number of rounds are fired with various elevations, generally a series of 10 rounds at 1°, 3°, 5°, &c., and the angle is calculated, by the subjoined formula, for each elevation.

The mean of the angles so obtained is adopted as the permanent angle of drift.

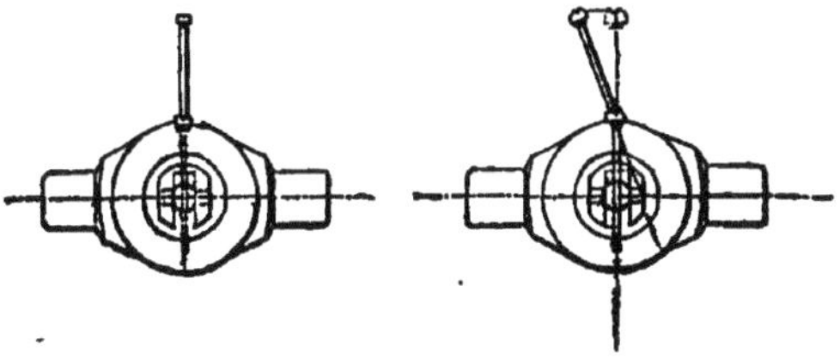

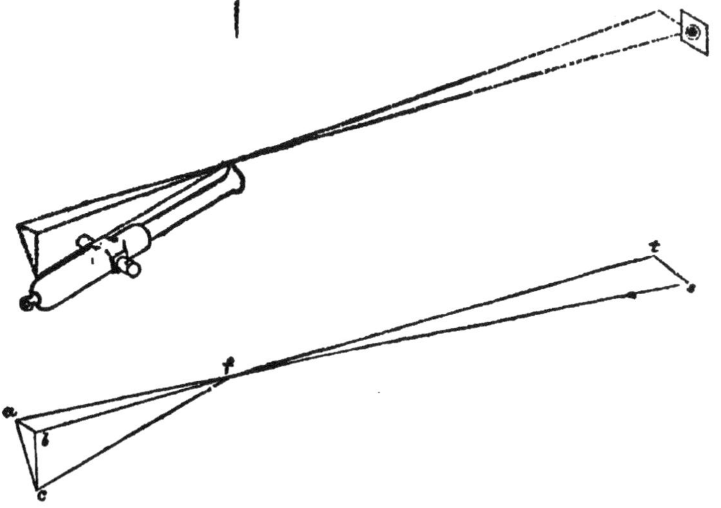

* Even should there be wind we have now, thanks to Lieut.-Col. Maitland, R.A., a formula by means of which we can without difficulty eliminate errors due to its effects in making the calculations founded upon the observed values of d and r. Vide "On the influence of the Wind on the Flight of Projectiles." Proceedings R.A.I. Vol. viii., No. 6.

The formula used for determining the angle for each range is

$$\tan\theta = \frac{\text{deflection}}{\text{range}} \times \text{cosec elevation,*}$$

which is proved as follows:—

In fig. above, let *bc* represent a perpendicular tangent sight, and *f* the foresight of the gun, then *bft* represents the line of sight, the gun being laid on the target *t*, with the angle of elevation $bfc = e$.

Suppose *s* to be the point where the shot falls, then *ts* measured at right angles to the line of sight, represents the deflection of the shot.

Join *sf* and produce it to *a*, draw *ba* at right angles to *bt* and join *ac*.

Now *a* is the point at which the head of the tangent sight should be placed, in order to compensate for the drift *ts*, and $acb = \theta$ is the *angle of* drift required.

Let the range $ft = r$ and the deflection $ts = d$.

Now by similar triangles

$$\frac{ab}{bf} = \frac{st}{tf} = \frac{d}{r}$$

But $ab = bc \tan\theta$ (*abc* being a right angle),
And $bf = bc \operatorname{cosec} e$ (*bcf* being a right angle),

$$\text{Therefore, } \frac{d}{r} = \frac{bc \tan\theta}{bc \operatorname{cosec} e}$$

$$\text{Or } \tan\theta = \frac{d}{r}\operatorname{cosec} e.$$

Note A to Chapter II.†

Description of Trials with various Systems of Rifling.

Commander Scott's gun was rifled in five grooves, which were shallower on the loading side than on the driving side, which was curved with a view to obtain a perfect centring for his shot. His rifling had a uniform spiral of one turn in 294 inches. Fig. 1. Commander Scott's system.

Fig. 1.

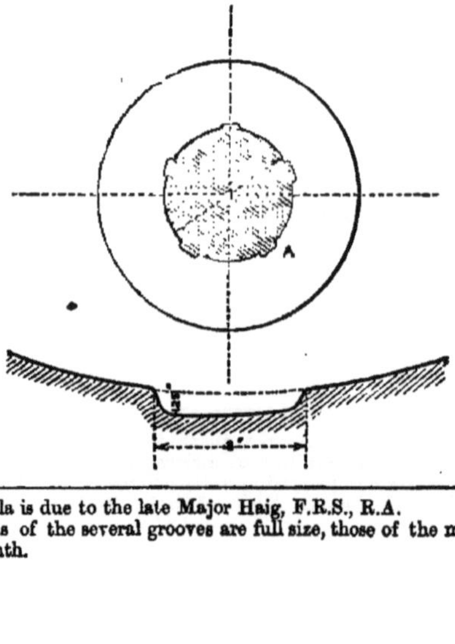

* This formula is due to the late Major Haig, F.R.S., R.A.

† The sections of the several grooves are full size, those of the muzzles are on a scale of one eighth.

CHAP. II.

His projectile at first had simply five iron ribs, with two very small copper studs inserted in the driving face of each, but afterwards the ribs were faced with zinc.

Mr. Lancaster's.

Mr. Lancaster's gun was oval-bored, the major axis being 7·6", and the minor axis 7". The rifling making one turn in 360". Fig. 2.

Fig. 2.

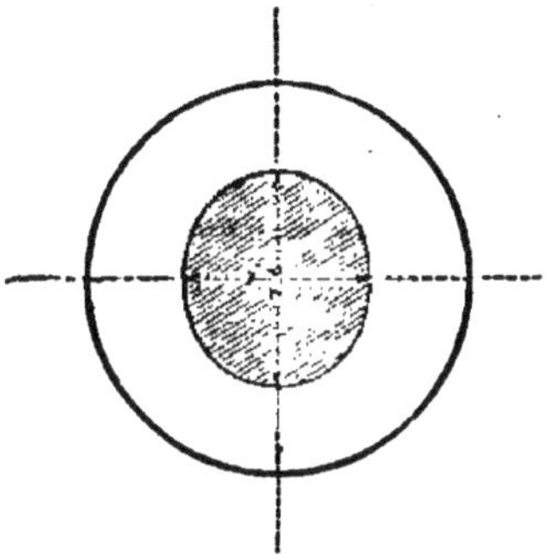

Messrs. Jeffery's and Britten's.

One gun was sufficient for Messrs. Jeffery and Britten, as their systems differed from one another only in manner of applying lead to the base of the projectile so that it might take the rifling.* This gun had 13 grooves 0·10" in depth, and 0·846" in width, and a turn of 1 in 805". Fig 3.

Fig. 3.

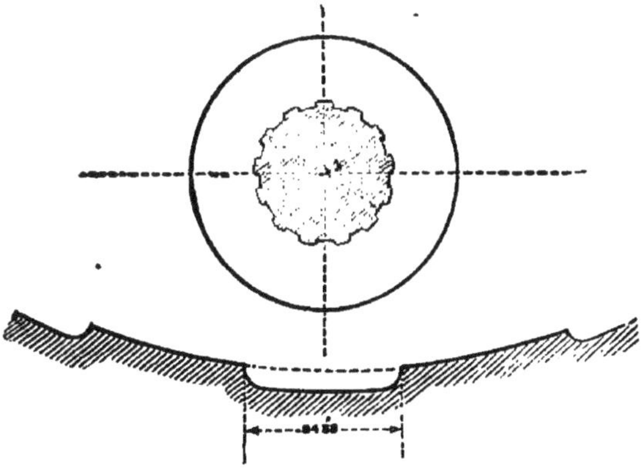

The French gun.

The French gun was rifled in three grooves 0·225" deep, 2·02" wide, the rifling gradually increased from nothing at the breech to 1 turn in 259" at the muzzle. The first batch of projectiles for the French gun had three large half-zinc studs in front, supported by an iron back, and three small ones behind, but as the experiments went on it was found expedient to adopt Major Palliser's proposal of changing the metal of the studs to gun metal, and of reversing the position of the studs by placing the smaller ones in front. Fig. 4.

* Mr. Britten's method of attaching lead coating chemically was adopted for the projectiles of the Armstrong B.L. guns, and has proved most satisfactory.

Fig. 4.

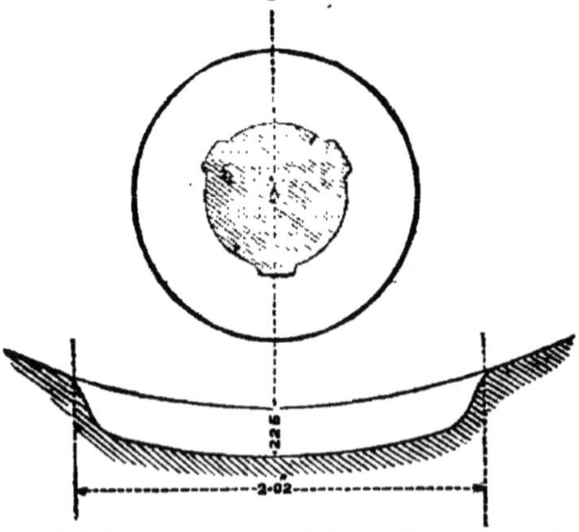

The shunt gun. The shunt gun had six grooves of the well known form; the spiral was uniform with one turn in 266″ or 38 calibres. The shot had 30 studs, *i.e.*, five for each groove. Fig. 5 shows section at muzzle.

Fig. 5.

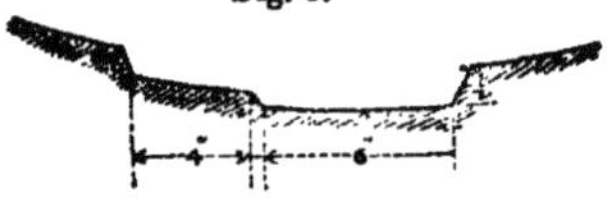

All the projectiles had hemispherical heads. The weight of each shell (filled) was 100 lbs., and of each shot 110 lbs., except that of the French shot, which weighed only 100 lbs.

A very short experience showed that the systems of Messrs. Jeffery and Britten were unsuited for heavy charges; large pieces of lead were blown off the shot, and the shooting was so wild as to throw these systems entirely out of the competition, which therefore was limited to those of Scott, Lancaster, the French and the shunt.

Recommendations of O.S.C. Experiments were carried on which tested these competitive guns in all the cardinal virtues of ordnance, and though the shooting qualities were alike, the Ordnance Select Committee in their final report, No. 3730, dated 1st May, 1865, recorded their unanimous opinion in favour of the so-called French system :—

(1.) "Because of the simplicity of its studding on the projectiles.

(2.) "The simplicity of the grooving of the gun, and

(3.) "From a disposition to admit the advantages of an increasing over a uniform spiral."

Fig. 6.

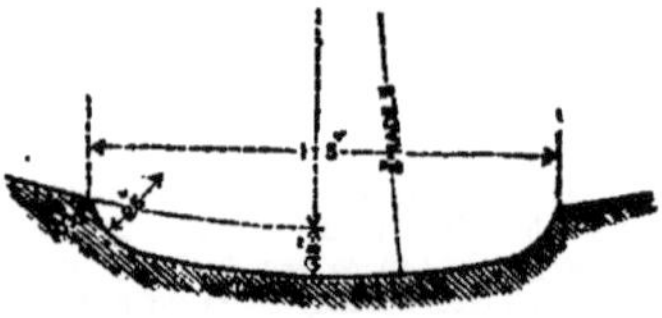

The modified groove. And further, the Committee recommended "that the heavy 7″ guns then in course of manufacture should be rifled in the same manner as the competitive so-called French gun, except that the width and depth of the grooves should be slightly decreased, and that 8″ and 9″ guns also should be completed with similar rifting. Fig. 6 shows a section of the modified groove.

CHAPTER III.

SMOOTH-BORE ORDNANCE AND STORES.

Rifled Ordnance rapidly replacing Smooth Bores.—Where S.B. are retained —Cast iron and bronze guns.—Wrought iron S.B. guns.—**Classes of S.B. Ordnance.**—Guns, carronades, howitzers, and mortars.—Gomer and cylindrical chambers.—Registration and designation.—Length.—Weight.—Preponderance.—Windage.—Natures of cast iron ordnance in the service.—**Bronze Ordnance.**—Inspection, proof, marking before issue, &c.—Marks showing nature of vents.—Venting.—Line of metal.—Quarter-sight line.—Line of horizontal axis.—Vertical line.—Quarter-sight scale.—Millar's sights.—Wood tangent scale of L.S —Small stores L.S. and S.S. Table of S.B. guns to be retained.—Table of S.B. guns to be abolished.

Rifled ord-nance replac-ing S.B.

In certain cases S. B. re-tained.

Rifled guns are rapidly replacing smooth-bore ordnance, yet we still have so many of the latter in our armament, both at home and abroad, that it is necessary for artillerymen to be well acquainted with such ordnance and the stores belonging to them. Moreover, although for the reasons given at p. 67 it has been found necessary to adopt rifled guns for our service armament, yet for certain objects, such as the defence of short flanks where range is limited, and generally speaking at close quarters, S.B. guns are still useful, and some are retained in all our large fortresses.

Manufacture of S.B. ordnance.

Cast iron guns in 1545.

S.B. ordnance, both of cast iron and bronze, were cast in moulds and afterwards bored out to the proper calibre. Cast iron ordnance were manufactured in England as early as 1545, but bronze pieces had been made in this country long before that date, for the founding of bronze was well understood prior to the art of smelting iron ores being perfected.

Brass foundry at Woolwich, 1717.

In 1717, the so-called brass foundry was established at Woolwich, and there our bronze ordnance have principally been made, some of the existing natures (the 4⅖ howitzer and Coehorn mortar for instance) having been manufactured from its earliest establishment.

Cast iron guns were supplied by contractors according to designs furnished to them, and were proved by Government officials.

Blomefield guns.

The oldest pieces still in our service were made between 1780 and 1822, when Sir Thomas Blomefield was Inspector-General of Ordnance. He instituted a rigorous proof, and improved the manufacture generally. Guns made after his designs may be known by the numerous architectural ornaments on the exterior.

Congreve, Dickson, Millar, Monk, Dundas.

General Sir W. Congreve, Sir A. Dickson, and Millar, Mr. Monk, and Colonel Dundas successively introduced improvements upon the Blomefield guns, but no new nature of cast iron or bronze S.B. service gun has been made since 1859, when the supersession of our smooth-bores by rifled guns commenced.

S.B. wrought iron guns, 1864.

In 1864, when guns *versus* armour plates became a serious question, the Admiralty proposed the construction of large S.B. guns of wrought iron for penetrating iron clad ships at close quarters. Two natures, 150 and 100-prs., were consequently made, being built upon the Armstrong coil system. Much more powerful results, however, than they could afford were obtained from the heavy R.M.L. guns adopted shortly afterwards, and their manufacture ceased. These two guns were the last two natures of S.B. ordnance introduced into our service, and the only description made of other materials than bronze or cast iron.*

Classification.

Classes of S.B. ordnance.

Our service S.B. ordnance are classified as follows:—

Cast iron	Guns. Carronades. Howitzers. Mortars.	Bronze	Guns. Howitzers. Mortars.

Guns. Guns are from 14 calibres and upwards in length, carronades about 7 calibres; they are adapted for both shot and shell.

Shell guns. Two guns however, the 10-inch and 8-inch, are made for shell fire only. These differ from other guns, in having conical (gomer) chambers and in being shorter and lighter in proportion to their calibre.

Carronades. Carronades have cylindrical chambers and differ much from other pieces, being short and tapering towards the muzzle, round which there is no swell, but a lip or rim projecting forward; they are without trunnions, but have a loop underneath by which they are secured to the carriage.

Howitzers and mortars. Howitzers, 5 to 10 calibres long, and mortars from 3 to 4 calibres, are chambered and adapted solely for shells, the former for so-called direct, and the latter for high angle fire. Their calibre is large compared with their weight.

The trunnions of mortars are, for convenience in high angle firing, placed at the breech end instead of near the centre of gravity.

Gomer chamber. All mortars and howitzers in the service have gomer chambers except the 4⅖-inch bronze howitzer, which is cylindrically chambered.

Registration. Difference between S.S. and L.S. S.B. ordnance are also divided into L.S. and S.S.; many pieces are common to both, *e.g.*, the 8-inch shell gun L.S. and S.S., others were only intended for one branch of the service. S.S. ordnance are always furnished with breeching loops, and a few also have housing blocks, otherwise they differ generally from L.S. in minor fittings only.

Designation. Smooth-bore pieces for firing solid shot are further designated by the weight of the shot in lbs. and the weight of the piece in cwts., as the 68-pr. of 95 cwts.; those for firing shell by the calibre in inches and weight as before, *e.g.*, the 8-inch mortar of 9 cwt. The 32-pr., 24-pr., and 12-pr. bronze howitzers are exceptions to this rule, and are distinguished like shot guns. When there is more than one pattern of the same calibre and weight some distinction should be specified; for instance, the 32-pr. of 25 cwt., length 6 feet.

Length. The length of all pieces, except mortars, is measured from the muzzle to back of base ring, and that of a mortar from rear of breech to face of muzzle along axis of piece.

Weight. The weights as given in the tables are termed nominal, because there is often a difference of two or three cwts. in pieces of the same nature.

Preponderance. Preponderance expresses the statical pressure on the elevating screw or coin.

---

* The 150-prs. are now obsolete. They have been returned into store, but are retained for the present. The 100-prs. are used by the Navy for drill purposes only.

CHAP. III.

Windage.

Windage—the difference between the diameter of the bore and the diameter of the shot*—allows room for ramming home the projectile when the bore is foul, &c. In old guns the windage is $\frac{1}{20}$th the diameter of the shot; but in those of more modern date it is much less, being only ·1 inch in field guns, and about ·15 inch in heavy guns. Windage should be as small as possible, for besides causing indentation of the bore and irregularity of flight, a great deal of the power escapes and is lost. In old guns this loss was computed to be equivalent to $\frac{1}{3}$rd or $\frac{1}{4}$th of the charge.

Natures of Cast iron Ordnance in the Service.

Ordnance in the service.

So many different pieces had been introduced into the service up to 1864 that our armament then embraced a very great variety of ordnance. In consequence we retained in our fortresses a number of guns and a vast accumulation of small stores of an obsolete pattern.

Table A.

To remedy this, two lists, A. and B. (pp. 65, 66, Tables III and IV), were made out in January 1864, showing the pieces to be retained in the service, and those which were to be abolished. These lists continue our official guide as to S.B. ordnance absolutely in the service. We will now go through the pieces mentioned in Table III, and remark on their specialities.

Cast iron ordnance. Shell guns.

There still remain some S.B. shell guns in our wooden ships and our fortresses. They have only two muzzle mouldings, while shot guns, with certain exceptions, have three.

10-inch.

The 10-inch shell gun weighs 86 cwts. The muzzle of this gun being too large for the ports of some ships, one of the muzzle mouldings was sometimes turned off in order to obtain a larger angle of training. Guns so treated are called L.M. (low muzzle), in contradistinction to the H.M. (high muzzle). The 10-inch should never be "double-shotted."

8-inch.

8-inch shell guns formed part of the siege train, and were much used for flanks or of permanent works. Those of 65 cwt. are now being converted into 64-pr. R.M.L. guns of 71-cwt. The 65-cwt. pattern is the one most used in both services. There are but few of the 54-cwt. pattern mounted in L.S. batteries, and none of the 60-cwt.

Shot guns. 68-pr.

The 68-pr., of 112 cwt., was the heaviest cast iron piece in our service; of them very few are left. There are, however, many of the 95-cwt. guns. They were much used as pivot guns and for sea faces of forts, and many have been now converted into 80-pr. R.M.L. guns L.S.

42-pr.

42-prs. are rare; a few may yet be found mounted in out-of-the-way batteries.

32-pr. of 63, 58, and 56 cwts.

32-prs. were formerly the principal armament of all classes of vessels, and hence we have several descriptions, varying in length and weight. The different amount of windage allowed from time to time has caused a considerable diversity in their calibres and ranges. The 63-cwt. gun is altogether for land service. The 58 and 56-cwt. are the patterns most commonly used in both services.†

Monk's A., B., and C. guns.

Monk's A., B., and C. guns still exist in Woolwich Arsenal, and in wooden ships. The A. pattern is sometimes found in garrisons, as are also the B. and C., but much more rarely. The 48 and 50-cwt. guns

* Properly speaking "windage" is the difference between the sectional areas of the projectile and the bore of the guns.

† The 58-cwt. is being largely converted, on the Palliser principle, into 64-pr. R.M.L., L.S.

are issued indiscriminately, mounted on the same carriages, and bracketed together in returns. The 48 and 50-cwt. guns, as well as those of 39 and 40 cwt., are to be found in inland works principally, and have been issued in large quantities to volunteers. 32-pr. of 48 and 50 cwts.

The 32-cwt. gun is a bored-up gun.

The 25-cwt. gun is the light 32-pr. It can be distinguished by having two muzzle mouldings. The 32 and 25-cwt. guns.

24-prs. are exclusively for land service, being garrison guns. The 48-cwt. has been issued extensively to Volunteers. 24-prs. of 20, 48, and 50 cwts.

A few 18-prs., both of the heavier natures and also bored-up guns, may still be found in the flanks of large works. 18-prs. of 42 and 38 cwt.

The three small natures of cast iron guns are used for saluting and drill purposes. 12, 9, and 6 pr.

Carronades are now to be found in flanks of a few works, and utilised for drill purposes (S.S.). Carronades.

The use of iron howitzers is limited to flanks, &c., where a very short range is necessary. Iron howitzers.

13 and 10-inch mortars, S.S., were originally intended for mortar vessels, but are now only used for coast defences. They have narrower chambers than the L.S. mortars of the same calibres, and are very much heavier. Mortars.

Bronze Ordnance.*

All the natures of bronze S.B. guns in the service are nearly alike, and resemble in exterior appearance the earlier cast iron guns. Bronze ordnance.

By an order, dated November 1859, a dispart patch is to be added to every bronze gun before issue. Guns.

The S.S. bronze howitzers are similar in pattern to the L.S., but the 24 and 12-pr. howitzers have a breeching loop, and the breech is rounded off.

The $4\frac{2}{5}$-inch howitzer still remains in the service, as well as the 3-pr. guns for colonial and mountain service. The howitzer is of exceptional construction, being very short, and having a cylindrical chamber. Howitzers.

The bronze mortars may still be useful in mountain warfare and in the advanced trenches of an attack. Mortars.

Inspection, Proof, Marking before Issue, &c.

The following were the tests applied in the R.G.F. to new cast iron pieces.

They were examined for flaws or holes, for concentricity of bore, dimensions, &c.; then two proof rounds were fired with a heavy charge and generally one service shot, pressed home with a junk wad or wooden wedges. The gun was afterwards proved by water pressure, and finally examined with a lamp passed down the bore. Fireproof. Waterproof.

Guns which endured the tests were weighed, marked, and registered. On the first reinforce were engraved the number by which the gun is registered in the R.G.F. books, the broad arrow, the exact weight of the piece in cwt. qrs. and lbs., and the year of proof, thus:— Marks on guns.

8736
↑
52–1–10
1864

* Most of the S.B. bronze guns and howitzers have now been sold as old metal, but some of these pieces still exist, especially in India and in our possessions abroad.

CHAP. III.

On the left trunnion were already marked the manufacturer's initials or the name of the foundry, the manufacturing number, and the year of casting.* Bronze ordnance have severally a register number engraved in Roman letters on some part of the piece, the foundry number being underneath, between the trunnions.

Venting or Bushing.

Prior to issue, a S.B. gun is vented, lined and sighted.

Bushing or venting.

Bushing iron guns was not the rule in our service early in this century, but in consequence of the enlargement of vent of the unbushed guns used at the sieges of 1812–13, experiments were carried out at Woolwich in 1813 as to the advantages of different bushes. Copper bushes answered best, and guns were ordered to be bushed, some with iron and some with copper.

In 1844 it was directed that wrought iron bushes only should be used,† but in 1855 this order again was cancelled, and copper bushes have been used since that date.

No better material for the purpose has yet been found, especially when it is hardened by hammering. Since 1855, cast iron guns (except 9-prs. and 6-prs.) have always been bushed before issue.

Bronze guns and howitzers of the present service natures have always been issued with copper bushes of different sizes and descriptions. Mortars are not bushed before issue, though some of the latter have been so subsequently.

When a gun is to be bushed for the first time, a cone vent is invariably used, but should the metal round the vent wear away in a gun so bushed, the cone vent will be replaced by a through one or otherwise according to the Regulations on that head, p. 314.

Process of venting cast iron guns.

The following is the process of venting a new gun: Throw the gun with one trunnion up to a convenient working position, and fix the drilling machine, place in the vent a stiff wire, to ascertain the rake or direction of the vent. This will show the position for setting the brace and drill. Drill right through the metal with the narrow set of drills, (as guns have been frequently destroyed by false drilling, wax impressions should be taken frequently during the process, to ascertain whether the workman is drilling straight,) and then with a larger set to within an inch of the bore, viz., to the spot where the thread ends (the thickness of metal may be conveniently found by using a vent scraper). Next finish the cone with the drill for the purpose.

Remove the drilling machine, and turn over the gun to prepare for tapping. In this case the seven-thread taps are always to be used. Tapping is a long and tedious process, requiring much care and skill, it is impossible to pass the first tap through, and then the second, and so on, but the thread must be brought to the right size gradually.‡

Where the thread ends there is usually a little metal thrown up

---

* This system of marking was not introduced until 1857, and old guns are not marked according to any regular system; in many the weight is engraved under the cascable. On bored-up guns the new weight and year of boring up are on the first reinforce. Carronades are marked near the elevating patch.

The Royal badge is on the first reinforce of most guns of all natures below the 68-pr., and bronze guns have in addition the monogram of the Master-General of the Ordnance on the chase.

† This was done because it was thought that a galvanic action was set up between the copper bush and the iron gun, which caused their corrosion. Experiments made in 1855 proved that this was not really the case, and the use of copper bushes was therefore resumed.

‡ It takes two men four or five hours to tap a gun.

which would prevent the copper bush from being screwed down properly, this burr must be removed with a conical rimer.

The hole in the gun is next cleaned with tow, and the copper vent, well oiled, is screwed in. The head should not be wrenched off as a fracture might occur below the surface of the metal of the gun.

The bush when properly fixed, will project about a quarter of an inch into the bore, and about two inches above the surface of the gun.

Wax composition.

Take an impression of the part in the bore with a mixture of:

Bees-wax	2 parts		Boiled together.
Soft soap	1 „		
Treacle..	1 „		

This will show whether the cone is well home, or whether there is a space left between the copper and iron.

If the bush is home, proceed to cut off the end in the bore. The instrument employed consists of a cutting tool supported by a metal head at the end of a long bar; the bar is kept in the axis of the bore by passing through a collar fitting into the muzzle, it is worked from side to side by two levers, being fed up by a small screw at the end of the frame; the spiral spring against the muzzle collar makes the knife work regularly. Care must be taken not to cut into the iron of the gun. It is probable that the end of the bush will not be cut off quite flush at first, so another impression is taken, and if necessary, the knife must be fed out with a small piece of tin, and the process repeated, as it is necessary that the copper and iron in the bore should be perfectly flush with each other. When this has been completed satisfactorily, remove that portion of the vent projecting above the surface of the gun by sawing it off about a quarter of an inch above the vent patch; chip a little copper away from the mouth of the vent to prevent it becoming choked when hammered, chisel it also at the edges, then hammer it well, chisel it off flush and open the mouth of the vent, then pass the set of rimers down one after the other and gauge, and if the gun is for S.S., rime out the mouth of the vent, tapering it from ·28″ at the top to ·22″ in a length of 1 inch. File the surface, take another wax impression of the inside, and if all is right the operation is finished.

Marks to show nature of vent.

Cast iron ordnance are also marked over the cascable according to the nature of vent they have.

C V
N
C } means copper vent, new gun, cone.

C V
C } indicates a cone vent, not new (gun).

C V „ a "through" vent.

The initial of the out station at which a gun is re-vented is added underneath.

I V on cascable means "iron vent."*

Lining.

Lining.

The line of metal† is obtained as follows:—

---

* Old guns may also be found marked CV / N / LO "copper vent; new; long cone: and TV "Through vent."

† This is a line extending from the base ring to the swell of the muzzle, and represents the intersection of the surface of the metal by the vertical plane passing through

CHAP. III.

The gun being levelled across the trunnions, a wood batten is placed in the bore, so as to project some distance from the muzzle. This batten is painted white on the upper surface, and bisected by a pencil line.

The upper surface is levelled transversely, and a T square being placed upon it, the position of the pencil line is squared up against the muzzle of the gun.

A wood straight-edge is placed on the top of the gun against the T square, and the edge of this straight-edge and the centre line of the batten are brought into the same vertical plane by aid of the eye. The line thus obtained is slightly marked on the swell of the muzzle, and also on the base ring. The T square is reversed, and the same operation gone through on the other side; if there is any error there will be two lines, and the mean is taken as the true one. The line of metal thus obtained is then cut permanently on the breech and muzzle.

Line of metal.

Quarter-sight lines.

The quarter-sight lines are next marked on both sides of the gun, at the breech and muzzle; they are parallel to the axis of the piece, but a little above it, so as to clear the trunnions and cap-squares when laying the gun.

Line of horizontal axis.

The line of horizontal axis is then cut on the breech, trunnion, and muzzle, on the right side of the gun. To obtain this line the gun is turned with the right trunnion vertical, and a similar process is repeated to that employed for finding the line of metal.

Vertical line.

Upon the right trunnion the line of horizontal axis is bisected, and a line drawn at right angles to it.

Quarter-sight scale.

Before the introduction of General Millar's sights, all cast iron guns were laid by means of a quarter-sight scale from 0 to 3° marked on the base ring on each side of the gun, starting from the horizontal line at zero. Such scales are now marked only on L.S. guns up to 32-pr. inclusive.

Sights.

Millar's sights.

The sights now used with cast iron ordnance* are Millar's sights. These consist of a fore or dispart sight of gun metal screwed on to the gun in rear of the trunnions, on what is termed the second reinforce, and of a half-round brass tangent scale† sliding in a gun metal block, which is secured to the breech of the gun by two screws.

Pieces of sheet lead are placed between the foresight and the gun metal block of hindsight and the gun, to assist in adjusting the sight accurately, and also to prevent the heads of the screws being broken off. In order to clear the breech of the gun, the scale of the hindsight is at an angle of 76° and not perpendicular.

Angle of sight.

The mode of adjusting Millar's sights is given in Chapter XII.

Short radius.

The distance between these sights is termed the "short radius," and is given in Table III for each nature of gun.

We find that at a certain point when elevating by these sights the muzzle of the gun begins to interfere with the line of sight, and it is necessary to use the muzzle notch as the foresight, and to employ another hindsight graduated for the "long radius,"‡ as the distance

Long radius.

---

the axis of the piece, when the trunnions are horizontal; it is marked on breech and muzzle. Similarly the line of horizontal axis is the intersection of the surface of the metal by the horizontal plane. It is marked on right side of breech, muzzle, and trunnion.

* Iron howitzers excepted, which are sighted in a similar manner to bronze guns.

† For S.S. this scale is hexagonal.

‡ The clearance angle is the angle of elevation obtained when the tangent sight is raised, so that the bottom of its notch, the top of the foresight, and the bottom of notch on muzzle are in line. It is the highest angle marked on the brass tangent sight, for if more elevation be given the muzzle would interfere with the line of sight.

Clearance angles, and the radii for the several natures, are given in Table III.

between the back of the hindsight and the notch on the muzzle is termed. For L.S. this other tangent scale is made of walnut-wood, and called wood tangent scale; it is graduated therefore from the clearance angle up to the extreme elevation (10° or so), the divisions being calculated to the long radius as the gun must be laid by the muzzle notch. It has also a degree scale from zero to the clearance angle, the same as that on the brass tangent scale, and there is a yard scale from pointblank to extreme range.

Wood tangent scale for L.S.

Clearance angle.

At the back of the scale is a brass staple which fits on the head of the brass scale when fully elevated, and at the bottom is a brass plate shaped so as to sit steadily on the hindsight block.

Long tangent bars for S.S.

For S.S. longer scales sliding in the same block as the tangent sight are used; they are either of brass or wood,* and are issued in sets of one, two, or three.

Tangent scales for bronze ordnance.

With bronze ordnance a half round brass tangent scale is used, which works in a socket drilled in the metal of the gun at the breech. This socket is fitted with a spring, and the scale has a small stud at the bottom which prevents it being removed altogether from the socket unless the spring is first taken out. The scale is clamped by a copper set screw, which is a separate store.

High head for S.S.

For S.S. the scale has a high head, so as to clear the head of the friction tube.

The notch on the muzzle of the piece serves as a foresight with bromze ordnance.

Quarter-sight lines are marked on both sides, and some old bronze guns have quarter-sight scales as well.

Wood side scales

For 32-prs. of 32 cwts. and 25 cwts., and also for 24-pr. bronze howitzers mounted on S.S. carriages with elevating screws, a wood side scale is used. This is graduated from 0 to 12° downwards, and 6° upwards. Elevation and depression can be given by means of it in connection with the ship's pendulum when smoke, &c. prevent the regular sights being used, the scale being cut so that when placed on the steps of the carriage and held upright, the zero of its gradation coincides with the axis of the gun when the latter is horizontal, the ship being on an even keel.

Graduated coins.

With other S.S. guns graduated coins are employed instead of this scale.

Small Stores.

Small stores.

In addition to sights and vents the following stores for S.B. ordnance are furnished by the R.G.F.

Priming iron or pricker ..	Specially for land service
Vent punches	Specially for land service
Lead aprons..	Specially for land service
Bit, vent	Specially for sea service.
Priming wire	Specially for sea service.
Friction tube pins	Specially for sea service.
Guide plates	Specially for sea service.
Ship's pendulums	Specially for sea service.
Vent plugs	Common to both services.
Wrenches for sight screws and friction tube pins.	Common to both services.
Spikes	Common to both services.

* They are supplied for the 10″ gun of 86 cwt., 8″ of 60 or 65 cwt., 68-pr. of 95 cwt., 32 prs. of 58 or 56 cwts. Long wood slides differ from the foregoing only in the material used.

CHAP. III.

L.S. stores. Priming iron or pricker.

The pricker is a rod of iron pointed at one end and with a ring at the other, for garrison service 12 inches long, and for field 7½ inches.

Vent punches.

Vent punches are for cleaning the vent from any hard substance which cannot be removed with the priming iron. They are of steel, and have a strong round head, so as to bear hammering. There are four sizes, varying in length from 3 to 14 inches.

Lead aprons.

Lead aprons are small pieces of sheet lead for protecting the tangent scale and vent of howitzers and bronze guns when mounted or in store. There are two sizes of aprons, large for iron howitzers, and small for bronze ordnance.

S.S. stores. Priming wire, and bit.

In the Navy the pricker is termed the "priming wire." A bit is also used, but it is not connected with the priming wire.

Friction tube pin.

The head of the S.S. quill friction tube is supported by a friction tube pin, which consists of a piece of steel threaded on the lower portion and formed above into a small pin. It is screwed into the gun to the left front of the vent in cast iron, and to the right front in bronze guns.

Guide plate.

To insure direct action the lanyard passes rearwards through a guide plate. This is a small iron plate with cross-heads on top and hole for lanyard, having also a slot near lower end for the screw which secures it to the sight block. That used with mortars is cylindrical and threaded at the lower end, and screwed into the body of the piece. The cross-head serves to loop the lanyard over.

Ship's pendulum.

A pendulum is used to show the angle at which the vessel is heeling over on the lee side, so that the necessary allowance may be made in elevation when laying guns by the wood side scale or graduated coin.

Vent plug.

A vent plug consists of a vulcanized disc of india-rubber, with a leather stem. It is employed for protecting the vents of mounted guns.

Wrench.

The wrench for sight screw and friction tube pin is a small iron instrument with four arms, one of which is a wrench for sight screws, another a turnscrew, a third a friction tube pin wrench, and the fourth a tommy.

Spikes.

There are two sorts of spikes, common and spring.

Common spikes.

The common spike is a conical piece of hard steel about three inches long. When it is desirable to disable a gun for some time, a common spike is to be hammered into the vent, and the top broken off. For the purpose of re-venting a gun thus spiked, two hollow drills are supplied, with the venting tools.

Spring spikes.

A gun may be temporarily rendered ineffective with a spring spike, which consists of a steel rod with a flat head at the top and a spring near the bottom, so that when the end has passed into the bore the spring acts, and the spike cannot be removed till the spring is pushed back. If the gun is likely to be recaptured, the spike should therefore be inserted with the spring towards the muzzle, and on this account there is a little notch on the edge of the head to show the side the spring is on, so that it may be pressed back by the rammer.

For L.S. there are four lengths of spring spikes, 11·4, 8·5, 6·3, and 3·6 inches. The Navy are supplied only with the shorter nature for field guns.

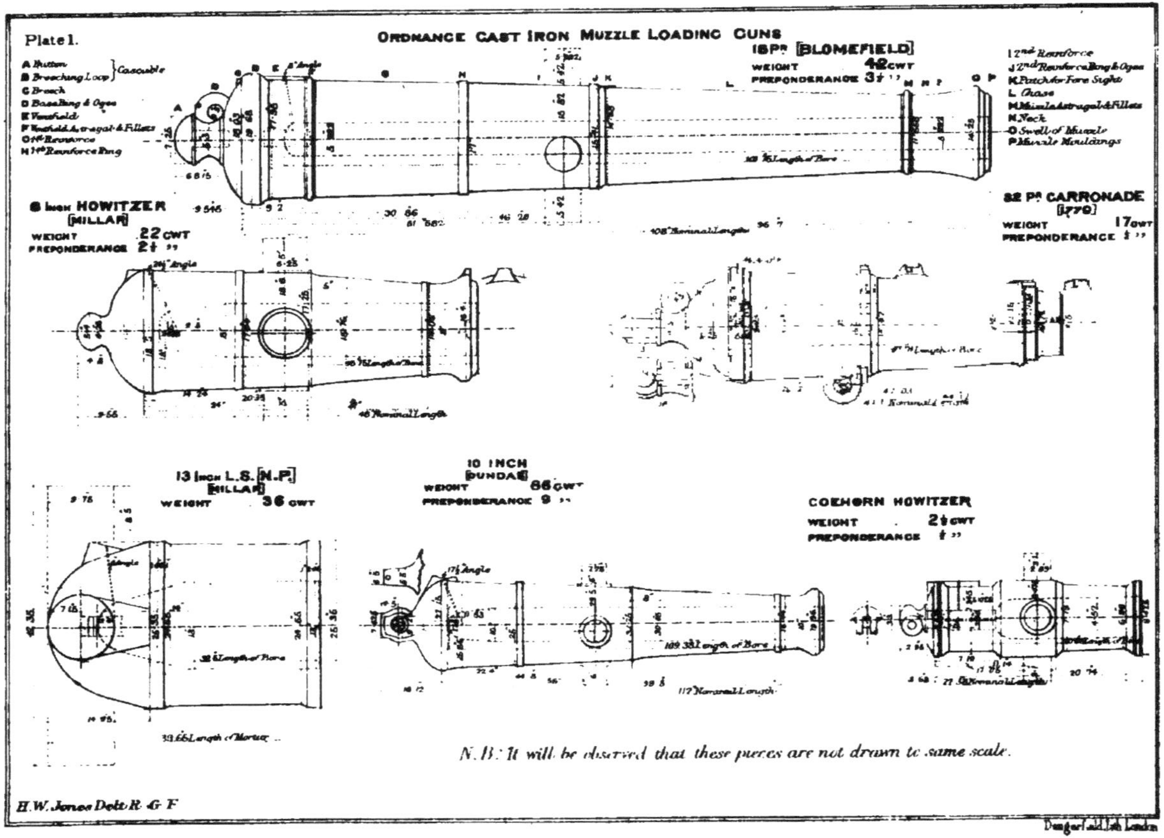
Plate 1.
ORDNANCE CAST IRON MUZZLE LOADING GUNS
18Pr [BLOMEFIELD]
WEIGHT 42 CWT
PREPONDERANCE 3½ "
A Button
B Breeching Loop
Cascable
C Breech
D Base Ring & Ogee
E Vent field
F Vent field Astragal & Fillets
G 1st Reinforce
H 1st Reinforce Ring
I 2nd Reinforce
J 2nd Reinforce Ring & Ogee
K Patch for Fore Sight
L Chase
M Muzzle Astragal & Fillets
N Neck
O Swell of Muzzle
P Muzzle Mouldings
8 Inch HOWITZER [MILLAR]
WEIGHT 22 CWT
PREPONDERANCE 2½ "
32 Pr CARRONADE [1779]
WEIGHT 17 CWT
PREPONDERANCE ½ "
13 Inch L.S. [N.P.] [MILLAR]
WEIGHT 36 CWT
10 INCH [DUNDAS]
WEIGHT 86 CWT
PREPONDERANCE 9 "
COEHORN HOWITZER
WEIGHT 2½ CWT
PREPONDERANCE ½ "
N.B: It will be observed that these pieces are not drawn to same scale.
H.W. Jones Delt R.G.F.
Dangerfield Lith London

CHAP. III.

TABLE III.

List A.

TABLE showing the Weights, Lengths, Calibres, Radii for Sighting, &c., of the CAST IRON and BRONZE ORDNANCE, which, according to the O.S.C., *List A.* are to be RETAINED in the Service.

	Nature and Service.	Nominal Weight.	Nominal Length.		Calibre.	Radii.		Clearance Angle.	By whom and when introduced.	Remarks.
						Short.	Long.			
		Cwt.	Ft.	In.	Ins.	Ins.	Ins.	Degs.		
Cast Iron Guns.	10-in. L.S. & S.S. ...	86	9	4	10·0	56·0	114·82	5 H.M. 6½L.M.	Dundas, 1846.	Two patterns.
	8-in. L.S. & S.S. ...	65	9	0	8·05	56·0	112·23	5½	Millar, 1834.	
		60	8	10	8·05	56·0	110·225	5	Millar, 1831.	
		54	8	0	8·05	48·0	100·715	5	Dundas, 1840.	
	68-pr. L.S. only ...	112*	10	10	8·12	59·8	132·4	4½	Dundas, 1841.	
	,, L.S. & S.S. ...	95	10	0	8·12	58·0	122·1	5	Dundas, 1846.	
	42-pr. L.S. only ...	84*	10	0	6·97	56·5	121·5	4½	Monk, 1843.	
		67*	9	6	6·93	54·3	118·0	4	Dundas.	
	32-pr. L.S. only ...	63	9	7	6·41	55·5	116·6	4	Millar.	
	,, L.S. & S.S. ...	58	9	6	6·375	55·0	117·895	4	Dundas, 1847.	
	,, L S. & S.S. ...	56	9	6	6·41	55·0	112·81	4	Blomefield.	
	,, L.S. & S.S., A. ...	50	9	0	6·375	50·0	109·25	5	Monk, 1838.	
	,, L.S. & S.S., B. ...	45	8	6	6·35	48·0	103·625	5	Monk, 1838.	
	,, L.S. & S.S., C. ...	42	8	0	6·35	46·3	97·62	5	Monk, 1538.	
	,, L S. only	48 & 50	8	0†	6·41	50·0	97·0	6½	Blomefield, Millar, & Dickson.	
	,, L.S. only	40	7	6†	6·35	40·5	85·1	7	Congreve ...	Bored up in 1830, from 24-pr. of 42 cwt.
	,, L.S. only	39	7	6†	6·375	43·0	89·8	5	Blomefield ...	Bored up in 1830, from 24-pr. of 40 cwt.
	,, L.S. & S.S. ...	32	6	6	6·3	39·0	78·0	6½	Blomefield ...	Bored up in 183?, from 24-pr. of 33 cwt.
	,, L.S. & S.S. ...	25	6	0	6·3	36·5	74·0	6	Dundas, 1845.	Has only two muzzl mouldings.
	24-pr. L.S. only ...	50	9	6	5·823	54·6	112·75	4	Blomefield.	
		48	9	0†	5·823	51·5	107·0	4	Blomefield.	
		20	6	0	5·823	34·0	70·75	5	Blomefield ...	Bored up from 12-pr., 22cwt.
	18-pr. L.S. only ...	42	9	0	5·292	51·5	107·0	4	Blomefield.	
		38	8	0†	5·292	46·5	95·75	4	Blomefield.	
	,, S.S. drill ...	20	6	0	5·17	34·0	70·75	5	Dickson ...	Bored up from 12-pr., 22cwt.
		15	5	6	5·17	30·0	65·25	5	Dickson ...	Bored up from 9-pr., 17 cwt.
	12-pr.	34	9	0	4·623	51·0	107·5	4	Blomefield.	
		33	8	6	4·62	48·5	101·2	4	Blomefield ...	For drill and saluting.
	9-pr.	28	8	6	4·2	48·0	101·25	4	Blomefield.	
		21	7	0	4·2	41·0	84·1	4	Blomefield ...	For saluting.
	6-pr.	17	6	0	3·668	35·5	72·0	4½	Blomefield ...	For drill or saluting.
	Carronades 68-pr. ...	36	5	4	8·05	30·0	—	—	1779.	
	Carronades 42-pr. ...	22	4	5	6·84	25·0	—	—		
	Carronades 32-pr. ...	17	4	0	6·25	23·0	—	—		
	Carronades 24-pr. ...	13	3	8	5·68	20·5	—	—		
	Howitzers 10-in. L.S. only	42	5	0	10·0	—	60·3	—	Millar, 1852	
	Howitzers 8-in. ...	22	4	2	8·0	—	48·0	—	Millar, 1852.	
	Mortars 13-in. S.S.	100	5	4	13·0	—	—	—	N.P., 1857.	
		100	4	5	13·0	—	—	—	O.P. or Blomefld.	
		81	3	2	13·0	—	—	—		
	Mortars 13-in. L.S. ...	36	3	4	13·0	—	—	—	Millar.	
	Mortars 10-in. S.S. ...	52	3	10	10·0	—	—	—	Blomefield.	
	Mortars 10-in. L.S. ...	18	2	5	10·0	—	—	—	Millar.	
	Mortars 8 in. L.S. ...	9	2	2	8·0	—	—	—	Millar.	
Bronze.	Guns 12-pr. L.S. ...	18	6	6	4·623	—	78·1	—		
	Guns 9-pr. L.S. ...	13	6	0	4·2	—	71·0	—		
	Guns 6-pr. L.S, ...	6	5	0	3·668	—	59·6	—		
	Guns 3-pr. L.S.	3	4	0	2·91	—	47·56	—		
		2½	3	0	2·91	—	35·5	—		
	Howitzers 32-pr. L.S. ...	17	5	3	6·3	—	62·5	—	Dundas, 1840.	
	Howitzers 24-pr L.S. ...	13	4	8	5·72	—	56·0	—	Millar.	
	Howitzers 24-pr. S.S. ...	13	4	8	5·72	—	56·0	—	Millar.	
	Howitzers 12-pr. L S. ...	6	3	9	4·58	—	45·0	—	Millar.	
	Howitzers 12-pr. S.S. ...	6	3	9	4·58	—	45·0	—	Millar.	
	Howitzers 4⅖-in. L.S. ...	2½	1	10	4·52	—	22·5	—	1738.	Coehorn.
	Mortars 5⅗-in. L.S. ...	1¼	1	3	5·62	—	—	—	1742.	Royal.
	Mortars 4⅖-in. L.S. ...	¾	1	1	4·52	—	—	—	1733.	Coehorn.

* Retained only until the few that now remain are superseded by Rifled Guns.
† These guns have been issued in large numbers for the Volunteer service.

CHAP. III.

TABLE IV.*

LIST (B.) of CAST IRON ORDNANCE to be ABOLISHED, but which are to be retained on the Works, if they are mounted at any station, until replaced by other pieces, which will be done when the carriages are worn out, if not sooner. (See par. 1,140 (*Changes in patterns,*) 1st January 1866).

	Nature.	Nominal Weight.	Length. ft.	Length. in.	Remarks.
		Cwts.	ft.	in.	
Cast Iron Guns.	10-inch	84	9	4	Millar.
		62	8	4	Millar. Only a few were made for experiment.
		57	7	6	Millar.
	8-inch	63	9	0	Millar.
		†59	6	9	Millar.
	68-pr.	88	9	6	Dundas.
	56-pr.	†97	11	0	Monk.
		†87	10	0	Monk.
	42-pr	75	10	0	Monk. Rarely met with.
	32-pr.	46	9	0	Bored up.
		41	8	0	Bored up.
		37	7	6	Bored up.
		25	6	0	Bored up.
	S.S.	25	5	4	Dickson.
	24-pr.	43	8	0	Blomefield.
		41	7	6	Congreve.
		40	7	6	Blomefield.
		38	7	6	Congreve.
		33	6	6	Blomefield.
		18	5	0	Dickson.
	18-pr.	40	9	0	Blomefield.
		32	6	10	Blomefield.
		27	6	0	Blomefield.
		22	7	0	Blomefield.
	12-pr.	29	7	6	Blomefield.
		24	6	0	
		22	6	0	Bored up from 9-pr.
	9-pr.	26	7	6	
		17	5	6	
	6-pr.	23	8	6	
		†22	8	0	
		21	7	6	
		19 or 20	7	0	
		18	6	6	
		11	4	10	
		6	3	6	
Cast Iron.	Carronades 18-pr.	†10	3	4	
	Carronades 12-pr.	† 6	2	8	
	Carronades 6-pr.	5	2	9	
	Howitzer, 24-pr. or 5½"	†16	3	4	Has a cylindrical chamber.
	Mortars, 8" Eprouvette.	8½	1	5	
	Mortars, 8" O.P.	8	1	10	
		9	1	9	

* This list is based on the rarity or inferiority of the pieces it contains.

† These pieces are to be found in the vocabulary of stores published by the Ordnance Store Department, because they still exist in the armaments.

CHAPTER IV.

BRIEF HISTORY OF THE RIFLED ORDNANCE IN THE BRITISH SERVICE.

Necessity for Rifled Ordnance.—Why rifled guns were not earlier employed.—Principal inventions with rifled pieces, 1615 to 1851.—Lancaster guns.—Krupp guns.—Comparative accuracy of S.B. and rifled pieces.—Experiments as to strengthening existing S.B. ordnance.—Armstrong and Whitworth systems.—**Armstrong System adopted 1859.**—64-prs. M.L. introduced.—Armstrong and Whitworth Committee, 1863.—R.M.L. *v.* R.B.L. guns.—Advantages in favour of R.M.L.—Manufacture of guns for India, 1868.—Bronze ordnance discontinued.—Special Committee of 1870.—Wrought iron M.L. gun with steel tube recommended.—Comparative value of English and foreign field guns, 1860-1.—**Abolition of the B.L. System.**—Introduction of **R.M.L. 9-prs. and 12-prs., 1870.**—French modified system.—16-pr. M.L.—Comparative trial of British and Foreign guns, 1872.—French trials.—25-pr. R.M.L.—Okehampton experiments, 1875.—New armaments of Germany, France.—Comparative value of English and Foreign field guns, 1876.—**Experimental 12-prs., 1876.**—Mountain and boat guns.—Guns of position and siege pieces.—Guns for secondary purposes of defence.—Heavy guns.—80-ton gun introduced, 1876.—Its projectile, charge, velocity, and penetrative power.—Resumé as to our service rifled ordnance.

Necessity for rifled ordnance, 1855.

THE general adoption of rifled small arms, about 1855, necessitated the introduction of rifled ordnance, in order that Artillery might still retain its superiority over Infantry, and remain as before the principal arm in the field, which certainly would not be the case if an enemy's skirmishers had the power of placing a battery *hors de combat* before its guns could be brought into action with effect upon the advancing columns from their limited range.

To supply this great want in warfare, involved a complete reformation in the architecture of Artillery, which had been almost at a standstill since the time of the Tudors, for although modifications had been occasionally made in the manufacture of ordnance, the general principles of construction remained unaltered. Anyone who examines the old guns in the Tower of London, or in the Museum of Artillery at Woolwich, may see that they are of the same genus as the smoothbores which still exist in our Service, and even notice some specimens quite as soundly and as artistically cast as any of those of the present century, nay more, he may infer that our existing S.B. guns can scarcely be superior to their prototypes in range-power, or susceptibility to rifling.

Why rifled guns were not produced earlier.

It is, however, worthy of note, that this stagnation in the construction of ordnance was not to be attributed to ignorance of the theories of gunnery, but to the backward state of metallurgy and mechanism, for

CHAP. IV.

professional as well as amateur artillerists have even at remote periods understood the value of rifled guns, but their endeavours to obtain them were rendered abortive by the want of suitable materials and proper machinery.*

Principal inventions with rifled pieces from 1615 to 1851.

* The following is a list of the principal home and foreign inventors and inventions up to 1858, as given by Major Stoney and Capt. Jones, R.A.

The names of the English inventors were selected out of nearly 200 whose proposals are described in "MS. notes on the various designs for elongated projectiles for smooth bore and rifled guns, which have been from time to time considered by the Ordnance Select Committee down to 1858," compiled from the records of the Committee by Lieut.-Colonel (now Major-General) A. G. Burrows, Royal Artillery.

For the foreign inventors an able paper by Captain R. A. Scott, R.N., published in 21st number of Vol. VI. "Journal of the Royal United Service Institution," was utilized:—

In the Arsenal of St. Petersburg is a gun 2¾ inches in diameter, and 62 inches in length of bore, which was rifled in nine grooves in 1615.

In 1661, the Prussians experimented at Berlin with a gun rifled with 13 shallow grooves.

In 1696, the elliptical bore was known, and had been tried in various parts of Germany.

In 1745, the date at which Robins was experimenting in England, the Swiss already possessed small rifled pieces.

1774, Experiments with elongated projectiles fired from a 6-pr. S.B. gun were carried on at Woolwich by the "Military Society." (This Committee existed previous to the formation of the "Board of Officers," which was succeeded by the Ordnance Select Committee.)

1776. Dr. Pollok proposed elongated shot for smooth-bore pieces.

1790. Mr. Wiggins made designs of a rifled gun and belted projectiles.

1803–1806. Proposals by Messrs. Davies, Barlow, Spencer, Eckhart, &c., were considered by the Board of Officers.

1816–1819. M. Ponchara, a distinguished French Artillery officer, was making various experiments with an old gun which he had rifled with 13 grooves.

1820. Captain Cullen, R.M. Artillery, proposed cylindrical shells filled with bullets without any bursting charge.

1821. Lieut. Croly, h.p. 81st Regiment, proposed *B.L. cannon and lead-coated projectiles.*

1823–32. Lieut. Norton proposed explosive shells, and a rifled gun.

1826. Experiments were made with some cylindro-conical percussion shells, designed by Lieut.-Colonel Miller, of the Rifle Brigade.

1833. M. Montigny of Brussels invented a breech-loading rifled piece.

1842. Colonel Treullie de Beaulieu first presented to the French Government his plan for rifling M.L. guns, with a few large grooves for studded projectiles, which was afterwards adopted in a modified form, and is now known as the French system.

1845. Major Cavalli, a Sardinian officer, invented a breech-loader (submitted to Ordnance Select Committee, 1850), rifled with *two* grooves, for a ribbed shot; his guns were used at the Siege of Gaeta in 1860.

1846. The Swedish Baron Wahrendorf proposed the system of using lead-coated projectiles with shallow-grooved breech-loaders. He also tried the Cavalli projectile, and rifling with guns closed at the breech on his own plan, whilst Lieut. Engstroem of the Swedish Navy affixed hard wood bearings or buttons to an iron projectile. Wahrendorf's and Engstroem's designs were submitted to the Ordnance Select Committee in 1855.

1852. H.R.H. Prince Albert proposed a concussion shell, Lord Clarence Paget a rifled projectile, Lieut.-Colonel Stevens, R.M. Artillery, a plan for rifling 13-inch sea service mortars; Mr. Mallet an improved form for rifling cannon shot and shell.

In 1853 proposals were submitted to the Ordnance Select Committee by Lieut.-Colonel Grant, Captain Norton, Captain Jodrell Leigh, Signor Verga, &c., &c., &c.

In 1854, by Major Parlby, Mr. Lancaster, Admiral Duff, Quartermaster-Serjeant Macbay, R.A., Major Parsons, Major the Hon. W. Fitzmaurice, Major Vandeleur, R.A., Lord W. Fitzroy, Mr. G. Nasmyth, Captain Anson, R.A., Mr. Hadden, Mr. (now Sir W.) Armstrong, Mr. Alfred Jeffery, &c., &c., &c.

1855. Captain Blakeley, R.A., patented his method of forming guns of an internal tube with cast iron or steel rings, heated and shrunk over it, and Sir J. Woodford, Captain Fowke, R.E., Messrs. Goddard, B. Britten, Underwood, Skelton, &c., and the Revs. J. Bramball and R. Potter, brought forward various designs.

CHAP. IV.

Such then being the state of the case it was fortunate for the ascendancy of Artillery that recent years should have seen immense progress in the science of metallurgy and in mechanical appliances. It is only of late that the manufacture of cast steel as a material for rifled ordnance has made rapid way whilst the difficulties which used to attend the forging of wrought iron in large masses were so great, that until the comparatively recent introduction of steam hammers it would not have been possible to forge our modern monster guns. We now have also machines so perfect and easily manipulated, that we are able to ensure the dimensions of our guns being true to gauge even to such minute limits as one-thousandth of an inch.

Owing to these facilities rendering the manufacture possible, rifled guns began to be introduced in large numbers into the Service in 1859, and since that date each year has seen great improvements in their manufacture.

Commencing in 1856 with the Lancaster rifling (which was soon abandoned), in a few cast iron guns we subsequently introduced (1859) R.B.L. on the Armstrong system for field, siege, and heavy guns. In 1864 the introduction of our present R.M.L. guns commenced, and we are still proceeding with the manufacture of these guns, but with ever increasing improvements in the construction of the guns themselves, in the methods of rifling, the natures of sights, and all other details connected with them.

In the following pages is given in detail a short historical précis as to rifled ordnance, field, siege, and heavy, showing how we have steadily and rapidly pushed forward in the path of progress.

Introduction of Rifled Field Guns.

Steps taken by the Continental powers.

In the lull which succeeded the Crimean campaign, Napoleon III. turned his attention to the rifled artillery problem, and came to the conclusion that the readiest mode of solving it would be to rifle his bronze field guns on Colonel Treuille de Beaulieu's plan, and at the battles of Magenta and Solferino, in 1859, some of these rifled field batteries were used by the French with great effect.

Rifled guns used by France in 1859.

About this date (1859) all the great powers, anxious to provide themselves with rifled field armaments, tried various systems, both of B.L. and M.L. guns, and the example of France in utilizing her S.B bronze guns for the manufacture of rifled field pieces was sooner or later followed by all the Continental powers, Prussia excepted; England, however, used other materials, while the Northern Kingdoms of Denmark, Sweden, and Norway retained cast iron.

Prior to the advent of rifled guns, it had been frequently proposed in Prussia to employ a lighter and stronger metal than bronze for field

---

1856. General Timmerhans of the Belgian Artillery invented a wad which by taking the rifle grooves gave rotation to the elongated shot.

1857. Mr. (now Sir Joseph) Whitworth submitted his system of ordnance.

In short, it appears from the records of the Ordnance Select Committee that up to 1855, experiments had been made with rifled guns and projectiles for half a century in this country, but without any satisfactory result. The projectiles tried were very numerous, most of them were intended to be fired from S.B. pieces, the necessary rotation being obtained either by the action of the gas in the gun, or of the air, on the peculiar shape of the shot.

The remainder were generally fired from special cast-iron 9-prs. rifled with four grooves, making a quarter turn in the bore. Specimens of those experimental projectiles are preserved in the Royal Arsenal; officers desirous of inventing a *new* shot or shell are recommended to examine previously this heterogeneous collection.

CHAP. IV.

guns, in order to increase their mobility. Between 1844 and 1855 several steel S.B. field pieces were manufactured for trial, both at Bochum and also by Messrs. Krupp at Essen, and on account of the good results given by these guns, an experimental committee actually prepared designs in 1857 for a service S.B. steel gun of 12 centimetre calibre, and to fire charges of between $3\frac{1}{2}$ and $4\frac{1}{2}$ lbs.

S.B. steel guns by Messrs. Krupp in 1857.

Smooth bores replaced in Prussia by steel R.B.L. in 1859.

When, therefore, it became evident that rifled guns must take the place of smooth-bores, and that a new armament had to be provided, Prussia was naturally ready to adopt for that armament a metal which had been proved, even for the old guns, to be better than bronze. The enterprising firm of Krupp completed in 1856 a 9c.m. (3·2-inch) rifled B.L. steel gun, according to the designs of an artillery committee, which was considered so satisfactory, that in 1859 steel was definitely adopted by the Prussian authorities as the material to be used for the new field guns; this material, as well as the B.L. construction on one system or another, has ever since been employed by this Power,* with partial exceptions.

Inadequate strength of bronze.

In England bronze was not considered of adequate strength. and the distinguished engineer, Sir W. Armstrong, having constructed a rifled gun of wrought iron as early as 1854, and having matured a satisfactory method of manufacture of B.L. guns, his system was adopted to satisfy the imminent necessity that existed for an immediate supply of rifled field pieces. His earlier guns were made entirely of wrought iron, but many of the later Armstrong field guns have steel barrels.†

Sir W. Armstrong's gun.

Experiments with the same, 1854.

Before adopting this system our own authorities carried on careful and extensive experiments with wrought iron rifled B.L. guns, constructed by Mr. (now Sir W.) Armstrong, whose system was brought to official notice in 1854, when a few guns were ordered for trial.

3-pr. B.L. gun, 1855.

The first of these, a 3-pr. was delivered in July 1855, and reported on favourably by the Ordnance Select Committee, who desired to make further experiments.

5-pr. and 18-pr. B.L. guns 1857.

A 5-pr. and 18-pr. on Sir W. Armstrong's system were subsequently tested at Newcastle and Shoeburyness respectively, and gave very good results, showing that they exceeded existing S.B. pieces of similar calibre in strength, power, and precision.

Experiments as to strengthening and rifling the existing S.B. ordnance, 1859 to 1865.

While these Armstrong B.L. guns were being tested, extensive experiments were being carried out, both to see if any safe method could be discovered of strengthening and rifling existing bronze and cast iron ordnance, and also to determine whether any better system of constructing rifled ordnance than that proposed by Sir W. Armstrong could be settled upon.

As was natural, however, in such a matter, Sir William Armstrong was not permitted to bear away the palm without a contest.

Various propositions for rifled guns were submitted to the Ordnance Select Committee, and in 1858 a special committee was appointed to report as to which was the best rifled gun for field service. This committee came to the conclusion that it was not expedient to incur the expense of trying further experiments with any except those of Messrs. Whitworth and Armstrong.

Armstrong and Whitworth rival systems.

A trial accordingly took place between guns rifled on the two systems, but at that time Mr. Whitworth did not propose any gun of his own construction, and had only rifled government blocks of bronze and cast iron, while the Armstrong breech-loading gun was complete in every respect.

* The construction, however, and mode of B.L. has often been changed.

† The experimental 3-pr. first tried was, however, an exception, having a steel barrel.

CHAP. IV.

The Armstrong R.B.L. system adopted, 1859.

No better nor cheaper system being found during the period of inquiry, and the necessity being urgent that we should have a rifled field equipment and so not be behind other nations, Government entered into a contract, in January 1859, with the newly-established Elswick Ordnance Company, and also commenced the manufacture of these R.B.L. guns in the Royal Arsenal, Woolwich. In February 1859, Sir W. Armstrong was appointed engineer of rifled ordnance, and in the following November he became also Superintendent Royal Gun Factories.

Service equipment of our artillery Armstrong R.B.L. guns, 1859–63

Our field artillery were soon equipped with Armstrong guns, and a number of the heavier natures were made both for siege and garrison purposes, and also for the armament of our fleet. These guns are still in the service, though many of them are kept in reserve: the different natures are given in the table at the end of Chapter VI.

The theoretical superiority of these R.B.L. pieces over the old S.B. field guns in power is shown by Table V. at the end of chapter.

Objections to the B.L. breech-screw system of Sir W. Armstrong.

This B.L. system, however, is wanting in simplicity and requires a great many implements for its repair. Further experience showed that serious accidents were likely to take place from various causes, such as the breech-screw being improperly screwed up, the vent-piece being weak or ill fitting, &c., and it is therefore probable that if we had been in the meantime engaged in a war, in which the B.L. guns had to be used in large numbers, such accidents would have considerably impaired their success; as it is, however, they have been only used on a small scale before an enemy, principally in China, New Zealand and Japan.

Behaviour of our R.B.L. breech-screw guns on active service.

In these wars the Armstrong B.L. guns answered very satisfactorily on the whole, both on land and at sea, and proved, as was to be expected, much more powerful than our old S.B. field and heavy guns, comparing well with the rifled guns of other powers at that period, as shown by the table at page 91. It was clear, however, that a less complicated system would be better adapted for general service.

Various changes in Continental systems.

Every Power in Europe which took an initiative in adopting rifled field artillery, had to purchase its experience in a similar manner. The French M.L. guns of 1864 were of a different nature from those used in the Italian campaign, while they were replaced in 1872 by the bronze B.L. Reffye guns, for which again new pattern steel B.L. guns of different calibres are now being substituted. The Austrians adopted the system La Hitte in 1859, changed it for Lenk's system in 1860, and this was changed again in 1862, while in 1876–7 a new armament of bronze B.L. guns was introduced. The Prussians adopted the Wahrendorf's system in 1860, and changed it for Wesener's in 1863, they afterwards adopted Krainer's, and since 1872 have equipped themselves with new and improved guns with Krupp's single B.L. wedge, so that since 1860 most foreign Powers have changed the pattern of their field guns at least twice over (vide Table VII. and VIII. at end of chapter.)

Armstrong and Whitworth Committees.

Sir J. Whitworth's system in 1863.

In 1863, Mr. (now Sir Joseph) Whitworth, having carried on a series of private experiments and perfected his system, obtained such good results that he demanded another trial.

Armstrong and Whitworth Committee, 1863.

A special committee was therefore appointed, 1st June, 1863 to examine and report upon the different descriptions of guns and ammunition proposed by Sir W. Armstrong and Mr. Whitworth.

The committee carried out careful and extensive competitive experiments with Whitworth 12-prs. and 70-prs., Armstrong 12-prs. and

CHAP. IV. 70-prs. breech-loaders, and Armstrong 12-pr, and 70-pr. muzzle-loaders; the 12-prs. having been chosen to decide the question for field artillery, whilst the 70-prs. were the best available representatives of heavy artillery, comprising siege, garrison, and broadside guns.

Whitworth M.L. guns. Both natures of the Whitworth guns were muzzle-loaders, and had his well-known hexagonal rifling, and mechanically fitting projectiles.

The 12-prs. were of solid mild steel (having trunnion-rings screwed on to them), with a hoop of the same material over the powder chamber.

The 70-prs. were of the same material, but consisted of an inner tube closed by a cascable screwed in, and strengthened by hoops forced on cold by hydraulic pressure.

Armstrong B.L. guns. The Armstrong breech-loaders were constructed with steel barrels, and with wrought iron coils superimposed as usual, but the 12-prs. had the ordinary breech-screw arrangement, and the 70-prs. were upon the wedge system.

Armstrong M.L. guns. His muzzle-loaders also had inner barrels of steel; they were rifled on the shunt principle,* for projectiles with soft metal studs.

Report made 1865. After a searching examination of important witnesses, and complete and comprehensive trials which cost some 35,000*l.*, the committee concluded their labours, on the 3rd of August, 1865, which is the date of their report.

The results of these experiments were very creditable to both inventors, especially as regarded the construction of their respective guns, each of which after firing about 3,000 rounds was only burst at last by abnormal means.†

Results in favour of the Armstrong M.L. guns. The report was on the whole most in favour of the Armstrong muzzle-loaders for the following reasons:—

"That the many-grooved system of rifling with its lead-coated projectiles, and complicated breech-loading arrangements, entailing the use of tin cups and lubricators, is far inferior for the general purposes of war, to both of the M.L. systems, and has the disadvantage of being more expensive both in original cost and in ammunition.

"That M.L. guns can be loaded and worked with perfect ease and abundant rapidity.

"That guns fully satisfying all conditions of safety can be made with steel barrels strengthened by superimposed hoops of coiled wrought iron, and that such guns give premonitory signs of approaching rupture; whereas guns composed entirely of steel are liable to burst explosively without giving the slightest warning to the gun detachment."

Conclusions arrived at by the committee. These remarks were not supposed to be limited to heavy guns, they apply with equal or greater force to field guns; in fact, the committee expressed their opinion that both Sir Wm. Armstrong's and Mr. Whitworth's M.L. systems, including guns and ammunition, were on the whole very far superior to Sir Wm. Armstrong's B.L. system for the service of artillery in the field.

---

* Vide p. 44 for description of shunt rifling.

† It may be noted that the Committee report upon the point:—

"After then these 12-pr. (8 cwt.) guns, viz., the Armstrong M.L. and B.L., and the Whitworth M.L. guns had each fired 2,800 rounds, the M.L. with 1 lb. 12 oz. of powder and 12 lb. shot, attempts were made to destroy them by firing increased charges of both powder and shot, with the following result. At the 42nd round the Armstrong B.L. gun split open, but did not burst; at the 92nd round the Whitworth steel gun *burst violently* into eleven pieces, while the M.L. Armstrong gun failed at the 60th round, one of the outer coils having cracked and fallen off without flying to pieces."

Committee of 1866.

Committee of 1866 in favour of M.L.

Another committee, of which Sir R. Dacres was president, was appointed in 1866 to inquire into the question of M.L. and B.L.; they came to the conclusion that "the balance of advantages is in favour of M.L. for field guns," because they are equal to breech-loaders in range and accuracy, and much superior to them in simplicity both of fittings and ammunition.

A few 12-pr. R.M.L. guns made in 1870.

At this date a few 12-pr. R.M.L. of 8 cwt. were made of wrought iron and steel, but as the expense would have been great and there was no urgency, their manufacture was not carried out on a large scale for issue instead of the service B.L. guns.

Committee of 1868.

Field guns for India, 1868.

Questions as to manufacture of guns in India.

Committee on Bronze field guns for India 1868–70.

The field armament of India consisted at this period (1868) almost entirely of S.B. guns, and it became necessary to provide her with a more efficient equipment. At this date we had ceased manufacturing R.B.L. and had not began to make R.M.L. field guns for issue. For India, however, it was finally decided to have a M.L. armament, and as it was considered advisable to make that country so far independent of England that her arsenals should be in a position to manufacture their own field guns, while there was an objection (on economical grounds) to sending out the expensive plant required for the manufacture of steel and wrought iron guns, a committee of officers was assembled in 1868 under the presidency of Major-General Eardley-Wilmot to carry out experiments with a view to recommending a R.M.L. bronze gun for Indian service. They first attempted to utilize S.B. 9-pr. guns, but had to abandon the idea, and subsequently R.M.L. cast at Woolwich gave such satisfactory results upon trial at Shoeburyness that in 1870 9-pr. bronze guns were definitely adopted for Indian service, while some were also issued to batteries at home.

Manufacture of rifled bronze guns commenced 1870, discontinued in 1871.

Time and practice during peace, however, developed so many defects* in these 9-prs. that they have now all been withdrawn from the service (vide p. 5). In external form, sighting, &c., these guns resembled the present service 9-prs.

Special Committee of 1870.

Special committee of 1870.

On the 25th July, 1870, a special committee of artillery officers, under Major-General Sir John St. George, K.C.B., as president, was appointed to carry out comparative trials with the 9-pr. bronze rifled guns (Indian pattern), and 9-pr. and 12-pr. B.L. guns of the home service, and to report fully on the whole subject, so that the question of B.L. and M.L. for our field guns might be definitively settled.

In their report dated 28th November, 1870, they say:—

"The Committee have no hesitation in giving the preference to the M.L. gun, both in respect to simplicity and facility of repair.

"If, as regards the question of endurance, the Committee are called upon to select between a structure of wrought iron and steel and one

---

* At the time of their adoption, the Supt. R.G.F., Colonel (now General) Campbell, strongly objected to the use of such bronze for rifled field pieces, vide p. 88 "Report of Field Artillery for India, 1868."

CHAP. IV.

of bronze, as tried by them, they unhesitatingly pronounce in favour of the former, whether the gun be a muzzle-loader or a breech-loader.

Results of trials between B.L. and M.L. field guns.

"It may be urged that considerations of manufacture favour the adoption of bronze for M.L. guns for India, but the same cannot be accepted as applying to this country, and taking weight for weight it is impossible to deny that far greater endurance will be attained by the present mode of construction than by the use of bronze.

"The service B.L. system, owing to the absence of windage, necessitates the employment of a mechanical arrangement for lighting the time fuze. This is effected by the employment of a detonator, which has proved to be highly sensible to climatic influences. The M.L. gun on the other hand has the advantage of being able to use an ordinary wood time fuze, which experience has proved to be little or not at all affected by climate.

"As regards cartridges, the B.L. guns have the great disadvantage of requiring the use of lubricators.

"In respect to other stores, such as percussion fuzes and projectiles, the Committee believe that, whether for breech or M.L. guns, there will be found little or no difference between them so far as regards their capability to bear the tests of travelling or climate.

"Judging from the results of the practice at Aldershot, the 9-pr. M.L. and the 12-pr. B.L. guns appear, in respect to shooting, to be much upon a par; the former being superior in point of shrapnel shell with time fuzes, the latter in point of segment shell with percussion fuzes, the 9-pr. B.L. gun being inferior to both.

A wrought iron R.M.L. with steel tube recommended.

"The advantages of simplicity, facility of repair, ease of working, rapidity of fire, original cost, and cost of maintenance, are in favour of a M.L. gun, and the Committee consider that these qualifications outweigh the important advantage of the superior amount of cover given to the detachments when entrenched and in the open, which a B.L. gun affords, and are therefore of opinion that, on the whole, a M.L. gun is the more efficient for war purposes; but they recommend that, if adopted for home service, they be made of wrought iron with steel tubes."

Resumé up to introduction of R.M.L. Field Guns.

Summary as to R.B.L. versus R.M.L. for field guns.

We see then that for field service our first rifled guns were R.B.L. pieces, built up on the Armstrong principle* between 1859 and 1864, that we soon found out that this system of breech loading had faults, and in consequence made very extensive experiments between 1863 and 1870 as to the comparative value of these B.L. guns and the various descriptions of M.L. field guns brought forward. The committees who carried out these trials were composed of distinguished artillery officers, and they all arrived at the conclusion that a M.L. gun was better than a B.L. for the reasons given above.†

Service construction.

As regards the construction or building up of the gun itself which is only affected indirectly by the question of B.L. or M.L., these committees were unanimous in recommending the built-up system of steel barrel and wrought iron jacket at present employed, instead of the use of bronze or of steel alone.

---

* The system of B.L. introduced with our service Armstrong guns, has nothing to do with the principles of the Armstrong construction of building up of a gun as pointed out at p. 31. Though we have abandoned B.L. guns, we still retain the Armstrong construction as modified by Mr. Fraser.

† Much hostile criticism has been spent upon our enthusiastic adoption and praise of the R.B.L. Armstrong system, and our speedy abandonment of the same in favour of R.M.L., but we have kept steadily to one mode of construction, and perhaps changed less in reality than most of the Powers.

CHAP. IV.

In consequence of these recommendations, no new R.B.L. field pieces have been made since 1864, but in 1871 we commenced the manufacture of the M.L. steel and iron guns with which our Field Artillery are now entirely equipped; these guns are built up of an inner barrel of steel, and a jacket of coiled iron carrying the trunnions.

Manufacture of R.M.L. field guns commenced in 1871.

Designs for a 9-pr. gun of 6 cwts. and of a 12-pr. of 8 cwts. had been prepared in 1866, the former being adapted for a boat and R.H.A. gun, and the latter for field batteries. A few of both were manufactured but not completed in 1868–69, and in 1870 these were rifled on the modified French system as 9-prs. of 18 and 21 calibres respectively, the former proved rather too short, and but few were made,* while of the 8-cwt. guns a number were manufactured both for field and boat service.

R.M.L. 9-pr. 6 cwts., and 12-pr. of 8 cwts. designed in 1866 and rifled in 1870.

Committee of 1869.

A committee of R.A. officers which carried out important experiments on Dartmoor in 1869† having strongly recommended that a field gun should be introduced suitable for high angle fire when necessary, and for throwing common shell with large bursting charges, the question was referred in 1870 to a special Committee‡ who after trials with guns weighing about 12 cwt. and of different calibres, finally settled upon the calibre of 3″·6 as the most suitable, and accordingly a 16-pr. gun of 3″·6 and of 19 calibres length of bore, was approved of for the armament of the heavy portion of our field batteries being, when introduced, the most powerful field gun in Europe.

Committee of 1869.

Introduction of the 16-pr. R.M.L. 1871.

Committee of 1871.

In 1871 a few Gatling mitrailleurs were introduced into the service for naval purposes after extended experimental trials had been carried out by a special committee. (Vide p. 381, Appendix, for further information as to this weapon.)

Gatling mitrailleurs, 1871.

Committee on High Angle Fire.

On the cessation of the Franco-Prussian War of 1870–71 in which the Prussian field guns played so important a part it was thought advisable by our War Office to test our new M.L. pieces in comparison with the Prussian breech-loader. Experiments were therefore carried on at Shoeburyness in 1872 by the Committee on High Angle Fire mentioned at p. 79, between a German 9-pr. B.L. of 3″·09 calibre, weighing about 6 cwt., and our own 9-pr. of 8 cwt. (now reduced to 6 cwt.) with highly favourable results for our own field gun.

Comparative trial of British M.L. and Prussian B.L. guns in 1872, by High Angle Fire Committee.

To quote once more from Captain Langlois' work, he says, talking of the metals employed for, and of the absolute construction of field guns:

"Prussia, in fact, after using steel pieces, came back to bronze in 1869; after the war of 1869–71, she carried out trials with bronze phosphor, and has lastly arrived at a steel ringed gun.

"France and Austria have so far preserved bronze, the first with a tendency towards steel, and the second towards phosphor bronze.

"*England alone has never varied;* she adopted ringed iron guns with steel barrels for the Armstrong system, and still preserves that metal (and a similar construction) for her so-called Woolwich guns."—*Les Artilleries de Campagne de l'Europe en* 1874, *par H. Langlois, Captain d'Artillerie.*

* Now only used for boat service, Mark I., 8S., vide p. 254.

† Major-General C. Dickson, C.B., V.C., President.

‡ On High Angle Fire, vide note p. 79.

CHAP. IV.

The Committee reported that the British gun had a decided superiority over the German, in length of range, rapidity of fire, power and ease of manipulation, while the German gun excelled somewhat in accuracy of shooting, though the difference in this point was but slight.

Throughout the trials, however, the breech arrangement of the latter required much attention, and the Committee considered that it required such constant care for its preservation, as in hard campaigning and severe weather it would be difficult to exact from any artilleryman whatever.*

French Experiments.

French official trials of field-guns in 1872.

Careful comparative trials were also made in the same year by the French authorities at Bourges and elsewhere, between the Woolwich 9-pr. and other field guns, the result being as follows according to the official report:

"Notwithstanding a few imperfections, the Woolwich material, taken altogether, constitutes a first-class system of field Artillery."

"The Woolwich wrought iron 9-pr. gun gave results which are not inferior to those of any gun actually in service in Europe. These results, however, can perhaps be surpassed. This is the end to be kept in view in the selection of a field gun, and which must be attained in the case of our adopting a breech-loader. This method of loading, since it possesses practical inconvenience, must give in compensation notably superior results to the best muzzle-loading gun in order to be adopted."

English Field Guns of this Date.

In 1872, then, we had every reason to be satisfied that our field artillery, even if slightly inferior in accuracy, was on the whole superior to that of all other nations in power.†

This will be evident upon inspection of the Table, p. 92 which shows the comparative values of the service field guns used by the Great Powers in 1872.

9-pr. 6 cwt. for the horse artillery and light field batteries.

The 8 cwt. 9-pr. proved somewhat heavy for horse artillery, and as experiment showed it was possible to use with this arm a lighter gun of 6 cwt. a piece of the latter weight and same calibre as the 8-cwt. gun with a length of about 22 calibres was introduced as a horse artillery gun in 1874.

Further trial showing also that the same charges and projectile could be used with this gun as with the 8-cwt. gun it was determined to arm both horse artillery and light field batteries‡ with 9-pr. guns of 6 cwt. which are about a calibre longer in bore, and give slightly more velocity than those of 8 cwt.

---

* The complications due to the breech arrangement was in fact severely felt during the war of 1870–71, when over 200 guns were rendered unserviceable in the field through the breech apparatus becoming defective. In the new German guns an improved and simpler breech closing apparatus has consequently been adopted, but the disadvantage of complication still remains. For every gun spare parts, tools, &c., have to be carried.

† In Captain Langlois' work on the Field Artillery of Europe, dated 1875, he states with regard to the armaments of the several Powers in 1870, "Two things are evident, one, that by breech loading, one may expect to obtain greater velocities and greater accuracy, thanks to the regular 'forcement' of the projectile. On the other hand, it must be acknowledged that no existing service breech-loader possesses over the Woolwich gun a superiority sufficiently great to make up for their complication."

‡ The 8-cwt. guns made are being turned down for S.S. Vide p. 253.

CHAP. IV.

Our present armament therefore for field service consists of 9-prs. of 6 cwts. for horse artillery and one half of the field batteries, the remainder of which (or heavy field batteries) are equipped with 16-prs. of 12 cwts. For boat and field marine service 9-prs. of 6 cwts. and 8 cwts.

16-pr. for heavy field batteries.

In 1871 a 25-pr. of 18 cwts. was proposed as a gun of position and introduced into the service in 1874. Its construction is similar to that of the field guns. Vide p. 78 for further account, under the heading of Siege Pieces.

25-pr. introduced 1874.

Our field guns proved very formidable field pieces during the experiments carried out at Okehampton in 1875, the 9-pr. of 6 cwts. up to 3,500 yards and the 16-prs. up to 4,000 yards range rendered the ground covered by them quite untenable by infantry, while at still longer range, cover, such as villages or houses, was destroyed by the fire of the 16-pr.

Experiments at Okehampton with 9-pr. and 16-pr. in 1875.

Progress, however, is as necessary in artillery as in all else, and it advances with no leaden foot. Since the Franco-German War, Germany has furnished herself with a new field equipment of more powerful guns, while Austria, France, Russia, and other Powers are making strenuous efforts in the same direction as will be seen from the Table at end of the chapter.

New or experimental armaments of Foreign Powers.

Experimental Field pieces.

Convinced as we are, notwithstanding numerous dicta to the contrary,* that we can increase the power of M.L. field guns in as great a ratio as the Germans have increased those of their breech-loaders, while retaining the undoubted and important advantage of simplicity we are now trying experimental M.L. field guns, which bid fair to surpass in power any field ordnance hitherto made.

Experimental M.L. 13-pr.

In the early part of 1875, in compliance with the wishes of the then Director of Artillery, Sir J. Adye, K.C.B., a 12-pr. R.M.L. gun of 8 cwts. was designed by the Royal Gun Factory to throw a 12 lb. projectile with a muzzle velocity of at least 1,500 feet per second.

After some preliminary trials with an old 9-pr. gun lengthened and chambered, three of these guns rifled on different systems were tested in 1876 giving very good results. The polygrooved system of rifling with a rotating gas check gave the best results, and was accordingly definitely adopted. Further experiments have led to the final adoption of a 13-pr. polygrooved gun, the M.V. of which will not be less than from 1560 to 1600 *f s.*

Certain minor details have not yet been finally decided upon. When these are settled only a short period need elapse before we can furnish our field artillery with a new and still more powerful equipment than that which they now possess.

Mountain and Boat Guns.

The history of these pieces is almost identical with that of field guns. When we ceased to manufacture R.B.L. guns, in 1863, we commenced to make these pieces also as muzzle-loaders.

Their history.

The 6-pr. Armstrong R.B.L., which had partially replaced the old S.B. mountain piece (vide p. 65), was in its turn superseded by the 7-pr. M.L. gun rifled on the French system. In 1865, a few mountain

7-pr. M.L. gun.

---

* Vide, for example, Captain Langlois' work already quoted.

CHAP. IV.

Conversion of 3-pr. S.B. bronze guns.

guns being ordered for Bhootan in haste, were made by turning down and rifling 3-pr. bronze S.B., and a small number of similar pieces were subsequently made; but steel now alone is used for these guns, which are manufactured from a solid ingot of that metal.

These 7-prs. have the French rifling (vide p. 46), and fire studded projectiles.

Steel. 7-pr. 150 lbs. 7-pr. 200 lbs.

Two natures are now being manufactured, one weighing 150lbs., for use in India, &c., and a later pattern of 200lbs. weight, introduced in 1873. This is a longer gun than the former pattern, and considerably more powerful. It is the piece principally employed for light boat service.

Guns of Position and Siege Pieces.

Ordnance for Direct or Indirect Fire.

Guns of position and siege pieces.

In the case of siege pieces, the same reasons which led us to discard the R.B.L. service guns for field service led us to do so in the case of these pieces also.

R.B.L. from 1859 to 1870.

The S.B. pieces originally used for siege service were replaced in 1859–63 by 20-pr., 40-pr., and 7-inch R.B.L., but since 1870, we have made 25-pr., 40-pr., and 64-pr. R.M.L.* pieces for direct and indirect fire siege guns,† to take the place of these R.B.L. pieces.‡

R.M.L. introduced 1870.

The introduction of the 64-pr. is described at p. 82.

40-pr. R.M.L. 1871.

The 40-pr. was introduced in 1871, and at that date a few of these pieces were made about 18 calibres in length of bore, but further experiments showed that an increase of length was advisable, and our 40-prs. for siege purposes have now very much the same dimensions as the 40-pr. R.B.L., but as they utilize a larger charge of powder they are more powerful guns.

25-pr. R.M.L. 1874.

The 25-pr. was introduced in 1874 to supersede the 20-pr. R.B.L. As far as power and accuracy of fire are concerned, we have good reason to be satisfied with these three last mentioned siege pieces.

Ordnance for Indirect and High Angle Fire.

Ordnance for high angle firing.

But besides ordnance for direct fire, siege trains must also include pieces to give an effective fire at high angles of over 15°, and for this purpose S.B. mortars remained in use in our own service, as in that of other nations§ long after rifled guns had been universally adopted for direct fire and for indirect or curved fire at angles under 15°; each year, however, their defects became more patent,‖ and as early as 1853, experiments were carried out with S.B. mortars rifled for the purpose on various systems, but without much success; these pieces were only capable of being fired at certain fixed angles; this necessitated variations in the amount of powder used, consequently a varying space in the chamber, and so to unequal action of the charge.

Experimental rifled mortars, 1853.

Fired at fixed angles.

---

* As explained in Chapter XI., we use 64-prs. also as garrison and naval broadside guns.

† Vide p. 257 as to constitution of present siege train.

‡ The 40-pr. R.B.L. guns, however, may perhaps be re-introduced as a siege gun.

§ In the war of 1870–71 the Germans employed many S.B. mortars, and in a German siege train, as at present constituted, of 400 pieces, we find a proportion of 120 S.B. (15c.) 6-inch mortars are included.

‖ At the siege of Fort Pulaski, during the American war of 1860, we are told that at a range of 1,700 yards more mortar shells missed than hit the fort.

CHAP. IV.

Various other proposals were brought forward from time to time,* but nothing substantial accrued from them. In 1864, Sir J. Burgoyne called attention to the great value of high angle fire against bomb-profs, and to the urgent necessity of increasing the range and improving the accuracy of our mortars; the O.S.C., however, then proposed that a piece should be tried of the nature of a heavy rifled howitzer capable of being fired at angles of elevation up to 30°. Finally three 68-prs. carronades were rifled and experimented with in 1866–67 and owing to the success of these experiments a 32-pr. bronze block was bored to 6″·3 calibre, and rifled with three plain grooves, so that the 64-pr. projectiles could be employed with it. The results obtained were very good, and it was submitted that such a piece should be added to our siege train. There were many objections to this, however, on account of the material, weight, and construction of the old S.B. 32-pr. bronze howitzer, it was proposed to bore up.

Proposals as to high angle firing. Sir J. Burgoyne, 1864. Rifled howitzers suggested. Experiments with 68-pr. carronades rifled, 1866–7. Results obtained.

Further experience having been gained as to high angle firing from rifled pieces during the war of 1866, the O.S.C. reported that year that the matter was ready for decision without further experiments; and in 1867 they suggested that for siege purposes an 8-inch rifled howitzer of about 50 cwts., should be made, and that for coast defence we should manufacture pieces capable of throwing shell of 200 lbs, and 300 lbs. and of the weights respectively of 4 and 6 tons. Designs were accordingly submitted in November, 1867, by Colonel Campbell, the Superintendent of the R.G.F. for rifled howitzers of 8, 9, and 10-inch bore, and of the weights above mentioned.†

Proposed 8-inch rifled howitzer, R.M.L. 8, 9. and 10-inch howitzers designed in R.C.F. 1867.

In 1870, a committee was appointed to consider whether it was advisable to convert 13-inch S.S. mortars, and also to report upon fire from rifled howitzers. The converted mortars did not give such good results as the rifled howitzers, so the idea of conversion was abandoned. Further experiments were carried out with rifled howitzers, and in 1872 it was finally decided that 8-inch rifled howitzers should take the place of S.B. mortars in our siege train; a number of these pieces were consequently manufactured in 1873–74; they are constructed with a steel tube and wrought iron exterior, and rifled on the Woolwich system.

Committee of 1870. Idea of converting S.B. mortars into 9 and 8-inch R.M.L. howitzers abandoned.

In 1874, this Committee being anxious to try a lighter nature of rifled howitzer capable of firing a shell of about 60 lbs. proposed that a howitzer should be constructed of 6·3-inch calibre. After considerable trial a piece of this nature was finally introduced into the service in 1878, the rifling being polygrooved. This piece is the first introduced into our service in which the stud system has been abandoned in favour of a rotating gas check. A definite stop is put in the rifling, by which the shot is always brought up at the same spot, in order that the charge may be burnt in a definite air space.

6·3 inch howitzer proposed, 1874; introduced 1878.

We thus have two natures of rifled howitzers in our siege train, viz.: The 8-inch howitzer of 46 cwt., and the 6·3-inch howitzer of 18 cwt. These two pieces should perhaps more correctly be termed rifled mortars, being short in the bore. We are now however manufacturing two howitzers, more properly so called, and of much greater length of bore, of 8-inch and 6·6-inch calibre respectively.

Service R.M.L. howitzers.

---

* Captain (now Sir W. Palliser) proposed rifled mortars in 1854, and in 1862–63, a 7-inch wrought iron mortar, proposed by Sir W. Armstrong, and rifled on the shunt principle was tried at Shoeburyness, but with indifferent results.

† These two latter do not come under the head of siege train ordnance but are intended for coast defences, &c. They are still being experimented with.

CHAP. IV.

Guns converted from S.B. for Secondary Purposes of Defence.

Introduction of converted guns.

Before going on to the history of our heavy guns it may be well to mention how and when guns converted from S.B. were introduced into the service. A detailed account of these pieces is given in Chapter X.

Strengthening existing C.I. ordnance, 1855.

Upon the introduction of rifled guns it was naturally attempted to utilize the existing cast iron ordnance by strengthening them to meet the increased strain due to the use of heavy elongated projectiles.

Major Palliser's proposal, 1863.

As early as 1855 various methods were tried to this end* but none of these succeeded well until in 1863 Major Palliser proposed to line the bored-out S.B. pieces with a wrought iron inner barrel; between this date and 1868 a number of guns so converted were tried, and in the latter year it was recommended by the O.S.C. that extensive conversion of certain S.B. guns on this system should be carried out in order to provide us at a cheap rate with rifled ordnance for secondary purposes of defence. Their recommendations were carried out in the case of certain guns mentioned in table p. 94, and we now possess over 2,000 of these converted rifled guns† which are employed principally for the armament of our smaller men-of-war, and for coast defences where the range is limited, or where none but small vessels of light draught could approach. Table IX. at end of the chapter shows how great an increase in power we have obtained by this mode of conversion.

Palliser system adopted, 1868.

Introduction of Heavy Rifled Ordnance.

Heavy R.M.L. guns.

We have so far followed the history of our rifled field, siege, mountain, and secondary guns of defence down to the introduction of the present R.M.L. armament, let us now see how our heavy‡ rifled guns were brought into the service.

Important as it is that our light artillery should be so well equipped that whether we have to fight in the open field, to attack a great fortress, or to take a mountain stronghold, our field, siege, and mountain guns shall prove all that can be desired, still more important it is that the heavy guns, which constitute the principal armament of our ships of war and coast defences should be the most powerful in the world.

Here we cannot afford to be behindhand, but must, as the greatest naval power, rather lead the way and retain our supremacy at sea, arming both ships and forts with guns more powerful than any which they will have to encounter.

Progress made since 1860.

Fortunately we are not wanting in this respect, but have since 1860 made such progress that whereas at that date our heaviest piece of ordnance for direct fire was the S.B. 68-pr. of 95 cwts., which, throwing round shot of 68 lbs. could not pierce 5 inches of iron at 200 yards,

---

* Amongst other modes tested were the strengthening the gun at the breech end with a wrought-iron jacket shrunk over, the use of a bronze jacket, of wrought-iron rings shrunk on, &c. Various tubes were also tried, made of steel, &c.

† Most of these pieces have been converted in the R.G.F., but some hundreds have been completed by the E.O.C. It will be evident, vide pp. 9, 10, that the inner barrel of these pieces is of too soft a material, and the outer casing too brittle to allow of heavy charges being used with them, but as a measure of economy it has allowed of our rapidly turning old S.B. guns into much more powerful rifled ordnance.

‡ Guns over 5 tons weight are designated *heavy* guns.

CHAP. IV.

we now have 100-ton guns capable of driving gigantic bolts of 2000 lbs. through more than 2 feet of iron at a distance of 1,000 yards.

Progress made since the middle of 18th century.

Let us trace this progress since the middle of the 18th century, from which date, excepting in the introduction of carronades,* hardly any advance was made in the construction of ordnance until 1830, when General Millar, R.A., the head of the Gun Factories at Woolwich, introduced the 8-inch and 10-inch shell guns, for direct fire. Ten years afterwards, *i.e.*, in 1840, Mr. Monk, chief clerk in the Department, proposed the 42-prs. and 56-prs.; and subsequently to this, Colonel Dundas, R.A., who succeeded General Millar, introduced 68-prs.—a great stride, as it was then thought, in the progress of heavy ordnance, and so indeed it was; but the supremacy of the 68-pr. was not of long duratión. It is no longer a heavy or a powerful gun, according to modern artillery ideas. It has been dwarfed by its big rifled brothers of recent date.

Carronades.

8-inch and 10-inch S.B. shell guns 1830.

42, 56, and 68-prs. 1840.

Lancaster.

In 1854 the first attempts were made to use heavy rifled guns by the employment during the siege of Sebastopol, of certain 8" and 68-pr., C.I. guns oval bored on Mr. Lancaster's plan. This attempt was not successful, the projectile occasionally jamming in the bore, and rupturing the chase, while the shooting was but indifferent.

R.B.L. screw, 1859.

In 1859, as already mentioned (p. 71), R.B.L. breech screw guns were introduced, of which the two heaviest natures made were the 40-pr. and the 7-inch or 110-pr. guns. In 1859–60, on service, especially when employed on board our men-of-war, many defects were found in this system, and in 1864, in consequence of the objections brought against his breech-screw guns, Sir William Armstrong introduced not only two natures of wedge-guns (40-prs. and 64-prs.)† as an improvement on the breech-screw arrangement in points of safety and simplicity, but also 64-pr. *muzzle-loading* guns with *shunt* rifling, and further proposed other shunt guns of larger calibre.

Wedge and shunt guns introduced. 1864.

64-pr. R.M.L. guns introduced, 1864.

In 1864, also, while further experiments were being carried on with wedge B.L. guns, the Admiralty requested that four 64-pr. (then called 74-prs.) should be completed as R.M.L. guns with a bore suited for S.B. 32-pr. projectile in case of emergency. Subsequently it was decided to complete all the remaining blocks prepared for wedge B.L. guns as 64-pr. R.M.L. on the shunt system, hence but few of this nature (the 64-pr. wedge R.B.L.) were manufactured and we have an exterior similarity between it and the 64-pr. R.M.L., Mark I. (vide Plate V.).

Necessity for heavy rifled M.L. guns.

Experiments, guns *v.* armour plates, 1860–64.

But although these rifled guns were more powerful than the S.B. 68-prs., they soon became insufficient for the modern requirements of naval warfare, for the power and precision of rifled guns and the growing use of shells which would burst on striking a ship's side and make a hole beyond repair, or having penetrated, burst between decks, dealing death and destruction around, and probably setting fire to the vessel, necessitated the use of iron-clad‡ ships. To penetrate these necessitated in turn the use of still more powerful guns, and then com-

* As to S.B. ordnance prior to that date, and as to carronades being introduced, vide p. 57, Chapter III.

† A small number of these guns were introduced into the service, but they are for the present in reserve; it is, however, in contemplation to utilize these pieces by improving their B.L. apparatus. Vide p. 159 for description.

‡ The power of direct shell fire was perceived only about 50 years ago; the French then adopted the Paixhans' shell gun, and General Millar, R.A., introduced our 10-inch and 8-inch shell guns; the new species of fire, however, was not much used until, through the impetus due to war, its development advanced *pari passu* with that of rifled ordnance. Nearly all the naval engagements which have taken place of late years furnish instances of the destructive effect of shells on wooden vessels:—

At Sinope during the Crimean war, the Turkish fleet was actually destroyed by

CHAP. IV.

menced those contests of guns versus armour plates, which are not yet at an end.*

As to the construction of these guns, the Armstrong coil system with its subsequent modification was readily accepted as the best, but

B.L. v. M.L.

the rival claims of muzzle loading and breech loading led to conflicts as severely contested as those of guns versus iron plates. The judg-

Opinion of the Armstrong and Whitworth committee.

ment, however, of the Armstrong and Whitworth Committee in 1865 in favour of M.L. guns was only in accordance—so far at least as heavy guns were concerned—with the preconceived opinion of our leading artillerists, for any B.L. arrangement with guns using the enormous charges required would of necessity be both cumbrous and readily liable to damage.

Manufacture of heavy B.L. stopped, 1862.

Our experience also in the wars against China and Japan in 1863 showed us the disadvantages of our heaviest 7-inch B.L. guns as to want of simplicity, &c., and in consequence no heavy guns of this nature were manufactured subsequent to 1860.

Trial of R.M.L. against iron plates, 1862-66.

Between 1862 and 1866 extensive trials were carried out against iron plates,† where both the 64-pr. R.M.L. already mentioned and other R.M.L. guns acquitted themselves well.

Meanwhile the manufacture of R.B.L. guns was entirely stopped, and Colonel Campbell, R.A. (now General Campbell, C.B., D. of A.), who succeeded Sir W. Armstrong as Superintendent of R.G.F. in 1863, had already begun to manufacture experimental R.M.L. guns built up on

Experimental shunt R.M.L. guns, 1863.

the Armstrong construction and rifled with the shunt rifling (the 64-pr. R.M.L. shunt was finally approved as a S.S. gun in March 1865) when the Admiralty proposed the construction of built up S.B. guns of large

100 and 150-pr. S.B. guns, 1864.

calibre to penetrate iron armoured ships at close quarters. Accordingly in 1864 two natures were constructed on the Armstrong principle, viz.: 100 pr. of 9″ calibre and 150-pr. of 10″·5 calibre. It soon appeared evident, however, that the so-called "racking effect" of the spherical

Good results obtained from 64-pr. M.L. shunt guns, 1864.

projectiles fired from such pieces would be powerless against the iron plating which each year became stronger. In fact such good results were obtained from the 64-pr. M.L. shunt gun, as well as from larger experimental guns on the same system of rifling and construction, that the Ordnance Select Committee suggested (Report No. 3558, 25/11/64) that the above two natures of S.B. guns should be also rifled on the shunt system.

They were found to be of too weak a construction for that purpose,‡

Paixhans' explosive shells fired from the Russian vessels "Constantine" and "Paris."

In the American war of 1860–61, the ironclad "Merrimac" blew up by shell the wooden vessel "Congress," having first made dreadful havoc of the crew; and the "Kearsage" sunk the "Alabama."

Lastly, at Lissa, the Italian vessel "Palestro" (partially plated) was destroyed by the Austrian shells.

* Shoeburyness, where these trials were carried out, is situated on the upper bank of the mouth of the Thames. It was purchased by Government for an artillery practice ground on account of the long range afforded by its far stretching sands. In 1849 the first gunnery practice took place there, and "Shoeburyness" then comprised only 8 acres of upland occupied by two officers and a gun detachment; but field by field was purchased as the science of artillery progressed, and there are now 216 acres of upland and 6,512 acres of sands, Government property. A straight row of pegs extends for 10,000 yards along the sands, and at the ebb of certain tides a range can be measured and the shot recovered up to 12,000 yards.

† Vide the valuable report on "The Penetration of Iron Armour plates by Steel "Shot, &c.," 1866, by Major W. H. Noble, Royal Artillery.

‡ They were too weak longitudinally for rifled guns, as pointed out by the Superintendent R.G.F., subsequently, indeed, to his giving this opinion, two of the 150-prs. blew out their breech during practice. The 150-prs. are obsolete, and have been returned into store. The 100-prs. are used for S.S. drill purposes only.

CHAP. IV.

and in 1864 it was proposed to make heavier pieces; a few 12-inch* and 13-inch guns were therefore ordered in 1864, and the latter (of which only four were made) were rifled with the shunt rifling.†

R.M.L. 12-inch and 13-inch guns ordered in 1864.

The shunt system of rifling had, however, some disadvantages, as pointed out at p. 45, and was soon succeeded (in 1865) by the so-called Woolwich system‡ which was adopted after a long series of competitive trials carried on with 7-inch built up guns rifled on various representative systems.§

Woolwich guns introduced, 1865.

Introduction of Woolwich Guns, 1865.

There was not much difference in the shooting qualities of the several systems, but in May 1865, the Ordnance Select Committee reported in favour of adopting the Woolwich system of rifling in which the groove as shown below is a modification of the groove then used with the French heavy guns.||

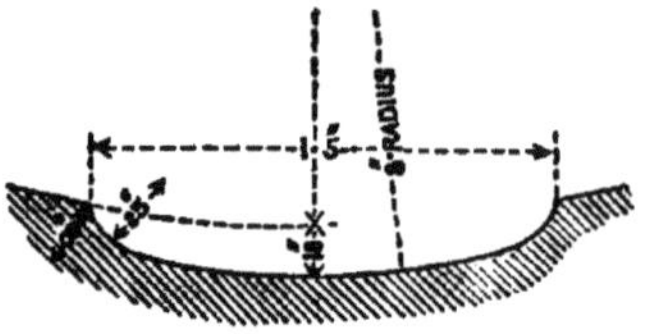

Shunt rifling abandoned.

The shunt system was therefore abandoned, and the M.L. 7-inch guns in course of manufacture were rifled on this principle, upon which all our heavy pieces since have been rifled.

Woolwich guns introduced, 1865. Definition of.

The 7-inch referred to, and introduced into the service in 1865, were the first of the so-called *Woolwich* guns, which then meant *wrought iron M.L. guns, built up on Sir W. Armstrong's principle, improved upon by hooking the coils over one another, and having solid ended steel barrels, rifled on the system shown above, for studded projectiles.*

8-inch and 9-inch gun, 1865.

The 8" and 9" guns (Woolwich) were also introduced in 1865, and a few 12" guns of 25 tons were completed by 1866, being built like the previous guns on what was called the "original construction" or that of Sir W. Armstrong (vide p. 28), where a forged breech-piece and a number of thin coils are used.

Original construction, 1863 to 1867.

This original construction, though it gave great strength and efficiency, was an expensive mode of construction,¶ and so far back as October 1864, the attention of the Ordnance Select Committee had been drawn to the desirability of obtaining some cheaper method. In 1863, General Campbell, C.B., then Superintendent R.G.F. applied himself

---

* These 25-ton guns were not tried until 1866, and were rifled on the Woolwich system, vide p. 283.

† The only guns rifled on this principle now in our service are certain 64-pr. R.M.L. of Marks I., II., and III., which were made prior to 1871, and which have not been re-tubed since then.

The obsolete 13"·05 gun, vide p. 286, was first rifled with shunt rifling, but as explained, were bored up and rifled with the Woolwich groove.

‡ For description, vide Chapter II., p. 45.

§ For further account of these guns, vide note at end of Chapter II., p.

|| The O.S.C. considered this system the best, both on account of the simple form of groove, the simplicity of studding the projectile, and the advantage it offered in allowing of the use of the increasing twist.

¶ The cost of the guns being in round numbers about 100*l.* a ton; that of our present construction is only about 65*l.*

CHAP. IV.

energetically to the question of decreasing the cost without decreasing the strength of our heavy guns. A coarser and cheaper iron was by his exertions obtained and used with much success in the manufacture of the bar iron for the coils.

Mr. Fraser's modification proposed in 1867.

Next came the question whether, in addition to using a cheaper material, the absolute construction of the gun could be improved, simplified, and also made less expensive. In April 1867, Mr. Fraser, the Deputy Assistant Superintendent, Royal Gun Factories, proposed a plan which was recommended by General Campbell to supersede the original or Armstrong construction.

The theoretical advantages and otherwise of the original, the partially modified, and the modified construction have already been discussed in Chapter II., p. 29, so we need only mention here that experimental practice proved that the two latter were as strong, if not stronger than the first mentioned, as was expected.

Retention of the forged breech-piece till 1866. Abolition of the same in 1867.

In the earlier Fraser guns the modified construction was only partially adopted, and the 64-pr., 7", 8", and 9", Mark II., few of which were made in 1866, retained the forged breech-piece. Since 1867, however, these natures, and indeed all our heavy guns, have been made entirely on the Fraser modification of the Armstrong principle and without any forged breech-piece.

Up to April 1867, therefore, all our heavy R.M.L. guns were made on the original construction, like the 9-inch gun, Mark I. (*see* Plate XIII.), and from that date up to August 1869, nearly all have been made like the 9-inch gun No. 368, or Mark III.—*i.e.*, consisting only of four parts, viz. steel tube, cascable, *B* tube, and breech coil.

80-ton gun.

The alteration which then took place in the manufacture for 9-inch guns simply consists in having a thinner steel tube and two coils on the breech (*see* Mark IV. and V., Plate XIII.), instead of one triple one. The higher natures are made in the same way, but have a "belt" in addition * and in the 80-ton gun we have another coil in the front layer of iron termed a 2 B coil, otherwise its construction is the same as that of the other natures except in some minor details (vide p. 170).

Heavy R.M.L. Guns since 1867.

10-inch guns, 1858.

So much as to the origin and present employment of the Fraser construction, with which we are quite satisfied, and which has given such excellent results in the 80-ton gun. To return once more to the history of our heavier guns: we see that in 1867 64-pr. 7-inch,† 8-inch, and 9-inch guns had already been introduced. Owing to the success of the 9-inch gun it was decided in 1868 to introduce a gun of 10-inch calibre into the service. A few of these powerful pieces (25) were made of the same pattern as the 9-inch gun, Mark III.; but the remainder, manufactured subsequently to August 1869, are of the construction of 9-inch, Mark IV. type (vide Plate XIII.).

Penetration of 10-inch guns.

We have a large number of 10-inch guns, both land and sea service; they are capable of piercing about 12 inches‡ of iron at 1,000 yards. Some of these pieces were constructed on the pattern of Mark II., 9-inch gun; but as above explained, all made since 1869 are on slightly different construction, both for convenience and economy of manufacture, as well as for sound theoretical reasons.

---

* See extracts from proceedings of the Ordnance Select Committee for 1864–9.

† As mentioned at p. 169, some of the 7" of 6½ tons have been reduced to still lighter guns of 90 cwts. for naval service against wooden ships, &c.

‡ According to Captain Noble's formula, vide Table, p. 356.

CHAP. IV.

As already mentioned, a few (4) 12-inch guns of 25 tons weight were proposed as early as 1864; these were made on the original construction and completed in 1866. Between 1866 and 1871 some other guns of this nature were manufactured on the same type as the 10-inch, Mark II. These guns are no longer being made, and if retubed are converted into 11-inch guns. The twist of rifling of these pieces is hardly sufficient, from 1 in 100 to 1 in 50; and they are not such accurate shooting pieces as most of our Woolwich guns.

12-inch guns of 25 tons. 1866-71.

An 11-inch gun was recommended in 1867 for trial in comparison with the 12-inch, in order "to determine what calibre and proportional length of bore were best adapted to a gun of from 23 to 25 tons weight." The question was decided in 1870 in favour of the smaller calibre, and since that date a number of these pieces have been made; their penetration at 1,000 yards would be about 13 inches of iron.

11-inch guns, 1870.

O.S.C. Proceedings, 1865, pp. 15, 94.

Penetration of 11-inch guns.

In 1870 also, as mentioned at p. 79, further experiments were carried out with regard to heavy R.M.L. howitzers meant to take the place of large S.B. mortars for coast defences and fortresses; these pieces 9-inch and 10-inch howitzers, are still under trial.

Experimental howitzers, 1870–76.

Committee on Explosives.

About this date much light began to be thrown upon the question of gun construction through the labours of the Committee on Explosives appointed in 1869,* to whose valuable scientific researches, carried on with the improved instruments described in Chapter XIII., we owe a very great increase of knowledge as to the pressures to which a gun is subjected in the powder chamber and other parts of the bore on the explosion of the charge. It was seen that by modifying the nature of powder used we could so diminish the pressure with very heavy charges that guns might be made of a size much larger than any we possessed, from which heavy projectiles could be fired with sufficiently large charges yet without straining the piece inordinately.

Committee on Explosives appointed in 1869.

Important results of their investigations.

In 1871, therefore, we advanced a step further by completing a 35-ton gun of 11"·6 calibre, but experiments having proved that this calibre is not suitable for the efficient combustion of 120 lbs. of "P" (pebble) powder, the proposed battering charge of the gun, it was decided to bore it out experimentally to 12", and to complete the first nine guns required at once by the navy to that calibre. This nature of gun, originally made for turret ships, proved rather short for its other dimensions; a pattern for a 38-ton gun about 3 feet longer in the bore was provisionally approved of in 1873.

35-ton guns, 1870.

38-ton guns 1873.

Experiments made by the Committee on Explosives with such a gun, bored out first to 12 inches and then to 12·5 inches, showed that the latter calibre was the most suitable of the two. It was consequently approved in November 1874, that the calibre of this piece should be 12·5 inches, the charge and projectile being provisionally 110 lbs. P². powder and an 800 lb. shell.

Further experiments by the Committee enabled them to report that with an 800 lb. shot and 130 lbs. of P² powder no excessive pressure was occasioned at any point in the bore when these charges were em-

* Colonel Younghusband, F.R.S. (President), Captain A. Noble, F.R.S., C.E., Captain W. H. Noble, R.A., F. A. Abel, Esq., F.R.S., Captain O. Malony, R.A., Captain Morgan Singer, R.N. (members), as originally appointed.

The constitution of this committee has been altered from time to time; at present it consists of: President, Major-General Younghusband, R.A., F.R.S.; Members, Captain Bridge, R.N., Colonel Fraser, R.A., Colonel Hay R.A., Captain A. Noble, F. Abel, Esq.; Captain C. Jones, R.A., Secretary.

CHAP. IV. ployed with the experimental gun. This gun was finally approved of in January 1875, and since that date a number of 38-ton guns have been manufactured both for the armament of our coast defences and men-of-war. Since 1875, further experiments have shown that the charge of this piece can be still further increased, and a corresponding increase in power obtained.

Heavy guns of Foreign Powers, 1876.

The absolute necessity, however, of our manufacturing larger ordnance in order not to be left behind by other nations was evident from the fact that experimental guns were at this time in process of manufacture by or for foreign navies, much exceeding in effective power our 38-ton gun.*

In addition to this, although the power of our guns had so far kept well up with the defensive power of iron shields, yet as the latter continued to increase in thickness it was evident we should have to acknowledge that guns were beaten unless we further developed their power.

Foreign iron-clads, 1873–76.

Gun required to pierce 20-inch of iron at 1,000 yards.

In 1873, indeed the Russians were hard at work upon a man-of-war, "Peter the Great," which was to carry 20 inches of armour; the Italian ships "Duilio" and "Dandalo" were to have 20 inches of iron plates; and other nations were moving in the same direction. We had no guns capable of piercing such a mass of metal, and in the same year the Director of Naval Ordnance applied to the War Office authorities for a gun capable of sending its projectile through 20 inches of iron at 1,000 yards; the Admiralty at the same time proposed a design for a turret ship capable of carrying four such guns.†

The data possessed as to the piercing power of guns is very incomplete (vide Chapter XIII., p. 332), so that the problem was not an easy one. At a meeting of the Heads of Departments, however, it was decided that an experimental gun should be manufactured to weigh about 75 tons, and to be 24 feet long in the bore, with a calibre of 14 inches, this calibre to be subsequently bored up to 15 and 16 inches successively in order to determine the proper calibre, rifling, charge, &c., &c.

80-ton gun ordered, 1874.

In the early part of 1874, the Admiralty acquiescing in the suitability of such a gun, the Secretary of State for War ordered its manufacture, the cost of the piece being estimated at about 8,000*l.* and its weight 80 tons.

The question of length offered no difficulties, thanks to the improved methods of loading by endless chain‡ or hydraulic rammer,§ and it was decided that the bore should be 18 calibres in length.

14·5-inch calibre.

Fire proof.

This experimental gun was manufactured in the usual manner (with the slight difference due to its size), having a steel tube and a few heavy coils. It was completed in September 1875, being bored up to 14·5 inches calibre rifled with 11 grooves, for studded projectiles, twist increasing from 0 to 1 in 35. It was proved with charges increasing from 170 to 240 lbs. and 1,260 lbs. projectile, giving with the last mentioned a M.V. of 1,550 f.s.

15-inch calibre and chambered.

Velocity and

Further experimental firing took place in December 1875, and the gun being bored up successively to 15 inches, chambered to 16 inches, and again bored up to 16 inches, was tested in 1876, with gradually increasing weights of shot and powder. With a projectile of 1,700 lbs. it gave as much as 542 f.-tons per inch of shots circumference at

---

* Vide Table p. 95.

† This vessel, the "Inflexible," is now in a very forward state.

‡ The method of loading guns by means of an endless chain, has been worked out in the R.C.D., and promises the most favourable results.

§ See an interesting account of this arrangement in the Proceedings of Institute C.E., by Mr. G. Rendell, of the Elswick Ordnance Co., to whom we are indebted principally for this admirable adaptation of hydraulics.

the muzzle, or a piercing power at 1,000 yards of over 21 inches of iron.

penetrative power.

16-inch calibre.

In the same year this gun was sent to Shoeburyness for trial as to range and accuracy and proved in that respect very satisfactory, the practice with shot of 1,700 lbs. and a charge of 370 lbs. P^2 powder, being very good. Thus in a series of rounds, at a range of about 4,700 yards, or nearly three miles, the mean error of direction was but 1·8 yards, and that of range under 15 yards.

In 1877, the tube, weakened as it was by the holes bored for a number of crusher plugs, gave way under the large number of heavy charges fired, and a slight crack appeared running some little way into the powder chamber. However as the defect did not in any way endanger the safety of the piece, the latter was further tested against iron plates, and then chambered to 18 inches. Several rounds have subsequently been fired against plates with still larger charges without developing the defect in the tube to any extent.

More guns of this nature are now in course of manufacture, they will have a calibre of 16 inches, an enlarged chamber of 18 inches in diameter, and a polygroove rifling.

While in the R.G.F. an 80-ton gun was being experimented with, the enterprising Elswick firm were manufacturing a still heavier R.M.L. piece of 100 tons for the Italian government. The first of these pieces was tested at Spezzia, in 1876.* So excellent were the results given by this gun that in 1878, when the political horizon was somewhat gloomy, our government considered it advisable to purchase 4 of these guns from Sir W. Armstrong and Co. These four guns will be completed in 1878, and will form a powerful addition to our artillery. They are built up on the Armstrong construction (as described at p. 28), but without a forged breech piece, but the inner tube of these guns consists of two steel tubes united together by means of a collar shrunk on. The system of rifling is polygrooved (28 grooves), the projectile being rotated by a copper expanding cup. The spiral of rifling increases from the breech to within a short distance from the muzzle, from which point it continues to muzzle with an uniform twist.

The competition between guns and plates still however continues, and the Italians are now constructing two ships which will be protected by steel armour, 27 inches thick,† put on in great blocks weighing from 60 to 70 tons. It is evident therefore, that it may be necessary before long to make much heavier pieces, than even those of 100 tons, and as soon as the necessity arises the manufacture could readily be carried out of guns, which would pierce armour plates 3 feet thick at the distance of a mile or more.

Résumé as to Service Guns.

Heavy guns.

To sum up in brief the history of our guns to the present date:

Our armament smooth-bore prior to 1859.

Lancaster gun, 1854.

Prior to 1859, our armament consisted exclusively of S.B. pieces, with the exception of a few rifled guns tried subsequently to 1854, such as those of Lancaster, which were not adopted, although a few S.B. guns rifled on Mr. Lancaster's principle were employed in the Crimean war without much success. Sir W. Armstrong's R.B.L. wrought iron (or steel and wrought iron) built up guns using lead coated projectiles, were adopted in 1859, and large numbers, roundly bout 3,500, of these pieces of calibres, varying from 2·5 inches to

* The Superintendent R.G.F., General Younghusband, C.B., attended these rials in an official capacity, and full details of this interesting series of experiments will be found in his report.

† Supposed to equal some 36 inches of wrought iron

CHAP. IV.

R.B.L. guns, 1859 to 1863.

7 inches, were manufactured between 1859 and 1863. We obtained by this means an armament surpassing at the time in power that of other nations, but found that although the general principles of construction were good, yet that the B.L. system itself was faulty.

R.M.L. guns introduced 1863.

In 1863, we therefore began tentatively to manufacture R.M.L. guns of similar construction, and after an extended series of experiments, in which every promising system of construction and rifling had a fair trial, we finally adopted in 1865 as the best then existing, the so-called Woolwich guns, where the guns were built up on the Armstrong principle, but were M.L. and rifled on a modification of the French system, using studded projectiles.

Original construction, 1865.

Fraser construction, 1867.

In 1867, the construction of these guns was improved and cheapened through the modifications proposed by Mr. Fraser, C.E. and carried through by General Campbell, C.B., R.A., (then Superintendent R.G.F.).

35-ton gun, 1871.

80-ton gun, 1876.

Since that date great numbers of heavy Woolwich guns have been manufactured, and as the means of defence became developed so the size of our guns has increased, until we arrived in 1871 at the 35-ton Woolwich Infant, which subsequently grew into the 38-ton 800-pr., while in 1876 we completed the 80-ton gun, throwing projectiles of 1,700 lbs.

In 1878, we added to our armament four 100-ton Armstrong guns throwing shot of 2,000 lbs., and guns have already been designed, and could readily be made in R.G.F. which would surpass the latter in power to as great an extent as they themselves surpass the 38-ton service gun.

In the 100-ton gun we see a polygroove system returned to with an expanding rotating cup. The same system will take the place of the stud system, also in our 80-ton guns. We have not only made these strides in the size and weight of our guns and projectiles, but of late great increase of power and accuracy of shooting has been given to the various natures of our heavy ordnance, by increasing the charges used, and by insuring that these charges are burnt in a definite and constant space or chamber in rear of the projectile.

Siege, field, and mountain guns.

With regard to medium, siege, field, and mountain guns, these also were S.B. of bronze or cast iron, as was the case with heavy guns until 1859, when the S.B. guns were superseded in a great measure by R.B.L. Armstrong pieces, with which our active army and siege train were entirely equipped. We ceased to make these R.B.L. guns in 1863, but it was not until exhaustive inquiries had been completed that in 1870 we carried out the manufacture on a large scale of our present R.M.L. siege, field, and mountain guns, which, like the heavy ordnance, are built on the Fraser modification of the coil construction, with steel tubes and wrought iron exterior coils, and which fire studded projectiles.

Converted guns.

Since 1868 we have also converted, as an economical measure, a very large number of S.B. cast iron guns into more powerful R.M.L. pieces on the Palliser principle.

For the last three years extensive experiments have been carried out with short howitzers or rifled mortars, as siege train pieces for indirect fire, and we have now a light and a heavy nature which afford an accurate and powerful fire. Two natures of long howitzers of a much more powerful description are now being manufactured, and our siege train will soon embrace a very formidable armament. Here, as with the heavier guns, the tendency is to adopt the polygroove system of rifling. Great improvements have also been made of late in increasing the power of our smaller guns by the use of gas checks, forward venting, the adoption of a definite powder chamber and increased charges. Our field artillery will soon be strengthened by the addition of a 13-pr., which will equal if it does not excel, the most modern improvement in field guns, that of the French.

CHAP. IV.

With the extensive and widely scattered possessions of this great empire of Great Britain, which has ships of war on every sea, and fortresses in every climate under the sun, it is evident that the replacing in a few years the whole of the armament of our immense fleet, and of our coast and land defences, would be too gigantic an operation to attempt, and we consequently find that we have four large classes of ordnance in the service: S.B. of cast iron and bronze of which thousands still remain in use though they are rapidly being replaced, besides a great number R.M.L. converted on the Palliser principle from S.B. ordnance; some 3,500 R.B.L. guns principally in reserve, and several thousands R.M.L. pieces of different descriptions. Our heavy R.M.L. guns, both in weight and number, surpass very much those possessed by any other nation, and constitute an armament of which we may well be proud; their manufacture is simple, their power and accuracy remarkable, and their cost comparatively small.*

Classes of Ordnance in the service.

We have in England carried out at a considerable expense more extensive experiments † than those ever attempted elsewhere, and are well repaid for an expenditure the results of which are so satisfactory.

So far as one can pry into the future, the history of our guns will be a successful history, being the outcome of patient perseverance, of scientific investigation, and of a steady disregard on the part of our artillery authorities of short-lived clamour for novelty, prompted by interest, panic, or an exaggerated estimate of the value of some foreign artillery.

Experiments necessary and well paid for by results.

Not to advance, however, is to go back in artillery as in other sciences, and on this account there is no doubt that we must ever continue to try and try again, without grudging too much a necessary expenditure upon those experiments which will be requisite as time goes on, if we wish the history of our guns in the future to show as steady and satisfactory a progress as that achieved during the last quarter of a century.

* *e.g.*, Our 80-ton gun will cost about 10,000*l.*; the 100-ton Armstrong guns cost about 16,000*l.*; while it is understood that the expense of the 57-ton gun of Krupp is 15,000*l.*

† Experiments, it may be observed, carried out in the face of day, as to the accuracy and reliability of which there can be no doubt.

Table referred to in p. 70.

TABLE V.

TABLE of COMPARISON between S.B. and R.B.L. SERVICE FIELD GUNS of 1860.

Nature of Gun.	Weight.	Calibre.	Length.					Charge.	M. V. f.s.	Value of log $\left(\frac{d^2}{W}\right)$	Weight of Common Shell filled.	Bursting Charge, Common Shell.	Common Shell.						Shrapnel.		REMARKS.
			Total.		Bore.		Chamber.						1,000 Yards.		1,500 Yards.		Energy per cwt. of Gun.		Weight of filled.	Number of Bullets.	
					Inches.	Calibre.							Velocity.	Energy.	Velocity.	Energy.	1,000 Yards.	1,500 Yards.			
	cwts.		ft.	in.			in.	lbs.			lbs.	lbs.	f.s.	f.t.	f.s.	f.t.	f.t.	f.t.	lbs.		
HORSE ARTILLERY.																					
6-pr. S.B. Bronze	6	3·67	5	0	57·5	15·6	—	1·5	1,484	0·2336	6·2*	—	712	22	572	14	3·6	2·6	5·0	31	It will be seen how very much superior the service R.B.L. guns were to the old S.B. in power, especially at long ranges.
9-pr. R.B.L. Iron	6	3·0	5	2	52·5	17·5	7	1·125	1,085	0·0217	8·56	0·4	846	42	780	36	7	6	8·25	42	
FIELD BATTERIES.																					
9-pr. S.B. Bronze	13	4·2	6	0	67·5	16	—	2·5	1,614	0·2511	9·3	—	726	34	568	21	2·8	1·8	7·87	54	
12-pr. R.B.L. Iron	8	3·0	6	0	61·5	20·5	8·5	1·5	1,300	1·9397	10·34	0·5	971	67	900	58	8·3	7·2	10·11	56	

* A solid shot of about the same weight was actually employed in obtaining the velocity.

Table referred to in p. 70.

TABLE VI.

TABLE showing approximately the COMPARATIVE VALUE of ENGLISH and some FOREIGN FIELD GUNS in 1860–61.

Nature of Gun.	Calibre.	Weight.	Total Length.	Length of Bore.		Powder Chamber.		Twist in Calibres.	Means of Rotation.	Powder Charge.	Shrapnel Shell.		Common Shell.		With Common Shell.						Remarks.
						Length.	Calibre.				Weight.	No. of Bullets.	Weight.	Bursting Charge.	Velocity at			Energy at			
															Muzzle.	2,000 Yards.	4,000 Yards.	Muzzle.	2,000 Yards.	4,000 Yards.	
ENGLISH.	Ins.	Cwts.	Ins.	Ins.	Clbrs.	Ins.	Ins.			Lbs.	Lbs.		Lbs.	Lbs.	f. s.	f. s.	f. s.	f. t.	f. t.	f. t.	Usually termed.
Iron, or iron and steel ... B.B.L. 12-pr. ...	3·0	8·5	72	61·3	20	8·5	3·2	38	Lead Coating.	1·50	10·7	56	11·75	0·50	1239	850	680	128	59	38	12-pr.
9-pr. ...	3·0	6·0	62	52·5	17	7·0	3·2	38		1·12	8·7	42	8·56	0·40	1035	641	558	64	24	19	9-pr.
AUSTRIAN.																					
Bronze, R.M.L. 15-pr. ...	3·4	9·8	66	57·8	14·5	—	—	22	Lead Coating	2.03	16·4	140	14·5	0·97	1125	809	647	127	66	42	8-pr.
8-pr. ...	3·2	5·2	55	47·7	15	—	—	22		1·16	9·0	80	8·0	0·45	1093	590	400	56	19	9	4 p..
FRENCH.																					
Bronze, R.M.L. 16-pr. ...	4·2	11·5	79	71·4	17	—	—	26	Studs ...	1·80	19·6	110	16·3	0·80	1007	705	540	114	56	33	8-pr.
9-pr. ...	3·4	6·5	63	55·1	16	—	—	26		1·21	10·0	80	9·0	0·42	1066	687	500	71	29	16	4-pr.
PRUSSIAN.																					
Steel, B.B.L. ... 15-pr. ...	3·6	8·6	80	71·2	20	?	?	51	Lead Coating	1·32	14·8	170	15·2	0·55	1065	776	618	119	63	40	9 centr.
10-pr. ...	3·1	5·8	86	70·6	22·5	?	?	51		1·10	9·6	80	9·6	0·37	1150	767	592	88	39	23	8 centr.
RUSSIAN.																					
Bronze, R.M.L. 28-pr. ...	4·8	15·9	—	74·4	15	—	—	50	Studs ...	3·16	28·5	105	28·3	0·88	1006	761	614	198	114	74	9-pr.
10-pr. ...	3·4	5·9	60·7	51·4	15	—	—	40		1·35	14·4	55	10·0	0·44	1050	677	530		82	19	4-pr.

CHAP. IV.

Table referred to in p. 70.

TABLE VII.

TABLE showing approximately the COMPARATIVE VALUE of ENGLISH and FOREIGN FIELD GUNS in 1872.

Nature of Gun.	Calibre.	Weight.	Total Length.	Length of Bore.		Powder Chamber.		Twist in Calibres.	Means of Rotation.	Powder Charge.	Shrapnel Shell.		Common Shell.		With Common Shell.						Remarks.
						Length.	Calibre.				Weight.	No. of Bullets.	Weight.	Bursting Charge.	Velocity at			Energy at			
															Muzzle.	2,000 Yards.	4,000 Yards.	Muzzle.	2,000 Yards.	4,000 Yards.	
ENGLISH.	Ins.	Cwts.	Ins.	Ins.	Clbrs.	Ins.	Ins.			Lbs.	Lbs.		Lbs.	Lbs.	f. s.	f. s.	f. s.	ft. t.	f. t.	f. t.	Usually termed
Wrought Iron, and Steel ... R.M.L. 16-pr. ...	3·60	12	75	68·4	19·0	—	—	30	Bronze Studs	3·00	17·2	128	15·8	1·00	1355	871	684	200	86	33	16-pr. 1871.
Wrought Iron, and Steel ... R.M.L. 9-pr. ...	3·00	8	68	63·5	21·2	—	—	30		1·75	7·9	63	9·0	0·50	1381	822	623	119	42	24	9-pr. 1871.
AUSTRIAN.																					
Bronze, R.M.L. 15-pr. ...	3·40	9·8	66	57·8	14·5	—	—	22	Lead Coating	2·03	16·4	140	14·6	0·97	1125	809	647	127	66	42	8-pr.
Bronze, R.M.L. 8-pr. ...	3·22	5·2	55	47·7	15·0	—	—	22		1·16	9·0	80	8·0	0·45	1093	590	400	66	19	9	4-pr.
FRENCH.																					
Bronze, B.B.L. 15-pr. ...	3·50	12·6	79·2	73·8	21·09		3·64	31	Lead Coating	2·50			15·2	0·88	1280	850	672	173	76	48	Canon de 7. 1870
Bronze, B.B.L. 11-pr. ...	2·95	9·1	—	78·7	26·67			30		2·13			10·8	—	1463	857	655	160	49	29	Canon de 5. 1870
GERMAN.																					
Steel, R.B.L. ... 15-pr. ...	3·6	8·6	80	71·2	20·00			51	Lead Coating	1·32	14·8	170	15·2	0·55	1065	776	618	119	63	40	9 C.M. 1868.
Steel, R.B.L. ... 10-pr. ...	3·0	5·8	76	70·6	22·50			51		1·10	9·6	80	9·6	0·37	1150	767	592	88	39	23	8 " 1868.
RUSSIAN.																					
Bronze or Steel ... R.B.L. 24-pr.	3·96	12·2	73	57·0	14·5			50	Lead Coating	3·25	28·0	210	24·3	0·87	1000	797	665	170	107	74	9-pr.
Bronze or Steel ... R.B.L. 13-pr.	3·42	6·77	70	60·7	20·0			41		1·50	14·5	105	12·6	0·44	1003	742	581	89	48	1	4-pr.

Table referred to in p. 70.

TABLE VIII.

TABLE showing approximately the COMPARATIVE VALUE of ENGLISH and some FOREIGN FIELD GUNS* in 1878 (including New and Experimental Pieces).

Nature of Gun.	Calibre.	Weight.	Total Length.	Length of Bore.		Powder Chamber. Capacity.	Powder Chamber. Ins. to lb.	Twist in Calibres.	Means of Rotation.	Powder Charge.	Shrapnel Shell. Weight.	Shrapnel Shell. No. of Bullets.	Common Shell. Weight.	Common Shell. Bursting Charge.	Double wall Shell. Weight.	Double wall Shell. Bursting Charge.	With Common Shell. Velocity at Muzzle.	Velocity at 2,000 Yards.	Velocity at 4,000 Yards.	Energy at Muzzle.	Energy at 2,000 Yards.	Energy at 4,000 Yards.	REMARKS.
ENGLISH.	Ins.	Cwts.	Ins.	Ins.	Clbrs.	Ins.	Cub. Ins.			Lbs.	Lbs.		Lbs.	Lbs.	Lbs.	Lbs.	f.s.	f.s.	f.s.	f.t.	f.t.	f.t.	Usually termed
W.I. and Steel } R.M.L. { 16-pr. ...	3·60	12·0	75·0	56·0	17·0	—	—	30	} Studs. ...	3·00	17·2	128	16·3	1·00	—	—	1366	871	684	200	86	33	16-pr. } 1871.
9-pr. ...	3·00	6·0	71·0	51·0	21·0	—	—	30		1·75	7·9	63	9·0	0·50	—	—	1390	824	625	119	42	25	9-pr.
13-pr. ...	3·0	8·0	87·9	84·0	28·0	110·38	35·5	{ 0 to 1 in 30	Cup gas check }	3·125	—	—	—	—	—	—	1580	—	—	—	—	—	13-pr. 1878.
AUSTRIAN.																							
Uchatius Bronze } R.B.L. { 16-pr.	3·43	9·6	79·2	57·4	17·0	—	—	45	} 4 copper rings.	3·30	15·4	163	16·1	0·40	16·1	0·40	1480	926	733	244	96	60	(8·7c.) or 9c. } 1876
10-pr.	2·95	5·9	76·8	56·2	19·0	—	—	45		2·09	10·3	105	9·5	0·22	9·5	0·22	1395	781	580	128	40	22	(7·5c.) or 8c.
FRENCH.																							
Bronze or Steel { Reffye, R.B.L. { 15-pr. ...	3·50	12·6	79·2	61·6	21·1	—	—	21	} Lead Coating	2·50	17·3	158	15·4	0·54	15·4	0·54	1280	850	672	173	76	48	Canon de 7. } 1872
11-pr. ...	2·95	9·0	78·0	62·9	26 7	—	—	21		2·10	12·5	166	10·8	0·40	11·0	0·40	1463	837	655	160	49	29	Canon de 5.
Steel { Du Bonge R.B.L. { 17½-pr. ...	3·54	10·43	98·4	89·7	25·3	176·6	40·4	25	{ 1 Copper ring }	4·4	17·6	—	17·6	—	—	—	1575	952	754	302	112	58	Can. de 90mm } 1877
13-pr. ...	3·15	8·46	84·0	76·6	24·3	133·5	40·4	25		3·3	13·2	—	13·2	—	—	—	1575	942	739	228	75	49	Can. de 80mm
GERMAN.																							
Steel ... R.B.L { 15-pr.	3·46	9·0	82·0	73·4	16·7	96·0	29·3	50	} Lead Coating	3·31	18·0	209	15·4	0·62	15·4	0·62	1456	903	710	226	87	54	9c. } 1873.
11 pr.	3·09	7·7	82·0	74·2	19·2	90·0	32·6	50		2·76	12·1	122	11·2	0·44	11·2	0·44	1526	896	691	180	62	37	8c.
RUSSIAN.																							
Bronze or Steel } R.B.L. ... { 24-pr.	3·96	12·2	73·0	57·0	14·5	—	—	50	} Lead Coating	2·70	30·0	210	24·0	0·88	—	—	1060	823	681	189	112	77	9-pr. } 1874.
13-pr.	3·42	6·8	70·0	60·7	20·0	—	—	41		1·50	14·5	105	12·6	0·44	—	—	1004	722	581	89	48	1	4-pr.
13-pr.	3·42	9·0	—	67·0	—	—	—	50		3·25	13·0	160	13·0	—	—	—	1558	889	679	219	71	42	{ Heavy 4-pr. Experimental.

* The data as to the foreign guns may not always be absolutely correct.

Table referred to in p. 80.

TABLE IX.

TABLE of COMPARISON between S.B. Cast Iron and R.M.L. Converted Guns.

Nature of Gun.	Weight.	Calibre.	Length. Total.	Length. Bore.		Charge.	M.V.	Value of Log. $\left(\frac{d^2}{W}\right)$	Weight of Common Shell filled.	Bursting charge Common Shell.	Common Shell. 1,000 Yards. Velocity.	1,000 Yards. Energy.	2,000 Yards. Velocity.	2,000 Yards. Energy.	Remarks.
	Cwt.	Ins.	Ft. in.	Ins.	Calibres.	Lbs.	f.s.		Lbs.	Lbs.	f.s.	f. tons.	f.s.	f. tons.	
32-pr. 58 cwt. S.B. ..	58	6·375	9 6	108·65	17·04	10	1900	0·2003	23·625	1·312	843	118	542	48	
$\frac{64}{32}$-pr. R.M.L. 58 cwt.	58	6·3	9 6	108·45	17·2	8	1245*	1·7814	64·0	7·125	1013	455	900	359	The comparative weights and M.V.s of the shrapnel fired from these guns bear about the same proportion to one another as shown in the case of the common shell.
8-inch S.B. 65 cwt. ..	65	8·05	9 0	105·27	13·07	10	1464	0·1096	47·812	2·562	806	217	556	102	
$\frac{\text{64-pr.}}{\text{8-inch}}$ R.M.L. 71 cwt.	71	6·3	9 0	103·27	16·3	8	1230*	1·7814	64·0	7·125	1006	449	895	355	Not only is the power of these pieces much greater than that of the S.B. from which they are converted, but their accuracy of fire is beyond comparison better.
68-pr. S.B. 95 cwt. ..	95	8·12	10 0	113·9	14·02	16	1810	0·0962	49·812	2·562	928	298	627	136	
$\frac{80}{68}$-pr. R.M.L. 5 tons..	100	6·3	10 0	113·25	17·9	10	1240*	1·6845	79·875	8·812	1042	602	938	488	

* By using gas checks and larger charges of Pebble powder an increase of velocity in the case of these guns of between 200 f.s. and 300 f.s. has been obtained, so that compared with the S.B. pieces they will in future be still more powerful than they are shown to be in the Table.

Table referred to in p. 86.

TABLE X.—Table showing approximately the WEIGHT, DIMENSIONS, CHARGES, &c., of the most powerful Service and EXPERIMENTAL HEAVY GUNS existing at the commencement of 1878.

Nature of Gun.	Calibre.	Weight.	Total length.	Length of Bore.		Twist in No. of Calibres.	Means of Rotation.	Enlarged Powder Chamber.		Cubic inches for lb. of Powder.	Powder Charge.		Common Shell.		Heaviest Projectile.	With highest Charge.									Remarks.
																Mean pressure on Chamber.	Velocity at,			Total Energy.			Thickness of Iron Plates pierced.		
								Length.	Calibre.		Weight.	Nature of Powder.	Weight, Full.	Bursting Charge.			Muzzle.	1,000 Yds.	2,000 Yds.	Muzzle.	1,000 Yds.	2,000 Yds.	1,000 Yds.	2,000 Yds.	
ENGLISH.	Ins.	Tns.	Ft.	Ft	Cals			Ins.	Ins.		Lbs.	P²	Lbs.	Lbs.	Lbs.	Tons.	f. s.	f. s.	f. s.	f. t.	f. t.	f. t.	Ins.	Ins.	
17·72 inch	17·72	100	32·9	30·25	20	Increasing from 1 in 150 to 1 in 50, 2·88" from muzzle remainder U.	Expanding Gas Check.	55·12	19·7	—	†	—	—	—	—	—	—	—	—	—	—	—	—	—	R.M.L. Steel and wrought iron.
16 " §	16·0	80	26·9	24	18		Do.	59·6	18·0	—	†	—	—	—	—	—	—	—	—	—	—	—	—	—	do.
12·5 " ...	12·5	38	18·75	16·5	16	Increasing 0 to 1 in 35.	Bronze Studs.	40·12	14·0	30	200	P²	800	43	800	21·0	1590	1460	1341	14022	11824	9975	20·4	17·2	do.
12·5 " ...	12·5	38	18·75	16·5	16	Do.	Do.	Not enlarged		25	130	P²	800	43	800	23·0	1451	1333	1229	11676	9856	8378	17·0	14·4	do
ITALIAN.																									
17·72 inch §	17·72	100	32·9	30·25	20	Increasing 1 in 250 to 1 in 50.	Expanding Gas Check.	—	—	—	†	—	—	—	—	—	—	—	—	—	—	—	—	—	do.
12·6 " ...	12·59	38	22·0	21·2	20	Increasing 0 to 1 in 45.	Copper rings.	—	13·1	30	150	do.	—	—	771	11·8	1411	1292	1189	10647	9921	7556	15·8	12·8	R.B.L Steel hooped.
FRENCH.																									
18·11 inch §	18·11	124	—	39·4	20·6	—	Copper ring.	—	—	—	—	—	—	—	2645	—	1640	1566	—	48810	44366	—	—	—	
13·4 " §	13·4	48	—	—	—	—	—	—	—	—	—	—	—	—	—	—	—	—	—	—	—	—	—	—	
12·6 " ...	12·6	38	20	17	16·1	Increasing 1 in 45.	Do.	—	—	—	139	Matteren.	630	37	770	—	1371	1257	1159	10054	8482	7185	—	—	R.B.L. Cast iron, steel hooped and lined.
GERMAN.																									
16 inch ..	15·75	70	32·8	28·5	22·0	U 1 in 45 ...	Do.	60·6	17·32	—	352	Prism.	—	—	1631	—	1640‖	—	—	—	—	—	—	—	R.B.L. 35·5 cents. steel hooped.
14 inch § ...	13·97	56	26·3	22·5	19·3	U 1 in 45 ...	Do.	—	—	—	280	—	903	66·5	1150	—	1565	1453	—	19460	16770	—	—	—	
12 " ...	12·0	36	22·0	28·9	19·0	U 1 in 70 ..	—	51·0	12·4	—	132	do.	566	32·6	650	—	1525	1385	—	10510	8672	—	—	—	30½ centr. steel hooped.
RUSSIAN.																									
12 inch ...	12·0	40	20·0	1·70	17·0	—	Lead coating.	—	—	—	114	—	—	—	650	—	1400	1273	1168	8830	7325	—	11·2	8·5	R.B.L. 30½ cents. steel hooped.

* The information contained in this Table has been collected from different sources, and may not in all particulars be quite accurate, especially as some of the guns are only experimental In case of much doubt the columns are left blank, so that they may be filled in when accurate information is obtainable, or when experimental pieces are finally adopted.

† Service charges not yet settled. ‡ Experimental.

§ These penetrations are calculated by formula (γ), p. 338, that of Lieut.-Colonel Maitland, R.A. It will be seen that the penetration is somewhat less than that given by the formula used by Major Noble, R.A., by which the Tables pp. 347–350 have been calculated. ‖ Estimated.

CHAPTER V.

PRINCIPAL OPERATIONS IN THE MANUFACTURE OF OUR RIFLED ORDNANCE.

OPERATIONS.

Machinery.—Steam hammers and their power.—**Manufacture.**—Steel Ingots.—Testing of Steel.—Testing machine used.—Manufacture of bars.—Coils and coiling.—Welding.—Solid forgings.—Trunnion ring.—Uniting coils to form a tube.—Shrinking.—Mode of crushing.—Manufacture of a jacket.—Centring.—Turning.—Boring.—Broaching.—Lapping.—**Rifling.**—Uniform twist.—Increasing twist.—Drilling.—Screw cutting.—Slotting and planing.—Viewing and gauging.

MACHINERY.*

Machinery.

In the Royal Gun Factories steam power is employed for driving the whole of the machinery, and also for working the heavy hammers, while hydraulic power is used in some of the heavy cranes and testing machines. Before proceeding to detail the various more important operations it is advisable to explain briefly some of the terms used in mechanics, in order to facilitate the description of the different machines used in the Department.

Communicating of motion from the engine to the machines.

Belts or bands.

Tooth-wheels.

Spur-wheels.

Bevil-wheels.

Mitre-wheels.

Steam engines are employed, as in most factories, to give rotation to long pieces of "*shafting*" or rods of iron which run the whole length of each shop, and carry a number of "*pulleys*" or wheels having a broad circumference or rim. Motion is transferred by means of "*belts*" or "*bands*" of leather from these to similar pulleys attached to each machine, this being the method usually employed for communicating motion from one axle to another at a distance from it. When axles are near one another "*tooth-wheels*" are used to transmit the power; these wheels being of different forms and sizes according to the relative positions of the axles and the work which the machine is required to perform. When the axles are parallel the ordinary "*spur-wheels*" are used, that is, wheels in which the teeth project radially from the circumference. If not parallel, wheels having the teeth formed on the surface of a cone instead of on the circumference of a circle, must be employed; these are called "*bevil-wheels.*"

"*Mitre-wheels*" are bevil-wheels of equal size whose axes are at

* For a fuller description of machinery, see "The Elements of Mechanism," and "Principles of Mechanics," by Professor T. M. Goodeve, M.A.; also "Workshop Appliances," by Professor Shelley, C.E.: which are some amongst the many useful manuals of recent date upon this subject. The details in this chapter are intended more especially for the instruction of Artillery Classes in the R.G.F.

right angles to one another, and are much used for changing the direction of motion.

Rack and pinion.

A "*pinion*" is a spur-wheel with a small number of teeth gearing into a larger wheel or into a rack. A "*rack*" is the name given to a straight piece having teeth projecting from its surface.

Worm-wheel.

A "*worm-wheel*" is a spur-wheel with oblique teeth shaped so as to gear with the thread of a screw, and the combination of the two is one method of obtaining increased power by the sacrifice of velocity. This arrangement is in general use in steam cranes, for should any part of the engine get disabled the screw becoming fixed prevents the lift running down. Another way of increasing power at the loss of velocity is also effected by using tooth wheels of different sizes; thus, if a wheel one foot in diameter with 10 teeth be made to drive another five feet in diameter and having 50 teeth, the latter will revolve only once while the former turns five times, but the force required to stop the large wheel will be five times that which works the small one, not taking friction into account. "*Cone-pulleys*" can also be used for varying the speed of machines.

Methods of increasing power.

Cone-pulleys.

Change-wheels.

Wheels of different sizes are also used to vary the rate of revolution of the work without reference to the power. For instance, in the screw-cutting lathe there is only one screw used as a copy, but a "*thread*" of any required "*pitch*" can be cut by introducing a series of "*change-wheels*" so as to alter the ratio of the revolution of the work to that of the copy.

Screw-thread.

A "*screw-thread*" is a projecting rim running spirally round the exterior or interior of a cylindrical surface; the former is called a "*male*" thread, and the latter a "*female*" thread, and the rim of each is shaped so as to correspond with the other. The "*pitch*" of a screw is the distance between corresponding edges of two adjacent threads measured parallel to the axis of the screw, and is the distance the screw travels when completing one revolution on its axis. In practice the pitch of screw bolts is usually estimated by the number of threads in an inch of length, thus, the screw of the present pattern copper vent bush is called "seven threads to the inch."

Pitch of a screw.

Generally right-handed.

Screws are generally *right-handed*, that is, the screw progresses when turned in the same direction as the hands of a watch. *Left-handed* screws are, however, sometimes used for special purposes, as in the instrument for taking impressions of the bores of guns.

Double-thread.

A "*double-threaded*" screw is one in which two parallel threads starting from opposite sides of the cylinder are wound round it side by side. The object is to increase the bearing surface of the thread without reducing the pitch, that is, the rate of progression. Thus, in a single threaded screw of one inch pitch the screw travels one inch for each turn, but if an intermediate thread be wound round the bolt the bearing surface will be as great as that of a screw whose pitch is only 0·5 inch, but the travel of the bolt will still be one inch. This kind of thread is therefore used in cases where considerable resistance to shearing is required, while the saving of time is also an object, as in the breech-screw of the 7-inch B.L. gun.

Conversion of circular motion into reciprocating, and vice versâ.

Now in order that the various machines may be capable of performing the work required of them, the rotatory motion derived from the steam engine must be changed into rectilinear and other motion, and nearly the whole of the combinations of machinery may be said to consist in different methods of converting *circular* motion into *reciprocating* motion, and vice versâ.

Cams.

The conversion of circular into reciprocating motion is most commonly effected by means of "*cams*" (the ordinary eccentric being only a particular form of cam), which name is given to a curved plate fixed

CHAP. V.

to, or a groove cut in, a revolving spindle, which plate or groove communicates motion to another piece by the action of curved edges; by varying the form of the plate or groove any required motion can be obtained.

Crank. Pawl and ratchet.

Rectilinear motion is converted into circular by the "*crank*," as in a steam engine, or by means of a pawl and ratchet-wheel, the pawl being carried on a jointed arm worked by a cam; the former gives continuous motion, while the latter gives intermittent motion such as is required in the "*feed*" of slotting and other similar machines.

Reversing motion.

The following is the method employed for reversing the motion of the boring or rifling machines.* The saddle carrying the boring (or rifling) bar works on a screw running the whole length of the machine. On the end of this screw, away from the gun, is fixed a bevil-wheel, into which two other similar wheels gear, the axis of the two latter being concordant and at right angles to that of the screw. One wheel is attached to a pulley by a hollow spindle, through which a smaller spindle passes, connecting the other wheel to a second pulley, so that each is capable of turning independently of the other. Between these two pulleys is a third, riding loose upon the spindles, and called an "*idle-pulley;*" the workman by shifting the band to this idle-pulley, can stop the machine without interfering with the rest of the shop, and by shifting the band to one or other of the "*fast-pulleys*" he can set the machine in motion in either direction, as the direction in which the screw will turn, and consequently the saddle travel, depends upon which bevil-wheel the power is passing through. When once set in motion the machine is self-acting. So soon as the saddle reaches one extremity of the bed a projection on it comes in contact with a stop fixed on a rod running along one side of the machine. This rod acts on the band through a series of bent or "*bell-crank*" levers, and shifts it from the one driving pulley to the other, passing over the idle-pulley. Thus the power is transferred to the other bevil-wheel, and the motion of the saddle reversed. On reaching the other end of the bed the projection from the saddle acts upon the rod in the opposite direction, the band is shifted back, and the motion again reversed.

Idle-pulley. Fast-pulley.

A simpler method of reversion, now much used, is by crossing the strap, which passes over one of the driving pulleys, and consequently motion is produced opposite to that which takes place when the uncrossed strap is on its driving pulley. Still another method is by passing the power alternately through a different number of spur wheels.

Nasmyth's steam hammers.

Steam Hammers.—The most powerful hammers employed in the Royal Gun Factories are Nasmyth's double acting steam hammers, that is, hammers in which the steam power is used not only in raising the block but also in driving it down upon the forging. In these the steam cylinder is fixed, and the falling weight is attached to the piston, but smaller ones are used constructed after Condie's patent, in which the piston is fixed and the cylinder moves, thus forming part of the falling weight. The piston is a double pipe through which the steam passes into the cylinder; when steam enters through one pipe it lifts the cylinder, and when through the other it drives it down on to the forging.

Condie's.

It is impossible to say absolutely what blow per square inch any particular hammer can give, as it varies with the height and size of the mass being forged, the pressure of the steam in the boiler, &c., &c.

Power of hammers used in the R.G.F.

The following table gives, however, a comparative statement of the equivalent in *foot-tons*† for each hammer, supposing the block to fall

* For a detailed description of the rifling machine, vide p. 114.

† A foot-ton being the blow struck by a ton falling through one foot.

the full stroke of the cylinder and the steam pressure to be about 50 lbs. on the square inch. CHAP. V.

TABLE XI.

TABLE showing the POWER of STEAM HAMMERS in the Royal Gun Factories.

Nominal Weight of Block.	Actual Weight of Block.	Diameter of Cylinder.	Length of Stroke.	Pressure of Steam on Piston.	Blow on Anvil.
	Tons.	Inches.	Feet.	Tons.	Foot-Tons.
Nasmyth's, 40 tons	40	55	10·5	53	977
" 12 "	16	38·25	7·66	25	314
" 10 "	14	37·625	8·41	25	328
" 7 "	8	29·125	6·33	15	146
Morrison's, 4 "	4	26·5	4.75	12	76
Condie's, 3½ "	3½	23·25	4·83	9·5	63

MANUFACTURING OPERATIONS.

Having thus briefly noticed some of the principal features in the machines used in the Gun Factories, we shall now describe the various manufacturing operations.

Raw materials used. The raw materials we employ are either steel or wrought iron, the former being obtained in ingots from contractors and the latter produced direct or else worked up in the Department as described further on.

Manufacture and testing of Steel Ingots.

Steel ingots for barrels. Steel ingots for the inner barrel of guns, or for the whole gun in the case of 7-prs., are supplied to the Royal Gun Factories by the contractors in the form of solid cylinders cast and afterwards forged under a heavy hammer.* Casting is necessary, not only for the purpose of obtaining a sufficiently large block of steel, but also for making the block homogeneous and uniform in structure. Forging or drawing out the cast block imparts to it the desirable properties of great solidity and density.

Casting. The mould is of cast iron and octagonal in shape, being smeared inside with some non-conducting substance, generally a mixture of black lead and oil. That for the 9-inch ingot is 5 feet long and 2 feet thick. After casting the steel is covered with ashes or other non-conducting substances, and allowed to cool very gradually; when cold a portion is cut off the top, and the lower end being the denser, is marked for the breech.

Forging. The block thus formed is drawn out by a series of heatings and hammerings which occupy several days, to a cylinder sufficiently long for an inner barrel, in which state it is sent to the Royal Gun Factories, where it is subjected to the following tests and treatment.

How tested. A slice is cut off from the breech end and divided into pieces for testing. Some of these are flat bars, 4 inches long and ¾ by ⅜ in section, and others are of the shape usually tested in the machine for tenacity and elasticity, viz., small cylinders 2″ long between breaking parts, and 0·533″ in diameter, having shoulders at each end by which they are fixed in the machine.

Bending test. Three of the former are marked respectively *S*, *L*, and *H*. One end of the *S* or soft (*i.e.*, untempered) piece is gripped in a vice, whilst the

* The ingots obtained from Messrs. Whitworth are forged by hydraulic pressure.

CHAP. V.

other end is hammered down towards it, to ascertain that the steel, by bearing this bending without cracking, is naturally of the mild quality required. The *L* and *H* pieces are raised to a low red and high heat respectively, immersed in oil, and when cold treated in a similar manner. The heat is judged by an experienced workman from the colour of the specimen.*

Test of tenacity and elasticity.

Whichever of these pieces shows the best fracture under the hammering test determines the heat at which the whole tube is to be toughened (see page 163). Should neither piece answer, others at intermediate temperatures are tried, and if all fail, the block is returned to the contractors; but some specimen having succeeded, as is generally the case, two of the remaining pieces—one in its soft state and the other toughened at the ascertained temperature—are tested for tenacity and elastic limit. The soft material should begin to stretch permanently at 13 tons per square inch, and break at 31 tons. The toughened piece should begin to stretch at 31 tons, and break about 50. The permanent elongation is also taken, but it is not considered necessary to lay down any limits in this respect.

Ingots, which pass all the foregoing tests, are accepted and are toughened in oil at the approved temperature previous to being put into the gun. The process of toughening is described in Chapter VIII.

Testing Machine.

Testing machine.

The machine used for testing metals in the Royal Gun Factories is shown below.

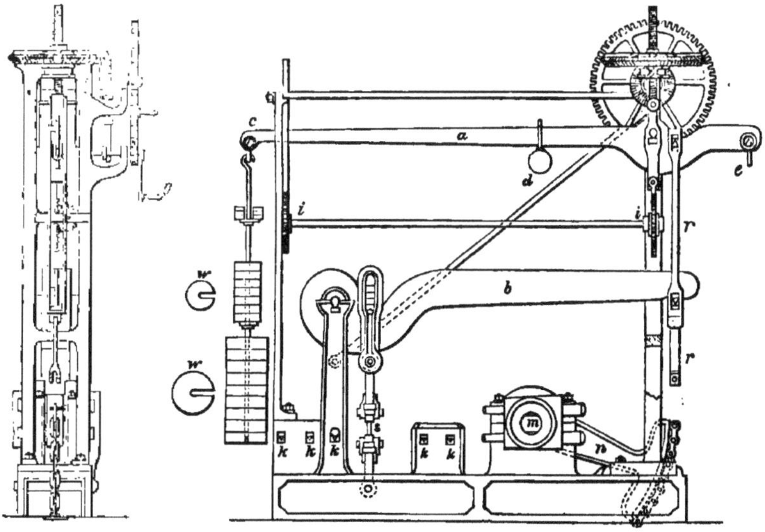

* It may seem remarkable that the colour at which a razor, a chisel, or a watch-spring must be tempered is definitely fixed, and yet that the heat for toughening a gun-barrel should vary in shade from a blood red to a bright cherry. Now, did the temperature depend alone on the amount of carbon in the steel, it would appear best to toughen every barrel at a bright heat, for the less carbonized, or in other words the milder the steel, the higher is the temperature at which it toughens most

Specimens of metals are tested by means of it both for elasticity and tenacity, while the copper plugs employed for crusher gauges (vide p. 347) are also tested as to amount of compression in the machine.

Description of machine.

The machine itself is of American invention, and consists of a combination of two levers *a* and *b*, which together give a purchase of 200 to 1; that is to say, 1 lb. applied to the end of the upper lever at *c* will exert a strain of 200 lbs. on the specimen at *s*, and as all the bearing points are hard knife-edges, on hard smooth surfaces, the friction is reduced to a minimum.

Transverse and torsional elasticities.

It is further provided with suitable bridles, holders, and other apparatus to test tensile, compressive, transverse, and torsional elasticities, and is adapted for experiments on the pressure required to punch or shear, and on the hardness or softness of bodies.

Prior to use the machine is adjusted as to balance by moving the small weight (*d*) on the upper lever, or by means of a weight attached at (*e*). Weights of the form shown (*w*) are placed upon a shelf on the iron rod *c* suspended from one end of the upper lever.

Testing for elasticity and tenacity.

We will suppose a specimen piece of form shown below (*K*) to be in its place, secured in the socket (*L*), which is fastened at its two ends to the bed of the machine and the short arm of the lower lever respectively.

The operator knows approximately what weight will be sufficient to try the specimen up to its limit of elasticity, and as that limit is approached he increases the amount of weight by only a little at a time.

Application of the weights.

As the weights are applied at *c* the specimen bar begins to stretch. In order to prevent the consequent inconvenience of the end of the upper lever dropping, there is an arrangement shown at *r* which allows us to raise or lower the fulcrum of the upper lever by means of a screw and train of wheels.

The same arrangement keeps a sliding wood stopper, close to the outer end of the upper lever, so that when the specimen breaks, the lever is caught at once, instead of being probably damaged by the effect of the jerk.

Temporary extension.

As the specimen stretches, an accurate measuring gauge pushed between the shoulders of the specimen shows the amount of extension. The operator removes the weights, and replaces them again as the limit of elasticity is approached, and on each occasion gauges the specimen carefully to see if there is any permanent elongation.

Elastic limit or permanent extension begins.

The determination of this point (the elastic limit) requires great accuracy and much practice on the part of the operator. Until this limit is reached, the extensions of the specimen for equal increments of weight are tolerably uniform (vide p. 3, and Fig., Chapter I.). but as soon as it is passed, the extensions increase rapidly for each increment of weight until the bar breaks.

---

satisfactorily: but it is a fact that the denser the steel is—*i.e.*, the more it is hammered in the process of drawing out—the less heat does it require for successful toughening; hence, each individual barrel must be tested. Doubtless experience will in time teach us the exact mildness and density suitable for a steel barrel, and the proper temperature to which it should be raised, before being immersed in the oil.

CHAP. V.

Great care necessary.

During the process of the test the operator has to watch with care the behaviour of the specimen, in order to note its general character, for by continuing to increase the weight upon the end of the lever gradually the characteristics of the specimen develop themselves more or less clearly.

Tenacity.

In the case of some metals the specimen will flow, or be drawn in the heart of the bar only, thus leaving a corrugated exterior surface from the crumpling of the outer skin; with other metals the flow is more uniform and the outside is comparatively smooth. If on the other hand the metal be hard and rigid, it may not be drawn out to any great extent but may break with very little reduction of sectional area and exhibit a high tenacity. If on the contrary it is of a soft and fluent nature, it will be drawn out to a considerably smaller section, and then will break at the point where the diameter is most reduced.*

The testing machine is equally suited for any other kind of metal; and in dealing with familiar materials, such as cast iron, copper, bronze, or other alloys, in order to arrive at their tensile properties, the same course is pursued as with wrought iron or steel.

Compression.

For testing compressibility of metals a socket of the form shown in Fig. 1 is employed.

TESTING METALS FOR COMPRESSION.

Fig. 1.

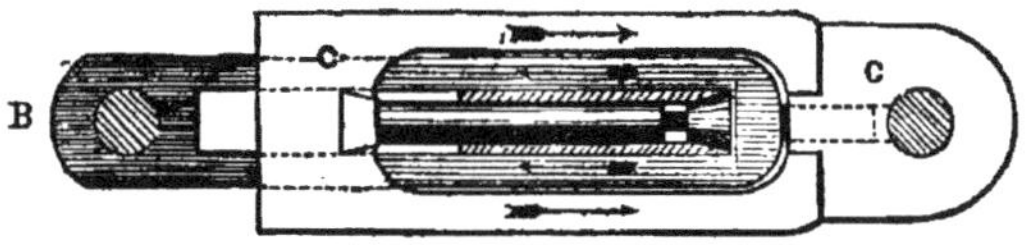

It consists of two parts, one sliding within the other, one of the parts (*B*) being attached to the lever and the other part (*c*) to the framing of the machine. The specimen for this purpose is in the form of a small cylinder; weights are applied to the end of the upper lever, producing a tension 200 times as great on the specimen as in the case of testing for tenacity.

Time not considered with this machine.

It will be seen that with this machine time is not taken into consideration, and in consequence another testing machine has been devised in the Royal Gun Factories, which it is hoped will enable us to test the compressibility of metals with regard not only to pressure, but also to time, within certain limits.

Manufacture of Wrought Iron.

Manufacture of wrought iron.

The wrought iron used in the Royal Gun Factories is either puddled iron (used principally for making bars) or scrap iron.

Puddled iron.

Puddled iron is made in the Department by burning out the carbon, &c., from cast iron obtained as pig iron, or by breaking up old guns, carriages, shot, shell, &c.

The reverberatory furnace.

The operation of converting this cast iron into malleable or wrought iron is called "puddling," and is carried out in a puddling furnace. This is of the description termed a reverberatory furnace; that is, a furnace in which a bridge of firebrick placed between the grate and the hearth prevents the contact of the coal and the iron which would be

* In fact the character of the specimen, and so of the metal, must be judged very considerably by the nature of fracture as well as by the actual fracturing weight.

CHAP. V.

detrimental to the latter, while the powerful draught generated by a tall chimney at the other side of the hearth induces the flames to play upon the metal with great intensity. The chimney is provided with a damper or lid, raised and lowered by a chain attached to a lever at the top, so that the draught may be regulated to a nicety, and stopped altogether when the hearth is temporarily empty.

Price's retort furnace, in use in R.G.F., provides for the heating of the air, and the heating and partial coking of the coal, before entry into the fire grate, by means of the waste heat of the furnace.

Puddling.

The cast iron scrap having been broken up under a steam hammer into pieces of a convenient size, a charge of between 5 and 6 cwts.* is placed in the furnace together with about 1 cwt. of iron scale or hammer scale, which is rich in oxygen † and aids in burning out the impurities.

The heat of the furnance soon melts the iron, which is well stirred by the workman and the carbon is so converted into gases which are carried off by the draught.

A liquid glass or slag is also formed, which is tapped and run out through a hole at one side from time to time.‡

As the carbon gets burnt out, the iron becomes pasty, and is collected by the puddler into large balls, which look in their white hot state not unlike loosely collected masses of snow. These balls are rolled up an incline on one side of the furnace bed and then removed from the furnace to a steam hammer, where the liquid slag entangled between the particles of iron is squeezed out, and the ball hammered into a rectangular block technically termed a "bloom."

Puddled blooms.

This iron is rather hard and brittle, and has to be heated and hammered or rolled if we want to give it fibre.

Scrap iron.

Scrap iron consists either of old wrought iron articles, bolts, nuts, screws, horseshoes, &c., which are purchased by contract, or of the shavings obtained in the Department from turning or boring wrought iron. The first gives the best fibre, but the latter is cleaner.

Drumming.

To work up the first mentioned it is drummed to get rid of rust, &c., and then piled on small pieces of board. The pile or charge is heated in a furnace to a white heat, and then hammered under a steam hammer into a bloom.

The Department scrap is thrown in loose instead of being piled as above, and then treated in a similar manner.

Manufacture of Bars and Coils.

Manufacture of bars and coils.

Most of the parts from which our rifled guns are built up are made by coiling a bar round a mandrel or core, and afterwards welding the folds together into a compact hollow cylinder; but certain portions, such as the trunnion ring and cascable, are made from solid masses of iron formed by welding together a number of slabs.

We will take first the manufacture of the bar and coil.

Bar iron made in R.G.F. since 1868.

Until the commencement of 1868 the bars used for coiling were supplied by contractors, but puddling furnaces and a rolling mill were then erected in the Royal Gun Factories, and all the bar iron required is manufactured on the spot from obsolete cast iron articles such as

* With double puddling furnaces the charges are of course much larger.

† Being the peroxide of iron, $Fe^3 O^4$.

‡ This contains some iron, but consists principally of the sand of the furnace bed or lining, mixed with impurities contained in the charge.

CHAP. V.

carronades, gun carriages, &c., in conjunction with the scrap iron before mentioned.

Flat bar.

Blooms of puddled iron are raised to a white heat, passed through the rolls, and made into a flat bar. Similarly blooms of contractor's scrap are heated, hammered and rolled, and made into a flat bar. The flat bars made from scrap iron are shorter, and may be distinguished from the flat bars made from puddled iron, by their irregular edges.

Piling.

Each of these flat bars is now cut into lengths, and arranged in a pile, with the bars of scrap in the centre, and puddled iron bars on top and bottom. The puddled iron is used for the exterior, as giving a cleaner surface, while the scrap has the stronger fibre. The pile is

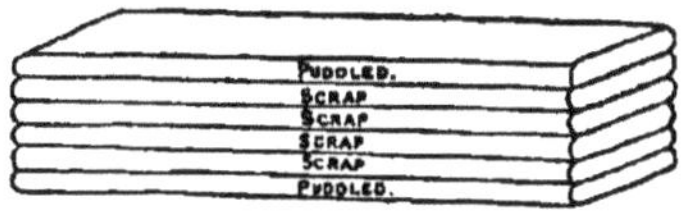

Rolling.

now raised to a white heat, and rolled into a long bar about 24 feet long, and varying in section from 2½ to 9 inches or more, according to the purpose for which it is intended.* Should the bar be required for the breech piece of a gun, or for the tube of a 64-pr. gun and guns converted on Major Palliser's system, it is cut into lengths, again fagotted, raised to a welding heat, and passed between the rollers; but in ordinary cases one rolling is sufficient.

Shape of bar.

A bar is always designated by the depth of its section. The section is slightly trapezoidal, in order that when the hot bar is wound round the mandrel, narrow side inwards, the spreading of the inside and the narrowing of the outside, natural to such a process, may be neutralized and no space left between the folds of the coils.

Testing quality of bar.

Samples of each week's work are tested for tenacity and elastic limit, the stretching weight should be about 12 tons, and the breaking weight 23 tons.

The comparatively short bars, made as described, have next to be joined together, to give us a bar of the length required for a coil.

Welding.

To weld two bars together, the ends must be scarfed down and placed from opposite sides in a furnace, from which, when they arrive at a white heat, they are withdrawn and welded under an adjacent steam

Why sand is used.

hammer, sand having been thrown on the hot bars (as is indeed customary in the case of all forgings) in order to clean the surface and prevent scale forming, by converting the superficial oxide into a liquid silicate, which will flow off of its own accord, or be squeezed out by the hammer. Another bar is welded on in a similar way, and so on, until a sufficient length is obtained for the required coil.

Coiling.

Furnace.

The bar to be coiled having the ends flattened down, is placed on trestle rollers in front of a long reverberatory furnace with a chimney at the far end, and grates along its sides.† A chain being hooked into an eye or hole in the far end, the bar is drawn by machinery into the furnace. When the bar arrives at a bright red heat, the end near the

---

* The bars from which the breech coil of 10-inch guns and upwards are made, are manufactured by forging, vide p. 107.

† The longest bar coiled in the Royal Gun Factories was 270 feet in length. The furnace is only 190 feet long, but should the bar be longer, the extra length is allowed to project through the end of the furnace outside, and is heated by fire placed on small iron trucks.

CHAP. V.

door is drawn out by means of the same eye, and attached to a pin, this end being cooled with water to prevent it tearing away with the weight of the bar. This pin is connected with a slightly taper iron roller or

mandrel, fixed across and in front of the door of the furnace. The mandrel tapers, in order to facilitate the removal of the finished coil. The apparatus is then put into gear, and the mandrel revolves, winding the bar round it. During the process, scales form between the folds, but their effect is almost entirely nullified by subsequent heating and forging, sand being used to assist in liquifying the oxide, as stated above. When the coil is formed, the fixed extremity is hammered off the pin, and water is poured on that end to cool it, in order that the folds there may not be opened out in the taking off of the coil. If the coil be large, a short iron bar is placed with one end resting on the ground, and the other end against the extremity which has been removed from the pin. The mandrel is then turned in the same direction as that in which it revolved when the coil was being formed, and the coil, being prevented from revolving by the iron prop, is loosened, and slips down towards the narrow end of the mandrel. The mandrel is afterwards raised by a crane, and the coil drops off. Small coils are hammered off, no water being used in this case.

Winding bar.

Detaching coil from mandrel.

If a double coil is required, a round bar is fixed by bearings at each end through the newly made coil when cold, and placed on the supports hitherto occupied by the mandrel; the second bar is then wound round the first coil in the same way that the first bar was wound round the mandrel, but in the reverse direction to break joints. A triple coil* would be formed by immediately winding a third bar around the second coil in the opposite direction. Thus the first coil would act as a mandrel to the second, and the second to the third, whilst the bar upon which all three revolve could be easily extracted when the triple coil is completed. The inner coils projecting at the ends, a little beyond the outer ones, in order that a close weld may be obtained at the interior of the cylinder.

Double coil

Triple coil.

The object of welding a coil is to unite the folds so that it may withstand the longitudinal stress. The operation is generally the same for

Object of welding coils.

* Triple coils are not employed now, and double coils but rarely, owing to the greater thickness of the bars from which coils are made.

CHAP. V.

Great care necessary with coils for inner barrel.

all kinds and sizes of coils, single, double and triple. When the coil is intended for an inner barrel, as in the case of cast iron guns converted on Palliser's plan,* the process must be very carefully performed; for if from the badness of the iron, or from dirt or grit between the folds, or from insufficient heating or undue hammering the welding becomes imperfect, the barrel will of course be unsound, and the powder gas will eat its way into the defective parts. Indeed, the difficulty of obtaining a perfect coiled tube was at one time so great, that two out of every three barrels were rejected, but excellent iron having since been procured for the purpose, the per-centage of defective tubes has been greatly reduced.

Heating a coil.

The coil is placed upright in a reverberatory furnace, for were it placed on its side, it should be turned over, in order to be equally heated all through, and moreover, drippings from the fire-brick which line the furnace, would probably fall from the roof in between the folds. If intended for an inner barrel, two furnaces are used; one is at a low temperature (termed a "blue light"), and when the coil arrives at a red heat it is brought out and transferred to the other, where it is brought to a welding heat. This is found to be more economical than placing the cold coil at once into a very hot furnace, and also prevents any injury to the iron which would result from so doing. In all cases of welding, it is necessary to "strike while the iron is hot," and that the surface to be joined should be perfectly clean, the white hot coil is therefore transferred from the furnace to the steam hammer as quickly as possible, and sand is thrown upon it for the reason before assigned.

Process of welding a coil.

The coil is first placed vertically under the hammer, and receives a few smart blows to weld the folds, it is then thrown on its side and being gradually turned, is hammered (or patted) all round to straighten it. It is then raised vertical again, and a punch or mandrel—rather over half the length, and a little larger than the interior diameter of the coil—is hammered down its own length, the coil is next placed on its side and hammered round, that half of its length thus being made very compact, and large enough to let the mandrel fall out. After this the coil is again raised vertical, and the mandrel is forced in the opposite end, and the process repeated.

The mandrels are of coiled iron and very hard.

The reason a long mandrel is not forced through the whole length of the coil is that it would tend to separate the folds.

The coil is replaced in the furnace for the second heating, and much the same process is followed, to render the ring more consolidated as well as more shapely; and if intended for an inner barrel, a fine mandrel is used to make the interior more perfect. If the coil is to be "faced," a flexible steel bar is used under the hammer to flatten the ends and prevent their being bell-mouthed.

Why water is used.

Before the coil is removed from the hammer, water is thrown over it, which, forming into steam, blows off the black scales and shreds when the work is good, but a black spot is left by the water if there is a bad part.

Loss in length in welding coils.

Coils lose in welding from one-tenth to one-third of their length according as they are thin or thick.

After welding, the cylinder is inspected by the gaugers as to size, shape, and soundness, and unless found satisfactory in all respects, is subjected to another heat.

All the tubes of which a gun is built are made from coils manufactured as above, but as we shall presently see, that more than one completed coil is often required to form such a tube.

---

* The process by which wrought iron tubes for guns made on the Palliser principle in America differ in several details from those described, as carried out in R.G.F.

Solid Forgings.

Building up with slabs. The porter bar.

Every solid forging, large or small, is made in the same way, namely, of slabs of iron successively welded together upon the end of a porter or carrying bar (a stout bar of wrought iron) which acts both as a lever and tongs in manipulating the work.

Formation of slabs.

To form a slab, departmental scrap (obtained in the shape of shavings during the processes of turning, boring, &c.) is thrown loose into the furnace, and hammered while hot into cubical blooms. These are again hammered into flat cakes, and several of these cakes being piled, heated, and welded together, constitute a slab.

Manufacture of a trunnion ring.

As an example of a solid forging we will take a trunnion ring.

The porter bar is heated and the end flattened or scarfed down on both sides, two slabs are placed on one face, and the end of the porter bar with the slabs brought to a white heat and welded together, the porter bar is turned over, slabs placed on the opposite face, and the same process repeated; one slab after another is added and welded in a similar manner, until we have a mass of iron of sufficient size.

Shaping and punching.

This mass is roughly hammered into the shape of a trunnion ring, the porter bar being in continuation of one of the trunnions. The roughly-shaped block so made, is converted into a ring by punching a small hole (or in the case of very large guns two parallel holes) through the centre with iron wedges, and then enlarging the same by taper oval mandrels increasing in size until the hole is sufficiently large.

The trunnion ring has to be heated between each punching, and the trunnions are then still further shaped. It will be seen that the fibre of the iron will run round the ring and along the trunnions exactly as we require it to do it for strength.

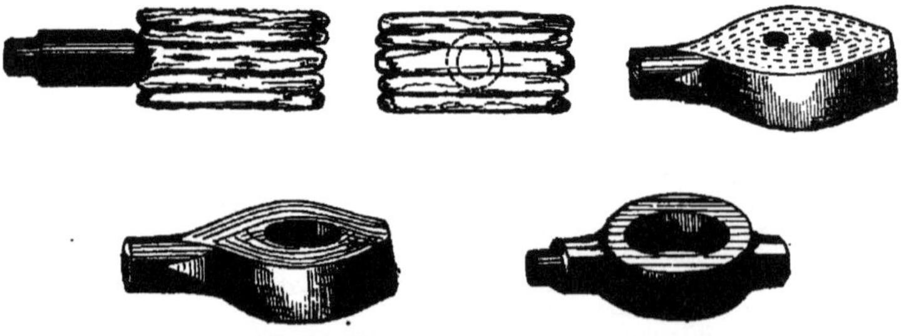

Shaving scrap.

Shaving scrap is used in making solid forgings for trunnion rings, as it gives us a good fibrous iron of uniform quality.

Binding slabs.

When we want a solid cylinder of iron, cross or binding slabs should be welded along the sides of those first welded to the porter bar, as in the case of the forging for a large cascable.

Manufacture of bars for breech coils, heavy guns.

To form the bars from which the breech coils of heavy guns are made, slabs are successively welded to the end of a porter bar scarfed to receive them, and the mass so obtained being again heated and scarfed down, another slab is added, and the forging drawn out to the shape and section of the bar required, and so the operation is continued until a sufficient length of bar has been made.

Why heavy bars are not rolled.

This mode of preparing the bars for the breech coil of very heavy guns is pursued, as by its means we obtain a denser and stronger material for that coil, where great strength is required.

CHAP. V.

Uniting two or more Coils to form a Tube.

Welding coils together end to end.

When a coiled inner barrel is required, several coils must be welded together, while the *B* tubes and breech pieces of heavy guns are composed of two united coils.

Preparation.

In both cases the coils must be faced (turned smooth at the ends) and reciprocally recessed; that is, a projection (spigot) is formed at one end of a coil, while a recess (faucet) is bored in the corresponding end of another coil. The height of the shoulder is a little greater than the depth of the recess in order that a close joint may be obtained on the interior. The recess is then expanded by heat and shrunk over the projection, so that the two coils are stuck sufficiently together to admit of their being put into the furnace for welding.

Process of uniting.

Thin coils.

If the tube is comparatively thin, as when intended for an inner barrel, the tube is put crossways through a furnace so constructed that intense heat acts on the joint while the remote ends project outside. When the joint arrives at a welding heat a stout iron bar is passed right through the tube, this bar is keyed up at one end, and by means of a screw-nut worked by a long lever at the other end, the two coils are welded or pressed together. The pressure slightly bulges the metal at the junction so it must be straightened under a steam hammer. Another coil is then added on in a similar manner, and so on till the tube is of the required length.

Thick coils.

Should the two coils be thick and strong, they are when shrunk together, heated in an ordinary furnace and welded gently together under a steam hammer

Shrinking.

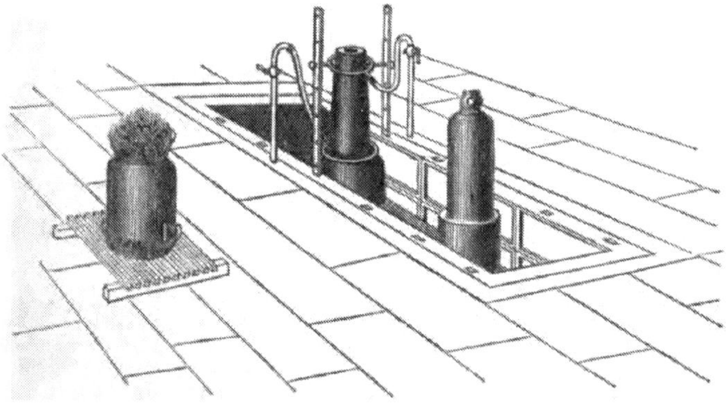

Double object of shrinking.

Shrinking is employed no only as an easy and efficient mode of binding the successive coils of a built up gun firmly together, but also for regulating as far as possible the tension of the several layers,

CHAP. V.

so that each and all may contribute fairly to the strength of the gun, vide p. 24, as to the theory of this construction.

The operation of shrinking.

The operation of shrinking is very simple; the outer coil is expanded by heat until it is sufficiently large to fit easily over the inner coil or tube (if a large mass, such as the jacket of a Fraser gun, by means of a wood fire for which the tube itself forms a flue; if a small mass, such as a coil, in a reverberatory furnace at a low temperature or by means of gas). It is then raised up by a travelling crane overhead and dropped over the part on to which it is to be shrunk, which is placed vertically in a pit ready to receive it.

Expansion of iron by heat.

High temperature not necessary.

How amount of shrinkage is limited.

The heat required in shrinking is not very great. Wrought iron on being heated from 62° F., (the ordinary temperature) to 212°, expands linearly about $\frac{1}{1000}$th* part of its length (the same amount of extension in fact as that due to its elastic limit or a stress of 12 tons per square inch of section); that is to say, if a ring of iron 1,000 inches in circumference were put into a vat of boiling water it would increase to 1,001 inches; and according to Dulong and Petit, the coefficient of expansion, which is constant up to 212°, increases more and more from that point upwards, so that if the iron ring were raised 150° higher still (*i.e.*, to 362°) its circumference would be more than 1,002 inches. Now it has already been shown† that no coil is ever shrunk on with so great a shrinkage as the $\frac{2}{1000}$th part of its circumference or diameter, for it would be strained beyond its elastic limit; allowing therefore a good working margin it is only necessary to raise a coil to about 500° Fahr.,‡ though in point of fact coils are often raised to a higher degree of temperature than this in some parts, on account of the mode of heating employed. Were a coil plunged in molten lead or boiling oil (600° F.) it would be uniformly and sufficiently expanded for all the practical purposes of shrinking, but as shrinkings do not take place in large numbers or at regular times, the improvised fire or ordinary furnace is the more economical mode and answers the purpose very well.

Heating a coil beyond the required amount is of no consequence provided it is not raised to such a degree of temperature that scales would form; and in all cases the interior must be swept clean of ashes, &c. when it is withdrawn from the fire.

Mode of cooling.

Cases where gas or water should be used.

With respect to the mode of cooling during the process of shrinking, care must be taken to prevent a long coil or tube cooling simultaneously at both ends, for this would cause the middle portion to be drawn out to an undue state of longitudinal tension. In some cases therefore water is projected on one end of a coil so as to cool it first. In the case of a long tube of different thickness like the *B* tube of a R.M.L. gun, water is not only used at the thick end, but a ring of gas or a heated iron cylinder is applied at the thin or muzzle end, and when the thick end cools the gas or cylinder is withdrawn from the muzzle and the ring of water raised upwards slowly to cool the remainder of the tube gradually.

Inner tube kept cool by a

As a rule the water is supplied wherever there is a shoulder, so that

* According to Lavoisier and Laplace the linear expansion of wrought iron on being heated from 32° F. to 212° is ·0012350 (*see* Phillips' Manual of Metallurgy, 2nd edition, p. 14), and according to Roy and Ramsden it is ·0012204 (*see* Ganot's Physics, translated by Atkinson, p. 205). Taking it therefore as ·001 for 150°, cannot be far wrong.

† See Chapter II., p. 26.

‡ The temperature may be judged by colour; at 500° F. iron has a blackish appearance, at 575° it is blue, at 775° red in the dark, at 1500° cherry red, and so on, getting lighter in colour, until it becomes white or fit for welding at about 3000°.

CHAP. V.

Flow of water always inside.

that portion may be cooled first and a close joint secured there; and water is invariably allowed to circulate through the interior of the mass to prevent its expanding and obstructing or delaying the operation; for example, when a *B* tube is to be shrunk on a steel barrel, the latter is placed upright on its breech end, and when the *B* tube is dropped down on it a continual flow of cold water is kept up in the barrel by means of a pipe and siphon at the muzzle. The same effect is produced by a water jet underneath, when it is necessary to place the steel tube muzzle downwards for the reception of a breech coil.

Manufacture of a Jacket.

Jacket.

The jacket or outer breech portion of all guns is made by welding together certain coils or tubes and a trunnion ring forged as described before.

Component parts.

It is usually composed of three parts, a breech coil, trunnion ring, and muzzle coil, although in many cases with heavy guns the latter is so short as to be a mere ring, and is sometimes dispensed with altogether, as in the 8-inch howitzer jacket, or 7-inch Mark IV, in which cases the shoulder on breech coil is made sufficiently long to extend to the face of trunnion ring.

How put together for welding.

The muzzle end of the breech coil is turned down to receive the trunnion ring, which is heated to redness and dropped over this shoulder.

While the trunnion ring is still hot, the muzzle coil is dropped down upon the front of the breech coil, through the upper portion of the trunnion ring. The trunnion ring thus forms a band over the joint, and in cooling contracts round the two coils, and grips them sufficiently tight to allow of the whole mass being placed bodily in a furnace, where it is raised to a welding heat.

Welding.

The heated mass is then placed on its breech end under a powerful hammer. Six or seven blows on the top suffice to amalgamate the three parts together; but to make the welding more perfect on the interior, as well as to obviate any bulging inside, a cast iron mandrel somewhat larger than the bore is forced down to within 20 inches of the breech end, a series of short iron plugs being used to drive it down. The whole is then reversed, and the mandrel is driven out with the same plugs, which have fallen out in the tilting over.

Breech Coil, Trunnion Ring and Muzzle Coil, ready for Welding.

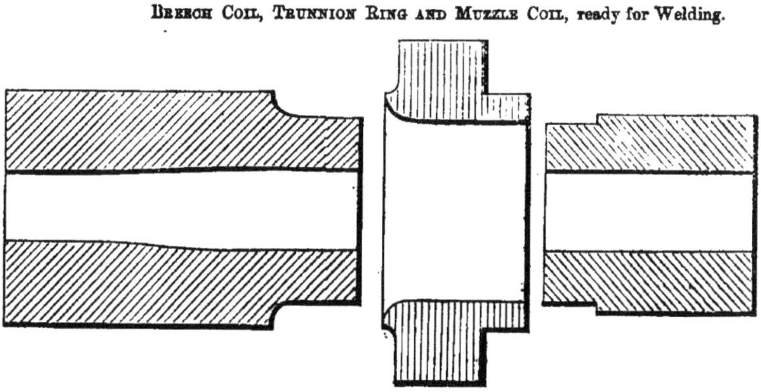

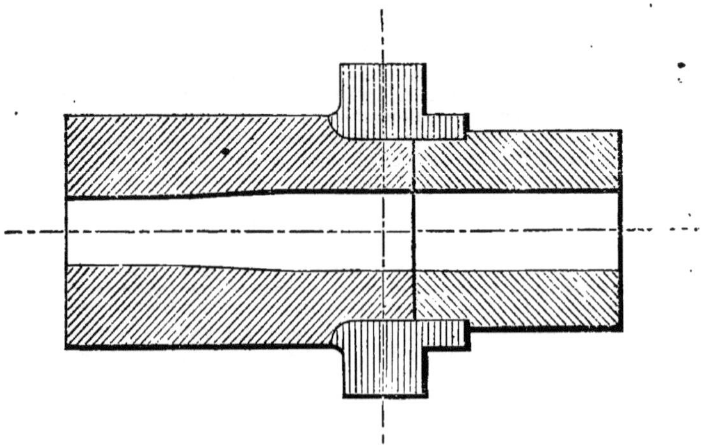

The *C* coil, (or jacket*) thus formed is turned, as to exterior, in powerful lathe, the belt removed in a slotting machine, and the jacket finally bored to the proper interior dimensions.

Machine Operations.

After the tubes, rings, and other portions of the gun have been made they have to be bored, turned, in some cases slotted, planed and shrunk together, while the inner barrel must be bored, broached, lapped, and finally rifled. These operations are shortly discussed below, as well as the processes of drilling, screw-cutting, and gauging. Machine operations.

Centring.

Previous to the first turning of any article, the axis must be found so as to centre it truly in the lathe. This is simply done in a solid cylinder by finding centres at each end with a pair of compasses; but in the case of a tube, bars of soft iron must be fixed across each end in order that the axis may be actually ascertained; the article being fixed accordingly in the lathe is turned truly cylindrical. Centring.

Large coils and jackets are centred as follows:—an iron spindle is passed through the interior and supported at either end, any flaws or defects on the interior are searched, and the spindle adjusted by means of a bar gauge so that the interior may be smooth and "clean" when bored to the proper diameter, the axis of the spindle being the axis of the bore; the ends of the mass are then whitened with chalk and a circle marked (with the spindle as centre) on each end by a scriber showing the amount to be bored out. The block upon which the cylinder is supported when in the turning lathe is adjusted to these circles, so that the exterior may be turned concentric to the interior.

* The jacket is sometimes termed the breech coil from its principal part when it comes immediately over the steel tube, as in the 7-inch gun, but it is called the *C* coil when it has between itself and the barrel a breech-piece of coiled iron, as in 9-inch guns and upwards.

CHAP. V.

Turning.

Turning.

Turning means cutting off the exterior surface all round. The machine generally used is a turning lathe on the ordinary principle, the work revolving as the cutter, fixed to a sliding saddle, moves along the side. In most machines two cutters work simultaneously, and rough or fine turning can be done as required. Turnings generally assume the shape of long ringlets and are again worked up into blooms.

Ordinary lathe used.

The size and power of the machines vary with the nature of the work. Those for turning the large breech coils of our heavy guns weigh each with foundations about 100 tons, and can turn off 7 inches in diameter in one cut. By arrangements whereby the motion can be sent into the machine through additional tooth-wheels of different diameters the speed can be greatly reduced and the power consequently increased.

Boring.

ROUGH BORING HEAD.

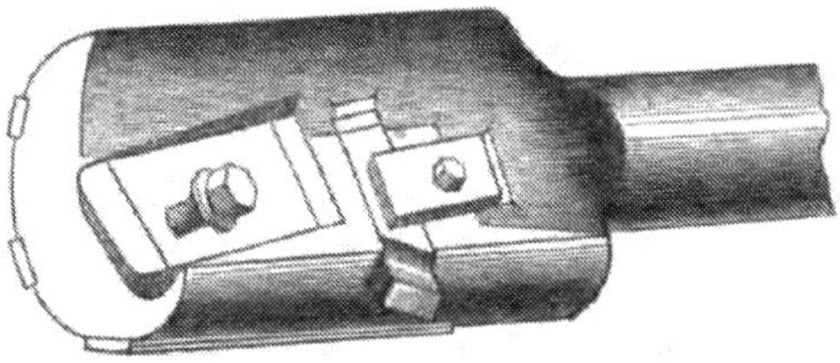

Boring.

The term boring is applied either to the process of reaming out the inside of a tube, or boring one out from a solid cylinder.

Rough and fine boring of an inner barrel.

Rough and fine boring of an inner barrel, as well as broaching and forming the chamber, are all effected in the same horizontal machines, the difference being in the shape of the boring-head and cutters. In this machine the barrel revolves, while the boring-head, guided by a supporting frame at the muzzle, simply progresses down the bore, being fed to its work by a long screw which passes through a nut in the sliding saddle to which the bar is fixed. The same effect would be produced did the tube move forward and the cutters revolve, but the shavings would obstruct the action by congregating at the bottom. The boring-head used for rough cutting is what is called from its shape, "a half-round bit," it has one pointed cutter set angularly and three steel "burnishers," or projections, to keep it steady in the bore.* This tool cuts away the metal in front of it and the same kind of head is used for cutting from the solid. In boring out a steel barrel for a 9-inch gun, 8¾ inches in diameter are taken out at each cut in segmental chips about ⅛-inch thick, and the operation takes 56 working hours, or about a week.

Rough boring head.

Fine boring.

Fine boring is performed with a round head having long cutters let in lengthways, and five lignum vitæ wood burnishers. The object of having the latter of wood, is that they may be set out further than the cutters, and so be cut to the same diameter as the tube on entering it,

* When the bore is to be a large one a second cutter is sometimes used, which is placed at right angles to the other in the same head.

as well as being expanded up to the full size of the bore by the water,* which is invariably projected down the bore; moreover were the burnishers of steel they might indent the bore. CHAP. V.

FINE BORING HEAD.

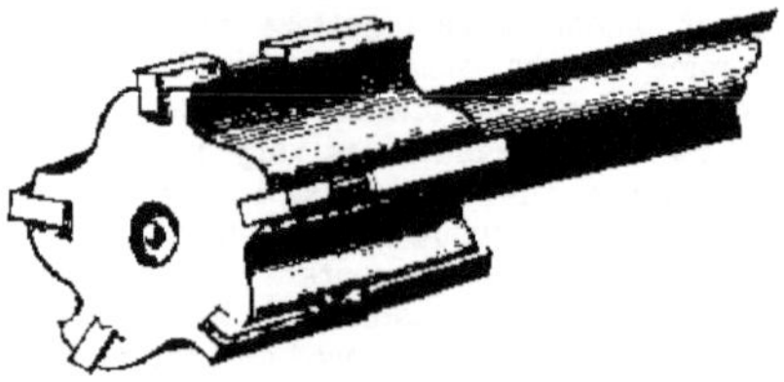

Machine for short and light tubes.

Short and light tubes, such as *B* tubes of R.M.L. guns, are bored in a horizontal machine, but on the opposite principle to that for inner barrels. The tube is fixed in a saddle capable of sliding along the frame of the machine, and the boring bar, through which are placed two adjoining and opposite cutters, is then passed through it and fixed at each end of the machine. The bar revolves while the saddle moves forward, and thus the boring is effected. This mode requires the bar to measure between the supports twice the length of the work to be bored, and the cutter to be in the middle of the bar; it is therefore unfit for long objects.

Large masses.

Large masses, such as the breech coil for a heavy gun, are bored in machines, somewhat similar to the preceding, but the work is stationary and the tool goes through a double motion of rotation and progression, the latter being given by means of a screw which passes lengthways through the side of the revolving bar, and is connected with the boring-head which in the shape of a ring slides along the bar; the screw is turned by a sun and planet wheel at the extremity of the bar.

The difference between the rough and fine cutters is generally that the former are deep and narrow, and the latter shallow and broad.

Trunnion rings.

Trunnion rings are both rough and fine bored in vertical machines, the work being stationary whilst the boring bar revolves and moves downwards.

Centring.

Large masses are centred by the exterior which has been previously turned.

Coiled wrought iron tubes for the barrels of 64-prs. and converted guns are bored before being turned, for if on examination of the bore a defect is detected which would condemn it, the labour and expense of turning is saved.

Centring of a tube.

Previous to the first boring of any tube the axis must be found as before described, and a bearing turned off at each end, by means of which it may be truly centred in the boring machine.

---

* More correctly soap and water. This mixture acts like oil in diminishing the friction on the tool, and it is much cleaner and cheaper; moreover the water keeps the tool cold and hard, and the soap prevents the formation of rust. It is always used in turning, boring, slotting, &c. iron, steel, and bronze.

CHAP. V.

Broaching.

BROACHING HEAD.

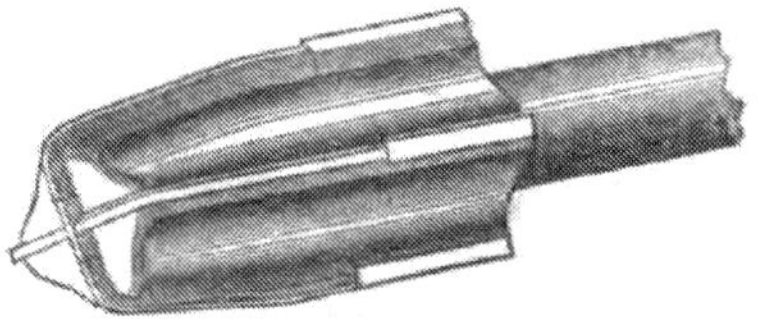

Necessity for broaching.

A broach is the name of the tool used in mechanical arts for making a taper hole or for perfecting a cylindrical one.

During the process of boring, the cutters wear from friction so that the tube is always slightly taper inside. This is of no consequence in an outer tube as the interior one can be turned to suit it, but an inner barrel must of course be truly cylindrical, hence broaching is employed to remove the taper.

Broaching-head.

The broaching-head is fitted with four long cutters fixed lengthways at right angles to each other and slightly tapering.

The cutters are carried round the front of the head and are shaped so as to finish the chamber of M.L. guns to the required form.

Process of lapping.

Lapping.

Lapping or the process of making any hard surface smooth and accurate by the application of polishing powders is used in the Royal Gun Factories for finishing off the bore of an inner barrel, for notwithstanding fine boring and broaching some little roughness and irregularity will exist. The process of rifling also often causes some of the metal on the edge of the grooves to be burred. Lapping is therefore performed both before and after rifling.

The operation consists of working (by means of levers attached to a revolving bar) a wooden head covered with lead and smeared over with emery powder and oil, backwards and forwards in the bore wherever the very accurate gauges indicate the necessity.

"Laps made of nearly every metal and alloy in common use have been more or less employed as vehicles for the application of several of the polishing powders, but of all laps those of lead slightly hardened and supplied with powdered emery stone render the most conspicuous service."*

Rifling machine.

Rifling.

One groove cut at a time.

This operation is performed by means of a rifling machine, which, though simple in its action, is not very easy to describe without using technical terms. This machine is horizontal, and the gun to be rifled is fixed in front of it, and in line with the rifling bar to which a stout head carrying the cutter is fixed.

---

* Holtzapffel's Turning and Mechanical Manipulation.

CHAP. V.

Only a single groove is cut at a time, and that as the cutter is coming down the bore bringing the chips of metal before it.

Grooves cut out roughly first.

All the grooves in the gun are first cut out roughly in succession, and then finely. The distance between the grooves is regulated by a disc fixed to the breech of the gun, having its periphery equally divided by as many notches as there are to be grooves. The gun is fixed each time by a pawl, and when a new groove has to be cut, is turned round to the next notch.

Rifling bar and head.

Rotation of the bar.

The gun remains stationary while the head carrying the cutter works up and down the bore, so it is necessary to make the bar to which the head is attached turn round more or less at the same time that it advances and returns, otherwise we should merely have a straight groove cut along the bore, instead of the spiral we require to give rotation to the projectile.

Gun metal head.

Sliding saddle.

The gun metal head in which the cutter is fixed fits the bore accurately by means of burnishers. It is fastened to a stout, hollow, iron bar termed the rifling bar. This bar is fixed to a saddle capable of sliding backwards and forwards on an endless screw. (For reversing motion of machine, see p. 98).

Motion of rotation.

Friction rollers.

Copying bar.

Angle at which set.

Alteration of angle.

Although the rifling bar is fixed to the saddle and moves with it, it can revolve independently of it, and towards the end furthest from the gun is fixed a pinion which gears into a rack sliding in the saddle at right angles to the bar itself; the outer end of this rack is fitted with two small rollers or friction wheels which run along a copying bar fixed to one side of the rifling machine; this copying bar is inclined at a certain angle to the side of the machine, and the greater this angle, the more the rack is pulled out by the friction rollers, and the greater the twist given to the rifling bar, and so to the grooves in the gun. The angle can be altered if required, and we can also take away the straight copying bar and use a curved one, as is done when a gun is to be rifled with increasing twist. By thus changing the copying bars, or their position, we can use a single machine for any description of rifling.

RIFLING HEAD.

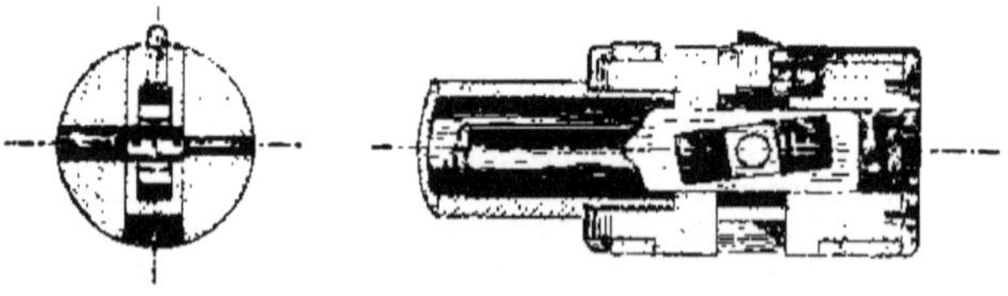

Cutting tool.

The cutting tool itself is of steel and works in and out of the head being drawn in or forced out by means of a cam attached to one end of an iron rod, passing through the hollow rifling bar.

Regulating motion of cutting tool.

The movement of the slide to which the outer end of the rod is attached (and consequently of the cutting tool) is regulated by another copying arrangement on the other side of the machine. This arrangement consists of two horizontal bars, one higher than the other, along which travels a weighted lever attached to a pinion which works the slide. When the rifling head is passing down the bore this weighted lever travels along the upper bar, but when the machine is reversed the lever is prevented by a small moveable piece from returning on the same bar, so that the weight falls over on the lower one, and in doing so draws back the slide and spindle and forces the tool out. By varying the form of the upper surface of this lower bar the depth

Varying depth of part of groove.

CHAP. V.

of the various parts of the groove can be regulated and altered as required.

Deepening groove.

To deepen the groove the outer end of the rod has attached to it a small wheel. the turning of which forces the bar forward and raises the cutting tool a certain distance. Therefore on the cutter being forced out as before, its height will be raised.

Calculation for copying bars.

In the case of a uniform spiral, the copying bar is straight edged, and the angle is easily calculated thus:—

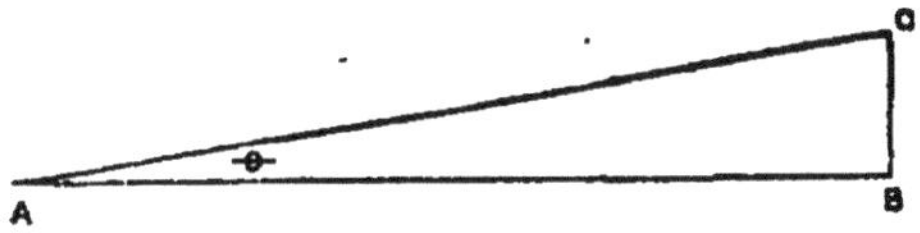

(1.) Uniform rifling.

Take AB equal length of rifling due to one turn, *i.e.*, the distance travelled by the projectile while it makes one complete turn upon its axis, and BC at right angles to it equal length of one turn, or the circumference of the bore, then AC will be the total length of spiral, and θ the angle of the rifling and of the tangent bar. Let n=number of calibres in which the shot makes one turn.

$$(a)\ \tan\theta=\frac{BC}{AB}=\frac{\pi\times\text{calibre}}{\text{number of calibres}\times\text{calibre}}=\frac{\pi}{\text{number of calibres}}=\frac{\pi}{n}.$$

For example, take the 7-inch gun, whose spiral is one turn in 35 calibres, then

$$\tan\theta=\frac{\pi}{n}=\cdot0897;$$
$$\therefore\ \theta=5^\circ\ 4'\ \text{nearly}$$

For shunt rifling, as in the 64-prs., there must be a corresponding shunt on the copying bar.

(2.) Increasing rifling.

For an increasing twist the edges of the bar must be curved accordingly; and as it is the property of the parabola to increase uniformly, we have adopted it as the curve of the rifling, for those guns having an increasing twist, its well known properties enabling us to construct without difficulty a copying bar for any required spiral.

Thus, suppose we want to make a copying bar for the 9-inch gun, which has an increasing twist from 0 at the breech to 1 in 45 at the muzzle, and in which n=45 and l the length of rifling=107·5 inches.

In the figure below let O represent the origin of co-ordinate axis, or that part of the bar corresponding to the bottom of the bore.

The distances measured along OM will correspond to abscissæ x, and those at right angles to the same to the ordinates y.

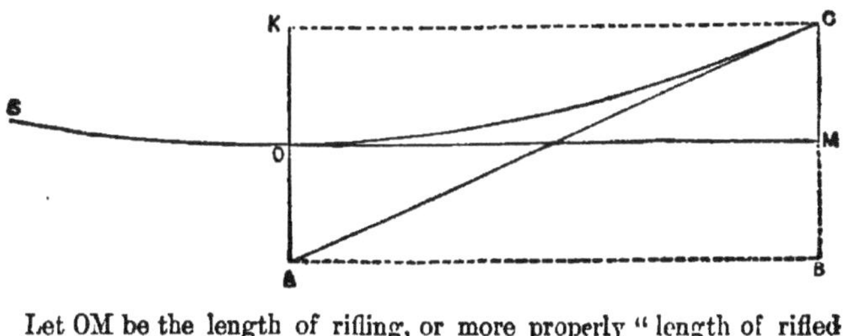

Let OM be the length of rifling, or more properly "length of rifled portion of the bore," and OC the curve required, or the absolute groove laid out, then MC or x at the muzzle is found as follows:—

$$* \; x=\frac{\pi y^2}{2nl} \text{ and at muzzle } y=l$$

$$\therefore \text{MC}=\frac{l}{2n}=\frac{3{\cdot}1416\times107{\cdot}5}{2\times45}\text{ inches}=3{\cdot}75\text{ inches.}$$

Preparation of copying bar.

In order to find the height $pq=x$ for any point on the curve, we proceed in a similar manner. Suppose $\text{O}p=y=50$ inches, then to find the height $pq=x$ of the curve at the point p we have $pq=x=\frac{\pi y^2}{2nl}=\frac{3{\cdot}1416\times(50)^2}{2\times45\times107{\cdot}5}={\cdot}81$ inch.

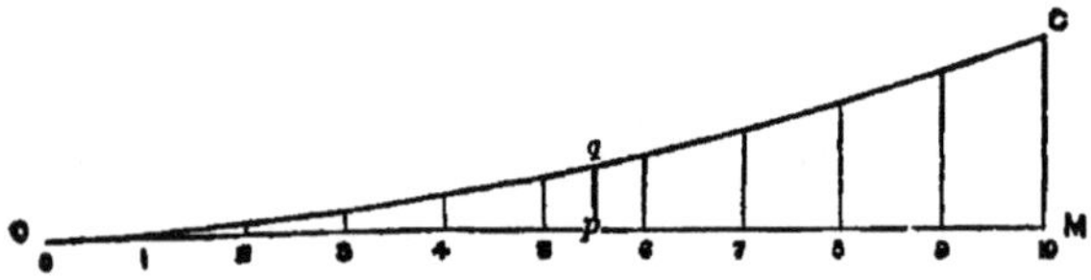

By this means the height of the curve can be calculated for any number of points, and by means of the various measurements the curve is traced on a steel bar, cast for the purpose, placed with one end resting on a plane table, and the other end supported at a height from the table equal to MC, after which the bar is planed or slotted to the proper size and shape.

If the pinion moved by the sliding rack and the bore of the gun are of the same diameter, the curves of the grooves and of the copying bar are the same; but if, as is generally the case (for, excepting the copying part, the same machine answers for all calibres and systems), the pinion and bore are of different sizes, the curvature of the bar will differ proportionately from that of the grooves. Thus, if the pinion is half the size of the bore, the bar must have half the curvature of the grooves.

Where the grooves commence, and how decided.

The grooves commence at a point about an inch below where the bottom studs on the projectile would reach with the lowest charge

* This is readily proved as follows: let us assume that OC is the required parabolic curve, and let OM drawn through the origin of the parabola perpendicular to its axis, represent the length of the bore to be rifled, $=l$ suppose. The tangent at C makes an angle with OM or AB equal to the angle of rifling at the muzzle, and cuts off (by the property of the parabola) OA on the axis equal to the abscissa OK.

$$\therefore \text{CM}=\text{MB}=\frac{\text{BC}}{2}$$

But from equation (a) p. 116.
$$\frac{\text{BC}}{\text{AB}}=\frac{\text{BC}}{\text{OM}}=\frac{\pi}{n}$$
$$\text{BC}=\frac{\pi\text{OM}}{n}$$
$$\text{BC}=\frac{\pi l}{n}$$
$$\text{and CM}=\frac{\text{BC}}{2}=\frac{\pi l}{2n}.$$

Further, as by the property of a parabola $\frac{x}{y^2}$ is a constant
$$\therefore\frac{x}{y^2}\text{ (for any point)}=\frac{\text{CM}}{\text{OM}^2}=\frac{\pi}{2nl}$$
$$\text{and } x=\frac{\pi y^2}{2nl};$$

By assuming different values for y, that is, different points at which we wish to calculate the curve of the bar, and substituting the proper values of π, n and l, we can find the x for that point.

CHAP. V.

employed, and are so placed that the two next the vent shall be equidistant from it.

The dimensions of the grooves and lands, and the twist of rifling for each nature of gun will be found in the tables (pages 160 and 291), whilst a full-sized drawing of section of the grooves is given under the description of the guns in the service in Chapter IX.

Drilling.

The cutting portion of the tool termed a drill is flat and shaped like a broad V.

Small holes such as those for friction tube pins, &c., are drilled by hand, the tool being turned by means of a brace or ratchet lever; but larger holes such as those for sights and vents are made with a drilling machine, the shape of the cutter is however the same.

The radial drilling machine.

The self-acting radial drilling machine is an ingenious piece of mechanism. The radial arm or "jib" carrying the drill spindle turns upon a bracket which can be raised or lowered as required. The motion to turn the drill spindle is derived from a cone-pulley behind the sliding bracket through mitre gear. The horizontal adjustment of the drill slide is done by means of a screw turned by a handle at the end of the arm. The vertical feed of the spindle is given by separate mitre gear in connexion with the horizontal shaft, as well as by a rack and pinion which can be worked by hand independently.

In drilling the holes for the vent bush and sights of guns, this machine is very convenient as the gun can be placed anywhere within the area covered by the radial arm, and the drill accurately adjusted in the proper position without moving the gun.

Screw Cutting.

Taps for screw-cutting.

A small internal screw thread, like that for vent bushes, is generally made by hand with taps.

A tap is a steel tool converted from a screw by the removal of part of its circumference in order to give the exposed edges a cutting action, and allow room for the chips to escape, while the circular parts which remain guide the instrument within the helical groove or hollow thread it is required to form. In the Royal Gun Factories, three or four taps gradually increasing in size, are generally employed to make a screw thread, but mechanics often use only one taper tap.

A small external screw is cut by means of a screw die.

The machines.

Large external and internal screw threads, such as the male and female cascable thread, are executed in screw-cutting lathes, in which the work revolves and the tool moves forward. The guide screw which governs the sliding motion of the tool is parallel to the axis of the work when fixed in the machine and connected with it by means of toothed wheels. The screw upon the work will be the same pitch as the guide screw if both revolve at the same rate, and this will be the case if there are only two change-wheels (or a train whose value is 1) with the same number of teeth in each. If it is desired to cut a screw with half the pitch of the guide, the work must be made to revolve twice as quickly as the latter, and so on for a thread of any pitch; each lathe is therefore supplied with a series of change-wheels, in order that a screw thread of any pitch required may be cut in it.

Slotting and Planing.

When the edges or surfaces of iron work cannot be readily reduced to proper size and shape by filing, a slotting or planing machine is employed as circumstances warrant. For example, the sides of a trunnion ring are planed, but its belt between the trunnions (which cannot be brought to shape in a turning lathe) is slotted, and the usual way of making any non-circular hole—such as the vent slot—is first to drill one or more round holes, and then slot or pare the edges to shape.

Slotting and planing machines.

Every slotting and planing machine is self-acting, that is, it has a double action, by means of which the cutter and the work move reciprocally, so that a fresh cut is made every time. Thus in the slotting machine (which cuts vertically), while the tool is being moved up, the work is advanced just sufficient for another slice, and in the planing machine (which cuts horizontally), while the work returns to be cut the cutter slides across the necessary distance.

In the large machine used for slotting the surface of the gun between the trunnions, the motion of the tool is obtained by means of a peculiar form of eccentric, giving a quick return. The table upon which the work rests can be fed either longitudinally, transversely, or circularly and the feed is obtained from a horizontal groove cam on the spindle which acts through a series of levers upon spur-wheels and screws in connexion with the table.

In the planing machines used in the department the reversing motion is similar to that used in the boring and rifling machines, but bevel-wheels of different sizes being employed, the table travels faster on the return than when moving forward against the cutter, thus economising time.

Viewing and Gauging.

The examining or viewing is a branch in itself, with foremen for each class of work, whose constant duty it is to examine every article carefully at the close of, and sometimes during each operation, to ascertain that it is manufactured properly, and to the correct dimensions.

Examinations during the several stages of manufacture.

Thus the flat bars which form a pile are examined as to manufacture and size before they are heated for rolling; the rolled bar is examined before it is welded to others; the long bar is examined before and after it is coiled; the coil after it is welded, faced, recessed, bored, turned, &c.

Working drawings.

Every gun, &c., is made from a working drawing, which shows the dimensions of the various parts, which latter must be exactly according to a sealed pattern deposited in the pattern room.

No manufacturing limits are supposed to be allowed; any departure from the correct dimensions is noted, and if considered of consequence is submitted to the Superintendent, who passes or rejects it as he may think fit.

The bore and internal surfaces, which have to be shrunk together, are measured by means of gauges and Whitworth's micrometers to $\frac{1}{1000}$th of an inch, which is the smallest measurement taken in the Department.

Gauges for interior.

A deviation of this amount in the bore can be directly ascertained by a bar gauge, provided with a self-registering micrometer scale. The bore of a B.L. gun is also tested at different parts with high and low cylindrical gauges. The former must not enter, the latter must pass through. Thus there are three pairs of gauges for a B.L. gun, viz., for the powder chamber, shot chamber, and the grip. The cylinder

CHAP. V. gauge for M.L. guns has studs* similar to those on the projectiles in order that the grooves may also be tested. From the care taken during the manufacture to have the inner barrel and all the super-imposed parts properly centred in the boring and turning lathes, the bore of a built up gun must be concentric with the exterior, and the trunnions must be in their proper position.

External gauging.

External dimensions are taken with callipers and measuring rods, but in this case $\frac{1}{100}$th of an inch is considered close enough.

After proof, guns are very carefully examined; this process will be found fully described at pp. .

Before any gun or article, which is a separate store, is issued for service, it is examined by the Assistant Superintendent, who is responsible for its exact accordance with the sealed patterns.

The sealed patterns.

These standard patterns are sealed by the Director of Artillery and Stores for the War Office, to govern future supplies; and no one is allowed to touch them but the officials of the R.G.F. in whose charge they are.

* The dimensions of the studs being about $\frac{1}{100}$th inch greater than the high gauge of studs on the projectile, and between $\frac{3}{100}$ths and $\frac{4}{100}$ths inch below the correct size of the groove itself. Vide table, next page.

TABLE XII.

TABLE showing the DIMENSIONS of R.G.F. CYLINDER GAUGES in comparison with those of BORE and GROOVES and R.L. SHOT GAUGES.

—	80-Ton.		12·5-Inch.		12-inch.		11-inch.		10-inch.		9-inch.		8-inch.		7-inch.		80-pr.		64-pr.		8-Inch Howitzer.		Mean Differences.
	Bore or Body.	Groove or Studs.	Bore or Body.	Groove or Studs.	Bore or Body.	Groove or Studs.	Bore or Body.	Groove or Studs.	Bore or Body.	Groove or Studs.	Bore or Body.	Groove or Studs.	Bore or Body.	Groove or Studs.	Bore or Body.	Groove or Studs.	Bore or Body.	Groove or Studs.	Bore or Body.	Groove or Studs.	Bore or Body.	Groove or Studs.	
Diameter of gun ...	16·000	16·400	12·500	12·900	12·000	12·400	11·000	11·400	10·000	10·400	9·000	9·360	8·000	8·360	7·000	7·360	6·300	6·590	6·300	6·530	8·000	8·360	
Diameter of R.G.F. cylinder gauge.	—	—	12·470	12·8775	11·965	12·377	10·970	11·3775	9·665	10·377	8·965	9·337	7·965	8·337	6·965	7·337	6·260	6·5575	6·265	6·4975	7·965	8·337	
High ring or cylinder gauge issued by R.L.	—	—	12·420	12·850	11·955	12·365	10·955	11·365	9·955	10·365	8·955	9·325	7·955	8·325	6·955	7·325	6·220	6·530	[6·220	6·470	7·955	8·325	
Difference between gauge of gun and grooves and that of high gauge studs and shot.	—	—	0·080	0·050	0·045	0·035	0·045	0·035	0·045	0·035	0·045	0·035	0·045	0·035	0·045	0·035	0·080	0·060	0·080	0·060	0·045	0·035	{ 0·045 { 0·035
Difference of diameters of gun and R.G.F. cylinder gauge.	—	—	0·030	0·0225	0·035	0·023	0·030	0·0225	0·035	0·023	0·035	0·023	0·035	0·023	0·035	0·023	0·040	0·0325	0·035	0·0325	0·035	0·023	{ 0·035 { 0·023
Difference between R.G.F. cylinder gauge and R.L. high ring or cylinder gauge.	—	—	0·050	0·0275	0·010	0·012	0·015	0·0125	0·010	0·012	0·010	0·012	0·010	0·012	0·010	0·012	0·040	0·0275	0·045	0·0275	0·010	0·012	{ 0·010 { 0·012

CHAPTER VI.

RIFLED BREECH-LOADING GUNS, THEIR MANUFACTURE, FITTINGS, AND STORES.

SECTION I.—MANUFACTURES.

R.B.L. Guns, original Construction.—Manufacture of 40 pr., 35 cwt.—Rifling.—Breech fittings.—Copper rings.—Chambering.—Bushing.—Vent-piece.—Breech-screw.—Tappet ring.—Lever and keep pins.—Indicator ring.—Proof of B.L. guns and fittings.—Processes after proof.—Lining.—Sighting.—Marking.

SECTION II.—SIGHTS, SIGHTING, AND STORES.

Tangent Sights, L.S and S.S.—Rectangular.—Hexagonal.—Barrel-headed.—Sliding leaf.—Graduations.—Set screws.—Moveable clamp.—**Trunnion Sights.**—Screw, and their adjustment.—Wood side scales.—Bearers, shot.—Bits, vent, Armstrong.—Bushes.—Clamps, moveable.—Collars, leather.—Crutches, iron.—Extractors, tin cup.—Eyes, elevating.—Implements, facing.—**Instruments taking impressions.**—Instruments, sighting.—Levers.—Machine, hand-rifling.—Patch, metal, elevating.—Pins, friction tube.—Pins, keep.—Pivot, steel for elevating arcs.—Plates, metal elevating.—Rings, copper vent-piece.—Rings, indicator.—Rings, tappet.—Saddles, metal.—Scale, wood side.—Screws.—Sights.—Sockets, metal.—Straight edges.—Vent-pieces.—Wrench, fixing elevating racks.—**Tables** of Fittings, Sights, and Stores, and of B.L. guns with steel or W.I. barrels.

SECTION I.—MANUFACTURE.

No R.B.L. guns have been manufactured since 1864, but we have so many yet in the service that all gunners should thoroughly understand their mechanism and know all about their stores and the repairs at times necessary, both for the guns and their fittings.

All R.B.L. service guns have been made on Sir W. Armstrong's original construction (as mentioned at Chapters II. and IV., pp. 28, 83), and are manufactured entirely of wrought iron * with the exception of a certain number having steel tubes (vide table, p. 143).

R.B.L. Guns of original construction.

This original construction, as already stated, consists of an inner barrel, a forged breech-piece, a forged trunnion piece, and one or more coils, according to the size of the gun. For example, the 6-pr. has one coil, and the 7-inch of 82 cwts. has six coils.

* The 3-pr. R.B.L. gun submitted by Sir W. Armstrong in 1855 had a steel barrel, but that material was abandoned owing to the difficulty of getting it of suitable quality at that time.

As a specimen of B.L. manufacture we may take the 40-pr. of 35 cwts. as shown in the diagram below. CH. VI. S. 1.

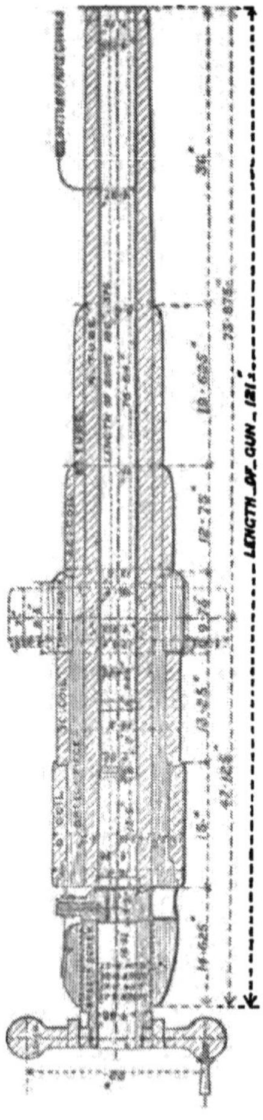

WROUGHT IRON B.L. GUN (BREECH-SCREW) 40-PR., 35 CWT., MARK I.

Scale $\frac{1}{24}$.

The inner barrel, or A tube as it is called, was made of coils joined together lengthways in the manner explained at p. 108, but a number of guns (vide table, p. 143), especially 7-inch 20-prs. and 12-prs., have inner barrels made of solid forgings, *i.e.*, solid cylinders of wrought iron afterwards bored out. These were adopted owing to the difficulty at one time experienced of making a coiled barrel free from defective welds, but they were soon given up, for although they presented a clean inner surface, the fibre which ran lengthways was in the worst direction for circumferential strength.

Solid forged barrels.

Why discontinued.

The breech-piece and trunnion ring were forged, and the number of coils required made, and the gun then built up by shrinking these parts

Breech-piece. Trunnion ring.

CH. VI. S. 1.

Shrinking the parts together.

over the A tube. In these guns the trunnion ring was not joined to other coils by forging, as in R.M.L. guns of present construction, but shrunk on separately.

Turning, &c.

After the gun was built up it was brought to the proper size and shape by turning, boring, &c., and the piece so completed, all but the rifling, and the preparation of the gun for breech fittings.

In this stage of manufacture we might close up the open breech end and make a muzzle loader out of the built up blocks, and indeed we know that certain blocks prepared for R.B.L. guns were afterwards so used for making R.M.L. 64-pr. guns, Mark I. (vide p. 81).

Rifling.

For the service R.B.L., however, the next operation was to rifle the block.

Grooves.

The groove is of the simple shape shown below, the driving side being radial to the bore.

SECTION OF RIFLING (Full size).

The number of grooves is considerable, and hence the system is usually called a poly-grooved or many-grooved system.

Lands.

The lands, or portion of the bore between the grooves, were made as narrow as possible so as to cut the more easily through the lead coating on the projectile.

Modes of Closing the Breech.

The two modes usually employed for closing the breech end are; (1) to press a block of metal firmly against the end of the barrel by means of a screw, or (2) by means of a wedge forced in behind it.

Vent-piece.

Breech-screw.

All our service B.L. guns use the first mode where we have a metal block called a vent-piece and a breech-screw.—In the wedge guns, however, which are now in reserve (p. 81) a stopper and wedge were employed instead.

Obturator.

Gas-tight joint or obturator.

In connexion with the vent-piece or wedge it is further necessary to have some arrangement for making the joint, where it meets the end of the chamber, quite gas-tight. As will be seen, we use for this purpose two copper rings (one on the vent-piece and another screwed into the gun),* or in the case of the 7-inch a ring of wrought iron and a tin cup.

Preparation for Breech Fittings.

Vent slot.

Water escape.

After being built up, the gun was prepared for the breech apparatus by having a slot cut for the vent-piece, and underneath this a circular hole, called the "water-escape" or "drip hole," was drilled through the gun. The breech end of the A tube was threaded for the breech-bush to be screwed in, while the end of the breech-piece was prepared in the same way for the breech-screw.

Chambering.

Grip.

Finally, the gun was chambered, *i.e.*, the powder chamber and shot chamber were bored out. The diameter of the first is greater than that of the shot chamber, which is also slightly conical. Immediately in

* The ill effect of a worn gas-tight joint is well shown on vent pieces which have suffered from neglect, through not having had the copper rings properly faced.

front of the shot chamber is the "grip," which is the smallest diameter of all, or calibre of the piece, in order that the projectile, which has a thin lead covering, shall, as it begins to move, have this covering compressed into the grooves of the rifling. From the grip to the muzzle the bore is slightly increased in diameter, to reduce the strain on the gun, until at the muzzle the bore is about 0·005 inch larger than the full calibre.

Bushing.

After chambering, the gun was bushed by screwing into the end of the barrel the ring or breech-bush already mentioned. This bush was faced to correspond to a ring on the vent-piece itself.

The breech bush can be easily repaired and refaced, and when worn out replaced by a new one. It is of copper, except in the 7-inch, which, though originally bushed with copper, are now bushed with iron.

Why tin cups are used.

With 7-inch guns, in order to make certain that there shall be no escape of gas when the breech-screw is home, a tin cup is used to complete the joint instead of a vent-piece copper ring. Copper bushing for these guns was not found suitable owing to liability to damage in so heavy a gun, and the copper bush was superseded by an iron bush of the same dimensions, 2" long, ½" thick; but as this occasionally shifted from want of sufficient bearing, a new pattern iron bush 3" long, ¼" thick was introduced. Guns bushed with this latest pattern * are called double bushed, and are marked D.B. on the right trunnion.

7" double-bushed guns.

Breech Fittings.

The breech fittings for breech-screw guns are:—

Vent-piece.	Lever and keep pins.
Breech-screw.	Indicator ring.
Tappet ring.	

The Vent-Piece.†

Vent-piece.

So called because the vent ‡ happens to go through it, is the block or stopper which closes the end of the breech before firing.

Vent channel.

Vent bushes.

The vent channel passes down through the neck to the centre of the block, and this vertical portion is partially bushed with copper; from the bottom of this portion a horizontal channel allows the fire to reach the centre of the cartridge. The bottom part is not bushed, as the vent-piece would be too much weakened thereby.

Material.

The material of vent-pieces has been changed frequently; wrought iron first, then steel, then Swedish iron, then steel toughened in oil, and lastly, Marshall's refined iron, have been employed. The material of which they are made is stamped on the back.

---

* Prior to the introduction of this new bush all 7-inch guns had been bushed either with thick iron or copper bushes. It is advisable to rebush with the thick iron bushes of O.P. any guns still having the copper bush, and should guns at out stations not be prepared for the double bush, and therefore not be marked D.B. on trunnion they must, if rebushed at an out station, be furnished with a new bush of the old pattern, as a gun can only be prepared in the first place for the latest pattern bush in an Arsenal. Vide p. 316. Guns marked D.B. would of course be rebushed with N.P. bush.

† This is not of necessity a part of this B.L. system. The fact of the vent being drilled through the vent-piece weakens the latter much, and is an expensive arrangement. At the time when R.M.L. guns were introduced experiments were being carried on as to venting R.B.L. guns through the body of the piece, with favourable results, but as shortly afterwards the manufacture of R.B.L. service guns ceased, it was not thought worth while to make any change in the system of venting, which would no doubt have otherwise been done.

‡ For weights of vent-pieces, vide p. 144.

CH. VI. S. 1.

The vent-pieces of all natures except the 7-inch and 40-pr. are lifted away from the gun after each round, while the former are placed either on the saddle or vent rest.

Component parts.

A vent consists of—

Body
Vent bush, copper (in two parts).
Vent-piece, copper ring (except 7-inch).
Cross-head.
Shackles.

There is a "beak" on the 7-inch and 40-pr. vent-piece, to prevent the "nose" on the face of the vent-piece being injured when the vent-piece is placed in the slot.

Bushing with copper.

The vent channel is bushed with copper at the upper end. The bush formerly consisted of three or four short plain pieces and one screwed piece on the top, but now one long plain piece is used instead of the short ones.

Copper ring.

The vent-piece next received the copper ring (vide p. 124), required to give a gas-tight joint when fitted against the copper bush. The face of the vent-piece itself has an annular undercut groove, into which the ring was forced.

SECTION OF RING.

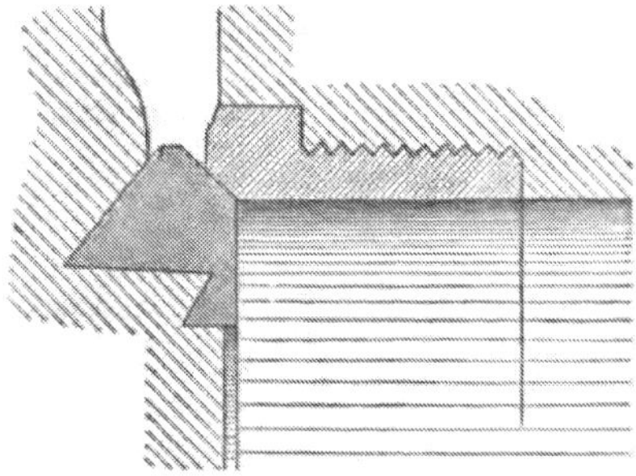

The ring was made from pure hardened copper, and has a dovetail corresponding to the groove in the vent-piece. In order to put * it on the vent-piece it was put into a screw press and the copper ring forced on, a small amount being left projecting to allow for refacing on service (vide p. 306).

Cross-head and shackles.

The cross-head and shackles were not put on till after proof. The cross-head is made from a block of scrap iron, and is so shaped that its projections rest on the slot when the vent is properly placed for being screwed up. Great pains were taken to get this correct to prevent the possibility of the vent-piece going wrong, so that it should rest on the slot whilst the vent-piece was being screwed up.

Vent-pieces without protection at the back.

The material, date of manufacture, pattern, and R.G.F. number are marked on the back; and the order is, that the latest pattern available should be used, while vent-pieces without a projection at the back should on no account be employed.

* On service, the breech-screw is utilized for this purpose, vide p. 309.

OBSOLETE VENT-PIECE. Scale 1 in. = 1 foot.

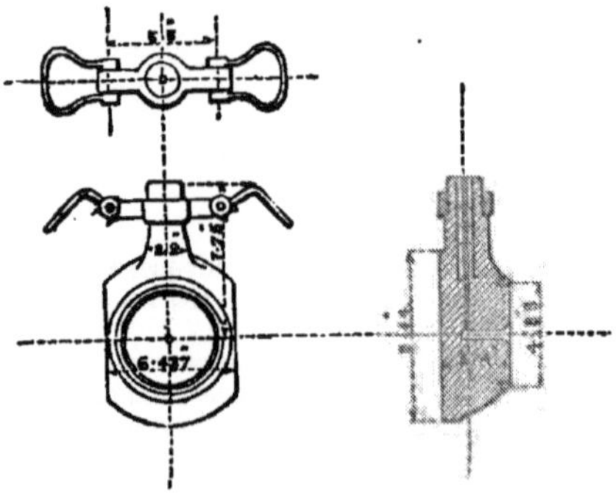

SERVICE VENT-PIECE.* Scale 1 in. = 1 foot.

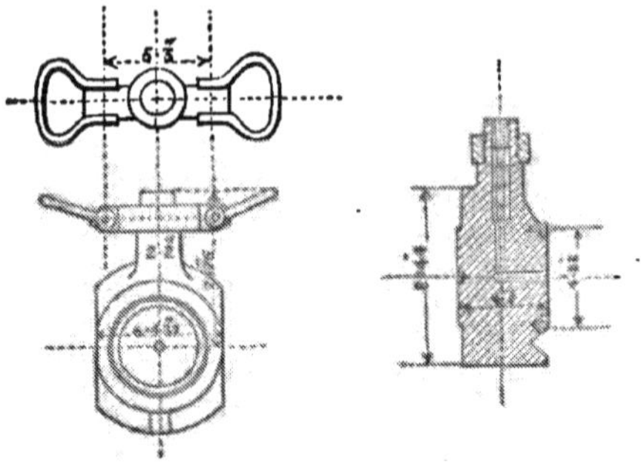

*Breech-screw.**

Is made of steel, toughened in oil for all B.L. guns except the 7-inch, which has an iron breech-screw with a steel face 6″ long screwed into it, an iron face being liable to "set up." The thread is called the "*V* bevelled" from the shape of its section, which is less liable to be jammed by grit or dust and also more easily worked than the square thread used in the early Armstrong guns. Some of the square thread breech-screws are still to be found in use. The ends of the thread are not left sharp but cut off at about half their thickness. The 7-inch breech-screw has a double thread; this allows of the screw being moved just twice as quickly, while giving the same amount of bearing surface. Breech-screw. Material. Thread.

The pitch of thread on the 40-pr. 32 and 35 cwt. are not the same, the former being 0″·9, the latter 0″·7, and consequently their breech-screws are not interchangeable for the same nature of gun. Pitch of thread.

The screws are marked with the nature of gun (but not for any Marks.

* For weights vide p. 144.

CH. VI. S. 1.

individual gun), with the register number, and also with the material of which they are made

The bore of the screw.

The bore of the screw is slightly less than that of the powder chamber, in order that the breech bush should not be damaged by the projectile when loading.

40-pr. Breech-screw, Tappet Ring, Levers, and Keep Pins.
Scale $\frac{3}{4}$ in. = 1 foot.

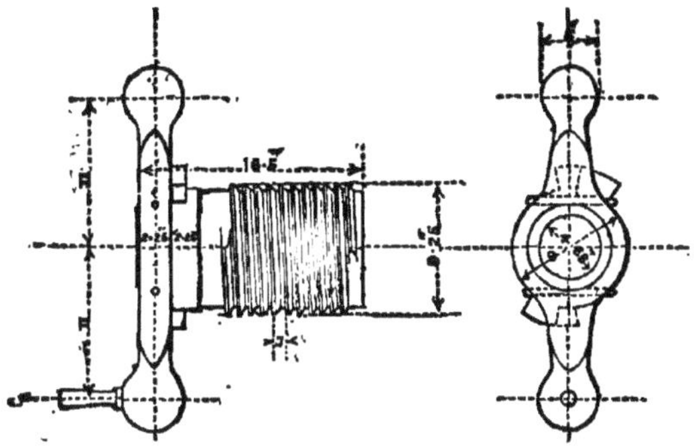

Tappet Ring.

Tappet ring.

Is octagonal in shape in the interior, and fits on a similar octagon on the breech-screw; hence it acts as a wrench to the breech-screw, the power being communicated through its projections from the tappets of the lever.

Lever and Keep Pins.

Lever and keep pins.

The lever fits on the breech-screw behind the tappet ring; it is free to revolve round the breech-screw, but is prevented from falling off by two keep pins which work in a cannelure. The lever is fitted with weight balls or accumulators, to give power in screwing up. The levers are all of wrought iron forged, the handles and tappets being forged separately. The small natures up to the 20-pr. inclusive have one handle, one weight ball, and one projection. 40-prs. have one handle, two weight balls, and two projections. 7-inch guns have two handles, two weight balls, and two projections. The levers have all a play of ·02 inch running round the breech-screw, and as there is no particular strain on them, they are not easily damaged; they are interchangeable for guns of the same calibre.

Indicator Ring.

Indicator ring.

Is a thin narrow ring of wrought iron fitted on the breech-screw in front of the tappet ring; on the internal circumference there are a series of grooves or feather ways, any one of which fits a "feather" on the breech-screw; it is so adjusted on the screw that when the vent-piece is properly screwed up, the raised line of brass or arrow on the ring and on the top end of the breech-piece must coincide, and then it can be seen at a glance whether the vent-piece is properly screwed up, or should it be dark this can be ascertained by the touch. As the copper rings and breech-screw are faced on service, the position of the raised line of brass must be altered to correspond by shifting the ring round on the breech-screw. CH. VI. S. 1.

Guns having indicator rings.

The 7-inch and 40-pr. are the only guns with which an indicator ring is considered necessary, and some 40-prs. of 32 cwt. do not use them, not having sufficient length of breech-screw.

RINGS, INDICATOR (for 7" B.L. Gun).

The shaded portions show the indicator ring.

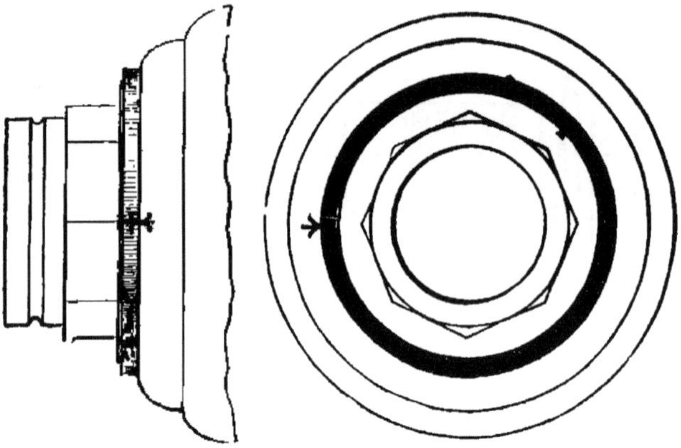

Top view, tappet removed. Rear view, tappet removed.

Proof of B.L. guns and fittings.

Proof.

A.B.L. gun was proved without its breech closing apparatus (spare being used) by firing six rounds of 1½ the service charge, and the service shot.

Vent-pieces and breech-screws underwent two service rounds.

After Proof.

The guns having passed proof were next marked, lined, sighted, &c., prior to issue.

Marks on the gun.

The weight of the gun, Royal monogram, and broad arrow are stamped in front of the vent slot, and the "mark,"* name of factory, date of proof, and register number on the left trunnion.

Sighting.

Vertical and horizontal lines were marked on the breech and muzzle, for the purpose of enabling the sighting plates to be adjusted (vide p. 310). Vertical and horizontal lines were also marked on the right side of the breech and right trunnion, and right side of the muzzle; but the field guns have not these latter marks.

---

* R.B.L. guns may be found on service with the "mark" omitted on the trunnion; this is accounted for owing to the order not being in existence as to their being so marked at the time of issue.

SECTION II.—SIGHTS, SIGHTING, AND STORES.

Sights.

Our B.L. guns being sighted on both sides have four sights, viz., two tangent sights* and two trunnion sights.

Process of Sighting.

Sighting.

In all natures, except the 7-inch, as the amount of metal at the breech end is small, the tangent sights work in sockets bored in a wrought-iron ring screwed on the breech of the gun, so that the sockets are inclined at an angle of 2° 16′ to the left, to compensate for the permanent deflection.

Permanent deflection.

Sockets.

In the 7-inch, holes are drilled in the metal of the breech of the gun itself at that angle, and in them are fixed gun metal sockets for the sights, similar to those described at p. 188 for R.M.L. guns.

The trunnion sight holes for drop sights are drilled perpendicularly, and in guns using screw trunnion sights the holes are threaded.

Tangent sights. Rectangular. Hexagonal.

The sights used are as follows :—For all except the 7-inch of 72 cwt. and the 12-pr., the tangent sight consists of a rectangular steel bar with a barrel head, or, in the newer pattern (introduced in 1867), with a gun metal head and plain sliding leaf. With the 7-inch of 72 cwt. and the 12-prs. the bar is hexagonal and of gun metal. The reason of this difference is that these guns were the first B.L. guns introduced, and it was then thought that the hexagonal gun metal bars would answer best, and afterwards when the flat steel bars were preferred it was not considered worth while to alter the few 7-inch guns of 72 cwt. in the service, whilst the cost and trouble of altering the numerous 12-prs. would have been considerable.

Tangent sights, L.S. and S.S.

L.S. sights have under the cross-head an "elevating nut," the circumference of which is graduated from 1′ to 10′, so that by turning it, any number of minutes of elevation less than 10′ can be obtained. This is considered unnecessary for S.S., which therefore have no nut; in all other respects the L.S. and S.S. tangent sights are identical.

Many B.L. sights of the older pattern, having so-called barrel heads in which the deflection leaf is placed (vide diagram next page), still exist. The second diagram shows the more recent pattern (according to which sights have been made since 1867), where the head is plain with a sliding leaf.

---

* For description of the different marks or patterns of these sights, vide Table, p. 146.

CH. VI. S. 2.

Barrel-headed Sights, with moveable Clamp.

Scale $\frac{3}{4}''$ = 1 foot.

R.B.L. (breech-screw) 40-pr., 35 and 32 cwt.

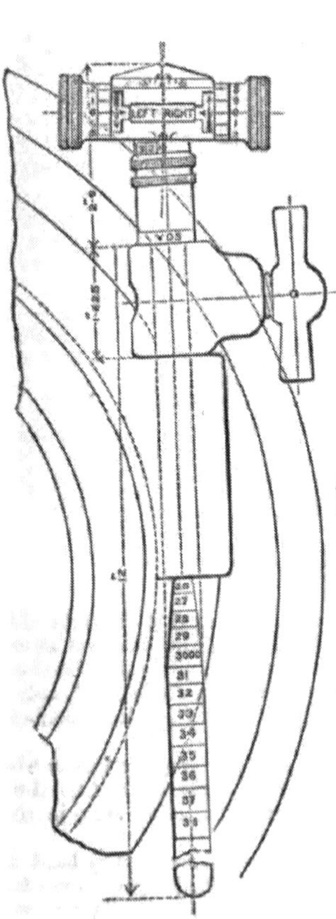

R.B.L. 7″ 82 cwt. (with gun metal socket).

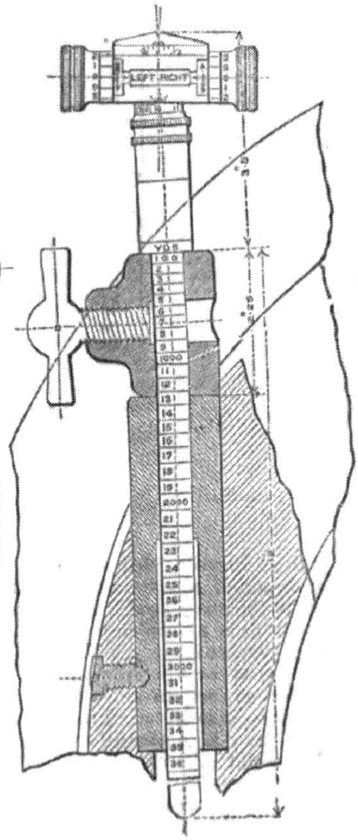

CH. VI. S. 2.

SLIDING-LEAF SIGHTS.

Scale 2½″ = 1 foot.

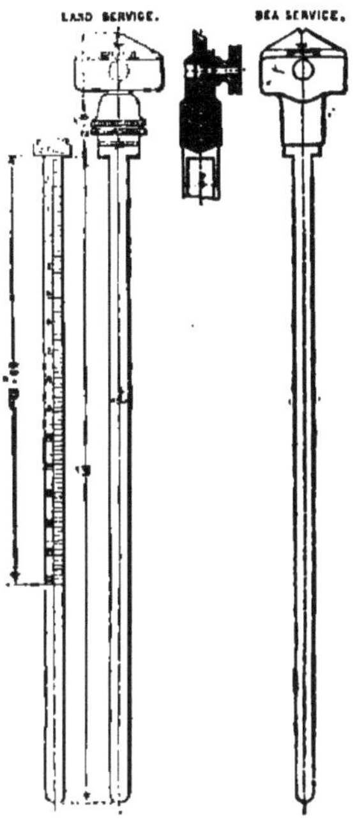

Barrel-headed sights. In the barrel-headed sights the leaf is traversed to the right or left by means of a screw worked by milled-headed thumbscrews at each end of the barrel head. This screw is of such a pitch that the thumbscrews make a complete revolution in traversing the leaf 10′, and the circumference being graduated from 1 to 10 any required number of minutes can be accurately given.

There are arrows on the barrel head showing the direction in which the screw is to be turned to give right or left deflection, and the deflection scale is graduated to half a degree on each side, in three divisions of 10′ each.

Sliding-leaf sights. In the present sliding leaf pattern the leaf is traversed by hand and clamped in the required position by a milled-headed screw on the front or muzzle side, and, though not quite so accurate as the old pattern, it has the advantages of lightness and cheapness, and is not so liable to become stiff in working. It is only necessary to use the deflection scale in order to allow for wind or similar causes of irregularity.

Graduation of tangent bars. The steel tangent bars are graduated on one of their narrow sides in degrees, and on the other in yards. Those for the 7-inch guns of 82 cwt. have also a graduation on one of their flat sides (right), showing the number of tenths of fuze corresponding to the range scale. Each degree is divided into six divisions of 10 minutes each.

Copper set screws. In the 20-pr. and lower natures the tangent sights are clamped at the proper elevation by means of a copper set screw, secured by chains to the piece. In the other two natures* the sight is fixed by a "moveable clamp,"† which permits of the sight being removed from the socket and taken to the light for adjustment during the operation of loading, a useful arrangement between decks or in casemates. CH. VI. S. 2. Moveable clamp.

MOVEABLE CLAMP.

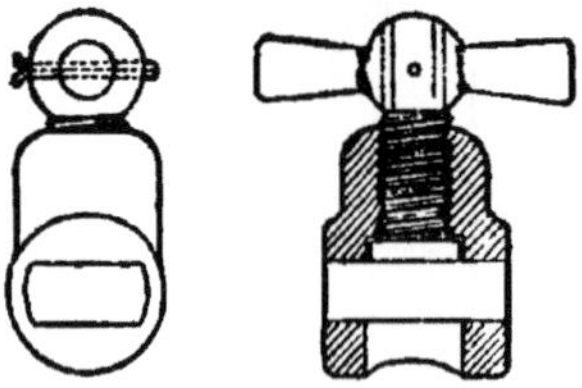

A list of different marks of sights for each nature is given at p. 146 and also in notes to the same the reasons are given on account of which alterations were made.

There are two kinds of trunnion sights, viz., the *screw* pattern and the *drop* pattern, as shown in diagram below. The former is used with field guns, *i.e.*, 12-prs., 9-prs., and 6-prs.; they are screwed into the gun so that they may not be shaken about and loosened when moving over rough ground. The latter is used with the higher natures. With the 7-inch of 72 cwts., however, a large screw fore sight is employed. Trunnion sights. Screw pattern.

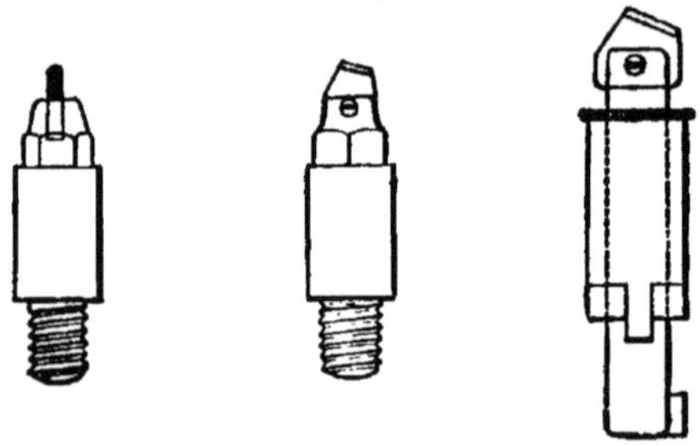

The advantages of the drop over the screw pattern are that the former can be easily removed for transit and afterwards replaced in its true position without any trouble; spare sights can also be carried ready to be placed in the gun without requiring any adjustment.‡ Drop pattern.

Spare drop trunnion sights are issued complete, and with the leaf finished, so that a new one may be put into the gun when required; but screw trunnion sights must be issued with rough leaves; hence the field guns are the only ones with which the process of adjustment has to be performed on service (vide p. 310). Adjustment.

---

* The 7-inch of 72 cwts. has its sights clamped by a fixed set screw.

† Vide note, p. 208.

‡ Screw trunnion sights are as far as possible made interchangeable, but on account of the great difficulty experienced in finishing off a thread with the very minute accuracy required in this case, some allowance would have to be made wherever there is a change of position. These sights can be used for either side, if circumstances should require it, only with the risk of a slight chance of error. With drop sights the case is different, there we have no screw, they can be finished off as desired, while any slight error in boring for the socket, &c., is made up for when the sockets are finally adjusted in the gun (see p. 181).

CH. VI. S. 2.

Both tangent and trunnion sights are marked for the nature of gun to which they belong, and all tangent sights (both L.S. and S.S.) are interchangeable for the same natures except between the L.S. and S.S. 20-prs.

Wood side scales.

Besides tangent sights, R.B.L. guns for S.S. down to 20-pr. inclusive, are supplied with wood side scales, giving 12° elevation and 6° depression, somewhat similar to those for S.B. guns. In the case of the 20-pr. 13 cwt. gun, however, when used on the decks of iron clad ships, the wood side scale is rectangular in section and has a moveable pointer similar to those employed with R.M.L. guns, and gives 20° depression and 12° elevation.

Moveable pointer.

Stores.

The stores for R.B.L. guns made in the Royal Gun Factories are as follows. A table of the same is given at the end of this chapter.

Bearers, Shot.

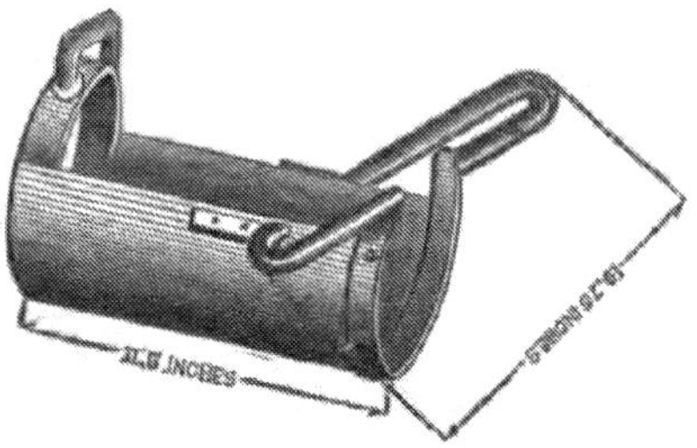

Shot bearer. § 919.

These shot bearers are of iron, with three handles covered with leather, and are so made that the projectile cannot be dropped from them. They are known as "Alderson's pattern." Two are issued for each 7″ gun.

Bits, Vent (Armstrong). Scale ½.

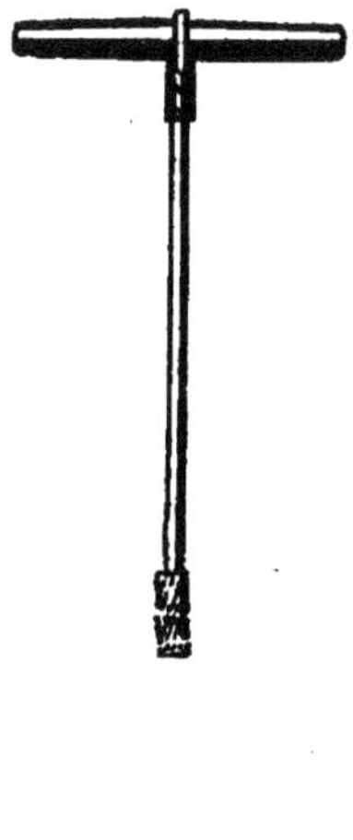

The Armstrong vent bit is of steel, with a cross-handle of wood (similar in shape to a gimlet). It is used to clear the channel of vent-pieces, should the copper bush become burred, and there is but one pattern for all natures. One to four 7″ or 40-pr. guns. For field service 1 per division. CH. VI. S. 2. Vent bit. § 842.

Bushes.

Bushes, breech, iron, thick, 7-inch.—Two inches long, half-inch thick, and is screwed into the breech end of the A tube of 7″ guns. Breech bushes § 399.

Bushes, breech, iron, thin, 7-inch.—Three inches long, quarter-inch thick, and is screwed into the thick bush. *Spare*, 1 per gun. § 467.

Bushes, breech, copper.—Screwed into the breech end of the A tubes of 40-pr. breech-screw guns and under, and projects ·08 inch so as to enable it to be refaced. *Spare*,* 2 per gun. §§ 526, 530. See also § 1145.

Bushes, copper, vent-piece.—*Spare*,* 1 to two 7″ guns; 1 to two 40-pr. for Garrison Service, and 1 for each gun for Field Service. § 1232.

Clamps, Moveable, for Tangent Sights.

The moveable clamps are of gun metal. See page 133. Clamp for sight. §§ 1144, 1357

Collars, Leather, for Breech Screws.

This collar fits on the breech-screw of 12-pr. guns, in rear of the breech. When the vent slot was widened for a thicker vent-piece some of the threads of breech-screw were exposed, and this collar protects that part. It can be pared as the copper is faced away. Leather collar § 845.

CRUTCHES, IRON. Scale ¼.

The vent-pieces of 40-prs. and smaller natures for sea service not affording room for a friction tube pin are fitted with a crutch, attached round the mouth of the vent channel by two screws. A slot is cut in the horizontal part of the crutch, through which the friction bar of the quill tube passes, thus the head of the tube is supported, and the liability to its being broken off prevented when pulling the lanyard. Crutch. §§ 27, 638.

---

* These copper articles should be carried in some place where they would not be liable to be injured, or put out of shape.

CH. VI. S. 2. With guns used for field marine purposes the crutch has two slots so as to enable them to be fired from the left side when on land.

EXTRACTORS, TIN CUP.

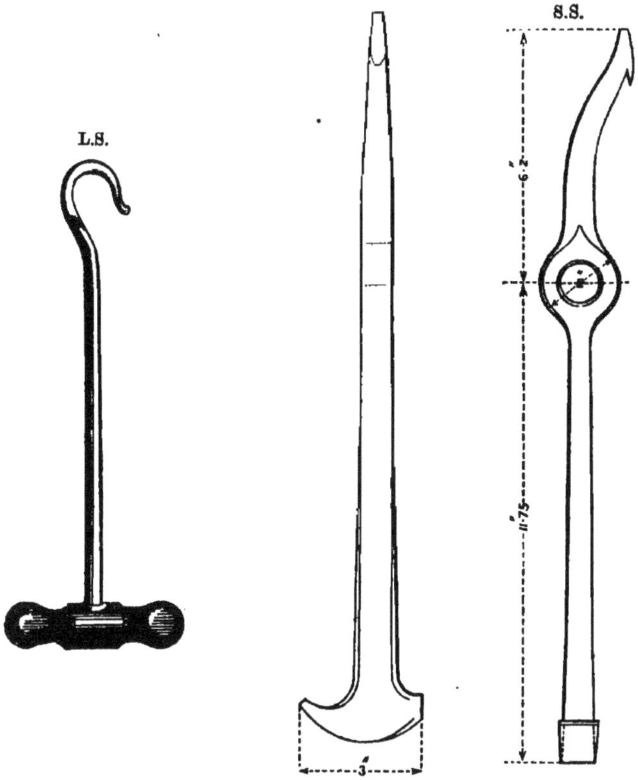

Tin cup extractor.

§ 733. The tin cup extractor for *L.S.* is an iron hook with a wooden cross-handle. One to each gun.

§ 1259. For *S.S.* it is an iron lever with a barbed hook at one end to extract the tin cups, whilst at the other end it has a curved-shape hammer used as a "lever lifting joint."

The above are only used with 7″ guns.

Eyes, Elevating (with Bolt, Washer, and Keep Pin)

Elevating eyes. §§ 173, 690, 711, 998. See also § 102.

The elevating eyes are of iron, and are screwed underneath the breech of L.S. field guns only. The 20-pr., 12-pr., and 9-pr. are double-headed, but the 6-pr. is solid-headed; the heads of the elevating screws will therefore respectively be single-headed and double-headed. CH. VI. S. 2.

GUIDE PLATES.

Scale ¼.

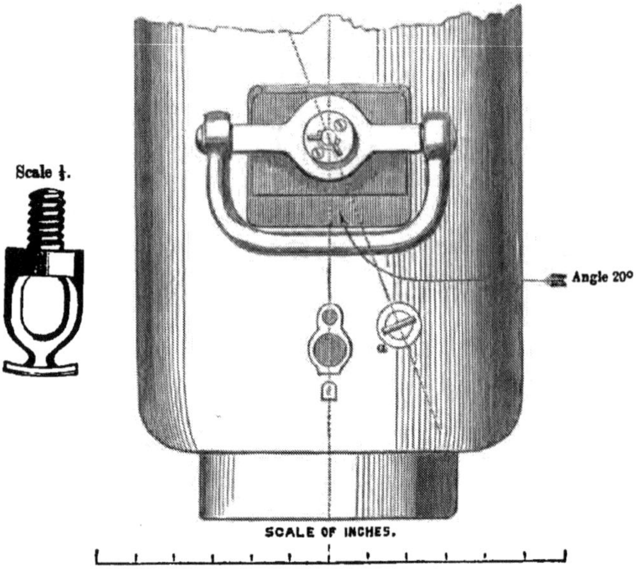

a Guide plate.

Guide plates are of steel, and one is screwed in at the right rear of the vent slot to guide the lanyard—which passes through it—direct on the quill friction tube. It has a cross-head on the top, to which a loop on the lanyard can be attached when the gun is loaded, and so prevent the gun being fired accidentally. The navy alone use it, as they fire their guns from the rear immediately the object is in line, and this guide plate enables the gunner to have a steady and direct pull on the lanyard while looking over the sights.

Guide plate. **§§ 476, 688.**

Implements, Facing.

For detail of facing implements see *Table* p. 321, 7" one per district, one in reserve; 40-prs. one per battery, one in reserve; other guns one per battery. The use of these instruments is described at p. 306.

Facing implements. **§§ 515, 516, 574, 575, 745, 1073.**

Instruments taking Impressions of Bores of Guns.

For taking impressions of bores of guns there are two sets of instruments. The small set (No. 2) is used for guns up to 20-pr.; the other (No. 1), from 40-pr. upwards, both M.L. and B.L. A set consists of a semi-cylindrical iron frame, about 2 feet long, connected with an iron

Instruments for taking impressions. **§ 1312. See also § 1625.**

CH. VI. S. 2. tube in such a manner that by screwing up a rod which passes through the tube the frame can be worked up or down.

Upon this frame a gun metal or iron plate, corresponding to each calibre of gun is screwed, and when an impression is required gutta-percha is spread on the plate, and by means of the rod is pressed against the defective part.

Plates are not necessary for the 6-pr. or 40-pr. guns, as the frames answer the purpose required.

Instruments, Sighting, Set.

Sighting instruments. §§ 1061, 1096.

Sighting instruments are for special issue only. See Ordnance Select Committee Proceedings, 1867.

Levers.

Levers.

Levers, breech-screw.—See page 128.

LEVER, IRON, FOR RELEASING VENT-PIECE. Scale 1 in. = 1 foot.

§769. See also § 1484.

Lever, iron, releasing vent-piece.—Is an iron crowbar about 2 feet 10 inches long, for prizing out the vent-pieces of 7-inch guns when they jam. One per gun.

MACHINE, HAND-RIFLING (for 6-pr. B.L. Gun. Scale 1½ in. = 1 foot).

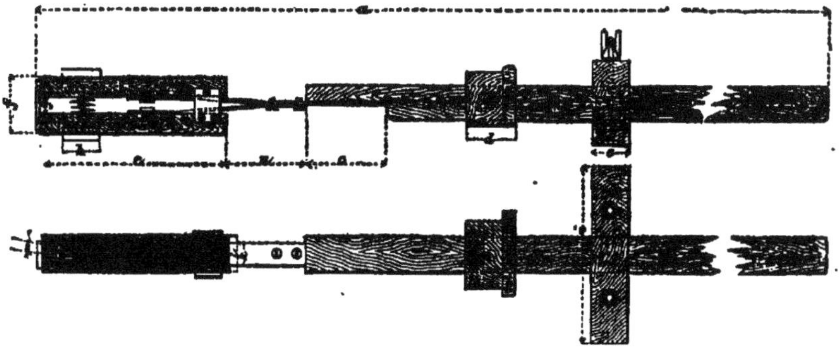

Hand rifling machine. §§ 973, 1096.

There is one hand-rifling machine for each nature of gun,* and they are nearly all alike in pattern. They are for the purpose of filing down any metal that may be turned up in the bore by the premature bursting of a shell, or other cause. The machine consists of—

1 Bar, working, (*b*) with cross-handle. (The 12-pr., 9-pr., and 6-pr. have no cross-handles.)	2 cutters, file (*h*).
1 block, guide (*d*).	1 distance piece, in halves, with bolts and nuts (*e*).
	1 head, filing, with springs (*c*).
	2 screws, fixing.

* That for the 12-pr. is the same instrument as that used with the 9-pr.

CH. VI. S. 2.

The filing head is of hornbeam wood, and is grooved like the barrel of the gun. The head is expanded by means of long springs, in the 7″ there are two, in the other natures one, and small files can be fitted on either side of the head which are kept up to their work in the bore by a spiral spring fixed beneath them.* One file is for the grooves and the other for the lands, and they are shaped accordingly; only one file at a time can be used. When required to be used the file for the lands or grooves (according to circumstances) is fixed and worked backwards and forwards until the bore is smooth. The guide block, which is placed in the muzzle, keeps the bar in the centre, and the distance piece is clamped on the bar at the ascertained distance of the flaw, &c. from the muzzle, thereby ensuring the action of the file on the proper place.

Patch, Metal, Elevating.

Elevating patch. §§ 717, 718. See also § 102

These patches are made of gun metal, and screw into the elevating eye holes of 12-pr., 9-pr., and 6-pr. guns when fitted for sea service. The 9-pr. patch has a steel screw.

Pins.

PIN, FRICTION TUBE.

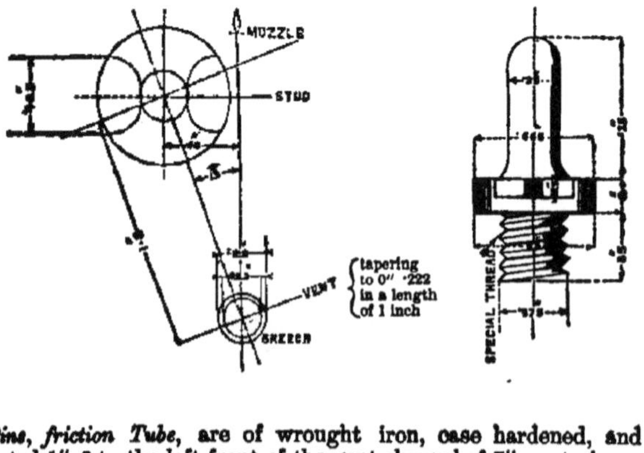

Friction tube pin.

Pins, friction Tube, are of wrought iron, case hardened, and are inserted 1″ ·3 to the left front of the vent channel of 7″ vent-pieces, for sea service only. The loop of the quill friction tube fits over the pin, and the mouth of the vent being rimed out to a depth of 1 inch, giving the tube a little play, ensures the pull of the lanyard acting upon the loop of the tube. The pin is now 1 inch long to prevent a liability of the loop slipping off. See *Crutches, iron.*

Lever keep pins.

Pins, keep, lever.—See page 128, *Spare*, one per 7″ and 40-pr.; two for other guns.

---

* The 7″ has no spiral spring, the double spring being sufficient for the purpose.

CH. VI. S. 2.

Pivots, Steel, for elevating Arcs. (*See Plates, metal elevating*).

Plates, metal, elevating.

Elevating plates. These stores are only required with the 20-pr. B.L. guns of 13 cwt. S.S. when mounted on wrought iron sliding carriages and slides, which guns will be fitted on both sides of the breech with the necessary screw holes. Only one plate is used, which is triangular, as in the case of the 40-pr. R.M.L.

Rings.

Vent-piece ring. §§ 472, 526, 350. See also § 528. *Rings, copper, vent-piece.*—Have a half dovetail on the inner side, which prevents the passage of the powder gas down between it and the iron of the vent-piece. It projects from the face of the vent-piece when new ″ ·05, so that it can be re-faced from time to time, as it is found to wear. Small channels are cut on the inside of the ring to allow the confined air to escape when placing it on the vent-piece. *Spare*, two per field gun; they should be carried in some place where they are not liable to be injured or put out of shape.

Indicator ring. §§ 790, 1083. *Rings, Indicator, for 7″ and 40-pr. B.L. Gun.*—Vide diagram and description at p. 129.

Tappet rings. *Rings, tappet.*—See p. 128.

Saddles, Metal.

Metal saddles. These metal saddles form the vent rest for the 7-inch guns, and are attached to the gun in rear of the vent slot by means of screws. The saddle for the 72 cwt. gun A is about 2¼ inches shorter than that for the B gun (the breech of the latter being longer than the former), and is fixed with four screws. All the other saddles have six screws. The saddles for the 82 cwt. gun are stouter made than those of the 72 cwt., and the position of the holes for fixing them is slightly different.

Scale, Wood Side.

Scale 1″ = 1 foot.

Wood side scales. § 1204. The wood side scale is used by the navy for broadside guns independently of all other sights, when the object to be hit is obscured by smoke, &c., thereby rendering the other sights useless. It must be used in connexion with the ship's pendulum, or other means employed to show the heel of the vessel, and the number of degrees of elevation or depression to be added or deducted from the required range, *i.e.*, if firing from the windward side, and the pendulum showed 3° of heeling over, *deduct* 3° from the correct elevation for the range of the object aimed at to allow for the inclination of the ship's deck.

These side scales are adjusted to the rear chock of the carriage with the zero notch coinciding with a point on the vertical line intersected by the horizontal line cut on the right side of the breech, the gun being horizontal. They are graduated to give 6° depression and 12° elevation, the radius being the distance from the centre of the trunnion to the point of intersection on the side of the breech.

The wood side scale for the 20-pr. of 13 cwt. or 15 cwt. is rectangular in section and has a moveable pointer similar to that used with R.M.L. guns. This scale gives a depression down to 20°, and from 0° to 12° of elevation. CH. VI. S. 2.

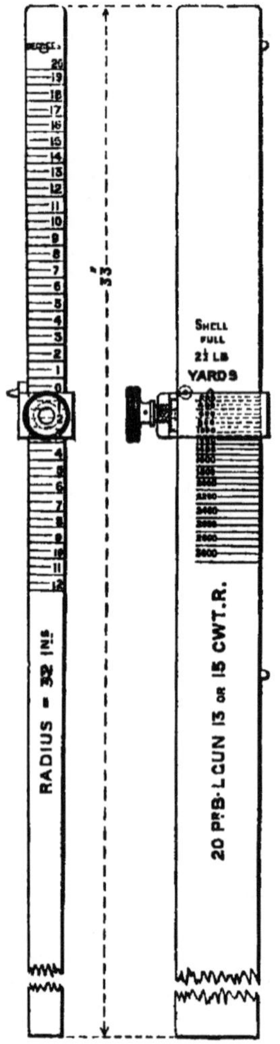

20-pr. R.B.L. Gun, 13 or 15 cwt. Scale, Wood, ½ size.

Screws.

Screws, breech.—See page 128.

Screws, copper, set, right-hand and left-hand.—Are used for the purpose of clamping the tangent sight, and pass through bosses in the Screws. §§ 935, 901–906, 1034, 829.

CH. VI. S. 2. tangent sight ring. These set screws for the 20-pr., downwards, are attached to the guns by a small chain.

Screws, fixing, plates, metal, elevating.—Wrought iron, and fasten the elevating plate to 20-pr. of 13 cwt.

§ 688. *Screws, fixing, crutch.*—Wrought iron, and attach the crutch to the top of the vent-piece.

Screws, fixing, metal saddle.—Wrought iron, and attach saddle to 7″ guns. *Spare* one per gun.

§ 485. *Screws, preserving.*—Wrought iron. Occupy the holes for the crutch, friction tube pin and guide plate, when the gun is used for land

§ 30. service, *and mounted.* When dismounted, those in the gun are removed and the holes filled with grease, to prevent the screw heads from being broken off.

Sights.

Sights.
§ 1480. *Sight, instructional, wood.*—Hexagonal in shape, and an enlarged model of the tangent sight. Rectangular are also issued on demand.

§§ 1143, 1476. *Sight, tangent.*—See p. 131.

§ 872. *Sight, trunnion.*—See p. 133.

Sockets, Metal.

Metal sockets.
§ 1481. These gun metal sockets are for the hind centre sight and tangent sight.—

Straight-Edges for testing Breech-screws, &c.

Straight-edge.
§ 1016. The straight edges are of steel, $18'' \times 1\frac{1}{8}'' \times \frac{3}{16}''$. See p. 303.

Vent-pieces.

Vent-piece. *For vent-pieces*, see p. 127. For field batteries, two vent-pieces per

§ 1185. gun are issued; one is carried in the gun, one at the side of the trail, and two in addition per battery. Garrison and siege batteries have three per gun.

Wrench, fixing elevating racks.

Wrench for elevating plates. Are used to attach the elevating plates and racks to the 20-pr. R.B.L guns when mounted on the decks of iron clads. Vide p. 211 for description of this store.

TABLE XIII.—Nature of Tangent Sights* and Particulars of Graduations for Rifled B.L. Guns.

Nature of Gun.	Permanent angle of Deflection of Sights.	Length of Radius in Inches.	Graduations for Degrees.		Graduations for Yards.			Remarks.
			Number.	Length in Inches.	Number.	Charge in lbs.	Length in Inches.	
7-inch B.L. 82 cwts.	2° 16′	45	15	12·057	3,600	11	8·286	
7-inch B.L. 72 cwts.		41·2	10	7·36	—	—	—	
40-pr. B.L., 32 and 35 cwts.		45	15	12·057	3,800	5	8·4319	
20-pr. B.L. 16 cwts.		36·2	12	7·636	3,500	2½	6·923	
20-pr. B.L. 15 and 13 cwts.		23·45	15	6·195	3,000	2½	4·2373	
12-pr. B.L., 8 cwts.		32·375	10	5·71	3,400	1½	6·033	
9-pr. B.L., 6 cwts.		23·45	15	6·195	3,000	1¼	4·044	
6-pr. B.L., 3 cwts.		23·45	15	6·195	3,000	¾	4·506	

* Besides tangent sights, the Rifled B.L. guns for S.S. down to 20-prs. inclusive, are supplied with wood side scales (vide p. 140).

Note.—The metal heads of the sights are not to be polished, as it would eventually destroy their accuracy.

TABLE XIV.—Table showing the Number of R.B.L. Guns in the Service, with the various kinds of Barrels.

Nature.	Barrels. Coiled.	Solid forged.	Steel.	Total.
Breech-Screw.				
7-inch 82 cwt.	690	179	5	883
7-inch 72 "	35	41	—	76
40-prs. 35 "	791	—	28	819
40-prs. 32 "	194	—	—	194
20-prs. 16 "	83	—	6	89
20-prs. 15 "	26	—	5	31
20-prs. 13 "	100	187	5	292
12-pr., 8 "	294	286	121	701
9-pr., 6 "	261	5	—	266
6-pr., 3 "	80	—	18	98

TABLE XV.

Weights of R.B.L. Vent-pieces and Breech Screws.

	Cwt.	qrs.	lbs.
Vent-pieces:—7-in.	1	0	24
40-pr.	0	2	3
20-pr.	0	0	27¾
12-pr.	0	0	15
9-pr.	0	0	14¾
6-pr.	0	0	8¼

Breech screw complete, with lever, tappet ring, and keep pins:—

	Cwt.	qrs.	lbs.	Pitch of thread. Inch.
7-inch	5	2	18	1·4
40-pr.	2	1	13	0·9—32 cwt. gun 0·7—35 "
20-pr.	1	0	8¾	0·5
12-pr.	0	2	12	0·5
9-pr.	0	2	6¾	0·5
6-pr.	0	1	10	0·5

TABLE XVI.

TABLE showing the FITTINGS and STORES for R.B.L. SERVICE GUNS.

Nomenclature of all Fittings and Stores made in the Royal Gun Factories for the RIFLED B.L. GUNS.	7-inch Gun. 82 cwt. Sea Service.	7-inch Gun. 82 cwt. Land Service.	7-inch Gun. 72 cwt. Land Service.	40-pr. Gun. Sea Service.	40-pr. Gun. Land Service.	20-pr. Gun. Sea Service.	20-pr. Gun. Land Service.	12-pr. Gun. Sea Service.	12-pr. Gun. Land Service.	9-pr. Gun. Sea Service.	9-pr. Gun. Land Service.	6-pr. Gun. Sea Service.	6-pr. Gun. Land Service.	REMARKS.
Bearers, shot*	1	1	1	—	—	—	—	—	—	—	—	—	—	Issued in certain proportions as required.
Bits, vent, Armstrong*	1	1	1	1	1	1	1	1	1	1	1	1	1	
Bushes, breech, iron, thick ...	1	1	1	—	—	—	—	—	—	—	—	—	—	
Bushes, breech, iron, thin ...	1	1	1	—	—	—	—	—	—	—	—	—	—	
Bushes, breech, copper	—	—	—	1	1	1	1	1	1	1	1	1	1	
Bushes, copper, vent-piece, sets ...	1	1	1	1	1	1	1	1	1	1	1	1	1	
Clamps, moveable, for tangent sights	2	2	—	2	2	—	—	—	—	—	—	—	—	
Collars, leather, for breech-screw ...	—	—	—	—	—	—	—	—	1	—	—	—	—	
Crutches, iron	—	—	—	1	—	1	—	1	—	1	—	1	—	
Extractors, tin cup, L.S.	—	1	1	—	—	—	—	—	—	—	—	—	—	
Extractors, tin cup, S.S. with lever lifting joint.	1	—	—	—	—	—	—	—	—	—	—	—	—	
Eyes, elevating	—	—	—	—	—	—	1	—	1	—	1	—	1	
Guide plates*	1	—	—	1	—	1	—	1	—	1	—	1	—	
Implements, facing, set	1	1	1	1	1	1	1	1	1	1	1	1	1	See p.
Instruments, taking impressions of bore of guns, No. 1 ...	1	1	1	1	1	—	—	—	—	—	—	—	—	
Instruments, taking impressions of bore of guns, No. 2 ...	—	—	—	—	—	1	1	1	1	1	1	1	1	
Instruments, sighting, set	1	1	1	1	1	1	1	1	1	1	1	1	1	
Levers, breech-screw	1	1	1	1	1	1	1	1	1	1	1	1	1	
Levers, iron, releasing vent piece*	1	1	1	—	—	—	—	—	—	—	—	—	—	
Machines, hand-rifling	1	1	1	1	1	1	1	1	1	1	1	1	1	
Patches, metal, elevating	—	—	—	—	—	—	—	1	—	1	—	1	—	
Pins, friction tube*	1	—	—	—	—	—	—	—	—	—	—	—	—	
Pins, keep, lever	2	2	2	2	2	2	2	2	2	2	2	2	2	
Pins, keep, pivot, elevating ...	—	—	—	—	—	1	—	—	—	—	—	—	—	For 20-pr. 13 cwt. guns when used on upper decks.
Pivots, steel, for elevating arcs ...	—	—	—	—	—	1	—	—	—	—	—	—	—	Do.
Plates, metal, elevating...	—	—	—	—	—	1	—	—	—	—	—	—	—	Do.
Rings, copper, vent-piece	—	—	—	1	1	1	1	1	1	1	1	1	1	Some 40-pr. of 32-cwt. have no room for an indicator ring.
Rings, indicator	1	1	1	1	1	—	—	—	—	—	—	—	—	
Rings, tappet	1	1	1	1	1	1	1	1	1	1	1	1	1	
Saddles, metal	1	1	†1	—	—	—	—	—	—	—	—	—	—	
Scales, wood, side	1	—	—	1	—	1	—	—	—	—	—	—	—	20-pr. 13 or 15 cwt.
Screws, breech	1	1	1	‡1	1	1	1	1	1	1	1	1	1	
Screws, copper, set, right hand ...	—	—	—	—		1	1	1	1	1	1	1	1	
Screws, copper, set, left hand ...	—	—	—	—	—	1	1	1	1	1	1	1	1	
Screws, fixing, crutch	—	—	—	2	—	2	—	2	—	2	—	2	—	The A pattern gun of 72 cwt. has four fixing screws.
Screws, fixing, metal saddle ...	6	6	6	—	—	—	—	—	—	—	—	—	—	
Screws, fixing, plate, elevating ...	—	—	—	—	—	3	—	—	—	—	—	—	—	
Screws, preserving, crutch*	—	—	—	—	2	—	2	—	2	—	2	—	2	
Screws, preserving, pin, friction tube*	—	1	1	—	—	—	—	—	—	—	—	—	—	
Screws, preserving, guide plate* ...	—	1	—	—	1	—	—	—	1	—	1	—	1	
Screws, preserving, plate, elevating ..	—	—	—	—	—	6	—	—	—	—	—	—	—	
Sights, instructional, wood	—	—	—	—	—	—	—	1	—	—	—	—	—	
Sights, tangent	2	2	2	2	2	2	2	2	2	2	2	2	2	
Sights, trunnion	2	2	2	2	2	2	2	2	2	2	2*	2	2	
Sockets, metal, tangent sight	2	2	2	—	—	—	—	—	—	—	—	—	—	
Straight-edges, for testing breech-screws and vent-pieces, &c.*	1	1	1	1	1	1	1	1	1	1	1	1	1	Issued in certain proportions as required.
Vent-pieces	1	1	1	1	1	1	1	1	1	1	1	1	1	
Wrench, fixing elevating racks ...	—	—	—	—	—	1	—	—	—	—	—	—	—	

* Universal patterns.
† Different patterns for A and B guns.
‡ The breech-screws for the 40-pr. of 35 cwt. and 40-pr. of 32 cwt. have threads of different pitch and therefore are not interchangeable.

NOTE.—All the above are interchangeable with guns of the same nature, excepting †‡.

TABLE XVII.

TANGENT OR SIDE SIGHTS OF R.B.L. GUNS

Muzzle.

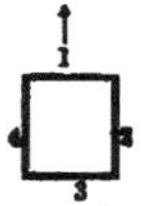

Mark of Sight.	No. of Face, vide Diagram.	7-inch. 82 cwt.	7-inch. 72 Cwt. Hexagonal Bar.	40-pr.	20-pr. 16 cwt.	20-pr. 13 and 15 cwt.	12-pr. Hexagonal Bar.	9-pr.	6-pr.
I.	1 2 3 4	§ 935. (2) *a* 0° to 11° *Blank.* C.S. Yards 3,600. Time of Flight—Secs. Common Shell, 12. *Fuze Scale C.S., 2″ o c*	§ 716. * (1) *a* 1. *Blank.* 2. *Blank.* 3. 0° to 10° 4. *Blank.* 5. 0° to 10° 6. *Blank.*	§ 901. (3) *a* 1. 0° to 12° 2. *Blank.* 3. Yards, 3,800 4. *Blank.*	§ 938. (3) *a* 1. 0° to 12° 2. *Blank.* 3. Yards 3,500 4. *Blank.*	§ 903. (3) *a* 1. 0° to 15° 2 *Blank.* 3. Yards 3,000. 4. *Blank.*	§ 829. (1) *a* 1. *Blank.* 2. *Blank.* 3. 0° to 10° 10′ 4. Yards 3,400. 5. *Blank.* 6. *Blank.*	§ 905. (3) *a* 1. 0° to 15° 2. *Blank.* 3. Yards 3,000. 4. *Blank.*	§ 906. (3) *a* 1. 0° to 15° 2. *Blank.* 3. Yards 3,000. 4. *Blank.*
II.	1 2 3 4	§ 1254. (4) *b* 0° to 15° *Blank.* C.S. Yards 3,600. Fuze Scale C.S. 2″·0.	§ 1254. (4) *b* 1. *Blank.* 2. *Blank.* 3. 0° to 10° 4. *Blank.* 5. 0° to 10° 6. *Blank.*	§ 1254. (4) *b* 1. 0° to 15° 2. Fuze Scale C.S. 2″·0 3. Yards 3,800 4. *Blank.*	§ 1254. (4) *b* 1. 0° to 12° 2. *Blank.* 3. Yards 3,500. 4. *Blank.*	§ 1254. (4) *b* 1. 0° to 15° 2. *Blank.* 3. Yards 3,000. 4. *Blank.*	§ 1254. (4) *b* 1. *Blank.* 2. *Blank.* 3. 0° to 10° 4. Yards 3,400. 5. *Blank.* 6. *Blank.*	§ 1254. (4) *b* 1. 0° to 15° 2. *Blank.* 3. Yards 3,000. 4. *Blank.*	§ 1254. (4) *b* 1. 0° to 12° 2. *Blank.* 3. Yards 3,000. 4. *Blank.*
III.	1 2 3 4	§ 1476. (5) *b* 0° to 15° *d* Shell Full 11 lbs. Fuze 24. Full 11 lbs. Yds. 3,600. *Blank.*		§ 1476. (5) *b* 1. 0° to 15°. *d* 2. Fuze 25. 3. Shell Full 5 lbs. Yards 3,800. 4. *Blank.*					

a Barrel head. *b* Sliding-leaf head. *c* In 1866 the "Time of Flight" was omitted and the "Fuze Scale" substituted on Mark I. tangent sight, 7″ of 82 cwt. gun.
d The fuze scale is graduated in even tenths, headings of yard scale altered, and weight of charge added. * It was *not* graduated in yards and fuze scale as stated in § 716.
(1) 1863, (2) 1864, (3) 1865, (4) 1866, (5) 1867.

CHAPTER VII.

DIFFERENT NATURES OF R.B.L. GUNS IN THE SERVICE.

R.B.L. Guns in the Service.—7-inch of 72 cwt.—7-inch of 82 cwt.—40-pr., 32 cwt.—40-pr., 35 cwt.—20-pr., 16 cwt.—20-pr., 15 cwt.—20-pr., 13 cwt.—12 pr., 8 cwt.—9-pr., 6 cwt.—6 pr., 3 cwt.—Wedge 64-pr., 61 cwt.—**Table of Dimensions**, &c. &c. of B.L. guns.—Number in the Service.

The following is a short description of each nature of R B.L. gun in the service, a list containing the whole of which will be found at p. 160.

There are two natures of 7 in. guns as shown below.

The heavier nature, weighing 82 cwt., was that first practically introduced into the service, a number of these guns having been issued in 1861.

From experience gained in the case of the 40-pr. guns it was thought advisable to manufacture these pieces with a strengthening coil over the powder chamber. Before their construction, however, was finally determined upon, a limited number of guns of a lighter description had been manufactured in part, and these, 76 in number, were subsequently utilized by completion as 7-inch guns of 72 cwt.

These latter guns are for L.S. only, and were issued to the service in 1863.

7-inch gun of 82 cwt. L.S. and S.S. consists of:—

A tube	Trunnion ring, and
Breech-piece and *B* tube.	Six coils.

7-INCH R.B.L. GUN, 82 CWT.—Scale ⅛ in. = 1 foot. § 935.

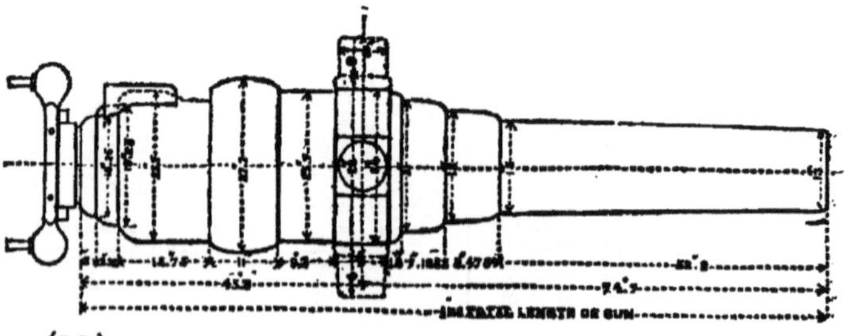

CHAP. VII.

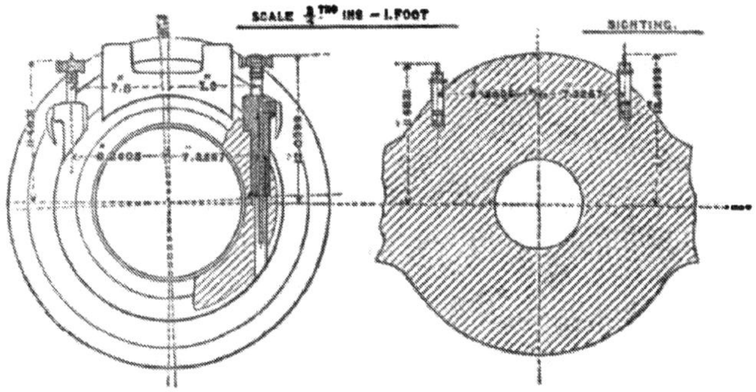

VENT-PIECE.*—Scale ¾ in. = 1 foot.

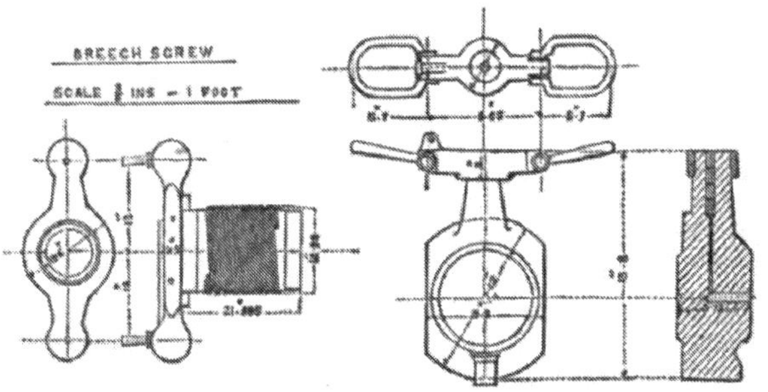

Parliamentary Report on Ordnance, 1863, p. 250.

It is used by the land and sea service, but not being sufficiently powerful for the penetration of iron plates, it has been replaced in the naval service by M.L. guns.

Report on Ordnance, 1862, p. 218. §§ 466-809.

It was first termed "100-pr.," but afterwards in 1861, when the weight of projectile was increased to 110 lbs., it was so designated 110-pr., and lastly, 7-inch, as at present.

§ 593.

7-inch of 72 cwt. L.S. consists of:

A tube or inner barrel.
Breech-piece and *B* tube.
Trunnion ring, and
Four coils.

§ 593.

There are two classes of this gun, respectively marked on the left trunnion A and B; they only differ in length of barrel and breech, A being 2″ longer in the barrel than B, and 2″ shorter in the breech.

§ 666-809.

In 1863 its designation was changed to "light" 110-pr., and finally, to 7-inch of 72 cwt.

* The projection shown on top of the Vent piece is obsolete.

40-pr. gun of 32 cwt.* cal. 4″·75 L.S. and S.S. consists of:— CHAP. VII.

A tube. § 901.
Breech-piece and *B* tube.
Trunnion ring, and
Three coils.

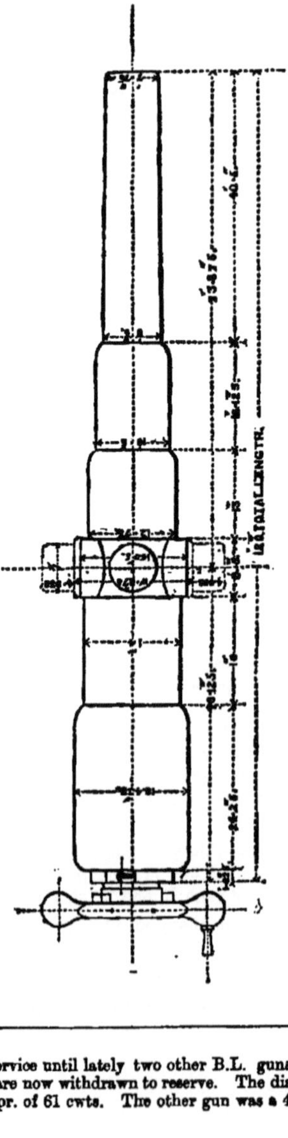

40-PR. R.B.L. GUN, 32 CWT.—Scale ¼ in. = 1 foot.

* There were in the service until lately two other B.L. guns with wedge breech-closing apparatus, which are now withdrawn to reserve. The diagram at p. 159 shows the heavier nature, a 64-pr. of 61 cwts. The other gun was a 40-pr. of 32 cwts.

CHAP. VII.

TANGENT SIGHT AND RING. Scale 2 in. = 1 foot.*

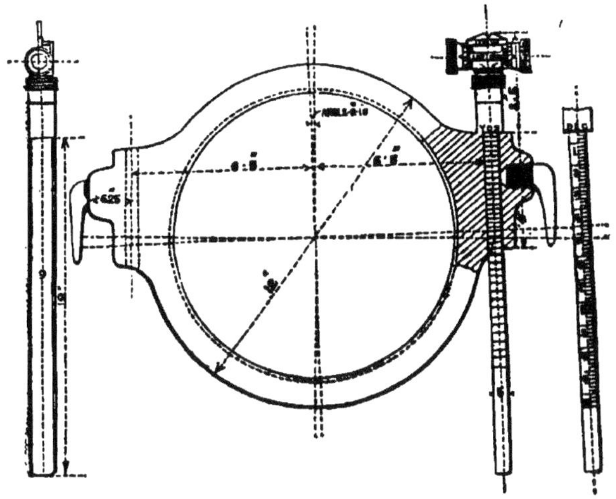

DROP TRUNNION SIGHT. Scale 2 in. = 1 foot.

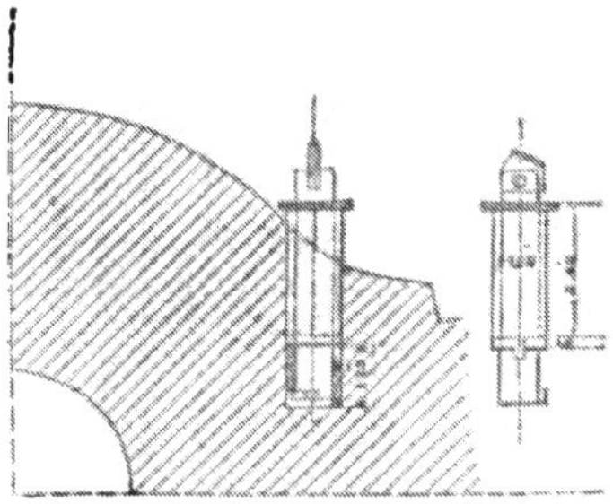

BREECH-SCREW. Scale ½ in. = 1 foot.

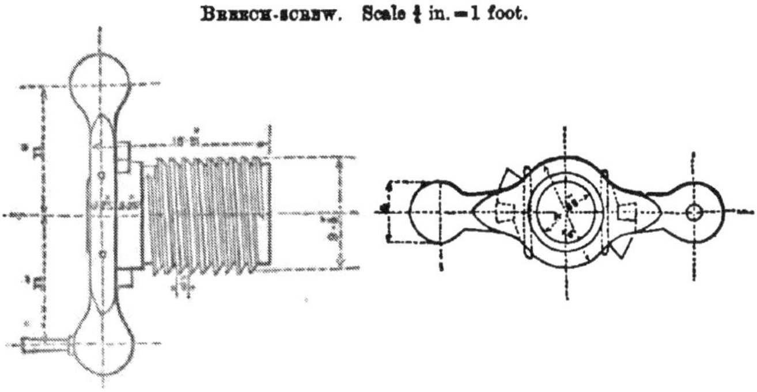

* The clamping screws shown in the diagram are now superseded by the moveable clamp.

Recommended in 1859, for the navy, as a broadside or pivot gun; it is also now used by the land service for batteries of position, siege and garrison purposes.

Report on Ordnance, 1862, p. 176.

A few of these guns have trunnions made of *cast iron*, and are known by the face of the trunnion being bored out in the centre.

This gun is sometimes termed the O.P. (old pattern) 40-pr.

40-pr. gun of 35 cwt. cal. 4″·75 L.S. and S.S. consists of the same number of parts as the 82-cwt. gun, but has a longer and a stronger breech-piece, which is unsupported behind the vent slot, and rounded off. § 902.

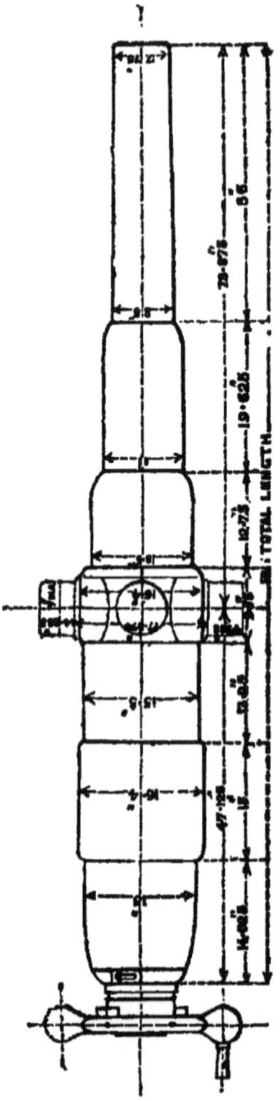

40-pr. R.B.L. Gun, 35 cwt. Scale ½ in = 1 foot.

It has a raised coil in front of the vent slot, and is known as the "G" pattern.

TANGENT SIGHT RING. Scale 2 in. = 1 foot.

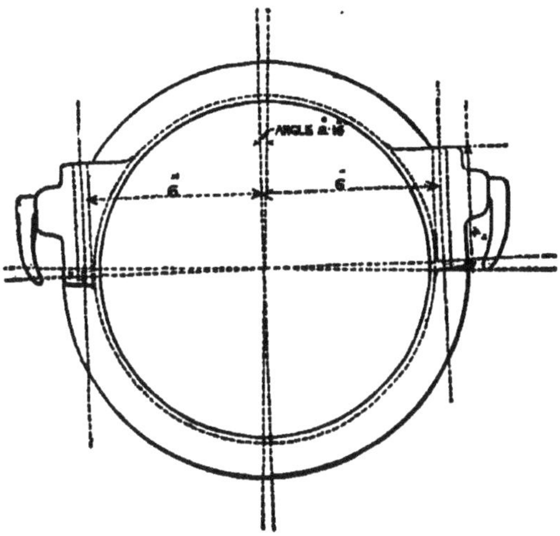

VENT-PIECE.* Scale 1 in. = 1 foot.

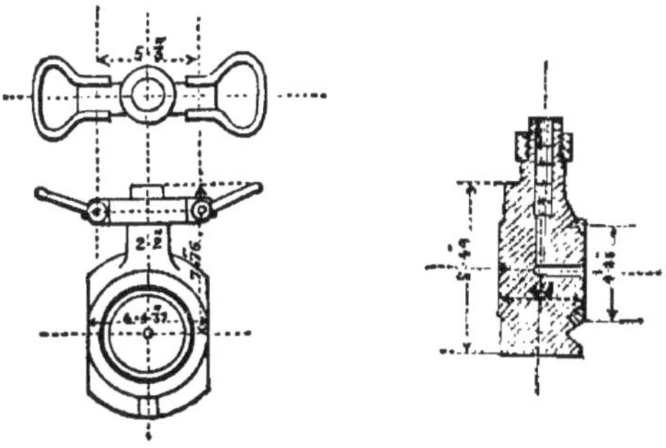

40-PR. BREECH-SCREW, TAPPET RING, LEVERS, AND KEEP PINS. Scale ¾ in. 1 foot.

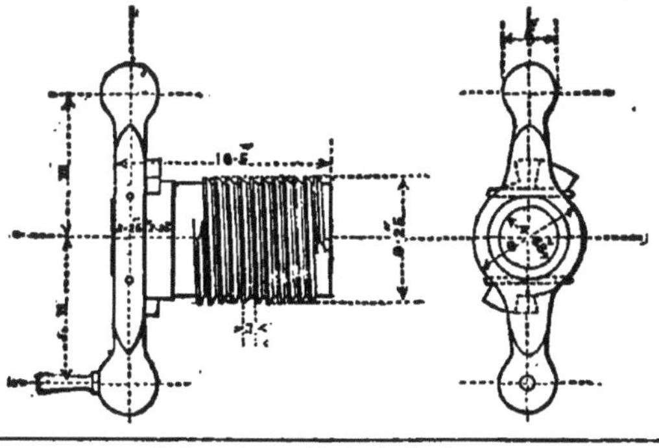

* Interchangeable for both 40-prs.

This pattern was introduced in 1860, more as a matter of precaution than from any symptoms of weakness in the lighter nature. It is used for the same purposes as the 32-cwt. gun, and the fittings are interchangeable with the exception of the breech-screw, which has a different shape and pitch of thread. Report on Ordnance, 1862, p. 218.

20-pr. gun of 16 cwt. cal. 3″·75 L.S. consists of:— § 998.

A tube.	Trunnion ring, and
Breech-piece.*	Five coils.

20-PR. R.B.L. GUN OF 16 CWT. Scale ⅜ in. = 1 foot.

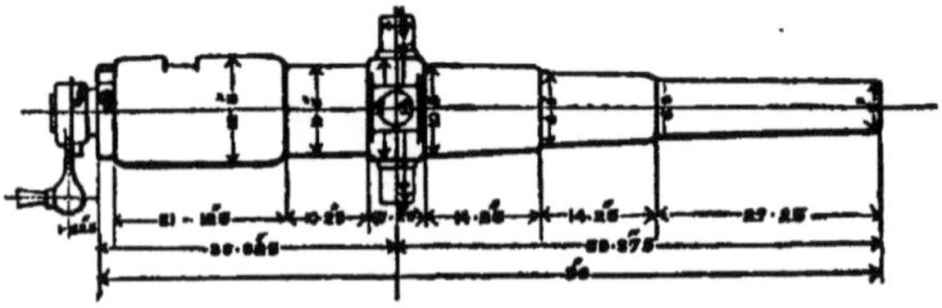

TANGENT AND DROP SIGHT. Scale 1 in. = 1 foot.

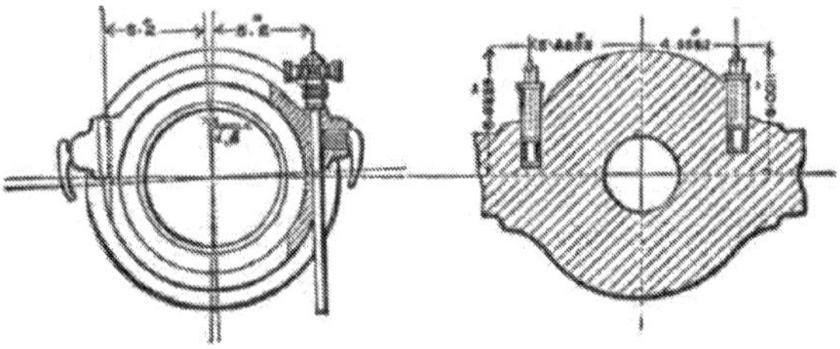

Recommended in 1859 (then 25-pr.) as a light gun of position, but subsequently it was resolved to use a lighter projectile; hence its alteration to 20-pr. It is used only for heavy field batteries of reserve. Report on Ordnance, 1862, p. 177.

20-pr. gun of 15 cwt. cal. 3″·75 S.S. consists of:— § 901.

A tube.	Trunnion ring, and
Breech-piece.	Three coils.

20-PR. R.B.L. GUN, 15 CWT. Scale ⅜ in. = 1 foot.

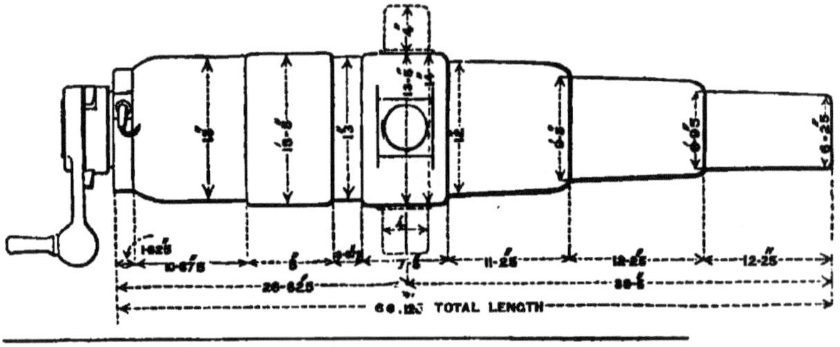

* Guns under 40-prs. have no *B* tubes.

CHAP. VII.

Report on Ordnance, 1862, p. 218.

Was adopted in 1859, for the navy, as a broadside gun for ships of the sloop class, in which there is not sufficient width of beam to work the land service 20-pr. Its external appearance is entirely different from that of the land service 20-pr., as it is 2½ feet shorter, has a raised coil in front of the vent slot, and the breech-piece is unsupported in the rear.

§ 903. **20-pr. gun of 13 cwt.** cal. 3"·75 S.S. consists of:—

A tube.
Breech-piece.
Trunnion ring, and
Four coils.

20-pr. R.B.L. Gun, 13 cwt. Scale ¾ in. = 1 foot.

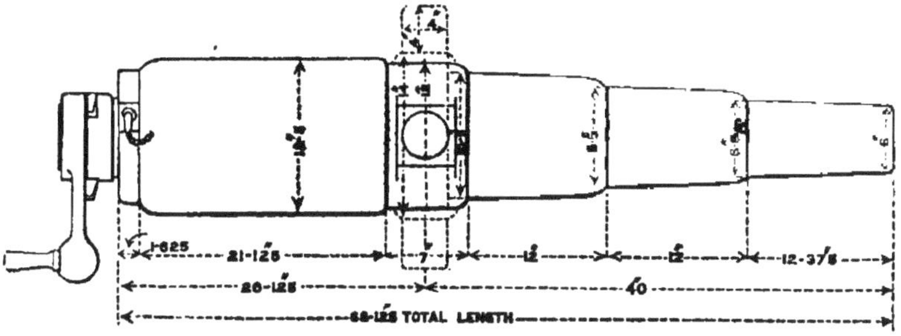

Tangent Sight and Ring. Scale 2 in. = 1 foot. Vent-piece.*

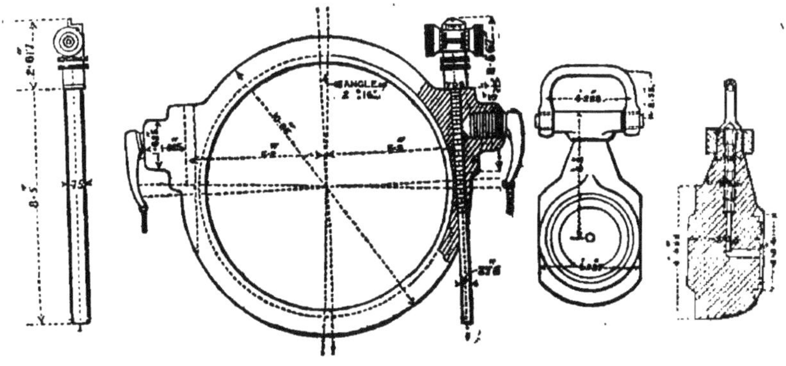

* Interchangeable for all 20-prs.

BREECH-SCREW.* Scale 1½ in. = 1 foot.

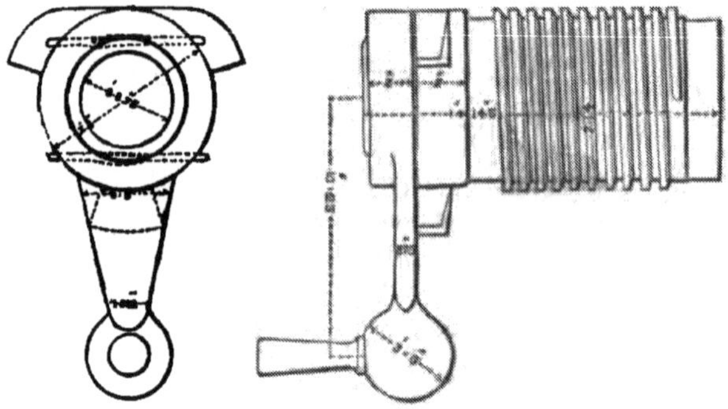

Recommended in 1859, for the navy, as a boat or field marine gun, and is termed the "pinnace gun." This and the 15-cwt. gun are much alike in appearance, but differ in their construction, the heavier pattern being stronger in every respect; neither of them shoot as well as the L.S. 20-pr. Report on Ordnance, 1862, p. 218; 1863, p. 155.

It is now used on the upper decks of iron clad ships for action at close quarters to repulse boarders, for firing at torpedo boats, &c., and is mounted on a carriage which allows 20° elevation or 30° depression. When so used it is fitted on the right side with a gun metal elevating plate, and steel pivot for elevating rack, similar to that used with heavy guns. § 1986.

12-pr. gun of 8 cwt. cal. 3" L.S. and S.S., and consists of:— §§ 939 and 829.

A tube.
Breech piece.
Trunnion ring, and
Three coils.

12-pr. R.B.L. GUN, 8 CWT. Scale ⅜ in. = 1 foot.

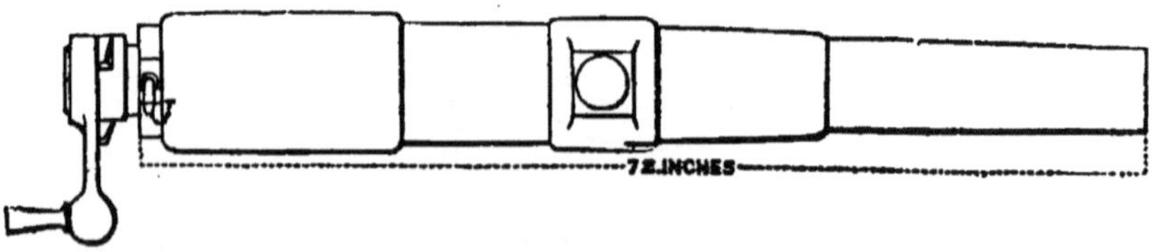

* Interchangeable for all 20-prs.

CHAP. VII.

TANGENT SIGHT AND RING. Scale 3 ins. = 1 foot.

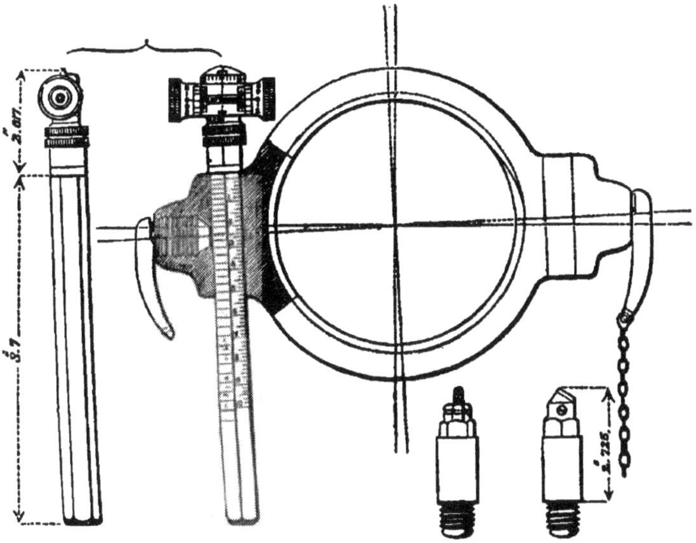

SCREW-SIGHT.

VENT-PIECE. Scale 1½ in. = 1 foot. BREECH-SCREW. Scale 1½ in. = 1 foot.

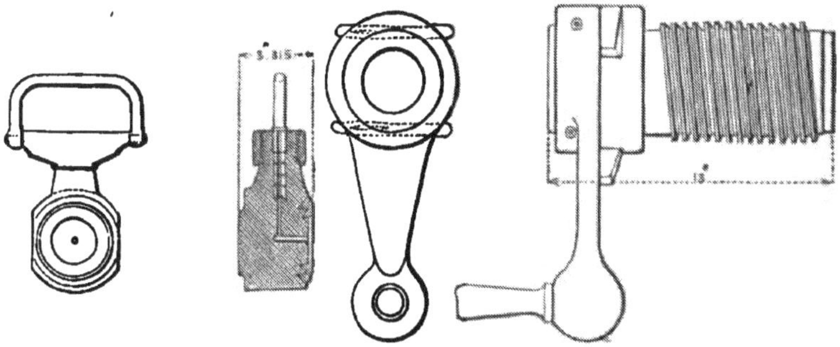

Report on Ordnance, 1862, p. 156. § 401.

This was recommended in 1858, for the land service, as a field battery gun, and subsequently adopted by the navy, as a boat or field marine gun; but the naval pattern was 12 inches shorter, and without the grip at the muzzle.

§ 829. § 939. § 826.

In 1863, an universal pattern was introduced for both services the L.S. pattern being altered. This necessitated a leather collar as a special store, for the vent slot was widened at its rear part to take a thicker vent-piece, causing a portion of the breech-screw thread to project behind. The leather collar can be pared round according as the breech-screw face wears on service.

The calibre of the 9-pr. gun being the same as this gun it can, on an emergency, use the 9-pr. ammunition, but not vice versâ, the 12-pr. ammunition being longer than the 9-pr. CHAP. VII.

9-pr. gun of 6 cwt. cal. 3" L.S. and S.S. consists of :— § 905.

A tube.
Breech-piece.
Trunnion ring, and
Three coils.

9-pr. R.B.L. Gun, 6 cwt. Scale ¼ in. = 1 foot.

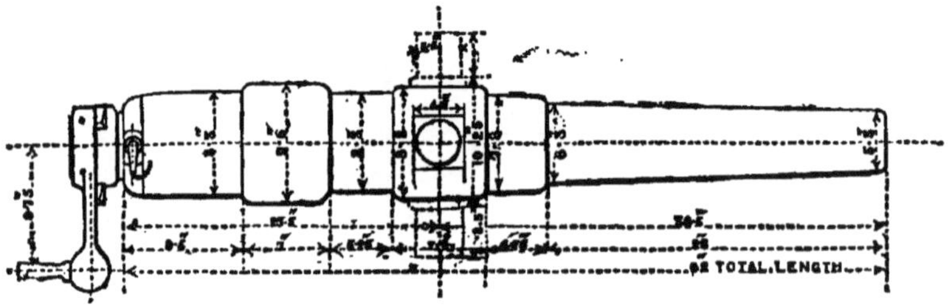

Tangent Sight and Ring. Scale 2 in. = 1 foot.

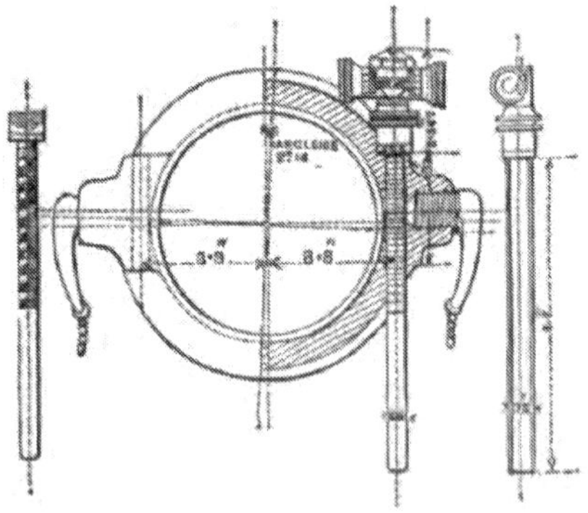

Screw Sight.
Scale 2 in. = 1 foot.

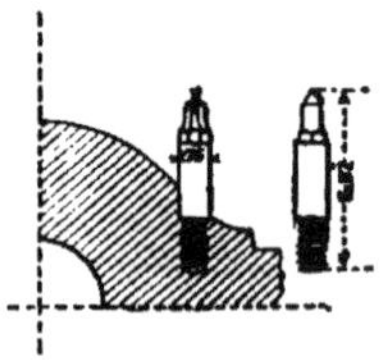

Vent-piece. Scale 1 in. = 1 foot.

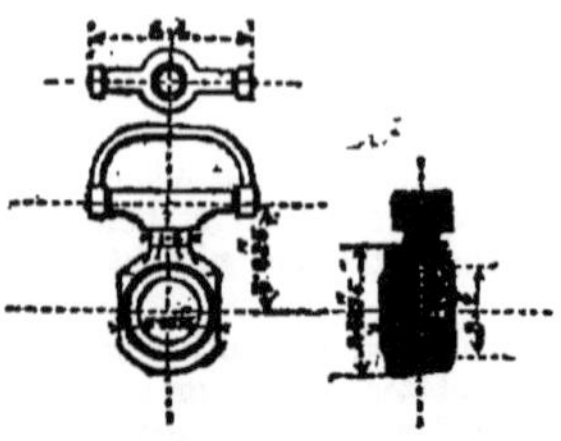

Breech-screw. Scale 1 in. = 1 foot.

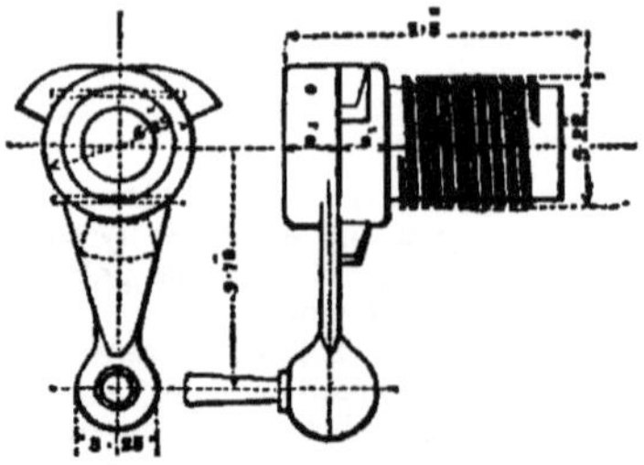

Introduced in 1862, for the Horse Artillery. The navy use it in some cases as a boat or field marine gun. § 474. § 529.

CHAP. VII.

§ 906. **6-pr. gun of 3 cwt.** cal. 2″·5 L.S. and S.S. consists of :—

A tube.
Breech-piece.
Trunnion ring, and
One coil.

6-PR. R.B.L. GUN, 3 CWT. Scale ½ in. = 1 foot.

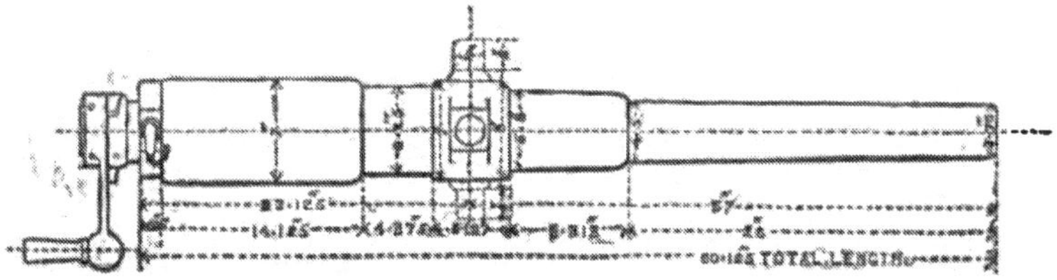

TANGENT SIGHT AND RING. Scale 2 in. = 1 foot.

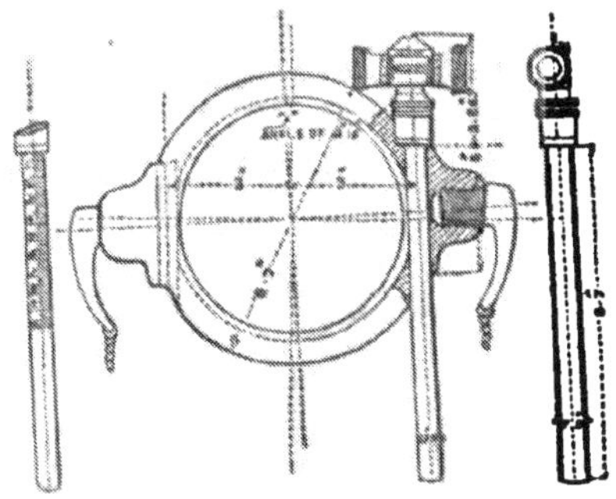

VENT-PIECE.
Scale 1 in. = 1 foot.

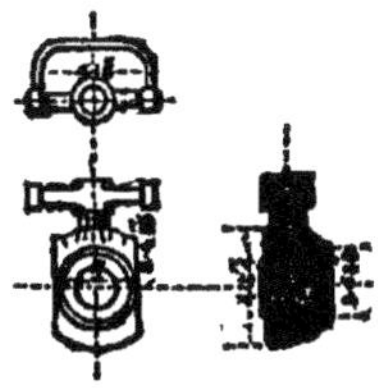

BREECH-SCREW, Scale 1 in. = 1 foot.

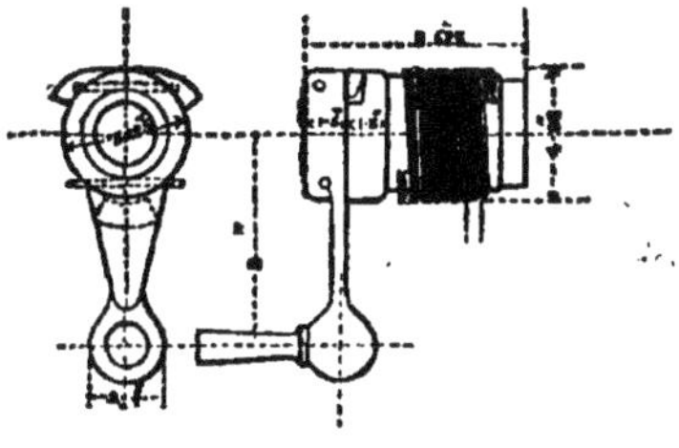

Report on Ordnance, 1862, p. 156.

Recommended in 1858, for mountain service, but, as it was considered too great a load for a mule, its use is restricted to colonial batteries; the navy employ it as a boat or field marine gun, where not replaced by R.M.L. 7-prs.

WEDGE GUNS.

§ 997. Wedge Guns. The two natures of wedge guns introduced in 1864. vide p. 81, have been withdrawn from service and placed in reserve, but in order to show the wedge system of breech-closing employed with them the diagram below is given.

64-PR. R.B.L. (WEDGE) GUN, 61 CWT. Scale ⅜ in. = 1 foot.

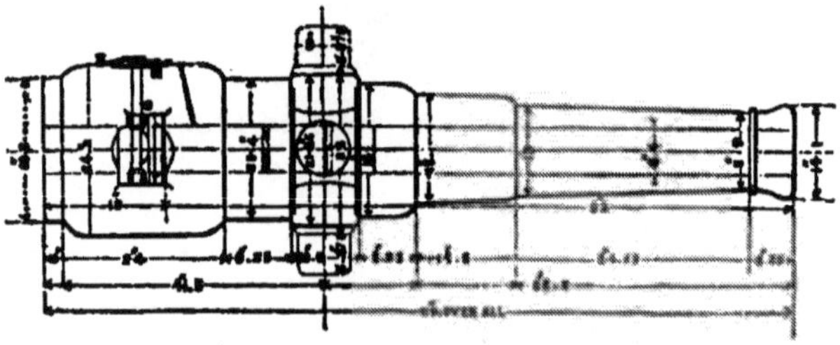

SIGHTING. Scale ¾ in. = 1 foot.

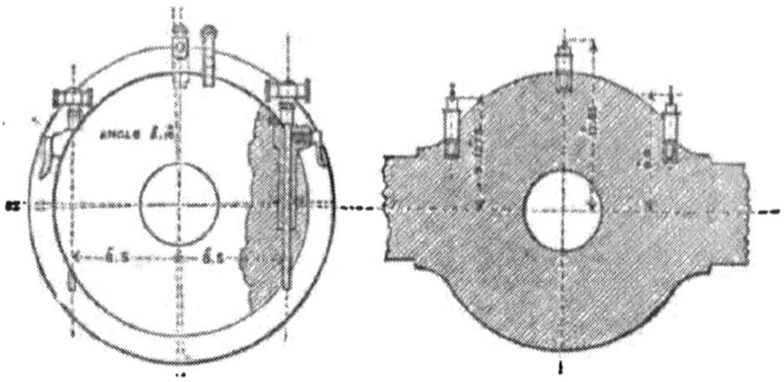

WEDGE. Scale $\frac{1}{10}$ STOPPER.

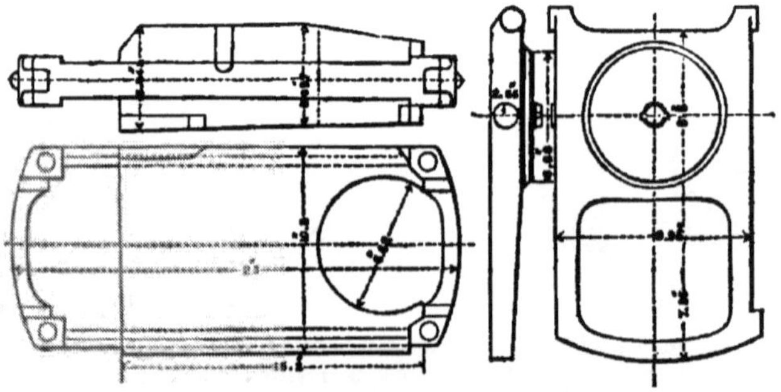

These pieces were built up and rifled like the breech screw guns, but had a horizontal instead of a vertical slot cut towards the breech end of the piece, and the breech was closed by means of a steel *stopper* supported by an iron *wedge* sliding in behind it.

Upon the face of the stopper was hung a tin cup to make the joint gastight.

TABLE XVIII.—Dimensions, Rifling, Service, &c. of Breech-loading Guns.

Nature, Weight, and Service.	Calibre.	Length. Barrel, Total.	Length. Powder Chamber.	Length. Shot Chamber.*	Length. Nominal.†	Length. From muzzle to axis of trunnion.	Length. From axis of trunnion to breech.	Diameter. Greatest over-charge.	Diameter. Muzzle.	Rifling (the many-grooved system). Twist.	Grooves. Number.	Grooves. Depth.	Grooves. Width.	Width of lands.	Preponderance of sealed pattern.	Remarks.
Breech-Screw.	"	"	"	"	"	"	"	"				"	"	"	cwt. qrs. lbs.	
7-inch of 82 cwt. L.S. & S.S. ...	7	99·5	16·0	9·0	120	74·7	45·3	27·7	13	1 turn in 37 cals.	76	·06	·166	·1233	6 3 16	The 7″ B.L. guns have been replaced by 7″ rifled M.L. guns of 6½ tons and 90 cwt. for S.S.
7-inch of 72 cwt. L.S. ...	7	97·5	14·25	9·0	118	71·25	46·75	24·7	13	Do.	76	·06	·166	·1233	7 3 27	
40-pr. of 35 cwt. L.S. & S.S. ...	4·75	106·375	13·5	7·0	121	73·875	47·125	16·4	7·75	1 turn in 36½ cals.	56	·06	·166	·1	4 3 0	
40-pr. of 32 cwt. L.S. & S.S. ...	4·75	106·375	13·5	7·0	120	73·875	46·125	16·438	7·75	Do.	56	·06	·166	·1	5 1 19	
20-pr. of 16 cwt. L.S. ...	3·75	84·0	12·0	6·0	96	59·375	36·625	12·5	6·0	1 turn in 38 cals.	44	·06	·166	·1	2 0 11	
20-pr. of 15 cwt. S.S. ...	3·75	54·125	11·0	6·0	66·125	39·5	26·625	13·5	6·25	Do.	44	·06	·166	·1	1 2 0	
20-pr. of 13 cwt. S.S. ...	3·75	54·125	11·0	6·0	66·125	40·0	26·125	12·5	6·0	Do.	44	·06	·166	·1	1 1 24	
12-pr. of 8 cwt. L.S. & S.S. ...	3	61·375	8·5	3·0	72	38·75	33·25	9·75	5·75	Do.	38	·045	·148	·1	1 3 3	
9-pr. of 6 cwt. L.S & S.S. ...	3	52·5	7·0	3·0	62	36·5	25·5	9·6	5·3	Do.	38	·045	·148	·1	0 2 26	
6-pr. of 3 cwt. L.S. & S.S. ...	2·5	53·0	7·0	2·5	60·125	37·0	23·125	7·0	3·75	1 turn in 30 cals.	32	·045	·148	·1	0 1 27	

* Exclusive of slopes in front and rear of shot.
† *i.e.* from face of muzzle to extreme end of breech, exclusive of breech-screw.

CHAPTER VIII.

MANUFACTURE OF PRESENT PATTERNS OF R.M.L. ORDNANCE.

CHAP. VIII.

All present Pattern Service Guns built up in a similar manner.—Converted guns excepted.—Armstrong or original construction.—Fraser construction.—Advantages of latter in manufacture.—**Building up of a 7-inch Woolwich Gun.**—Steel tube, operation of toughening.—*B* tube, how prepared.—Breech coil or jacket.—Different parts of.—Shrinking the several parts together.—8-inch gun manufactured as 7-inch.—Difference in construction of 9-inch.—Of 10-inch guns and upwrads.—Of 80-ton gun.—Construction of rifled howitzers.—Manufacture of 64-pr. guns.—Siege pieces.—Construction of our field guns.—7-pr. steel guns.—**Processes after building up and before Proof.**—Forming gas escape.—Chamber in Woolwich guns, 40-pr. and upwards.—Rifling.—Venting for proof.—**Examination and Proof.**—Proof charges for heavy guns.—**Processes after Proof.**—Lining.—Sighting.—Service venting.—Marking and adjusting fittings.—Painting and final issue.

All R.M.L. pieces of the patterns now being manufactured (except converted guns) are built up* in a similar manner. That is to say, over a barrel (or *A* tube) of steel are shrunk a certain number of coils or other pieces of wrought iron according to the size of the gun, while a cascable is screwed into the end of the coil which comes over the breech end of the tube and fitting closely against the solid end of the barrel, supports it firmly. *Manufacture of latest pattern R.M.L. ordnance.*

In 25-pr. guns however and smaller natures, the solid end of the steel tube projects beyond the breech end of the jacket and is turned down to a cascable. Such guns therefore have no cascable screw and also no gas escape. Vide p. 245. *Cascable screws.*

The smallest piece we manufacture—the 7-pr.—is made out of a single block of steel and is therefore an exception to the above.

Until April 1867 all our rifled M.L. guns were built up like the B.L. guns—of wrought iron coils shrunk together successively on Sir William Armstrong's original plan. The plan proposed by Mr. R. S. Fraser, of the Royal Gun Factories, was then adopted; but manufacture on the original construction did not cease altogether until March 1868. *Difference between the "original" and present construction.*

Mr. Fraser's plan is, as stated in a previous chapter, an important modification of the original method, from which it differs principally in *Fraser construction.*

---

* Guns of patterns made previous to 1869 were manufactured on the Armstrong or modified Armstrong construction, vide p. 28, or had some other points of difference. The details of their construction, showing how they differ from the present pattern, are given in Chapter XI., and are shown also in the Plates at the end of the book.

CHAP. VIII.

building up a gun of a few large and comparatively heavy coils instead of several short ones and a forged breech-piece.

Small number of parts.

For example, in addition to the steel barrel and cascable, a "Fraser" 12·5-inch R.M.L. gun has only four separate parts, viz., the breech coil or jacket, *B* tube, the 1 *B* coil and breech-piece, whereas the 7-inch R.M.L. gun of original construction has a forged breech-piece, a *B* tube, a trunnion ring, and six coils—nine distinct parts—which are shrunk on separately (see Mark I., Plate IX).

The formation of a heavy coil is a simple forge operation, but great expense is saved by its means, as there is much less surface to be bored and turned, for each coil having to be made as smooth as possible, and at the same time true to gauge (to a thousandth of an inch), it follows that it must be cheaper to have a few thick ones in lieu of many thin ones. For the same reason there is also less waste of material; for although the turnings are afterwards worked up into bars, iron in its scrap state is only worth one-third of its forged value.

Moreover, time and labour are also saved in having fewer pieces to move from workshop to workshop; for instance, in the case of a gun of original construction, when a coil was shrunk on, the mass had to be moved from the shrinking pit to the turning lathe, and turned down for the next coil, and so on, coil by coil, until the gun was built up; but in the Fraser construction only two or three separate shrinkings are required.

Reduced cost of present construction.

From the circumstances, combined with the employment of cheaper iron, a Fraser gun can be made more cheaply than a gun of the same nature as originally manufactured, while the experiments which were carried out previous to the introduction of this construction clearly prove that guns of this pattern are at least quite as trustworthy and serviceable as those of the original pattern.

"Marks" of guns at present manufactured.

Up to 1869 Fraser guns were made on the same type as the 7-inch gun, Mark III. Now, however, 7-inch,* 9-inch and heavier guns have been made with the wrought iron over the breech in two layers of coils. The "Mark"† of each calibre of gun approved for future manufacture is as follows, viz.:—For 7-inch guns, Mark IV. 8-inch, Mark III.; 7-inch of 90 cwt. Mark I.; for 9-inch guns, Mark V.; for 10 inch,‡ 11-inch, and 12-inch, (25-ton guns,) Mark II.; and for 12-inch, 35-ton guns, Mark I.; 12·5-inch 38 tons, Mark I.; 64-pr. 64 cwt., Mark III.; 40-pr. 35 cwt., Mark II.; 25-pr. 18 cwt., Mark I.; 16-pr. 12 cwt., Mark I.; 9-pr. 8 cwt., Mark II.; 9-pr. 6 cwt., Mark II.; 7-pr. steel 150 lbs., Mark III.; 7-pr. steel 200 lbs., Mark IV.

Details of Manufacture of Woolwich Guns.

Manufacture of a heavy gun, 7-inch.

We will now proceed to the details of the construction of a Woolwich gun, and will, for the sake of convenience, take the manufacture of a heavy gun, and, as a good example, that of a 7 inch, and then see how the manufacture of the larger and smaller natures differs slightly from the same.

---

* The construction of the 8-inch gun still remains as in 1869.

† In November 1867 the word "Mark" was substituted for "Pattern." See § 1545.

‡ It is probable that a new pattern will shortly be adopted for 10″ and 11″ guns.

7-inch Gun, Mark IV.

This gun consists of:—

An inner barrel or tube of toughened steel (*A* tube).
A *B* tube.
1 B coil or belt.
Breech-piece.
C coil or jacket.
Cascable.

Steel tube.

The steel for the tube is received from the contractors in the form of a solid ingot, which is rough turned, care having been taken to fix it truly central in the lathe by means of a chuck at the muzzle and the centre at the breech. In this operation a lip or collar is formed at the muzzle to facilitate the lifting of the tube in and out of the furnace and oil bath; the slice for testing is also cut off the breech end during the rough turning. The tube remains in this state until the result of the required tests (which are described in Chapter V.) are known. Should the ingot not be rejected the manufacture proceeds. *A* tube.

The block is next bored roughly from the solid. The boring head is the ordinary shaped "half-round bit" with one pointed cutter set angularly, and three steel burnishers. After this, the conical chamber (which is in all Fraser guns) is roughly formed by means of a cylindro-conoidal head with one long cutter and six steel burnishers, two on the taper part and four on the cylindrical.* Rough boring and turning.

The tube is now ready for toughening in oil. This operation consists in heating the roughly bored tube to the approved temperature in a vertical furnace, and then plunging it bodily into a bath of rape oil, in which it is allowed to cool. Toughening in oil.

The tube is lifted by a crane, and placed in a perpendicular position in an upright furnace; an iron coil, larger in diameter than the steel tube, is placed upon the fire bars at the bottom of the furnace for the tube to rest upon; beneath this iron coil is placed a piece of plate iron, to prevent the cold air coming in contact with the steel, and in order to obtain an uniform temperature at the extreme end of the steel tube the iron coil is filled with wood ashes.

After the steel has acquired the proper temperature throughout, the crane is brought over the furnace; the cover of the latter is removed, and the block of steel is drawn out and placed in a large iron tank full of oil. The heated steel sometimes causes the surface oil to take fire, which is extinguished by closing the covers at the top of the tank. The tank has a water space around it through which a supply of cold water permeates for the purpose of keeping the oil cool.†

---

* Rough and fine boring, forming the chamber, and broaching, are all effected in the same horizontal machine, the difference being in the shape of the boring head and cutters.

† A full description of this process is given in "The Management of Steel" by Mr. George Ede of the Royal Gun Factories.

CHAP. VIII.

Effects of toughening in oil.

The process of toughening not only warps the steel a little, but sometimes causes the surface to crack. The barrel must therefore be slightly turned and bored to make it straight inside and outside, as well as to remove any flaws that may have been generated. This second boring (performed with a cylindrical boring head, fitted with five long edged cutters and five wood burnishers) increases the diameter to 6·6 inches. By this the cracks are generally removed, but lest there should be any dangerous flaws the steel barrel is subjected to the following water test:—

Water test after toughening.

The tube having been recessed on the face for a gutta-percha ring, and inside the muzzle for a leather cup, is fitted with these, placed in a horizontal hydraulic press, and screwed tightly up between two cast iron heads by means of two strong wrought iron bars extending from head to head, and portions of which are threaded and furnished with nuts, worked by a long spanner. The tube is then filled with water from the main, through a hole in the head and the leather cup, the pipe of the press fixed into the hole, and the pump set to work by steam. The pressure on the interior is shown by two indicators, one vertical and one horizontal, so as to check one another When 4 tons per square inch is indicated, the pressure is withdrawn, and if no flaw has been detected by moisture on the exterior, the tube is considered safe and sound. The barrel is left in this state until the *B* tube is ready to be shrunk over it.

All steel tubes treated alike.

The steel tubes for every piece built up, and that from which a 7-pr. is entirely made are treated in exactly the same manner, except as to the amount of pressure applied.

The Breech Piece.

Breech piece. Two coils welded together.

The breech-piece is composed of two single coils united together end for end.

The two coils, being made and welded in the usual way, they are faced and reciprocally recessed, and then united together endways by expanding the faucet of one coil by heat, and allowing it to shrink round the spigot of the other. This fastens the two coils sufficiently together to admit of the cylinder thus formed being placed upright in a furnace, whence, when it arrives at a white or welding heat, it is removed to a steam hammer, and receives on its end six or seven blows which weld the joint completely. The rear portion of that coil for the breech end is formed of a thicker section of bar, so that after welding, the exterior having everywhere been hammered to the same diameter, there may be sufficient thickness of iron in the interior portion at breech end to allow for the formation of the female screw for the cascable.

The breech-piece is next rough turned, during which process a rim is formed on its exterior for the convenience of lifting the tube in "shrinking." After this, it is rough and fine bored in the same horizontal machine, and a female thread cut in the breech end for the reception of the cascable screw.

Measuring interior prior to shrinking.

The interior of the breech-piece having been brought to the degree of smoothness requisite for close contact with the steel barrel, is gauged every 12 inches down the bore. To the measurements thus obtained, the calculated amount of shrinkage is added; a plan is then made out according to which the exterior of the *A* tube (or rather that portion of it on which the breech-piece is to go) must be turned down, in order that it shall be larger than the bore of the tube by the required amount of shrinkage at the respective parts.

The plan (as illustrated by the annexed drawing), is made on a slip of paper, and together with a corresponding series of accurately measured horseshoe gauges, is furnished to the turner, who turns down the breech end of the *A* tube accordingly.

PLAN OF *A* TUBE.

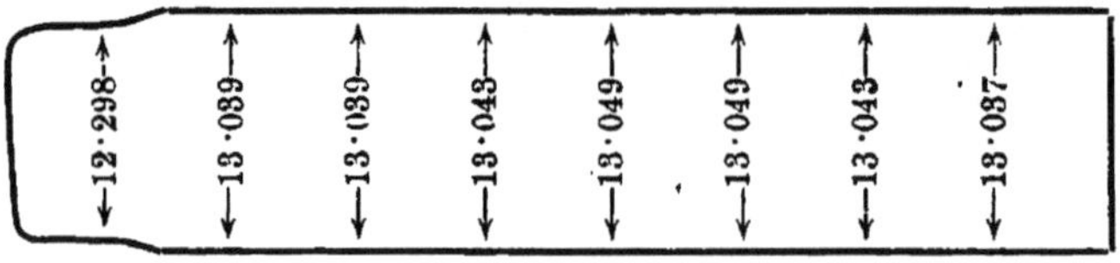

The reason an *inner* tube is turned to suit an exterior one, instead of the latter being bored to suit the former is, as previously stated, that it is much easier to turn than to bore to very exact dimensions, on account of the great command which the operator has over the turning lathe, and the facility he has of testing his work by gauges, and correcting it by means of emery powder and oil.

1 *B Coil or Belt.*

The 1 *B* coil is composed of a thick single coil, which after welding is rough turned. It is then rough and fine bored, gauged every 12 inches, and a plan made out, so that the centre portion of the *A* tube, and the front portion of the breech piece may be turned down accordingly.

B Tube.

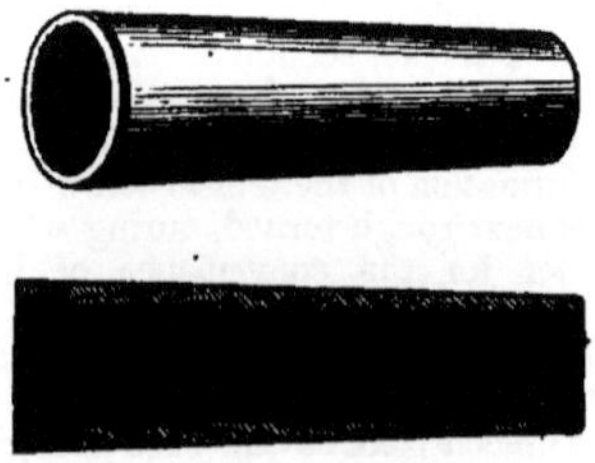

CHAP. VIII. The *B* tube is composed of two single and slightly taper coils. These coils having been made and welded in the usual manner, are united endways (vide p. 108). The *B* tube so formed is rough turned, during which process a rim is formed near the muzzle end, for the convenience of lifting the tube in shrinking. After this it is rough and fine bored in the same horizontal machine. The interior is gauged, and a plan made out to guide the workman in turning down the muzzle portion of *A* tube, and the front of the 1 *B* coil.

*The Jacket.**

Jacket. The jacket is composed of a breech coil, and a trunnion ring, made and welded together as follows:—

Breech coil. The breech coil* is formed of a single coil, and in order to weld its folds, it is placed in a furnace and brought to a welding heat, whereupon it is rapidly transferred to a powerful hammer, and receives a few smart blows on its upper end, which close the folds longitudinally. A mandrel somewhat larger in diameter being then forced down, the coil turned on its side, and well hammered all round to make it dense, and also to weld the three layers together. It is replaced in the furnace for about four hours, and the same process repeated at the breech end, but with a smaller mandrel.

When cold, the ends are faced and the coil is turned down at the muzzle end to form a shoulder for the reception of the trunnion ring.

Trunnion-ring. The trunnion ring is made—as described at p. 107 like all wrought iron trunnion rings, namely, of slabs of iron consecutively welded together on the flattened end of a porter bar, and gradually formed into a ring by means of, first, a small iron wedge, which is driven through the centre and punches an oval hole, and then by a series of taper mandrels increasing in size, which make the hole sufficiently large and round. The trunnion ring has to be heated for each punching, and the occasion is utilized to hammer the trunnions roughly into shape, one of them being in continuation of the porter bar. Eventually the ring is cut off from the bar by means of strong blunt hatchets of steel hammered through it. After this it is roughly bored out.

The Jacket having been built up from these pieces is turned in a very powerful lathe

It being impracticable to turn down the trunnion portion in a lathe, it is slotted smooth in a self-acting vertical machine with a double motion, one of which moves the jacket round for a fresh cut at every stroke of the tool which the other works up and down accordingly.†

The trunnions themselves have yet to be turned down to shape; so the jacket has to be moved for the purpose to another machine, a *break* lathe, in which it is made to revolve on the axis of the trunnions while the cutters act on their surface.

The jacket is next rough and fine bored in a machine like that used for the *B* tube, but more powerful, and the front portion is recessed to fit over the rear portion of 1 *B* coil.

---

* The jacket is often termed the breech coil from its principal portion, when it comes immediately over steel tube. With guns having a coiled breech-piece, the jacket is termed the C coil.

† Since 1874 the trunnion shoulders of 10-inch guns and upwards have been rounded off.

Building up the Gun, or shrinking the parts together.

Shrinking the parts of the gun together.

The steel barrel and breech piece being prepared for one another, as described, are shrunk together in this manner:—The breech piece is placed on a grating, and heated for about two hours by means of a wood fire, for which the tube itself forms a flue, until it is sufficiently expanded to drop easily over the breech end of the steel barrel, which is placed upright in a pit ready to receive it. The breech piece is then raised, and the ashes, &c., being brushed from the interior, it is dropped over the steel barrel by a travelling crane overhead. During the process of shrinking, a stream of cold water is forced up the steel barrel, to keep it as cool as possible.

The cascable is now screwed in

Cascable.

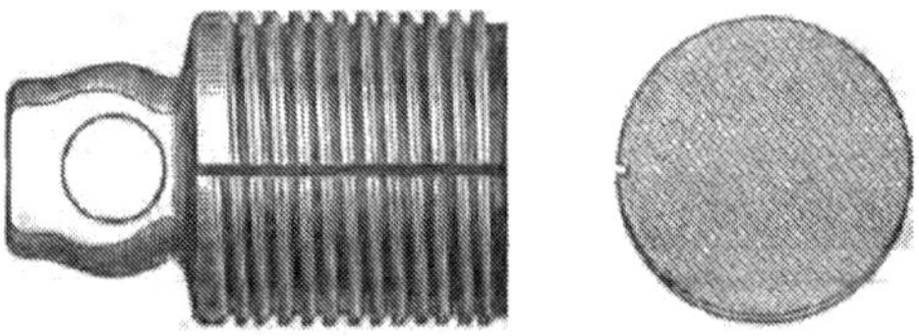

The cascable is made of the best scrap iron. It is first forged into an oblong block, then turned cylindrical, and a bevel thread cut on it. The outer end is partially turned, and a hole is drilled for the purpose of screwing it into the gun, which hole is subsequently enlarged into the loop.

Loop or button.

In the case of the 35-ton and larger guns, and with R.M.L. howitzers, the outer end of the cascable is turned into a plain button.

Screwing in the cascable

This operation of screwing in the cascable requires great care, for the front of it must bear evenly against the end of the steel barrel, and in order that this may be the case, the end of the tube is smeared with red lead and the cascable screwed in tentatively, then unscrewed again, and filed down on the prominent parts, which are indicated by the presence of the red lead. This is repeated several times, until the equal distribution of the lead on the front shows that it bears evenly against the steel barrel.

Gas escape.

At this stage, one round of thread is turned off the end of the cascable, so that there may be an annular space there, which in connexion with a channel cut along the cascable and across the thread, will form a gas escape, or tell-tale hole, in case the steel barrel should split. The channel is about $\frac{3}{8}$th inch broad, and extends $\frac{1}{10}$th inch below the thread. In all guns made before the 1st September 1869, the channel comes out directly under the loop; but in guns made since that date, it will be found at the right side, where it may be more easily noticed. The channel ought to be kept clear, and should the barrel split, gas would be seen issuing from the hole; it is therefore

CHAP. VIII. advisable, in case there should be any suspicion concerning the gun, to keep an eye on this hole, and to cease firing should it give warning.

When at length the cascable fits properly, it is finally screwed in, and to prevent its moving, a hole 2¼ inches long, and ¾ inch in diameter is drilled and tapped through the breech-piece, and into the cascable in a slanting direction on the left side, and a plug is screwed in to prevent any chance of the latter turning round.

The *A* tube and breech-piece thus shrunk up are placed in a lathe, and the front portion of the breech piece, and the centre portion of the *A* tube are turned for the reception of the 1 *B* coil. The half formed gun being next placed, standing on its breech end in the shrinking pit, the 1 *B* coil is heated and shrunk on in the same manner.

The muzzle portion of the *A* tube and front portion of the 1 *B* coil having been turned for the reception of the *B* tube, that tube is now shrunk on.

During the process of shrinking, a stream of cold water is poured into the steel barrel, to keep it as cool as possible, the water being supplied and withdrawn by a pipe and siphon at the muzzle. A ring of gas or a heated cylinder is placed round the muzzle or thin end of the *B* tube, to prevent its cooling prematurely, whilst a jet of cold water plays on the other end, which it is desirable should grip first; were both ends allowed to contract simultaneously, the intermediate part of the tube would be drawn out to a state of longitudinal tension, and weakened accordingly.

The partially formed gun having now a complete layer of coils over

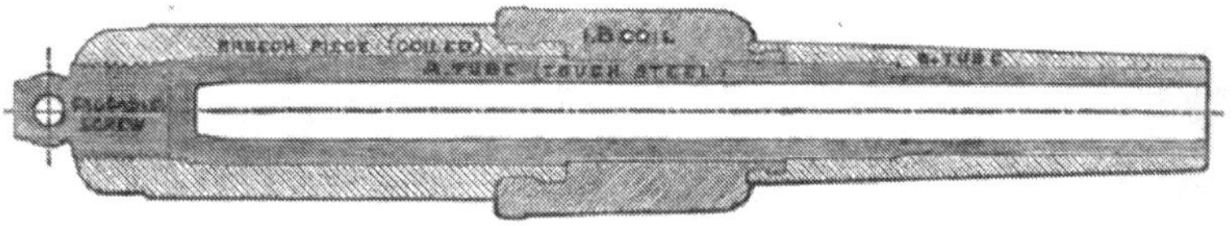

its steel tube is removed to the workshops, and its exterior is turned for the reception of the jacket.

Having been replaced on its muzzle end in the shrinking pit, the jacket is heated for about 10 hours, and shrunk on in the same manner as the *B* tube; the jacket is however (being nearly of the same thickness throughout), allowed to cool naturally, and to keep the interior cool, cold water has to be forced up, into the bore of the gun by a jet found which the muzzle rests.

8-inch Gun, Mark III.

Consists of same parts as the 64-pr., Mark III., (vide p. 172).

9-inch, Marks IV. and V.*

Consists of:—

A tube (toughened steel).
B tube.
Coiled breech-piece.
Jacket (*C* coil).
Cascable.

---

* Mark V. is of exactly the same construction as Mark IV., from which it only differs in the position of the axis of the trunnions being 0″·875 further back.

The A tube is prepared up to the point of shrinking, as already described. CHAP. VIII.

The coiled breech-piece consists of two coils united, and being finish bored, and a thread cut in the breech end for the cascable, is shrunk on the *A* tube, after which a shoulder is formed on its muzzle end.

The B tube is manufactured like that already described, except that a recess with a hook is cut in the breech end for the purpose of joining it to the coiled breech-piece.

The cascable is screwed in before the *C* coil is shrunk on in this and all guns of similar construction, the cascable thus gaining the advantage of compression due to shrinking.

For these 9-inch guns, and for all heavier natures, the rifling &c. of the piece is completed before the jacket is put on, before the gun is made too heavy to be moved from one workshop, or machine, to another without inconvenience.

The C coil or jacket is composed of a breech coil, trunnion ring, and muzzle coil welded together, and being finish bored and turned, is shrunk on over the coiled breech-piece.

The remainder of the operations are similar to those for a 7-inch gun.

7-inch Gun, 90 cwt., Mark I.

Consists of much the same parts as the 9-inch gun, Mark V.,* described below, viz. :—

A tube (toughened steel).
B tube.
Coiled breech-piece.
Jacket (*C* coil).
Cascable.

Reduction of 7-inch 6½-ton guns to 90 cwt.

Only a few of these guns have been manufactured as new guns, but a considerable number have been made by reduction from the 7-inch S.S. gun of 6½ tons, vide p.

In both cases the jacket is made of a single coil and a trunnion ring, but in the reduced guns the breech-piece is a part of the old jacket which remains after removing the trunnions and turning off the outer portion.

In the new guns of this nature, on the other hand, the breech-piece consists of two coils united.

Removing the old cascable.

With reference to the reduction of these guns from 7″ of 6½ tons, it may be noted that it is not easy to get the cascable out of the latter. In order to do so advantage is taken of the expansion of the breech-piece during the process of shrinking on the new jacket.

10-inch gun, Mark II., *and higher natures of those manufactured since* 1869, *up to the* 80-*ton Gun,*

Consists of :—

A tube (toughened steel).
B tube.
1 *B* coil (belt).
Coiled breech-piece.
Jacket (*C* coil).
Cascable.

---

* The reduced 7-inch R.M.L. is described here as it consists of similar parts to the 9-inch R.M.L.

CHAP. VIII.

The manufacture of these guns is similar to that of 7-inch, Mark IV., except that as before mentioned, these guns are rifled, &c. before the jacket is shrunk on, to save the labour of shifting the whole weight from one machine to another.

Ring coils.

In making the *C* coil (or jacket) of such guns, in addition, a very short muzzle coil (technically called a "ring coil") is placed inside the trunnion ring and upon the end of the breech coil and welded with these two latter.

The 16-inch R.M.L. Gun of 80 Tons.

This gun consists of:—

A tube (toughened steel).
B tube.
2 *B* coil (in continuation of *B* tube).
1 *B* coil (or belt).
Coiled breech-piece.
Jacket (*C* coil).
Cascable.

Its manufacture is similar to that of the 10-inch, Mark II, and upwards, except in the following particulars:—

2 *B* coil.
1 *B* coil.
B tube.
C coil.

The first layer of tubes shrunk over the steel barrel consists of four instead of three pieces, on account of the great length of the gun, an additional tube called the 2 *B* coil being placed between the 1 *B* coil or belt and the *B* tube.

As the *C* coil or jacket is of very massive proportions, it is thus constructed:—Four single coils are prepared from bars of large section and two joined together to form a tube which is turned; the other two are then shrunk over it, and the whole welded into one mass.

To the breech coil so made, the trunnion ring and short muzzle or ring coil is added as usual, and then all are welded together into the solid jacket. The section of the bar from which the coils mentioned are made, and the proportions altogether are so large that the mode of constructing the breech coil by shrinking one tube over another, as described, has been adopted both to ensure a fairly homogeneous mass and for convenience of manufacture.

The remainder of the manufacturing operations with this piece prior to proof are similar to those described in the case of smaller guns.

Cascable.

The cascable will probably terminate in a plain button for L.S. like that of the 38-ton gun, but a hole will be bored in the direction of the axis of the cascable, and threaded for the reception of a strong screw bolt and shackle, the latter may be used in lifting the gun, and afterwards removed so that the total length of the gun may be as small as possible.

The 80-ton gun is the heaviest gun which we have as yet made, though there appears no manufacturing reason why guns of twice or four times its size and weight should not be made in the same manner and of equal comparative strength.

Construction of Natures below 7-inch.

Construction of guns lighter than 7-inch.

We will now take in detail the construction of R.M.L. pieces which are lighter than the 7-inch, and of patterns now being made.

10-inch Rifled M.L. Howitzer of 6 tons (Experimental).

The 10-inch howitzer consists of :—

A tube (toughened steel).
Coiled breech-piece and *B* tube.
Jacket, (*C* coil).
Cascable.

The construction of this piece is slightly exceptional, for the coiled breech-piece and that part which in most guns is termed the *B* tube, are united together before being shrunk on to the steel barrel, and so form a single tube which extends from breech to muzzle.

This mode of construction is adopted for convenience in manufacture, the two coils of which this breech-piece and chase are made being short and comparatively thick coils.

The *A* tube being prepared, the coiled breech-piece is finished bored, the thread for cascable screw cut, and a shoulder left for a corresponding shoulder on the steel tube to abut against. *A* tube.

The coiled breech-piece and *B* tube (in one) are shrunk on to the *A* tube, and the cascable fitted and screwed in, a gas channel being formed in the usual manner. Coiled breech-piece and *B* tube.

The exterior is next turned down for the reception of the jacket, two shoulders being left, one a little in front of the trunnions, and the other towards the breech end.

The jacket or *C* coil is made of one single coil and a trunnion ring, manufactured and welded together as usual, and then shrunk on from the breech end. It hooks over the two shoulders left on the coiled breech-piece, so binding the gun together more firmly in a longitudinal direction. This and the remainder of the operations are similar to those described in the case of the 7-inch gun. *C* coil.

The cascable terminates in a plain button having no loop. Cascable.

8-inch Rifled M.L. Howitzer of 46 cwt., Mark I.

This piece consists of :—

A tube (toughened steel).
B tube.
Jacket.
Cascable.

It is built up and completed in a similar manner as the 7-inch gun, Mark IV. described in Chapter VII. Breech coil.

In turning down the *A* tube a shoulder is left over which the *B* tube hooks when shrunk on. *A* tube. *B* tube.

The cascable has no loop, but terminates in a plain button. Cascable.

6·3-inch Rifled M.L. Howitzer of 18 cwt.

In construction this howitzer exactly resembles the 8-inch already described, and it consists of the same number of parts. Construction

CHAP. VIII.

The 64-pr. Rifled M.L. Gun, Mark III.

This gun consists of:—

A tube (toughened steel).
B tube.
Jacket.
Cascable.

The jacket consists of a double coil, a trunnion ring and a single coil welded together. The *B* tube is first shrunk over the muzzle end of the steel tube, the jacket fitting directly over the *A* tube and rear portion of *B* tube, which is cut away for the purpose.

Steel barrel.

Guns of this nature made since April 1871 have solid ended steel tubes. Before that date 64-pr guns had wrought iron tubes as explained in remarks on different natures, p. 262.

The 40-pr. Rifled M.L. Gun of 35 cwt., L.S. Mark II.

This gun consists of:—

A tube (toughened steel).
B tube.
B coil.
Jacket.
Cascable.

In construction this gun differs from the 64-pr. in having an additional *B* coil between the jacket and *B* tube; this mode of construction is adopted as the *B* tube would be inconvenient to manufacture were it sufficiently long to extend from the muzzle to the jacket, as that part of the exterior is naturally rather thin in a gun of this size.

The 25-pr. Rifled M.L. Gun of 18 cwt. L.S. Mark I.

This gun consists of:—

A tube (toughened steel).
B coil.
Jacket.

Cascable.

This gun has no *B* tube nor cascable screw, and so stands midway in construction between the heavier guns and our field pieces, the chase consisting in part of the steel tube unstrengenthed and the cascable being turned down from the projecting breech end of the steel barrel itself.

A tube before tempering.

Before tempering the inner tube of steel a hole 1 inch diameter and some inches deep is bored in its solid end (so as to reach within about 2¼ inches of the bottom of the bore), in order to obviate any chance of its splitting at that part during the operation of toughening. This hole is subsequently filled in by a screw plug when the tube has been toughened and proved.

16-pr. Rifled M.L. Gun of 12 cwt. L.S. Mark I.

This gun consists of:—

A tube (toughened steel).
Jacket.

This gun consists of two parts only, viz., a toughened steel tube, and a jacket, composed of two single coils and a trunnion ring welded together.

The cascable is cut out of the solid end of the steel tube, as in the 25-pr., and the chase of the gun for a distance of 30½ inches from the muzzle is entirely of steel, the tube being thicker at that part.

9-pr. Rifled M.L. Gun of 8 cwt. L.S., Mark I.

This gun consists of:—

A tube (toughened steel).
Jacket.

This gun is identical in construction with the 16-pr., except that there is a swell at the muzzle and a dispart patch. This swell is cut out of the solid steel, except in a few of the guns first made, in which it consists of a wrought iron ring screwed on. These can be known by the small fillet which runs round the chase where the iron ring ends.

9-pr. Rifled M.L. Gun of 8 cwt., S.S., Mark II.

This gun is constructed in exactly the same manner and of the same parts as the 9-pr. R.M.L. gun of 8 cwt., L.S., Mark I., except that there is no swell on the muzzle.

The 9-pr. Rifled M.L. Gun of 6 cwt., S.S., Mark I.

This gun is identical in construction with the 16-pr., and differs from the 9-pr. of 8 cwt. only in weight and length, and in having no swell at the muzzle.

9-pr. Rifled M.L. Gun of 6 cwt. L.S., Mark II.

In construction this gun is similar to the 9-pr. of 8 cwt., L.S., Mark I. It is, however, both lighter and longer, the bore being of greater length than that of the latter piece by 2½ inches.

It has a swell on the muzzle and a dispart patch.

The 7-pr. Rifled M.L. Gun (steel) of 150 lbs. L.S., Mark III.

This gun is made out of a solid block of steel (see Plate I), rough bored and shaped, then toughened in oil, and afterwards finished in the usual manner.

7-pr. Rifled M.L. (steel) of 200 lbs., L.S. and S.S., Mark IV.

Like the 7-pr. of 150 lbs. weight, this gun is made out of a single block of steel, which has, however, no projection at the muzzle like the block for the former, but only a slight patch to be formed into a dispart patch subsequently.

CHAP. VIII.

Processes before Proof.

Processes before proof.

We now come to the various processes which the 7-inch gun we have taken as a type undergoes after it has been built up and before it has been proved with the modification necessary in the case of other guns. The following are the operations :—

(1.) Engraving the Royal cypher.

(2.) Fine boring.

(3.) Second rough cutting of chamber.

(4.) Finished boring.

(5.) Broaching of bore, and finishing of chamber.

(6.) Lapping.

(7.) Rifling.

(8.) Temporary venting.

8-inch and smaller pieces. 9-inch and upwards.

In the case of the 8-inch and smaller guns, all of these processes are performed, as stated above, after the gun is completely built up; but with 9-inch and higher natures they are carried out before shrinking on the jacket, with the exception, of course, of the temporary venting, and engraving the royal cypher. We thus avoid having to move about from one machine to another during these processes so heavy a mass as would otherwise be the case.

Engraving.

(1.) Her Majesty's monogram is engraved in front of the vent. In the case of the 6″·3 and 8″ howitzers, the monogram is cut on upper surface of chase, to avoid interfering with the elevating plane, the outline being marked on the gun by means of a perforated brass plate, rubbed over with charcoal.

Fine and finish boring.

(2.) The gun is next removed to the boring mill, where it is fine bored to 6″·9.

Chambering.

(3.) The chamber at the bottom of the bore is next roughly bored out with the same boring head as before. In the case, however, of guns having enlarged powder chambers as in the 80-ton gun and in the new 13-pr. field pieces, a different nature of tool is required for the operation.

Shoulder on *A* tubes.

In all R.M.L. guns, 40-pr. and upwards and also in the 8-inch and 6·3-inch howitzers, the end of the bore has a conical chamber to allow of a shoulder (as shown below) being cut on the outside of the steel tube, which abutting against the coiled breech-piece helps to distribute the longitudinal strain over the latter.

Why necessary.

With the present construction this shoulder is required, because the layer of iron next the tube being a coil, it would be necessary, in order to retain the same diameter of cascable as that used with guns having forged breech-pieces, to make the interior of the whole coil much smaller than is now required, and afterwards to bore out a large portion of it for the reception of the steel tube. This would cause a considerable waste of labour and material, consequently in these guns the cascable screw is made considerably larger in diameter, that is of nearly the same diameter as the exterior of the tube, leaving only a small shoulder about half an inch broad.

The cutting of this shoulder weakens the steel tube at that part, as shown by dotted lines "*a b*."

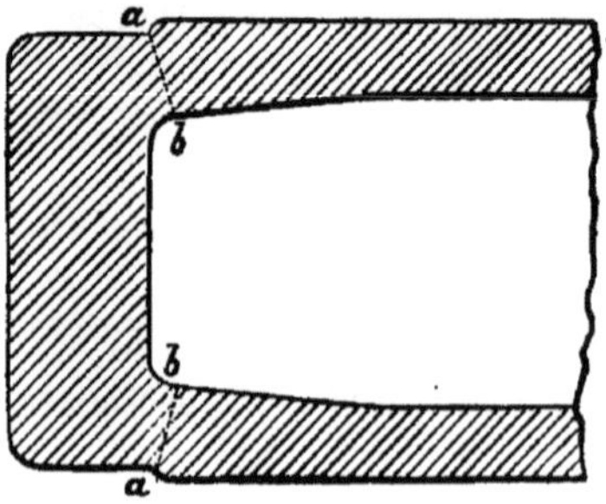

In guns of the original construction, viz.,

7-inch,	of	6½	tons,	Mark	I.
7	"	7	"	"	
8	"	9	"	"	
9	"	12	"	"	
12	"	25	"	"	

this shoulder was not necessary, as they had forged breech-pieces, and the diameter of the cascable screw was made about equal to the calibre of the gun (see 7-inch gun, Mark I, Plate IX), thus affording a strong shoulder in the breech piece against which the end of the *A* tube abuts, thereby reducing the strain on the thread of the cascable screw.

In the 25-pr. and smaller natures the bottom of the bore is merely rounded off, as such a shoulder is not required, for these guns have no cascable screws, the *A* tube projecting at the breech end, and the steel tube, which is thicker at the chase than breech end, has a large shoulder abutting against a recess cut in the jacket and *B* coil, so as to take the longitudinal thrust.

(5.) The finished boring to 6″·997 is then performed.

The fine boring and the finished boring are effected with the boring head used in the second rough boring.

Finish boring. (6.) In each boring the cutters wear a little during the operation, so that the bore becomes slightly taper towards the breech. This is of no consequence in an outer tube, as the exterior of the inner one can be turned accordingly, but the bore of the gun must be truly cylindrical, so broaching is employed; that is, boring the barrel by means of a cylindro-conoidal head, fitted with four long cutters at right angles to one another, and slightly tapering. The cutters are edged on the front as well as on the side, as the chamber is also finished off at this time, and for this latter purpose there is also a centre cutter for the end of the bore. **Broaching.**

Lapping. (7.) In order to make the bore absolutely true, lapping must be finally resorted to, the bore being at the same time brought up to its correct diameter of 7 inches. In this operation no cutter is used, but a wooden head, covered with lead and smeared over with emery powder and oil, is worked up and down those portions of the bore which are indicated by the gauges as imperfect.

Rifling. (8.) The gun is now taken to the rifling machine, and the grooves cut as described at p. 114, the form of groove depending, of course, on

CHAP. VIII.

the nature of the piece. With Woolwich rifling the grooves commence about an inch lower down the bore than the rear stud of projectile reaches when home, and they are so arranged that the two grooves nearest the vent are at equal distances from the latter.

Number of grooves.

The number of the grooves varies. With 7-inch guns and all lower natures only three grooves are used, with 8-inch four, 9-inch six, 10-inch seven, 11-inch, 12-inch 12·5-inch nine. In the case of polygrooved guns, the number of grooves varies from 10 in the 13-pr. to about 33 in the 80-ton gun.

Depth of groove.

Width of grooves.

For studded projectiles the depth varies from 0″·1 in the smaller pieces to 0″·2 in 10-inch guns and upwards, and the width from 0″·6 in the 7-pr. to 1″·5 in the heaviest guns. For rotating gas checks the depth is from ·05 in 13-pr. to ·1 in 16-inch gun, and 0·5 to 1 inch in width respectively.

Splay of grooves.

Grooves are widened at the muzzle in 10-inch guns and upwards, in order to facilitate loading, the loading side being cut away to a breadth of 2½″, tapering down to the ordinary width at 2″ from the muzzle. This change was introduced in October, 1871.

According to order dated 15/1/76, in all R.M.L. guns, 10-inch and upwards, of future manufacture, and in guns of these natures which pass through the Royal Gun Factories for repair, the splay of the loading sides of the grooves will be increased, so as to remove nearly the whole of the lands at the muzzle, in order to facilitate the entrance of the projectile into the bore. The angle of splay varies slightly for each nature of the gun; that for the 12″·5 gun of 38 ton is shown in the diagram below.

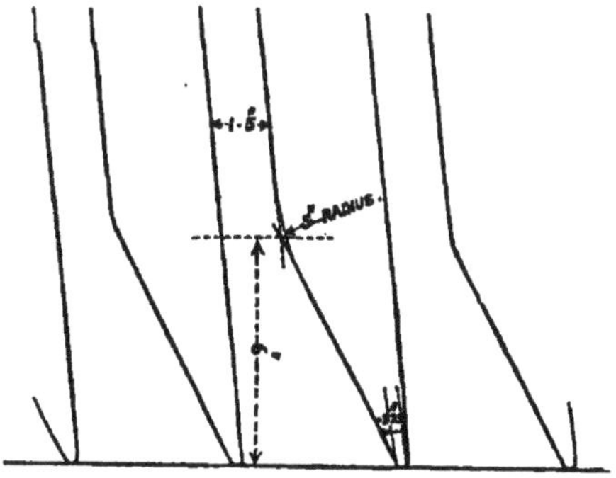

A similar splay is given in the R.M.L. howitzers.

Chamber.

As a rule, about two calibres in length is left plain or unrifled for a powder chamber. The unrifled part should be as long as possible, for grooving tends to weaken the barrel.

Bell-mouthing.

All 10″ guns and upwards of future manufacture, as well as those passing through the department, have a conical portion formed at muzzle end of bore for a distance of 1·5 inches to facilitate loading.

BELL-MOUTHING FOR GUNS OF 10-INCH AND UPWARDS.

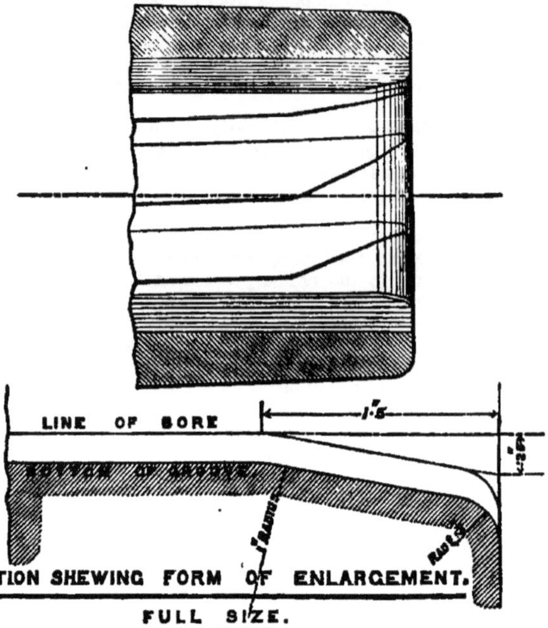

SECTION SHEWING FORM OF ENLARGEMENT.

FULL SIZE.

(9.) Previous to the 23rd January, 1868, rifled M.L. guns were left unvented until after proof, at which they were fired by means of electric wires passed in at the muzzle. Since that date, all guns are drilled and tapped before proof,* and fired through a removable cone vent,† which is unscrewed after proof and replaced by the permanent vent; the object of this is to prevent the proper vent being strained by the large proof charge. Venting.

The cone of this removable vent is about 0"·025 smaller than the service pattern, but after proof the cone in the gun is broached out to the proper size. Proof vent.

The gun is now ready for examination and proof.

Examination and Proof.

All guns are minutely examined before proof, and gutta-percha impressions are taken of the whole length of the bore in four quarters. The bore of all guns of 9-inch calibre and upwards is also accurately gauged every three inches. Gutta-percha impressions taken before proof, and bore gauged.

The object of testing a gun before issue is to make quite sure that it is strong enough to bear not only the strain caused by firing the Object of testing a gun.

* As to position of vent and nature of bush, vide pp. 48, 50.

† Except the 7-prs., which are proved with their service vent.

CHAP. VIII.

ordinary charges, but also any unusual strain which can possibly occur caused with service projectiles and charges.

Proof. | Light guns.

For this purpose the smaller natures are proved by firing two rounds of 1¼ the highest service charge and the service projectile.*

Heavy guns.

With guns above the 8-inch however, the proof consists of one round, with battering charge and two rounds of proof charges which differ for each gun, but are always less than 1¼ the highest service charge. This is done because when the powder charge becomes very large, a pressure upon small increase of powder gives a considerable increase of comparatively the chamber of the gun.

Proof charges.

The proof of guns, 9-inch and upwards, firing Pebble powder or cubical powder, consists of one round with battering charge and two with proof charges; all with service weight of shot.

The following are the charges for proof of 9-inch guns and upwards, viz.:—

	Proof.		Battering.	
	lbs. Powder.		lbs. Powder.	
38-ton gun	180 P^2 at 30 c. in. to lb.		160 P^2 at 30 c in. to lb.	
35 „	115 P.	Service ramming.	110 P.	Service ramming.
25 „	95 „		85 „	
18 „	75 „		70 „	
12 „	58 „		50 „	

Water test. Object.

After proof rifled M.L. guns are tested by having water force pumped into the bore, the pressure being 120 lbs. on the square inch. This test was instituted for guns with wrought iron barrels, having *loose* ends to ascertain that the breech was perfectly closed, for which purpose it is still used with the *converted* guns. It is also continued in guns having solid ended steel barrels, to make sure that the end has not been split at proof.

Impressions after proof compared with those taken before.

Gutta percha impressions of the bore are taken after proof, and the bores of heavy M.L. guns are again gauged. The impressions taken after proof are compared with those taken previously to ascertain whether any defect of a serious character has been developed, and whether slight ones have perceptibly increased. If such should appear to be the case, the gun is subjected to five more rounds with service charges, and if after that the defect appears unimportant the gun is passed.

Further test if necessary.

Impressions of defects.

The impressions of any such defects, however, are cut off, and the position in the gun marked on the back, after which they are registered and preserved for future reference.

In addition to comparing the impressions, the expansion of the bore at the seat of the charge is ascertained by comparing the gaugings before and after proof. This expansion seldom exceeds a few thousandths of an inch, but it may be greater in guns having coiled barrels.

Should the gun pass proof it is now to be prepared for actual issue by fitting with vent, sights, &c.

It has therefore to undergo the processes undermentioned.

Processes after Proof and before Issue.

Processes after proof.

(1.) Lapping.
(2.) Obtaining preponderance and weight.
(3.) Lining.

---

* 64-pr. Mark III. (with steel tubes) will be proved with 15 lbs. powder and 90 lbs. projectile, to cover a charge of 12 lbs. which may be exceptionally used.

(4.) Sighting.
(5.) Venting.
(6.) Marking, and the "marks" denoting pattern.
(7.) Fixing on elevating plates and small fittings, sloping sides of and completing cascable, cutting planes on piece, &c.
(8.) Painting and lacquering, and final inspection.

All the above processes, except the last, are performed in the same workshop (the sighting room), and generally, but not necessarily, in the exact order given.

Lapping.

(1.) Every gun is lapped after proof, for the purpose of removing any little burs which may be thrown up on the edges of the grooves by the heavy proof rounds.

Preponderance.

(2.) "Preponderance," means the pressure which the breech portion, when the gun is horizontal, exerts on the elevating arrangement.

To ascertain its amount, the gun is supported at the trunnions by steel bars placed beneath them, and is brought horizontal by means of long handspikes in the bore. A Kitchener's weighing machine (like that ordinarily used at railways for weighing luggage) is then placed under the breech, and a block of wood fixed upon it, touching the gun underneath midway between the elevating points. The handspikes being then removed from the bore, the pressure on the block is indicated on the arm of the machine, and gives the preponderance of the gun.

Up to 1869 the preponderance of 7-inch guns was about 3 cwt., of 8-inch 4 cwt., and of 9-inch 5 cwt.; since that date, however, 9-inch guns and upwards have been made without any preponderance, anything under 3 cwt. being considered nil. With smaller natures it differs, the amount for each piece is given in Table, p. 291.

Actual weight.

The actual weight of each gun is taken by means of a strong steelyard to the short arm of which the gun is slung by the trunnions. But with very heavy guns the finished portions, *i.e.*, the jacket, &c., are weighed before they are shrunk together, and the total weight is obtained thus with sufficient approximation.

Lining and sighting.

Line of metal.

(3.) The object of lining is to enable the sights and elevating plates to be adjusted. The line of metal is the first line required, and is obtained as in cast iron guns, by finding the axis of the gun and a line in the same vertical plane along the top of the gun, but the process is much more accurate. The gun is placed on a horizontal iron table, and being levelled across the trunnions and along the bore, is carefully scotched up.

Instead of using a wooden batten to find the axis, a centring block, capable of being pressed out so as to fit tightly in the bore, is pushed home to the breech end. From the very centre of this block, a silk thread is extended through a plate on the muzzle to an iron upright (plumbed) stand, some feet in front of the gun, which is furnished with a plumb line, so that it can be adjusted to be truly vertical. The stand is moved to the right or left until the thread passes through the centre point on the muzzle plate.

A "breech gauge," provided with a vertical slide, having been fixed horizontally on the cascable, another silk thread is stretched from the stand to the breech slide so as to pass through a point in the muzzle plate in the same vertical plane as the lower thread, and just high enough to clear the breech of the gun. This gives the position of the line of metal, which is marked for about 1½ inch in length at the extreme end of the cylindrical part of the breech.

Vertical and horizontal lines.

Vertical and horizontal lines are marked on the face of the muzzle along slots in the plate, and short horizontal lines are also marked on the right side of the muzzle, and on both sides of the breech, by means of a scribing block, the moveable arm of which is adjusted to the horizontal slot, the block resting on the table. These lines are useful for the purpose of adjusting the sights, elevating plates, &c.

On the right trunnion vertical and horizontal lines are also cut, except in 25-prs. and smaller natures. The vertical line on the right trunnion enables the gun to be laid point blank (or brought horizontal) at any time without the aid of sights, while the horizontal line on the right trunnion is for use when firing at angles of depression.

Drilling for sights. (4.) The gun is placed under a radial drilling machine. The breech gauge and muzzle plate (the same as used for lining) are then attached, and the gun is levelled to the angle at which the tangent sights are to be inclined to the left, as the machine drills vertically. This brings the

Tangent sights. right sight higher than the left, and the right tangent sight socket nearer the vertical axis than the left. Two silk threads are stretched at one side of the gun from the breech gauge to the muzzle plate, and at the width of the socket apart. The given distance of the tangent sight socket from the line of metal being ascertained (by a gauge), the arm of the machine is brought over the spot and the hole drilled completely through the breech so as to allow of the water and turnings in the after processes to escape.* The drills, &c., work between the threads which answer as a check.

In subsequent borings the drills are not carried through. This operation is repeated on the other side.

The corners of the sight recesses are rounded off $\frac{1}{8}$th inch, to prevent any injury to the thin edge when moving the gun.

Centre hind sights. The hole for the centre hind sight is drilled in the same way and at the same angle, but only of sufficient depth to admit of the sight.

Socket holes for trunnion sights. For the trunnion sights the holes are drilled in a similar manner, but as they are not to be inclined at an angle the gun must be previously relevelled with the trunnions horizontal; the distance from the centre of the tangent sight holes to the centre of the trunnion sight holes is accurately measured (by a gauge) according to the radius at which the gun is to be sighted.

Fitting sockets. The bearing for the tangent sights in the metal of the gun being long and liable to rust, gun metal sockets are fitted in by hand and afterwards fixed by side screws. There is also a gun metal socket and clamping screw for the centre hind sight.

Trunnion sight sockets. Adjustment. For the drop trunnion sights, vide p. 188, gun metal sockets are fixed in the bottom of the hole. These sockets are carefully adjusted by means of gauges, &c., until a pattern drop sight fits into them accurately. This adjustment corrects any slight error due to boring or otherwise, of the socket hole, and as the sockets are fixtures in the gun, it is ensured that any drop sight will answer with them when they have been so adjusted.

In the howitzers, 16-pr. and smaller guns, the holes for the screw trunnion sights or fore sight are bored and threaded.

Venting. (5.) The drilling and tapping of the hole for vent bush has already been completed, as described, p. 177, for the temporary vent used at proof. All that remains is to rime out the cone to the proper size, and to vent the gun with the service bush.

8-inch under 9-inch and upwards. Guns below the 9-inch are vented in a similar manner to S.B. ordnance as far as the operation of venting is concerned, vide p. 60, but with the 9-inch and upwards the bush is screwed in after the cone has

---

* With 35-ton guns and upwards, the vertical hole for the sight is drilled only to a depth of about an inch greater than the length of the tangent bar. From the bottom of this hole a channel is bored to the interior, having a slight slope downwards.

been cut to about the proper length in a lathe. An impression is then taken in the bore and the amount of projection marked on the part of the bush in the bore. The bush is screwed out, the bottom of cone turned off as far as necessary and the operation repeated. When finished the bush projects in the bore about 0″·075*

Marking. Set of tangent sights.

(6.) In addition to the marks made in lining and the Royal cypher before mentioned, the broad-arrow and actual weight are stamped in front of the vent, and the angle of set of tangent scale is also stamped on the gun.†

Letter D.

With 9-inch guns and upwards the letter D is also stamped in front of the socket when the latter has been deepened for the lengthened centre hind sight (vide p. 187).

Two parallel lines are cut across the vent field to indicate the unrifled space. That in front of the vent denotes the end of the rifling, and that in the rear of the vent the end of bore. These lines enable us to mark the sponge or rammer staves for the exact distance from the muzzle to the end of bore or rifling, and show us size of powder chamber, when the end of the rifling is used as stop for projectile (vide p. 46).

Centre of gravity. Half weight line.

Lines are also cut on the top of the gun to denote the position of the centre of gravity and the point at which the sling must be placed in order to take half the weight, the gun being at the same time slung at the cascable.‡

The material of the inner barrel (for example FIRTH'S STEEL) is stamped on the face of the muzzle, as is also the number of the barrel as entered in the registry of manufacture.

On the left trunnion are§—the initials R.G.F., or otherwise of the factory where the piece was made, the register number of the gun, the numeral signifying its pattern, and the year of proof. The register number is that by which the gun is registered in the department records; it indicates also the number of that nature manufactured.

Heel scales.

Those 64-prs. which fire 10 lb charges, as well as 7-inch and upwards (except the L.S. 7-inch of 7 tons), have a heel scale on the cascable for use with the Wood scale for

* At present the reventing of these guns is performed as described at p. 315, with the tools issued for the purpose, the part projecting into the bore being cut off by a knife as with S.B.

† All guns made, or which have passed through the department after 24/6/75, will be found with this angle marked, but others are without it.

‡ See § 1936, List of Changes, 1st September, 1870.

Only 8-inch guns and upwards are marked with the lines indicating centre of gravity and half weight; and as some of the following natures have been issued without these lines the respective distances are given:—

Nature of Gun.		Distance of Centre of Gravity from Muzzle.	Distance of Half Weight from Muzzle.
8″ M.L. guns, 9 tons,	Mark I.	86·45	36·4
„ „	Mark II.	87·85	39·2
„ „	Mark III.	87·75	39·0
9″ M.L. guns, 12 tons,	Mark I.	90·55	34·1
„ „	Mark II.	90·5	34·0
„ „	Mark III.	90·9	34·8
„ „	Mark IV.	90·55	34·1
„ „	Mark V.	90·585	34·17
10″ M L. guns, 18 tons,	Mark I.	109·75	49 5
„ „	Mark II.	109·55	48·35

§ Except with 7-prs., where these are found on the right trunnion.

CHAP. VIII.

S.S., as described at p. 195. This scale is marked by means of a template, and then cut by hand.

Screw holes for fittings.

(7.) Screw holes have to be bored for the screws which secure the different fittings to the gun, index plates, guide plates, &c., according to the nature of the piece.

The positions for such are carefully obtained, by means of accurate gauges, from the lines already marked on the gun.

Gun metal plates for elevating racks.

The position of the gun metal plates for the elevating racks being measured, and the holes drilled and tapped by hand, the plates are firmly attached to the gun by means of a screw at each corner. They are also marked with the number of gun to which they have been adjusted.

Until November 1871 many guns, 7-inch to 9-inch, were furnished with two studs screwed into the face of the muzzle, for supporting the shot bearer during the process of loading. At the above date, however, they were abolished; such guns are provided with preserving screws for these holes.

S.S. fittings.

All heavy L.S. guns are drilled and tapped for the guide plate and friction-tube pin, the holes being filled by preserving screws; thus these guns can be made available for sea service should occasion require.

The preserving screws in the friction-tube pin holes also answer the purpose of indicating the position of the vent by the touch, during night firing.

Shaping the cascable.

To bring the cascable to the approved shape, its sides are sloped towards the rear, except when it ends with a button.

In 25-prs. and smaller natures the cascable is slotted for reception of head of elevating screw.

To prevent the handspikes slipping when working the gun, the breech was formerly scored underneath at each side. This is no longer required, as heavy guns are now elevated by means of elevating racks.

Preparing for derricks.

9-inch guns and upwards must also be prepared for index plates and muzzle derricks.

Trunnion studs.

38-ton L.S. guns are also prepared for the trunnion studs, described at p. .

Quadrant planes.

On Howitzers. the elevating plane is cut by hand, and the two levelling planes by means of a small planing machine.

Painting and lacquering.

(8.) The exterior of the gun being well cleaned, receives one coat of Pulford's magnetic paint, and the bore is lacquered. This only applies to 25-pr. guns and upwards; 16-pr. and smaller natures are browned (vide p.) and the bore left unlacquered.

The gun having been inspected, and found in exact accordance with the sealed pattern, is now ready to be issued for service; and when it has been provided with the fittings described in the next Chapter it is issued by the R.G.F. to the Commissary General of Ordnance, Woolwich, complete.

CHAPTER IX.

SIGHTS, FITTINGS, AND STORES OF R.M.L. ORDNANCE.

CHAP. IX.

Sights.—Six used with 64-pr. and upwards.—Four with smaller natures, except 9-prs. and 7-prs., which have only two.—**Special Sights.**—Other means used for obtaining elevation and direction.—**Tangent Sights.**—Various natures.—Difference between L.S. and S.S.—Centre hind sights.—New pattern for 9-inch guns and upwards.—Trunnion sights and centre fore sights.—Drop pattern.—Screw.—Turret sights.—Moncrieff sights of two descriptions.—Chase sights, when used.—Wood scales.—Index plates and readers.—Clinometer.—Quadrant.—Experimental hanging scales.

Fittings and Small Stores.—Bearers for shot.—Brackets.—Clamps.—Derricks.—Guide plates.—Pivot pieces and elevating plates.—Prickers.—Trunnion studs.—Wrenches.—Table showing different marks or patterns of tangent and fore sights for Heavy guns, Medium, Siege, Field, and Boat or Mountain guns.—Table of wood scales.—Table of fittings and small stores.

Sights.*

64-pr. guns and upwards have six sights, *i.e*, two tangent sights or side sights, one centre hind sight, and three fore or trunnion sights. The trunnion sight used in conjunction with the centre hind sight is usually termed the "centre fore sight." Sights. 64-pr. upwards.

The 40-pr., 25-pr., and 16-pr. have two side or tangent sights and two trunnion sights; while the 9-prs. and 7-prs., are central sighted only, having one hind sight and a small screw fore sight on the muzzle. 40-pr. and under.

All the fore sights are of the drop pattern (p. 188), except in the case of the 16, 9, and 7 prs., with which screw fore sights of different patterns are used. Fore sights.

With the rifled 8-inch and 6·3-inch howitzer exceptional sights are employed. Exceptional sights.

Special Sights.

Besides the ordinary sights named above, we shall find that with certain guns special sights are used at times. These are:— Special sights.

---

* By order dated W.O. 21/12/76, any sights passing through the department and *altered but not brought up to the latest sealed pattern*, are to have the numeral of the mark to which they are assimilated stamped after the numeral of their original mark, thus, I. (IV.).

CHAP. IX.

(1.) Turret sights.
(2.) Moncrieff „
(3.) Chase „

In addition to sights which can be used both for laying and giving elevation to a gun, the following arrangements are sometimes employed for giving elevation only, viz.:—

(4.) Wood scales.
(5.) Index plates and readers.
(6.) Clinometers or quadrants.

Again, guns are sometimes laid as to line of fire alone by the following*:—

(7.) Hanging scales for howitzers and siege guns.
(8.) Graduated arcs on racers used in conjunction with the index plates.

Hind Sights.

Hind sights. All hind sights, whether side or centre, must be inclined at a certain angle as explaned in Chapter II., or else have some other arrangement by which to make up for permanent deflection, as in the howitzers, where a long deflection leaf is employed; but the fore sights are always put in vertically.

We have already seen (p. 180) how guns are prepared for these sights, and how the sockets, &c. are fitted; we can now, therefore, go on to the description of the sights themselves and to the details of the different patterns in the service.

Ordinary Service Sights.

(a.) *Tangent Sights.*

Tangent. sight. The tangent or side sights for heavy guns consist of a rectangular steel bar rounded off on two sides and having a gun metal head, in which slides a gun metal leaf.

Slow motion screw. Excepting L.S. sights made before 1871, which have a slow motion screw for giving minutes of elevation, as with smaller pieces, these sights have a plain head.

For S.S. sights the slow motion screw has never been employed.

Depth of notch. The notch in the deflection leaf through which the sight is taken is now 0″·15 deep;† it was formerly 0″·06, the depth of notch still used for siege and field guns.

The diagram, *A*, p. 185, represents the description of tangent sight‡ now employed with heavy guns and also 64-pr. guns, except 64-pr. siege guns, which have L.S. slow motion screw, &c.

Graduations. The graduations of the several faces differ in the several guns and also in the different marks or patterns, as shown in pp. 212, 225, and explained in notes to the same.

---

* The Collimator described in Manual of Artillery Exercises, p. 90, is practically obsolete.

† The object of making the notch deeper, and of doing away with the slow motion screw in the case of L.S. sights for heavy guns, was to make them interchangeable with S.S. sights.

‡ With the 38-ton gun sights, however, both the L.S. shallow notch and slow motion screw are reintroduced. For S.S., these sights would not be used, such guns being mounted in turrets, and therefore the question of interchangeability between L.S. and S.S. does not arise, as in the case of sights for other heavy guns.

TYPES OF TANGENT SCALES. Scale $\frac{1}{3}$.

A	B	C
10-inch Gun.	40-Pr.	7-Pr. L.S. and S.S.

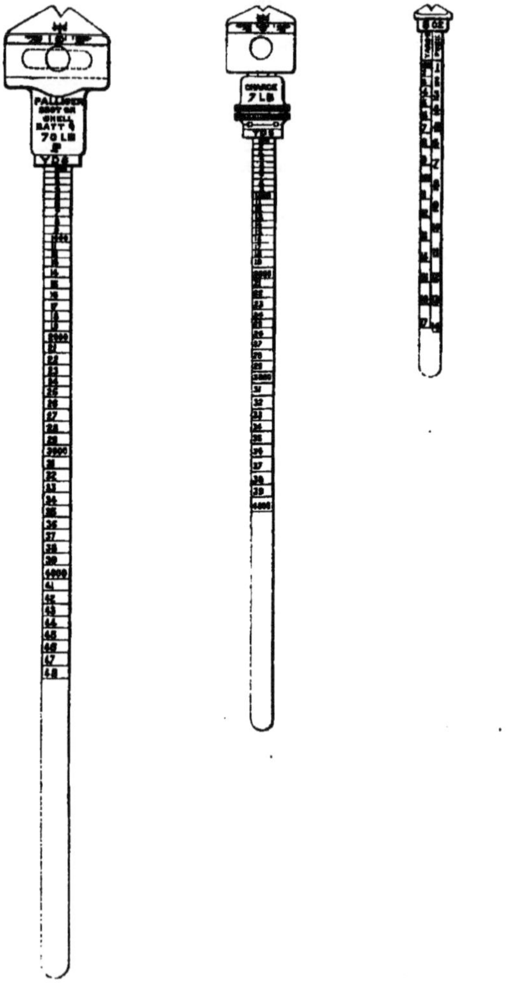

For 64-pr. guns with steel tubes, and smaller guns down to the 9-pr (*i.e.*, all siege and field guns having side sights), the tangent sights consist of a similar steel bar with gun metal head and sliding leaf (*B*); but in these a slow motion screw is always used, and the notch in the leaf is only 0″·06 deep as already mentioned. 64-pr. and under.

The hind sights of 9-pr. guns consist of a bar of steel, which for L.S. has a cross-head of steel and a sliding gun metal leaf, and for S.S. is 9-pr. guns.

CHAP. IX.

quite plain, without any cross head or leaf for deflection. As these sights are graduated for the long radius—from breech to muzzle—they have no slow motion screw for giving minutes of elevation, the sights themselves being graduated for every 3 minutes.

7-pr. guns. The hind sights for the 7-pr. are similar to those of the 9-pr. S.S. (*C*).

Long and short sights. In both 9-pr. and 7-pr. two tangent sights are provided, one long and one short. The long sights are of tempered steel, and have to be employed instead of the short sights for certain elevations. If always in the gun, they would, from their length, be very liable to damage.

The short sight is made of a length about equal to the thickness of the gun at the breech.

The different marks or patterns of the tangent sights for siege and field guns are given at pp. 220, 221.

Howitzer sights. For the 8-inch and 6·3 inch howitzers one tangent sight is placed vertically in the piece and provided with a long steel cross-head and

SIGHT, TANGENT.

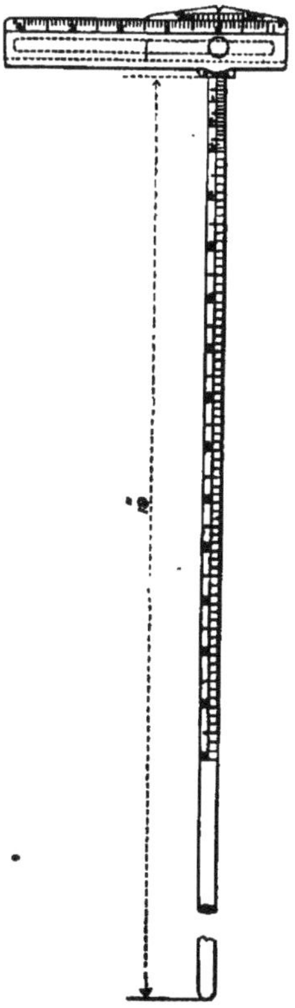

SIGHT, MUZZLE.

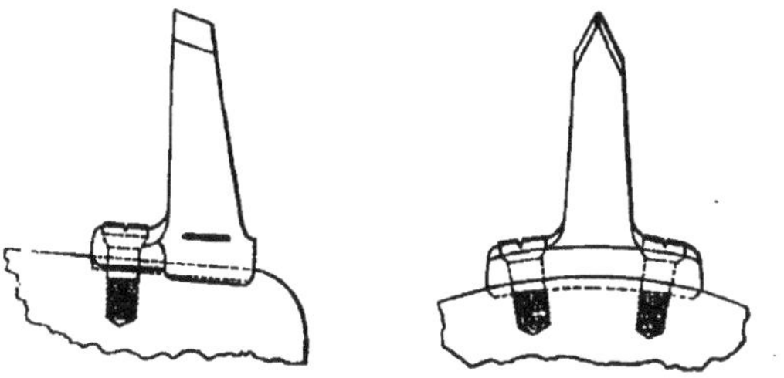

SIGHT, FORE.

deflection leaf. It is used for short ranges in conjunction with a short screw sight on the trunnion ring, and for elevation over 8° with a sight secured by screws near the muzzle, as shown above.

CHAP. IX.

When under cover so that the mark to be hit cannot be seen, the howitzer would be laid as to elevation by the quadrant or clinometer (vide p. 199), and as to direction by the hanging scales described at p. 200.

*Centre Hind Sights.**

Centre hind sights.

A centre hind sight consists of a hexagonal bar of gun metal,† (Fig. 1) which has a plain head in all the older patterns, and also in the most recent for 8-inch guns and downwards; but 9-inch guns and upwards will for the future be provided with lengthened centre hind sights, as in (Fig. 2). These lengthened sights are not only longer but have also a gun metal head, with a sliding leaf for giving deflection.

9-inch guns and upwards, lengthened centre hind sights.

FIG. 1.

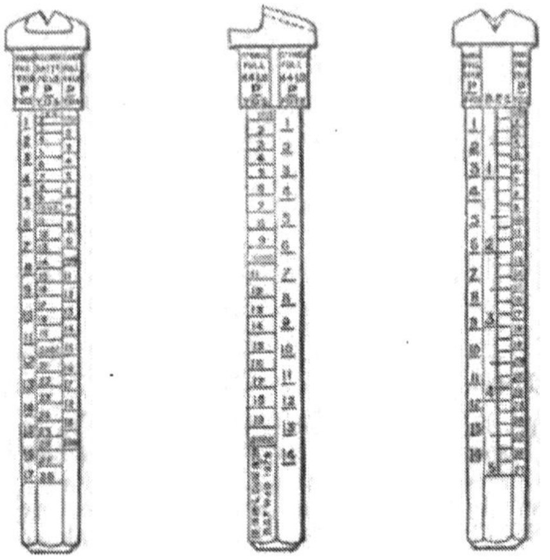

This alteration (introduced in 1874) was adopted for those heavier pieces of ordnance, 9-inch and above.‡ which at 2,000 yards and over would be formidable against iron clads.

Deepening holes, long centre hind sight.

These new sights will not be supplied immediately to all guns, but the latter will be prepared for them by deepening the hole for hind sight. Vide p. 379, Appendix, for this operation, for list of tools used, and for instructions for deepening the sockets at out stations. The letter D is stamped in front of the socket, after the operation.

* Although the tangent sights employed with 9-pr. and 7-pr. guns are central, yet this term above is not applied to them, but only to the centre hind sight used with the 64-pr. guns and upwards, which have three hind sights.

† With 12" of 35 tons and 12"·5 of 38 tons the centre hind sight is precisely the same as the side tangent sight.

‡ When these guns are mounted behind iron shields it is not always possible to use side sights.

CHAP. IX.

FIG. 2.

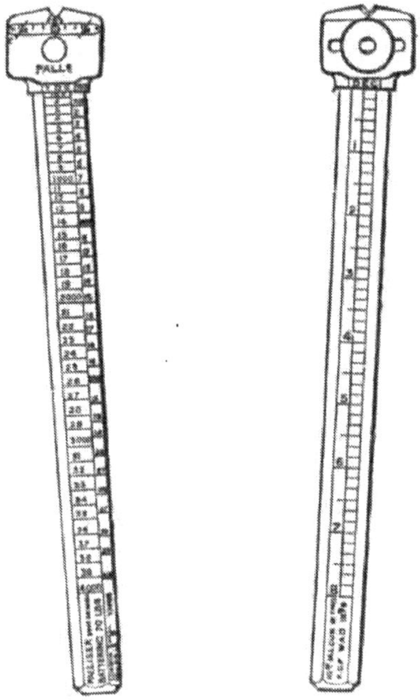

Trunnion Sights, Centre Fore and Muzzle Sights.

Trunnion sights, 64-pr. and upwards. Centre fore sights, 64-pr. and upwards. Trunnion sights, 40 and 25-prs. Drop sight. Component parts.

With 64-pr. R.M.L. guns and upwards, which all have 3 rows of sights, two trunnion sights of the "drop" pattern are employed, and one centre fore sight of the same pattern, but shorter.*

The 40 and 25-pr. guns having but two rows of sights, are provided with two "drop" trunnion sights, similar to those already mentioned.

These fore sights are termed "drop sights" as they can readily be dropped into the socket, and as easily removed. The sight itself consists of a pillar and collar of gun metal, a small steel leaf, and a screw for fixing the leaf.

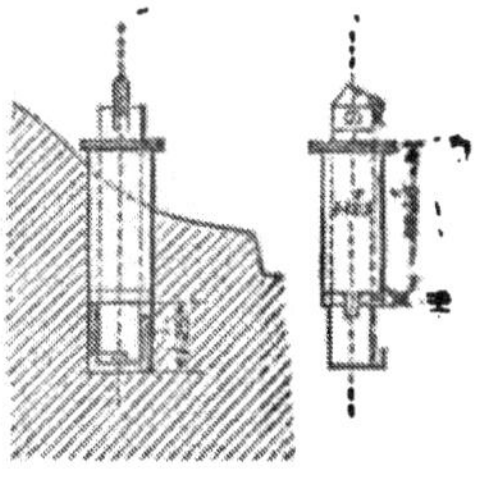

* The radius or distance between the centre sights is slightly greater than that between a side sight and the corresponding trunnion sight, vide Table p. 212.

Sockets. There is a socket of gun metal, into which the sight is secured by means of a double bayonet joint. This socket is permanently fixed in the gun, as explained at p. 180, and adjusted by means of a gauge sight.

To remove a drop sight. To remove the sight the collar must be raised, and then the pillar moved round a quarter of a circle.

As the plane surfaces of the socket and sight can be very accurately fitted, these sights are made interchangeable with the sight leaves already prepared. This is not the case, as we shall see, with screw fore sights.

With the 16-pr., two "screw" trunnion sights are employed, being the only R.M.L. gun so sighted.

Screw sights. Component parts. These are of steel, and consist of a pillar of steel threaded towards the lower end, and furnished with a steel sight-leaf secured to the top of the pillar by means of a small screw.

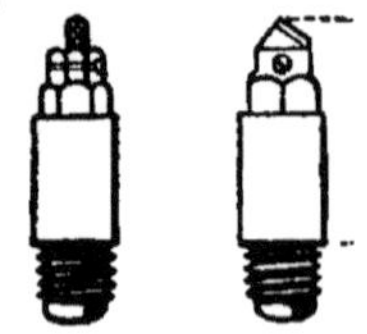

Why spare screw trunnion sights are issued. As it is difficult to end a screw thread very accurately, these sights must be carefully adjusted in each case, and spare sights are issued with rough leaves.

R.M.L. howitzers and guns below the 16-pr. have but a single, central row of sights.

In the 9-pr. L.S. the muzzle sight consists of a small sight, screwed into a recess cut into the projection on the muzzle.*

MUZZLE SIGHTS.

A	B	C
9-Pr., 8 cwt., L.S.	9-Pr., 8 cwt., S.S.	9-Pr., 6 cwt., S.S.

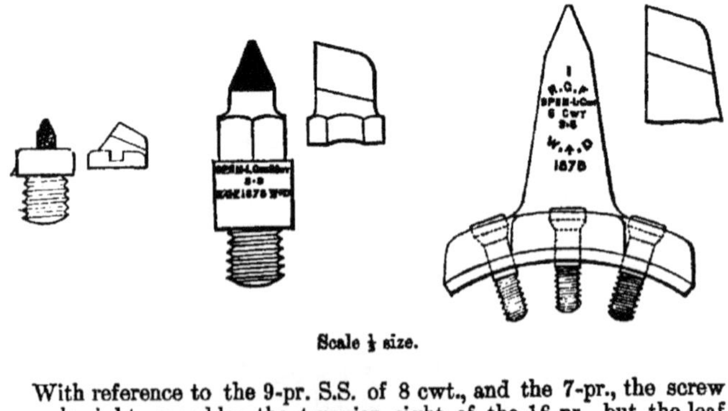

Scale ½ size.

With reference to the 9-pr. S.S. of 8 cwt., and the 7-pr., the screw muzzle sight resembles the trunnion sight of the 16-pr., but the leaf forms an integral portion of the sight.

---

* So as to protect the sight from damage when the gun is being dismounted.

CHAP. IX.

In the case of the 9-pr. S.S. of 6 cwt., there is a high sight fixed on to the muzzle by three screws, as is the case also with the 8" and 6"·3 howitzers. These latter have in addition a small screw centre fore-sight, fixed on the trunnion ring.

For the adjustment of rough leaves of screw sights, vide p. 312.

Special Sights or Means of adjusting the Elevation.

1. *Turret Sights.*

Guns in moveable turrets are mounted upon carriages moving on fixed slides, so that direction must be given to them by traversing the turret itself, which is furnished with means of obtaining the correct line of sight.

Each turret has a number of "man-holes," centre, intermediate, and slide or wing, through any one of which the captain of the turret can raise his head to look along the sights and lay the turret.

A set of sights for each man-hole.

For each man-hole a fore sight and hind sight* are so adjusted that a vertical plane through any pair of sights will be parallel to a vertical plane passing through the axis of the gun and slide. When the hind sight is at zero and the ship on an even keel, the plane passing the notch of the hind sight and the top of the fore sight is a horizontal plane. The sights are also of sufficient height to allow of the line of sight clearing the edge of the turret when the hind sight is at its maximum height. A second set of sights is provided for considerable angles of heel.

To obviate the necessity of the captain of the turret exposing himself to the enemy's fire while laying the turret, a reflecting arrangement has been fitted to the wing man-holes of some turret ships, so that he can direct the traversing and obtain the proper line of sight while under cover. This arrangement consists of two mirrors, one fixed inside the turret, the other while in use is secured upon the outside, and in rear of the man-hole. This latter mirror works on a hinge, and is readily adjusted to any required angle by a lever or hand wheel in the turret acting on a system of bell-crank levers. When not in use the mirror can be brought inside the turret and slid along the roof, where it is secured so as not to be in the way nor liable to damage.

The line of sight is reflected from the outer mirror upon that fixed in the turret, whence the captain of the turret may safely lay upon the object.†

For giving the necessary elevation or depression to turret guns a wood scale is in all cases used.

Two corrections in elevation are at times required:—1st, to make up for the ship's heel, and 2nd, to make up for the gun being raised bodily from one step to another when mounted on muzzle pivoting carriages.

The first correction is thus made:—The captain of the turret observes how many degrees above or below zero are given by the hind turret sight when laying on the object; this number of degrees is added to or taken from the proper elevation given by the wood scale as follows:—The position of the clamp on the wood scale is not altered, but the pointer on the clamp is applied to the given number of degrees above or below zero on the heel scale marked on the cascable.

---

* The hind sight is a rectangular bar of steel with a gun metal head and sliding leaf. On one side is a ratchet in which gears a small pinion worked by a hand wheel. The fore sight is of a special drop pattern.

† This arrangement will not be adopted in future ships.

7-inch gun carriage, Mark I. Description of sight.

The second correction is made by means of a simple arrangement in the wood scale,* so that the latter can be shortened or lengthened to correspond to the height through which the gun is lowered or lifted when placed on the several steps.

2. *Moncrieff Sights.*

In addition to the ordinary sights, two special sighting arrangements are employed with guns mounted upon Moncrieff carriages.

(1.) That used with 7-inch guns mounted on Moncrieff carriages, Mark I., consists of a skeleton gun metal bracket secured by screws to the trunnion ring; the sight-notch is a small V-shaped pin of metal sliding up or down in the skeleton fore sight frame, and clamped by a thumbscrew, and the graduation is in hundreds of yards. To elevate the gun the V must be, of course, lowered.

MONCRIEFF'S FORESIGHT.

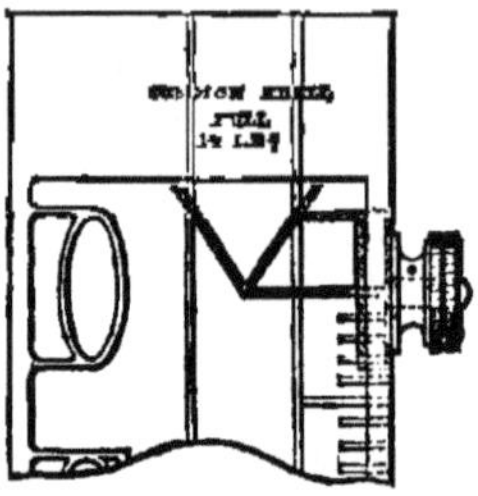

The hind sight is a mirror in a gun metal bracket screwed to the top of the gun, near the breech end, as shown in Fig. below :—

Wrought Iron R.M.L. Gun, 7-inch 7 tons.

Sketch, showing position of Moncrieff Sights for M.I. carriage.

Scale $\frac{1}{12}$.

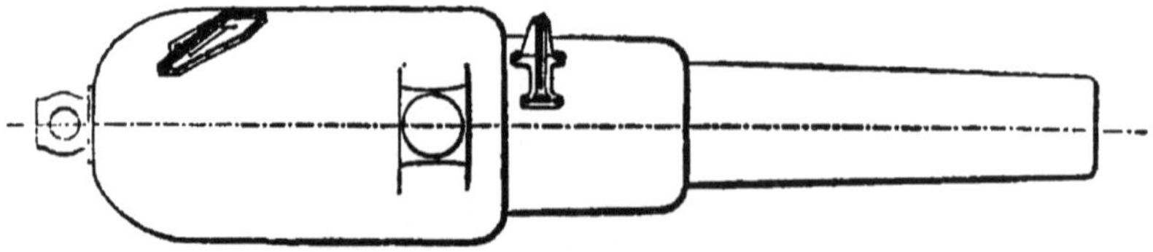

The woodcut shows the service position of the sights when finally adjusted on this particular gun with carriage, Mark I.

The mirror has cross lines upon it, and in order to lay the gun on an object, the V is clamped at the proper elevation and the gun traversed until the object itself and the bottom of the V are reflected together on the intersection of these cross lines.

7-inch or 64-pr. gun

(2.) The system of sighting employed with 7-inch R.M.L. guns of 7 tons, mounted on carriages, Mark II., and for 64-pr. guns on carriages,

---

* A gun metal tube with a button at the lower end slides in and out of a socket in the scale, the button resting on a small plate secured to the bottom plate of the carriage.

CHAP. IX.

mounted on carriage, Mark II.

Mark I., consist of two mirrors and are used without a fore sight. One of these mirrors is secured to the end of the right trunnion of the gun by a circular bracket, and the other is fixed in a sliding bracket or frame which can be moved along a gun metal arc screwed upon the lower part of the elevator, and which is graduated in degrees.

To lay the gun by these sights it is necessary that the object aimed at and the intersection of the cross lines upon the upper mirror should correspond with the intersection of the lines upon the lower mirror, the latter being clamped on the bar at the proper angle of elevation.

To fix the sights, Mark I. carriage.

To fix the sights on the 7-inch gun, carriage Mark I., first take out the preserving screws and thoroughly clean the fitting surfaces of both gun and sight brackets; then screw on sights (using the thick screws for the reflector), taking care that the screws are sent firmly home, otherwise the shock of firing will loosen the sights, destroy the adjustment, and perhaps break the glass.

7-inch M.L. Gun, mounted on Moncrieff Carriage (Mark II.).

Sketch, showing position of Reflecting sights.

Scale ¼ in. = 1 foot.

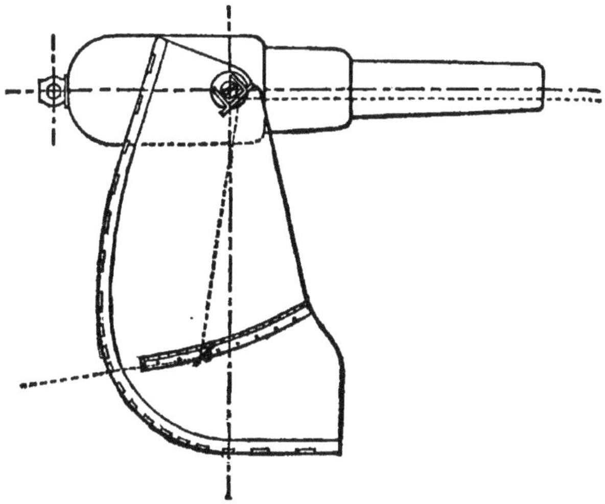

When the sight brackets are firmly fixed, lay the gun accurately by the tangent sights (at zero) on a fixed object sufficiently far off to render the lateral distance between the two systems of sighting inappreciable; then observe whether the intersection of the cross lines (*i.e.*, the centre of the top of the T) upon the reflector is in line with the zero of the fore sight and the object, also whether the vertical line (the upright of the T) on the glass coincides with the centre line of the fore sight, indicated by the extremities of the longer line. If both these conditions are fulfilled, the sight is in adjustment; if not, the glass must be shifted.

To move the glass, unscrew the metal strips round the face of the frame, and take out the narrow frame cushions of india-rubber or other elastic material, and also the wood packing, leaving the glass in the frame; then adjust the glass by paring down the old pieces of wood, and wedging up with fresh ones, as may be required; when correctly and firmly fixed, replace the cushions, and screw in the metal strips again; the sight will then be ready for use.

There are four fine lines on the edges of the mirror, and four on the frame; the relative positions of these should be accurately noted after the final adjustment of the sight, so that any shifting may be readily detected.

Mark II. carriage.

The same course should be pursued with carriages Mark II. as to cleaning the fitting surfaces, &c., and fixing the sights to the gun and elevator, as in the case of Mark I., already described.

Run the elevator up to its firing position, set the lower sight at zero on the curved bar, and clamp it. After this lay the gun by the ordinary tangent sights, also at zero, upon a distinct object, and observe whether the object itself and the intersection of the cross lines on the top mirror coincide with the intersection of the similar lines upon the lower one, also whether the vertical and horizontal lines on the one mirror strictly coincide with those on the other. If these conditions are fulfilled the sights are in adjustment, if not, the glasses must be shifted as described in instructions given for sights with Mark I. carriage.

3. *Chase Sights.* § 3120.

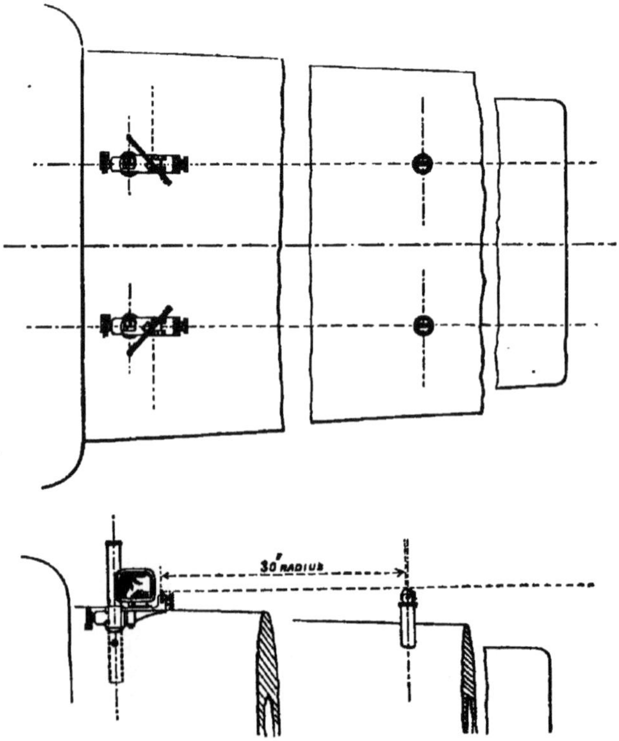

CHAP. IX.

Chase sight.

§ 3120.

In consequence of the small size of the ports in certain works, armed with iron shields (*e.g.*, Plymouth Breakwater), it is found that the ports foul the ordinary sights when the guns are traversed.

Various modes of getting over this difficulty have been proposed by placing sights on the chase of the piece, and it has been approved of that for 12″·5 guns of 38 tons* a mode of sighting should be adopted which had been brought forward by Colonel Inglis, R.E.

The system embraces a fore sight and a tangent sight with reflector, Both are fixed on the chase of the gun at a distance apart of 30 inches, and a set being furnished for each side of the gun.

Fore sight. The fore sight is an ordinary drop sight, fitting into a gun metal socket let into the chase.

Hind sight. The hind sight consists of a rectangular steel bar, graduated in degrees on its rear face, and fitting into a socket in the chase. Upon this bar slides up and down a cross-piece of steel, the direction of which is parallel to the axis of the gun; upon the muzzle end of this cross-piece is a vertical sight-leaf with a notch for use as a back sight, the leaf being capable of sliding laterally to give deflection up to 30″ R. or L.; it can be clamped as usual by a set screw.

Between this leaf and the tangent bar there is a vertical axle on the

CHASE SIGHT FOR HEAVY GUNS.

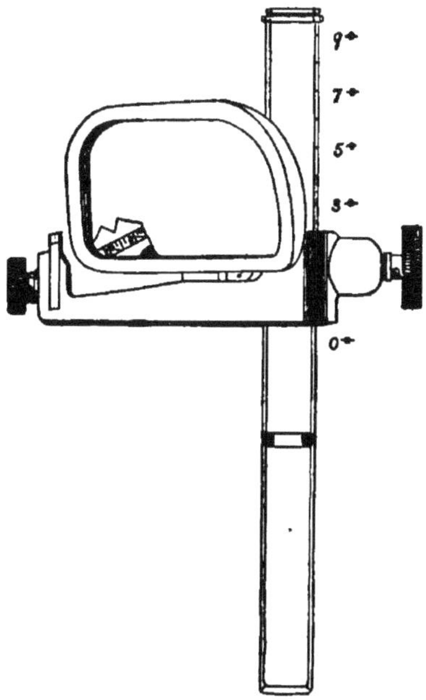

* *i.e.*, Mounted on "muzzle pivoting" carriages.

cross-piece (the axis of which is in line with the sights), and upon this axle a mirror is fixed which can revolve freely in a horizontal plane and partially in the vertical plane also. The cross-piece carrying the back sight leaf and mirror can be clamped at any required elevation ou the tangent sight bar by means of a powerful set screw. CHAP. IX.

To use this sight the man laying the piece stands in front of the trunnions with his back to the port, and having clamped the cross-bar at the necessary elevation, lays the gun by traversing until the object, the notch on back sight, and point of fore sight are reflected together upon the mirror, which he can adjust to suit his own position, or else the mirror can be clamped at an angle of 45°, when the person laying will have to shift position.

4. *Wood Scales.**

For naval service a *wood scale* is used in connexion with the ship's pendulum or director, for giving elevation or depression when the object aimed at cannot be seen from the gun. The scale is square in section, and is graduated for degrees and yards, both for full and battering charges. It is provided with a moveable slide, fitted with a pointer and clamping crew. Wood scale.

The gun, when on its carriage, and run out to its firing position is to be trained on the beam, and laid perfectly horizontal when the ship is upright. If the ship is heeled over the gun should be elevated or depressed to compensate for the heel. The lower end of the scale is then to be shortened if necessary, so that when the scale is held in a perpendicular position, with the foot resting on the slide or carriage, the zero of the graduations on the rear face of the cascable may correspond with the pointer of the clamp, when the latter is set to zero on the wood scale. Adjustment. §§ 1477–8.

The clamp is set at the required elevation or depression or at the graduation in yards for required range on the scale, and the gun then elevated or depressed until the degree on the cascable "From" or "To," corresponding with the heel of the ship coincides with the pointer on the clamp. Use.

Those used with heavy guns mounted on muzzle pivotting carriages have also an arrangement by which they can be lengthened or shortened (vide p. 190).

The graduations on this scale are computed with a radius equal to distance between the rear face of cascable and the axis of the trunnions.

* The various marks, &c. of these scales are given in Table at p. 228.

CHAP. IX.

MANNER IN WHICH THE WOOD SCALE IS USED.*

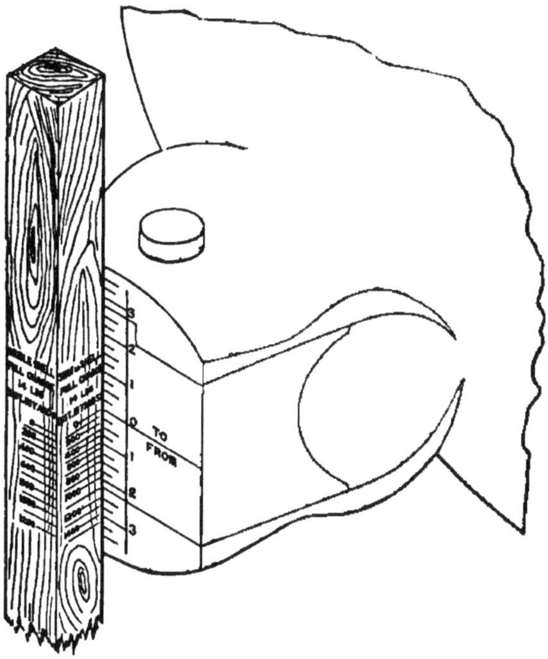

* The scale shown in this drawing is not fitted with the moveable slide and clamping screw used in the latest patterns, which is shown in the woodcut opposite.

Wood Scale (8-inch R.M.L. Gun). Scale.

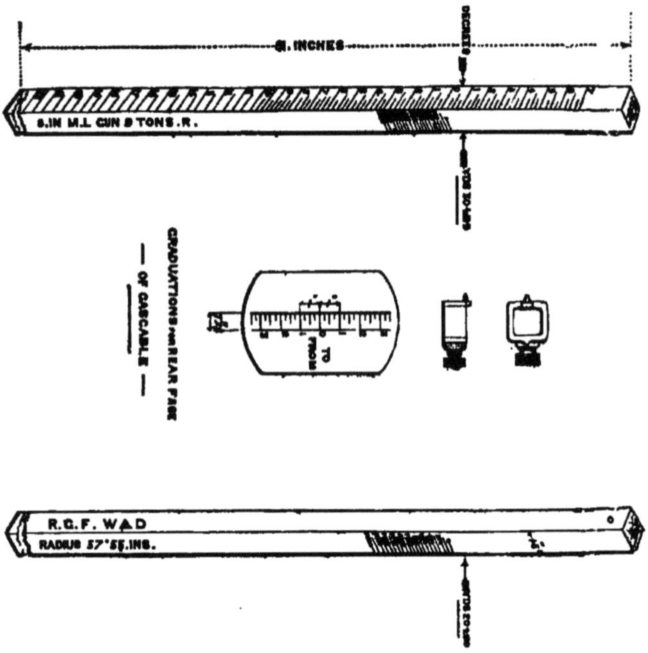

When the radius is above 40 inches the graduations on the tangent scales are calculated for each degree, and therefore increase in length for the higher elevations. Under 40 inches radius the scale is calculated for the highest elevation and divided into degrees of equal length.

5. *Index Plates and Reader.*

These articles have been introduced for service with guns of from the 9-inch to the 12-inch of 25 tons, mounted in casemates or behind iron shields, for the purpose of giving the necessary elevation, when the object aimed at cannot be seen from the gun. They are used in conjunction with the arc graduated for racers, by which the required training is given. **For 9-inch guns and upwards.**

The index plate is a gun metal arc, secure to each side of the gun near the breech end by two screws. These arcs are graduated for 10° elevation down to 6° depression, and a gun-metal pointer termed "reader for index plate" is secured to the carriage, so that the number working the handle of the elevating gear can see when the required elevation is given. There are two patterns, Marks I. and II. **§ 3226.**

Mark II. is used with 10″ guns on low carriages and platforms. (See "Treatise, R.C.D., 1874," p. 216.) **§ 1212, 1575.**

CHAP. IX.

10-INCH R.M.L. GUN, 18 TONS. INDEX PLATE AND READER.

½ Size.

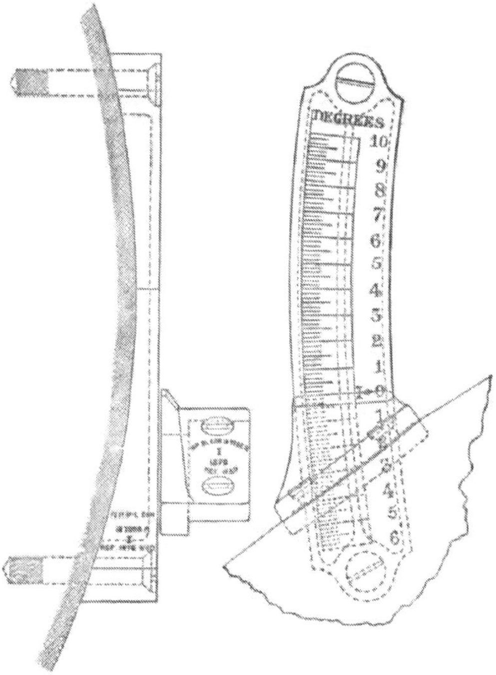

For 12·5-inch and 12-inch guns.

In the case of 12·5-inch guns of 38 tons and 12-inch guns of 35 tons, the elevating arcs will be graduated, in order that they may be employed, in conjunction with pointers fixed on the elevating clamp brackets. Index plates are consequently not used with these guns.

6. *Clinometer.**

§ 1444. This instrument, which was introduced into the service in 1867, can be used for giving the angle of elevation. It is issued as part of the equipment for field guns.

It consists of a 12-inch boxwood rule, with a quadrant marked on the joint and a spirit level set in one edge.

A few useful scales and memoranda are marked on the faces.

7. *Quadrant.**

Quadrant. A quadrant of the description shown below was also introduced for

* These stores are not made in the R.G.F., but obtained by contract.

the same purpose in 1867. This pattern was introduced to replace an older one which had previously been employed for these pieces of ordnance.

For the rifled howitzer lately adopted a new and improved form of quadrant will (probably) be employed, with a spirit level showing, in addition, whether the trunnions are horizontal or otherwise.*

NEW QUADRANT FOR LAYING HOWITZERS.

½ Full Size.

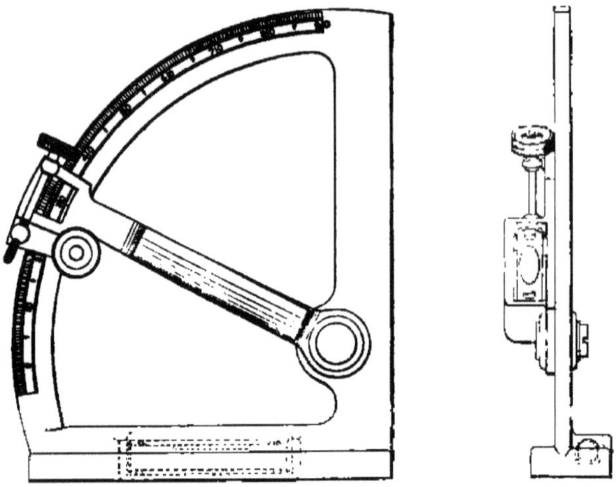

This instrument will be applied to steps cut on the surface of the piece.

---

* The importance of having some such arrangement for eliminating the errors due to difference of level of trunnions with sights in fixed positions in the gun, will be seen from the following examples. Of course for purposes of experiment where extreme accuracy is desirable, and delicacy of instrument of comparatively little importance, such a sight as that described would be particularly useful.

Let R = the range.
E = angle of elevation.
β = angle of inclination of trunnions, with horizontal plane.
δ = deflection due to the inclination of the trunnions.

Now it can readily be proved that $\delta = R \cdot \tan E \cdot \sin \beta$.

With the 16-inch 80-ton gun, let us suppose that when firing at an elevation of 7° with range of 4,694 yards, one trunnion was ¾ of an inch higher than the other from platform slightly giving away; then, as length of axis passing through both trunnions is 106 inches,

We have in this case $\delta = 4694 \cdot \tan 7^\circ \cdot \sin \beta$.

$$= 4694 \cdot \tan 7^\circ \cdot \left(\frac{0 \cdot 75}{106}\right)$$

Therefore $\delta = 3 \cdot 6$ yards nearly, whereas the measured deflection with five rounds as above, was only 1·8 yard.

We see, then, that the error due to a very slight inclination of the trunnions, would be just double that of the gun as fired.

CHAP. IX.

8. *Hanging Scales.*

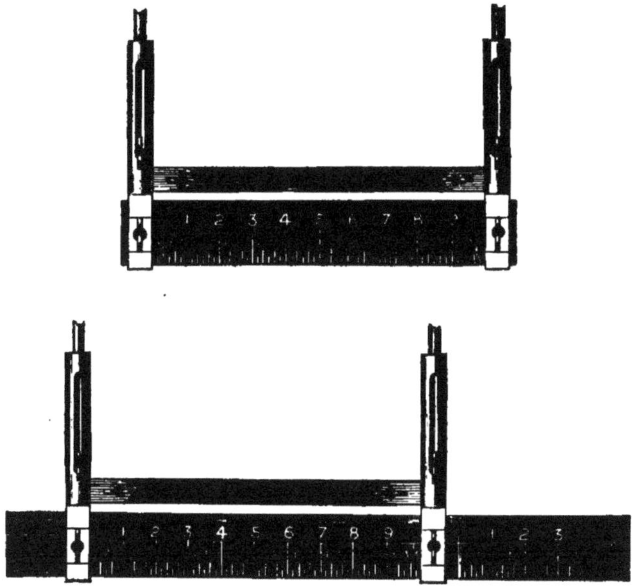

Laying apparatus for siege ordnance.

This is a special arrangement employed for laying siege pieces when under cover, so that the object to be aimed at cannot be seen. The elevation in such case is given by means of (a clinometer or) quadrant, as before explained.

Hanging scales.

The scales consist of two frames, one hanging from the axletree of the gun, and a second parallel to the former hanging from the trail, the frames being at right angles to the axis of the gun and swinging freely on the supporting hooks and eyebolts, so that they always assume a vertical position when in use.

When the piece is in action and the frames hanging down, the distance between their lower edges varies with the nature of carriage measured in the vertical plane passing through the axis of the gun.

Front frame.

For the front frame two eyebolts are screwed into the underside of the axletree at equal distances from the centre. Two suspending rods of iron are hung by hooks on their upper ends from these eyebolts.

Suspending rods.

Each rod slides in a gun metal socket, and has a scale of three inches at its lower end, and a set screw, by means of which the rod can be clamped to length required.

Distance piece.

A "distance piece" is secured to the rods above the caps, completing a rigid frame, the upper portion of which is the axletree itself, the sides being the two suspension bars and the lower side the "distance bar."

The frame swings freely on its upper side in a vertical plane, the bearing surfaces of the suspension bar hooks and the eyebolts in the axletree being knife-edged.

The front "scale bar" fits through thel ateral slots in the caps, and is therefore parallel to the distance bar. It is capable of lateral movement in the frame to the right or left, and can be clamped by a thumb-screw in each cap. **Scale bars.**

The scale bars are rectangular and of gun metal. They are graduated in degrees from 0° to 10°, each of which is subdivided into smaller parts corresponding to 10 minutes. The scales are enamelled and fixed to the bars by means of screws.

How the scales are prepared for use. When required for use the hooked ends of the suspension bars are inserted into the eyebolts under the axletree, and adjusted so that the scale bar when passed into the caps may nearly touch the platform with its lower edge.

The lower ends of the suspension bars are thus connected by the distance bar, and the scale bar placed in the caps.

Rear frame. For the rear frame two eyebolts are screwed into the side of the trail, to which the frame is suspended.

This rear frame consists of the same parts as the front frame, but the supension bars are of course shorter.

The scale bar is rather longer than that of the front frame, and in addition to being graduated like the latter has a scale on the rear face for giving deflection up to 4°.

How to use the scales. In order to use the scales, obtain the line of fire in direction (by pointing rods or otherwise), and mark its prolongation along the length of the platform. The division on the first scale cut by this line is noted, and the piece traversed until the same line cuts a division on the rear scale equal to that on the front.

Deflection scale. Deflection is given by pushing the rear scale to the left, the number of degrees being given by the graduation on the right end of the rear face.

Captain French's laying Apparatus.

This arrangement also allows of laying guns, when under cover, or when the object cannot be sighted from the gun. It consists of two graduated steel bars or sights (fore and hind) adapted to fit the service sight sockets. These are provided with two horizontal bronze arms about 8 inches in length, graduated into intervals of 10 minutes, and a sliding leaf with notch or pointer moves along each arm.

The line of fire, having been obtained, is marked by pickets or plumb line at the rear of the battery, and at the point where this line cuts the graduated arm of the fore sight, the sliding leaf is clamped. The trail is then traversed until the corresponding graduation on the arm of backsight is also in the line of fire, the number laying the gun, standing with his back towards the parapet.

FITTINGS AND SMALL STORES.

We now come to the other fittings and stores for R.M.L. guns, besides their sights. These are all shown in Table, p. 381, in alphabetical order.

Bearers, shot or shell.

These are for use with the L.S. 9-inch, 7-inch, and 80-pr. guns, and the 8-inch howitzer. § 1320.

SHOT BEARERS, L.S. 9-inch R.M.L. Gun.

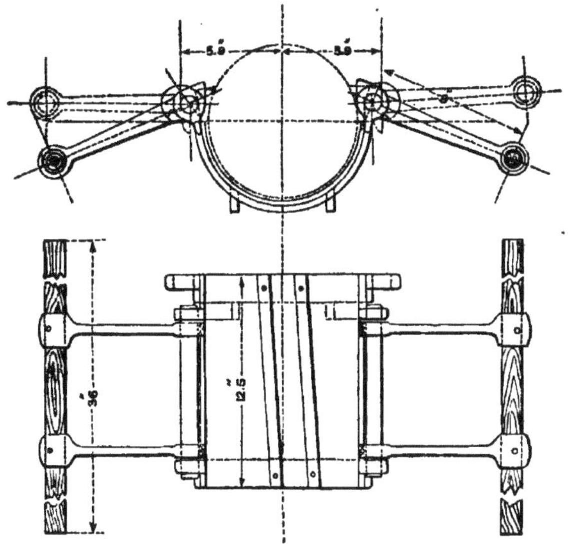

7-inch R.M.L. Gun.

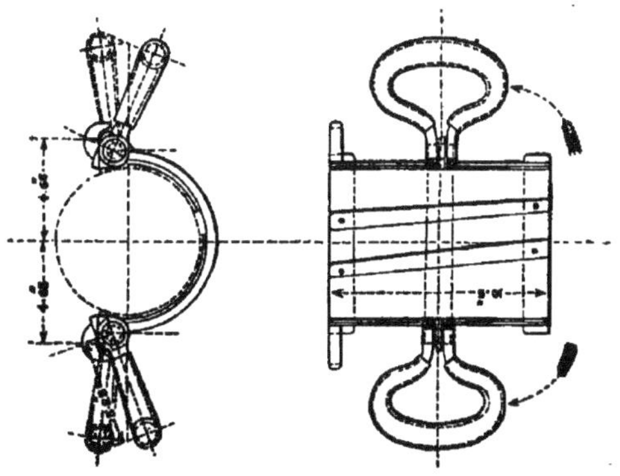

§ 2207. The hooks on the short bearers shown in the cut, have been removed. The handles of the 9-inch are now made of iron tubing.

Studs to support the shot bearer were formerly screwed into the muzzle of heavy guns before issue, vide p. 182.

Bolt, iron, elevating eye complete with washer and keep pin.

For 25, 16 and 9-pr. guns and consists of a bolt, washer, and keep pin, and are interchangeable, except the 9-pr. of 6 cwt. bolt which is shorter than the others.

Bolts, steel, for Pieces Pivot Mk. II.

For overbank carriages.

Brackets, metal.

Attach the trunnion and centre fore sight sockets to the converted cast iron guns. See p. 241. §§ 1752, 2066, 2220.

Clamps, metal, for Wood Scales.

Are used with R.M.L. heavy guns for S.S. and slides on the wood scale. It has a pointer and is used to give elevation or depression in connexion with the heel scale on the cascable. See p. 195. § 1478.

Clamps, moveable, for Tangent sights.

Are of gun metal and clamp the tangent sights at the required elevation, see p. 133. They are used with 25-pr. R.M.L. guns and upwards.* §§ 1144–1857.

Derricks.

This fitting is now provided for 9-inch L.S. guns and upwards, for the purpose of raising the projectile to the muzzle of the piece. It is used only for guns not mounted behind iron shields. § 3007.

It is made of bronze, and consists of a *band* (A) and *derrick* (B), as shown by the drawing. The band is fixed round the chase of the gun at a distance of from seven to twelve inches from the muzzle, according to the nature of gun, and the derrick is secured to it by two screw bolts. The derrick has a "bridge piece" (*b*) which rests on the top of the chase, supporting the fore part of the derrick, which projects over the muzzle the amount required for convenience of loading. To the loop (*c*) would be hooked the double block of the small tackle by means of which the projectile is raised.

The band when adjusted in the proper position is secured to the gun

---

* These clamps are only interchangeable for certain natures of guns, viz.:—
1 is 2·25 inches in height for 7″ B.L., 82 cwt.; 100 pr. S.B. 9″ of 12 tons 11 and 12″ of 25 tons R.M.L.
2 is 1·625 inches in height for 64-pr. 64 cwts. R.M.L., and $\frac{80}{68}$-prs. and for 40-pr. R.B.L. Mk. I. sight with barrel head.
3 is 1·425 inches in height for all other R.M.L. guns, except as below, and also used for 40-pr. R.B.L. Mk. II. and III. sights.
4 is 1·125 inches in height, for 40-pr. and 25-pr. R.M.L.
5 is 1·125 „ for 8″ and 6″·3 Howitzers.

CHAP. IX.

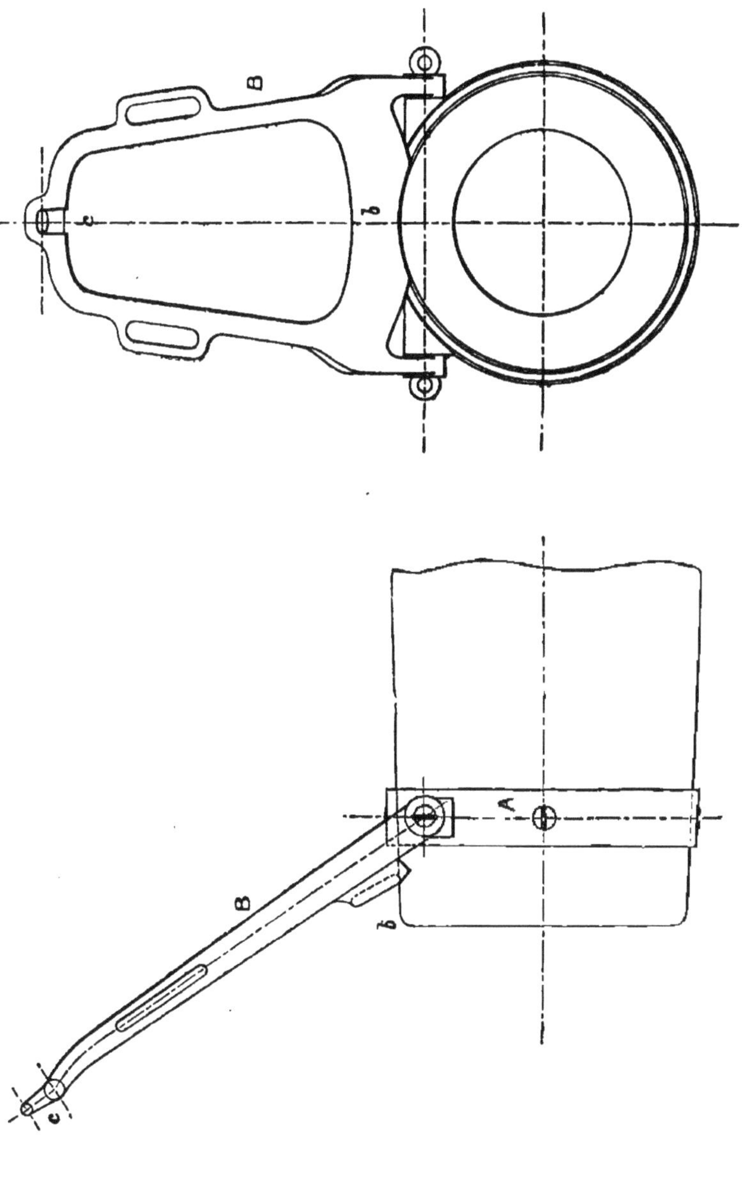

MUZZLE DERRICK. Scale 1 inch = 1 foot. ORDNANCE WROUGHT IRON R.M.L. 12·5-Inch 38 Tons.

by four screws. Instructions for adjusting and securing these derrick bands are given at p. 379, Appendix II. CHAP. IX.

Guard, metal, vent.

For S.S. 10-inch guns when mounted on gun boats of the "Blazer" class, to prevent injury to the men from the flash of the vent when the gun is fired.* § 2118.

These are issued to the Navy and are fitted by their artificers to the guns.

Guide plates.

Guide plates are of steel, and one is screwed in at the right rear of the vent to guide the lanyard— which passes through it—direct on the quill friction tube. It has a cross-head on the top, to which a loop on the lanyard can be attached when the gun is loaded, and so prevent the gun being fired accidentally. The Navy alone use it, as they fire their guns from the rear immediately the object is in line, and this guide plate enables the gunner to have a steady and direct pull on the lanyard while looking over the sights. §§ 476–688.

Scale ½.

Plate, preserving, muzzle sight.

This covers the screw holes for the muzzle sight of the 9-pr. 6 cwt. S.S. gun and the 8″ and 6″·3 howitzers when the sight is not on the piece.

Plate, metal, elevating.

See pieces, pivot.† § 1435.

Pin, iron, Friction Tube.

The friction tube pin is screwed in 1·3 inch to the left front of the vent and a spare hole is made adjoining it, lest the pin should be broken off, leaving its stump in the first hole. The leather loop of the S.S. quill friction tube is placed over this pin, to prevent the tube coming out or breaking when the lanyard is pulled, whilst to ensure direct action, the lanyard is passed through the *guide plate*, which is screwed into the gun in rear. § 1141.

---

* The method in which these guns are mounted, bringing the axis of the gun lower than usual. 8-inch guns on board the "Sultan" are also so fitted, the guards being slightly altered to allow of an electric tube being entered into the vent.

† The latest pattern for 10″ and upwards are termed "Countersunk." They have a square projecting boss fitting into a recess cut in the metal of the gun.

CHAP. IX.

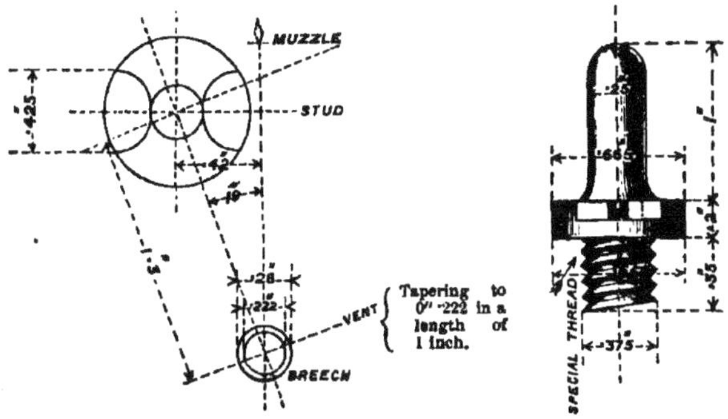

Pins, iron, keep, pivot, elevating.

§ 1435. See Plate, metal, elevating.

Pivots, steel, for elevating arc.

§ 1435. See Pieces Pivot.

Pieces, Pivot.

§§ 1435, 1647. DETAILS OF A PIVOT PIECE, FOR 9-IN. R.M.L. GUNS, 12 TONS.

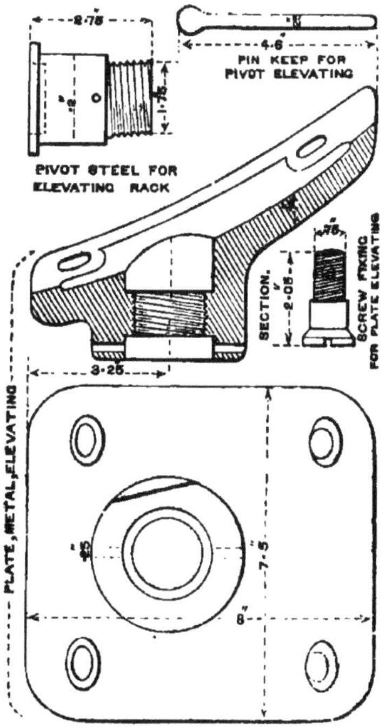

Scale ½ size.

CHAP. IX.

A pivot piece* consists of the following parts: a metal plate, steel pivot for elevating rack, keep pin for pivot, and screws for fixing the plate.

The metal plates are right and left handed (excepting 38 and 35-ton guns which have two plates, both attached to one side of the breech of the gun by six screws each and termed top and bottom).

Pivot pieces serve to connect the gun to the elevating racks on the carriage; they are attached to the guns by means of a wrench, vide p. 211.

To fix the elevating rack the keep pin is removed, the steel pivot unscrewed, passed through the hole in the elevating rack, and once more screwed into the plate, after which the keep pin is again inserted.

A pivot piece for an overbank carriage, consists however of the following parts: Pivot piece, overbank carriage.

A plate, metal, elevating, a bolt, steel, with keep pin and screws for fixing the plate.

Two pivot pieces are attached to the gun, one near the breech, marked B, the other towards the front, marked M.

The ends of the elevating rack are attached to the pivot pieces by means of the bolts.

ORDINARY WROUGHT IRON R.M.L. 40-PR., 35 CWT. MARK II.

PIECES PIVOT COMPLETE (FORE AND HIND). MARK II (for Overbank carriage).

Scale ½″ = 1 foot.

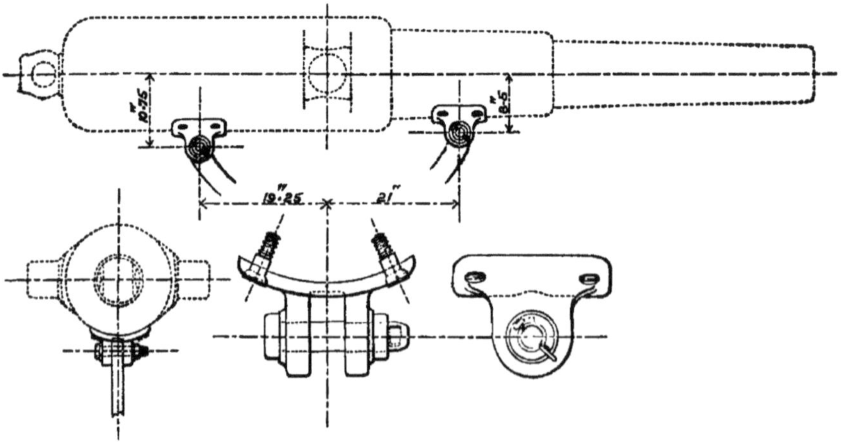

---

* The following are the marks of pivot pieces (dependent on the form of the plate metal elevating) for the different natures of guns :—

Mark I for 12″·5 M. I; M. II. for 12″·5 M. I. mounted on small port carriages and is central only. M. I. or II. for 12″ of 35 tons; M. I. or II. for 12″ of 25 tons M. II.; M. I. or II. for 11″ M. II; M. III. for 11″ M. II (Alexandra and Temeraire class); M. IV. for 11″ M. II. Barbette of Temeraire to suit Armstrong elevating arrangement); M. I. or II. for 10″ M. II.; M. III. for 10″ M. II. on small port carriages; M. IV. for 10″ M. I. (has a strenghtening boss); M. I. for 9″ M. I.; M. II. for 9″ M. III.; M. III. for 9″ M. II., IV., or V; M. IV. for 9″ M. I. fitted for Sultan class; M. I. for 8″ M. I.; M. II. for 8″ M. III.; M. I. for 7″ M. I.; M. II. for 7″ M. III.; M. I. for 7″ of 90-cwt.; M. I. for 64-pr. M. II.; M. I. for 64-pr. M. III.; M. II., for do. (for overbank carriage); M. I. for 40-pr. M. I.; M. I. for 40-pr. M. II; M. II. for do. (for overbank carriage); M. I. for 8″ howitzer; M. I. for 25-pr. Mark II for 25-pr. (overbank carriage); M. I. for 20-pr. R.B.L. 15 cwt., M. I. for 20-pr. R.B.L. 13 cwt. § 3067.

CHAP. IX.

Plates, index.

These plates are of gun metal and for use with L.S. 9-in. to 12-inch of 25-ton guns and are right and left handed, being attached to the side of the breech of the gun.

They are used in conjunction with a pointer or reader attached to the carriage, vide p. 198. See Table, p. 231.

Prickers, priming.

They are to prick the cartridge and are of four different lengths, viz., 29-inch, 23-inch, 17-inch, and 7½ inch. For the future all are to be made of steel.

Reader for index plate.

§§ 1306, 1478. See Index plate. Are of gun metal and attached to the carriage, and are right and left handed.

Scales, wood.

§ 1204, 1752. Are used in connection with the heel scale for S.S. guns, vide p. 196.

Scales, wood, side.

The 64-pr. guns of 71-cwts. for S.S. are provided with these scales, which are used in a similar manner to those for smoothbore guns.

Screws, copper, set.

Right and left hand for 16-pr. guns, and serve to clamp the tangent sight.

Screws, fixing and preserving.

All fittings, &c. are attached to the guns by means of "fixing screws," and when such fittings, &c. are removed, the holes are filled up by "preserving screws."

Sights.

See pp. 183–195, and Table p. 212.

Sockets, metal.

§ 1481. The centre, hind, and tangent sights work in these sockets which are let into the gun.

§ 3245.

Spanner, box, Fore Sight.

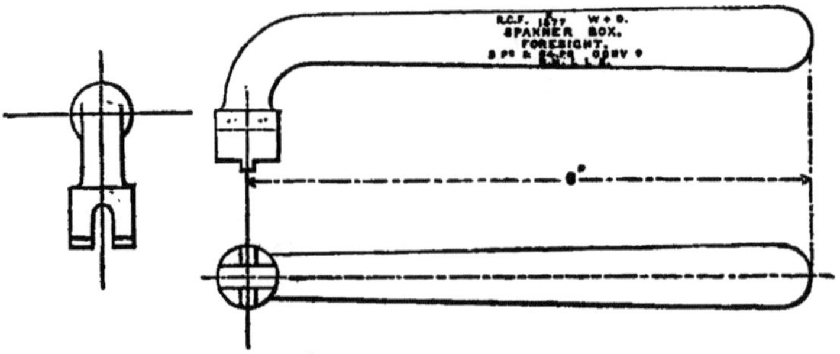

This store is used for both the 64-pr. converted gun of 58 cwt. and 9-pr. L.S. for unscrewing the small muzzle sight.

Spanner, iron removing Trunnion Sight, 16-*pr.* § 3246.

For removing screw trunnion sights of 16-prs.

Spikes.

Common and spring are of steel, and serve to disable guns in the field. Vide Table p. 231.

The Common spike is 3 inches long, and of universal pattern.

The Spring spikes are issued in three different sets, according to natures of guns for which required.

Studs, trunnion.

In order to facilitate the mounting and dismounting of very heavy guns in casemates, Colonel Inglis, R.E., proposed in 1874 that studs should be screwed into the trunnions of 38-ton guns, which would afford points by means of which the gun could be raised with the aid of powerful hydraulic jacks and a beam with wrought iron loops fitting over the studs. These studs were tried that year and found to answer well, in consequence of which their employment for 38-ton guns fo L.S. was approved of in 1875.

The gun is prepared for them by boring into the face of each trunnion, the holes being about 4 inches in depth and 3·5 inches in diameter, and tapped for about three inches. The studs themselves are of wrought iron of the shape shown below, and when screwed home project 3·75 inches.

20-ton hydraulic jacks would be used in conjunction with them.

CHAP. IX.

TRUNNION STUDS.

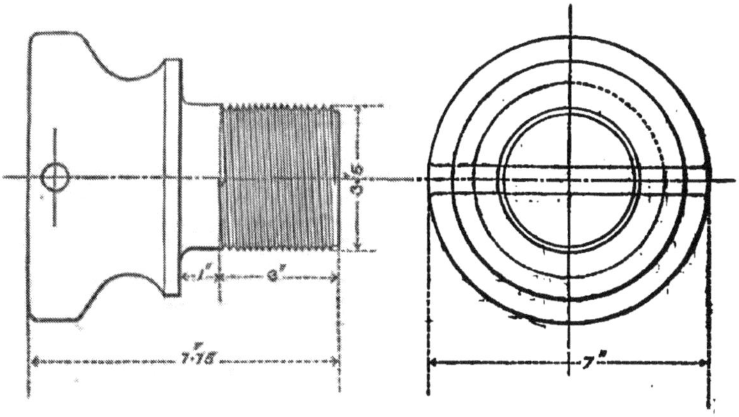

Washer, iron, for trunnions of 12·5 inch M.L. gun of 38 tons for mounting and dismounting.

This washer is used in connection with trunnion studs, being screwed into position between the studs, and face of trunnions. It prevents slipping of chain or rope when such chain or rope passes round the trunnions.

Wrench for fixing Elevating Rack, Index Plates, and Muzzle Derricks.

MARK II.

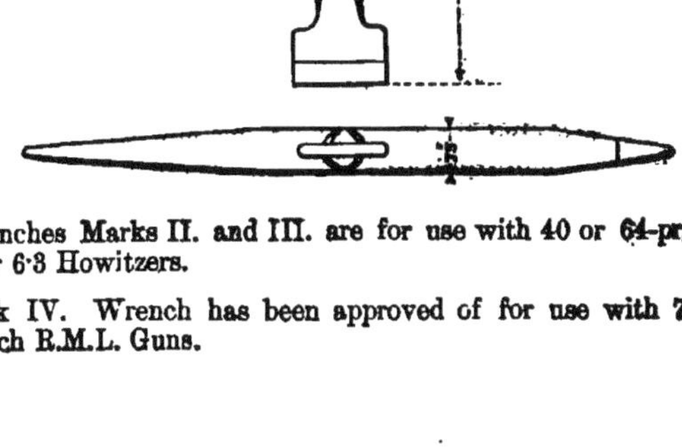

Wrenches Marks II. and III. are for use with 40 or 64-pr. R.M.L., 8 in. or 6·3 Howitzers.

Mark IV. Wrench has been approved of for use with 7 inch to 12·5 inch R.M.L. Guns.

WRENCH, MARK IV., WITH CLAMP COMPLETE FOR REMOVING ELEVATING PIVOTS.

¼ size.

MARK IV.

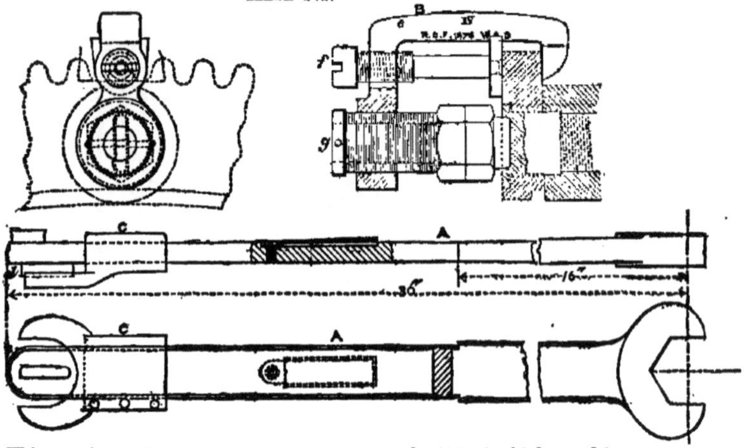

This consists of two separate pieces, one only (A) of which need be employed to remove the pivot piece when the elevating arc is not secured to the gun, but both (A and B) are necessary should the arc be attached.

Part A is a strong wrought iron bar with a wrench at one end, and at the other a projecting stud to fit into the slot in the face of the pivot piece. Upon it slides a jaw (*c*) (kept in the centre by a spring when not required).

To use the wrench in removing a pivot piece preparatory to fixing on the elevating rack, place the stud (*d*) into the slot on the pivot piece, slide the jaw (*c*) over the pivot piece behind the head, and then unscrew.* The jaw prevents any slipping and keeps the stud firmly fixed in the pivot piece.

To remove the pivot piece when the elevating rack is attached by it to the gun, we must use A and B. B is also of wrought iron, and has a bent portion (*e*) to fit over the elevating arc, the screw (*f*) securing it firmly in position; the screw (*g*) has a spindle passing through its centre, on the end of which is a stud for fitting into the pivot piece.

On the screw and spindle where they meet are cut hexagonal faces, § 1213.
and the wrench fits over both together.

To remove the pivot piece, fix B to the arc, screw up (*g*) until the stud is well home in the slot of the pivot, then apply the wrench (A) to the hexagonal surfaces, and unscrew the stud.

Wrench, Pin, Friction Tube.

This wrench has four arms, which are used for the following purposes:—

Box wrench for friction tube pin.
Spanner for sight and fixing screw
Tommy for guide plate.
Screw-driver for sight screw.

* The keep pin being of course removed.

CHAP. IX.

TABLE XIX.

NATURE of LATEST PATTERNS of TANGENT SCALE and CENTRE HIND SIGHTS* and particulars of GRADUATIONS for RIFLED M.L. GUNS.

Nature of Gun.	Pattern of Sight.	Permanent angle of Deflection of Sights.	Length of Radius in Inches.	Graduations for Degrees. Number.	Graduations for Degrees. Length in Inches.	Remarks.
17·72-inch, 100 tons	I.	—	—	—	—	—
16-inch, 80 tons	I.	—	—	—	—	—
12·5-inch, 38 tons	I.	2° 10′	72″	10	12·695	Side or centre.
12-inch, 35 tons...	I.	1° 25′	"	10	12·695	Side.
			72	10	12·695	Centre.†
13·05-inch, 23 tons	—	Nil.	45	13	10·388	Side } No yard graduations.
			45·1	5	3·936	Centre
12-inch, 25 tons	II.	30′	54	12	11·478	Side.
			59·95	5	5·249	Centre.
	III.	30′	59·95	10½	13·5	Long centre hind sights ‡
11 inch, 25 tons	I.	2° 26′	54	12	11·478	Side.
			59·95	5	5·249	Centre.
	II.	2° 26′	59·95	10½	13·5	Long centre hind sights.‡
10-inch, 18 tons	II.	1° 10′	54	12	11·478	Side ... } The side sights of L.S. guns are placed 2″ nearer the vertical axis than S.S. guns.
			60	5	5·249	Centre ...
	III.	1° 10′	60	8	10	Long centre hind sights.‡
9-inch, 12 tons	III. } IV. }	44′	45	15	12·057	Side ... ... } Existing sights to be altered for P. powder according to instructions below.§
			45·1	5	3·936	Centre ...
	V.	44′	45·1	7½	8	Long centre hind sights‡
8-inch, 9 tons	II.	28′	38	15	10·05	Side ... ...
			38·1	5	3·35	Centre ...
7-inch, 7 and 6½ tons ...	III.	3°	38	15	10·05	Side ... ...
			38·1	5	3·35	Centre ...
7-inch, 90 cwt.	I.	44′	38	15	10·05	Side.
			38·05	5	3·35	Centre.
80-pr. (converted) 5 tons ...	I.	19′	38	15	10·05	Side.

* Besides tangent sights, the Rifled M.L. guns for S.S. have wood scales used in conjunction with heel scales on the cascable, whilst 64-pr. M.L. are supplied with wood side scales (giving 12° elevation and 6° depression) similar to those for S.B. guns.

† This is a third tangent sight with steel bar.

‡ Vide p. 187.

NOTE.—The metal heads of the sights are not to be polished, as it would eventually destroy their accuracy.

§ The face of 7-inch sights graduated for yards with battering charges R.L.G. will be planed down, and a fresh scale showing yards for battering charges P. powder engraved instead. The faces of 8-inch and 9-inch sights graduated for yards with battering charges of R.L.G. powder will be cancelled by two lines drawn across the scale, and a fresh scale showing yards with battering charges P. powder will be engraved on the blank face opposite.

TABLE XIX.—*continued.*

Nature of Piece.	Pattern of Sight.	Permanent angle of Deflection of Sights.	Length of Radius in Inches.	Graduations for Degrees.		Remarks
				Number.	Length in Inches.	
64-pr. 64 cwts.	V.—VI	2° 16′ 2° 50′*	38	15	10·05	Side.
64-pr. 64 cwts.	IV.	2° 16′	38·1	5	3·35	Centre.
64-pr. converted 71 cwts. ...	IV.	2° 16′	38	15	10·05	Side.
64-pr. converted 58 cwts. ...	III.	2° 16′	38 1	5	3·35	Centre.
40-pr. 34 cwt.	I.	1° 20′	30	12	6·376	Side.
40-pr. 35 ,, ...	II.	1° 20′	36	12	7·651	Side.
25-pr. 18 cwt.	I.	0° 53′	30	12	6·376	Side.
16-pr., 12 cwt.	I.	1° 50′	24	12	5·101	Side.
13-pr., 8 cwt.	I.	...	...	... ...	... ...	Short } Central Long } sighting.
9-pr. Wrought Iron 8 cwt.† ... Mark I.	II.	1°·30′	66	6 12	6·936 14·028	Short } Central Long } sighting.
9-pr. Wrought Iron 8 cwt. ... Mark II.	I.	1° 30′	65	6 12	6·831 13·815	Short } Central Long } sighting.
9-pr. Wrought Iron 6 cwt. ... Mark I. ...	I.	1°·30′	54·6	7	6·703	Central sighting.
9-pr. Wrought Iron 6 cwt. † ... Mark II.	III.	1° 30′	68	5 12	5·948 14·456	Short } Central Long } sighting.
7-pr. 150 lbs. (Steel) ...	II.	3°	24·2	10 20	4·28 8·81	Central sighting. Do. Wood.
7-pr. 200 lbs. (Steel) Mark IV....	I.	3°	36·4	8 12	5·116 7·736	Short } Central Long } sighting.
7-pr. 200 lbs. (Bronze) ...	I.	3°	33·8	8 17	4·708 10·241	Central sighting. Do. Wood.
Howitzers. 6·3 inch 18 cwt. ...	I.	Nil.	Short 20·5 Long 49	15	11·635	Central.
Howitzers. 8 ,, 46 ,, ...	I.	,,	Short 27 Long 54	15	13·054	Do.
Howitzers. 6·6 ,, 36 ,, ...	...	...	...	...	...	
Howitzers. 8 ,, 70 ,, ...	...	...	...	...	...	

* Those with steel tubes firing 10 lb. charges.
† A brass plate showing the range table is attached to left bracket of carriage.

TABLE XX.

TANGENT SCALE OR SIDE SIGHTS OF HEAVY R.M.L. GUNS.

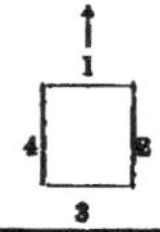

Mark of Sight.	No. of Face. Vide Diagram.	100 Ton.	80-Ton.	12"·5 38-Ton.	12" 35-Ton.	12" 25-Ton.	11" 25-Ton.	10" 18-Ton.	9" 12-Ton.	8" 9-Ton.	7-inch. 7 or 6½ Tons.	7-inch. 90-cwt.
I.	1	—	—	0° to 10°	0° to 10°.	0° to 12°.	§ 2198.* 0° to 12°.	0° to 12°.	0° to 15°. d	§ 1477.† 0° to 15°.	§ 1231.‡ 0° to 15°.	§ 2549.§ 0° to 15°.
	2	—	—	—	C.S. Full, 85 lbs. P. Yds. 4,500. Fuze 20.	C.S. Full. Fuze 24	C.S. Full. 60 lbs. P. Yds. 4,500 Fuze 20	S. Full. 40 lbs. Fuze 30	Fuze 2·65	Shell Full. 20 lbs. Fuze 30.	Fuze.	—
	3	—	—	—	Pal. S. or S. Battg. 100 lbs. P. Yds. 4,800.	,, Yds. 4,000	Pal. S. or S. Battg. 85 lbs. P. Yds. 4,800	Full. 40 lbs. Yds. 4,000	S. or S. 30 lbs. Yds. 4,000	Full. 20 lbs. Yds. 4,000	C.S. 14 lbs. Yds. 4,000	C.S. Full. 14 lbs. Yds. 4,000
	4	—	—	—	—	C.S. Battg. ,, 4,000	—	Battg. 70 lbs. P. Yds. 4,800	—	Battg. 35 lbs. P. Yds. 4,800.	—	C.S. Full. Fuze 20.
II.	1	—	—	—	—	§ 2198.* 0° to 12°. d	—	§ 2198.* 0° to 12°. d	§§ 1254, 1476.† b 0° to 15°.	§ 2198.* d 0° to 15°.	§ 1437.† c 0° to 15°.	—
	2	—	—	—	—	C.S. Full 55 lbs. P. { Yds. 4,000 Fuze 24	—	C.S. Full. 44 lbs. P. { Yds. 4,000 Fuze 30	Shell Full. 30 lbs. Fuze 27.	C.S. Full 20 lbs. { Yds. 4,000 Fuze 30	C.S. Full. 14 lbs. Fuze 20.	—
	3	—	—	—	—	Pal. S. or S. Battg. 85 lbs. P. Yds. 4,800	—	Pal. S. or S. Battg. 70 lbs. P. Yds. 4,800	Full. 30 lbs. Yds. 4,000	Pal. S. or S. & C. S. Battg., 35 lbs. P. Yds. 4,800.	C.S. Full. 14 lbs. Yds. 4,000.	—
	4	—	—	—	—	C. S. Battg. 85 lbs. P. { Yds. 4,800 Fuze 32	—	C.S Battg. 70 lbs. P. { Yds. 4,800 Fuze 34	Battg. 50 lbs. P. Yds. 4,800.	C.S. Battg. 35 lbs. P. Fuze 33.	Battg. 30 lbs. P. Yds. 4,800.	—

III.	1	—	—	—	—	—	—	—	§ 2198.* *d* 0° to 15°.	—	§ 2198.* *d* 0° to 15°.
	2	—	—	—	—	—	—	—	C.S. Full 30 lbs. { Yds. 4,000 Fuze 27	—	C.S. Full. 14 lbs. { Yds. 4,000 Fuze 20.
	3	—	—	—	—	—	—	—	Pal. S. or S. & C.S. 50 lbs. P. Yds. 4,800	—	Pal. S. or S. & C. S. Battg. 30 lbs. P. Yds. 4,800.
	4	—	—	—	—	—	—	—	C.S. Battg. 50 lbs. P. Fuze 32.		C.S. Battg. 30 lbs. P. Fuze 27.
IV.	1										
	2										
	3										
	4										
V.	1										
	2										
	3										
	4										

a. Barrel head sight.
b. Has sliding leaf head, fuze scale in even tenths, headings of yard scale altered, and weight of charge added. In 1866 this system was adopted.
c. Fuze scale in even tenths, and projectile and charge added.
d. Owing to the introduction of P. powder regraduation was necessary; the depth of notch increased, and the graduations made uniform.

* Introduced in 1871. † 1867. ‡ 1866. § 1873.

TABLE XX.—Tangent Scale or Side Sights of Heavy R.M.L. Guns—*continued.*

Mark of Sight.	No. of Face. Vide Diagram.	100 Ton.	80-Ton.	12"·5 38-Ton.	12" 35-Ton.	12" 35-Ton.	11" 25-Ton.	10" 18-Ton.	9" 12-Ton.	8" 9-Ton.	7-inch.	
											7 or 6½ Tons.	90 Cwt.
VI.	1											
	2											
	3											
	4											
VII.	1											
	2											
	3											
	4											
VIII.	1											
	2											
	3											
	4											

TABLE XXI.

Tangent Scale or Side Sights of Medium R.M.L. Guns.

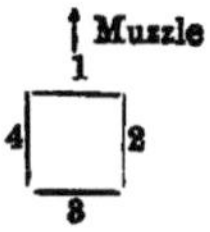

Mark of Sight.	No. of Face on Diagram.	64-pr. of 64 or 71 cwts.	$\frac{80}{68}$-pr. 5 tons. (Converted.)	Remarks.	Should be issued with.
I.	1 2 3 4	§ 1032. 0° to 15°. *a* *Blank.* *Blank.* *Blank.*	§ 2220. 0° to 15°. S.F. 10 lbs. Fuze 20. Full do. Yds. 3,500. *Blank.*	*a* Barrel head. No yards graduation owing to range not having been determined. Introduced in 1865.	Nil.
II.	1 2 3 4	§ 1143. § 1254. 0° to 15°. *b* Fuze scale, F, 2·0. Yards, 3,600. *Blank.*		*b* Sliding leaf head, 1866	Nil.
III.	1 2 3 4	§ 1476. § 1752. 0° to 15° *c* S. Full, 8 lbs., Fuze 27. Full „ Yds. 3,600. *Blank.*		*c* Fuze scale graduated in even tenths, heading of yard scale altered, and weight of charge added. 1867. Adopted for the converted 64-pr. 8-inch gun, 1869.	Nil.
IV.	1 2 3 4	§ 2198. 0° to 15° *d* S. F. 8 lbs. Fuze 27. Full „ Yds. 3,600. *Blank.*		*d* In 1871 a uniform system of graduation was introduced, and the depth of the notch was at the same time increased, as is the case with this sight.	A.B. E.
V.	1 2 3 4	§ 2842. 0° to 15°. *e* *Blank.* Full, 10 lbs. Yds. 5,000. „ „ Fuze 34.		*e* Has a slow-motion screw for giving minutes of elevation being required for L.S. 64-pr. with steel tubes for siege purposes, and has also the shallow notch. 1875.	D.
VI.	1 2 3 4	§ 2993. 0° to 15°. *f* Red[d]. 8lbs. { Yds. 3,600. / Fuze 27. Full, 10 lbs. Yds. 5,000. „ „ Fuze 34.		*f* Has no slow-motion screw, being used for S.S., but has the deep notch, and is graduated for reduced charges. 1876.	A. B.C. E.
VII.	1 2 3 4				

CHAP. IX.

TABLE XXI.—Tangent Scale or Side Sights of Medium R.M.L. Guns.—*continued.*

Mark of Sight.	No. of Face on Diagram.	64-pr. of 64 or 71 cwts.	$\frac{80}{68}$-pr. 5 tons. (Converted).	Remarks.	Should be issued with.
VIII.	1				
	2				
	3				
	4				
IX.	1				
	2				
	3				
	4				
X.	1				
	2				
	3				
	4				

Note.—§ 2938 (1.) There are at present six classes of the above-mentioned guns, viz.:—

A. 64-pr. of 64 cwt., Mark I. (S.S.)
B. „ „ „ II. (S.S.)
C. „ „ „ III. (S.S.)
D. „ „ „ III. (L.S. siege train).
E. „ 71 cwt. converted, Mark I. (L.S. and S.S.)
F. „ 58 cwt. „ I. (L.S.)

(2.) From 1865 to the present date there have been introduced into the service for these natures of guns six patterns of tangent sights and five patterns of centre hind sights.

(3.) Tangent sights of Mark I. when returned into store from H.M. ships now in commission will be condemned, together with those in store at home and abroad, and sent to Woolwich as opportunities offer.

(4.) Tangent sights of Marks II. and III, at present on board H.M. ships will remain as they are until the vessels are paid off. Those in store, together with those returned from ships from time to time, will be sent to Woolwich for alteration to suit the guns for which sights of Mark IV. can be issued. The altered sights will resemble those of Mark IV., but will not be brought in all respects to pattern. They will be distinguished by having the Mark "IV." within parentheses, added to their original numeral; thus II. (IV.) and III. (IV.).

TABLE XXII.

TANGENT SCALE OR SIDE SIGHTS OF SIEGE, FIELD, MOUNTAIN, AND BOAT GUNS.

CHAP IX.

TABLE XXII.—Tangent Scale or Side Sights,

Mark of Sight.	No. of face on diagram.	64-pr. 64 cwt. Mark III. steel tubes.	40-pr.		25-pr.[1]	16-pr.	* 13-pr.		* 9-pr. 8 cwt. L.S.I.		* 9-pr. 8 cwt, S.S.II.	
			I.	II.			Short.	Long.	Short.	Long.	Short.	Long.
I.	1	a	§ 2478 b 0° to 12°	§ 2854 b 0° to 12°	0° to 12°	§ 2221 0° to 12°	—	—	§ 1920 § 2067. —	—	§ 2743 0° to 7°	7° to 12°
	2	—	Fuze 26	Fuze 26	—	Fuze 28	—	—	—	—	Fuze 17	—
	3	—	Charge 7 lbs Yds. 4,000	Charge 7 lbs. Yds. 4,000	Yards 4,000	Yards 4,000	—	—	—	—	Yards 2,600	—
	4	—	— [d]	— [d]	— [d]	— [d]	—	—	0° to 6° [e]	6° to 12°	— [d]	—
II.	1	a	—	—	—	—	—	—	§ 2352 § 2636 —	—	—	—
	2	—	—	—	—	—	—	—	Yards 2,400	Yards 3,500	—	—
	3	—	—	—	—	—	—	—	0° to 6°	0° to 12°	—	—
	4	—	—	—	—	—	—	—	Fuze 16	Fuze 20	—	—
III.	1	a	—	—	—	—	—	—	—	—	—	—
	2	—	—	—	—	—	—	—	—	—	—	—
	3	—	—	—	—	—	—	—	—	—	—	—
	4	—	—	—	—	—	—	—	—	—	—	—
IV.	1	a										
	2	—										
	3	—										
	4	—										
V.	1	§ 2842 0° to 15°										
	2	—										
	3	Full 10 lbs. Yds. 5,000										
	4	Do. Fuze 34										
VI.	1	a										
	2	—										
	3	—										
	4	—										

[1] Introduced in 1875; [2] 1873; [3] 1876;

a. See table of sights for medium guns.
b. These sights are not interchangeable, owing to the different radius with which these guns are sighted.
d. No sight of this pattern has been sealed or issued.
e. The deflection head on the sight gives 1° left and 4° right.

SIEGE, FIELD, MOUNTAIN, and BOAT GUNS.

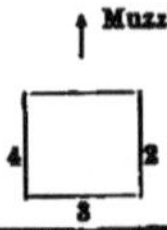

* 9 pr. 6 cwt. S.S. I.		* 9-pr. 6 cwt. L.S. II.		* 7-pr. 200 lbs. Bronze II.		* 7-pr. 150 lbs. (steel II. and III.		* 7-pr. 200 lbs. (steel) IV.		* Howitzers.			
Short.	Long.	Short.	Long.	Short.	Long.	Short.	Long.	Short.	Long.	8-inch of 46 cwt.	6·3 inch.	6·6 inch.	8-inch of 70 cwt.
0° to 7°	0° to 12°	§ 2683 —	— [d]	§ 1935 0° to 8°	§ 1943 0° to 17°	§ 1506 [g] 0° to 10°	[d] —	§ 2498. 0° to 8°	8° to 12°	— [c]	—		
F. 15	F. 20	—	—	D.S. {Y F}	[h] Yards 1,200 Fuze 17	—	—	12 oz. Yards 1,800 Fuze 13	Yards 2,400 Fuze 19	—	—		
Yards 2,400	Yards 3,300	—	—	C.S.Y.	[h] Yards 2,800	—	—	8 oz. Yards 1,700 Fuze 14	Yards 2,200 Fuze 19	0° to 15°	0° to 15°		
—	—	—	—	C.S.F.	[h] Fuze 26	—	—	D. shell. Yards 4,000 Fuze 12	Yards 900 Fuze 19	—	—		
—	—	§ 2682 § 2855 —	—	—	—	§ 1717 0° to 10°	10° to 20°						
—	—	Yards 2,100	Yards 3,500	—	—	—	—						
—	—	0° to 5°	5° to 12°	—	—	—	6 oz. Yards 2,000 Fuze 21						
—	—	Fuze 13	Fuze 20	—	—	—	D. shell 4 ozs. Yards 1,100 Fuze 17						
—	—	§ 3071	[f]										
—	—	Yards 2,100	Yards 3,500										
—	—	0° to 5°	5° to 12°										
—	—	Fuze 12	Fuze 20										

[a] 1869; [b] 1872; [c] 1870; [d] 1874; [e] 1868; [*] 1867.
f. This pattern sight has its fuze graduation longer than the previous pattern.
g. This was for Mark I. gun whic was side sighted.
h. These graduations added in November 1871.
[*] These tangent sights are central though not termed central hind sights.

CHAP. IX.

TABLE XXII.—TANGENT SCALE OF SIDE SIGHTS, SIEGE, FIELD,

Mark of Sight.	No. of face on diagram.	64-pr. 64 cwt. Mark III. (steel tubes)	40-pr.		25-pr.	16-pr.	* 13-pr.		* 9 pr. 8 cwt. L.S.I.		* 9-pr. 8 cwt. S.S II.	
			I.	II.			Short.	Long.	Short.	Long.	Short.	Long.
VII.	1											
	2											
	3											
	4											
VIII.	1											
	2											
	3											
	4											
IX.	1											
	2											
	3											
	4											
X.	1											
	2											
	3											
	4											
XI.	1											
	2											
	3											
	4											
XII.	1											
	2											
	3											
	4											

MOUNTAIN and BOAT GUNS.—*continued.*

* 9 pr. 6 cwt. S.S. I.		* 9-pr. 6 cwt. L.S. II.		* 7-pr. 200 lbs. Bronze II.		* 7-pr. 150 lbs. (steel) II and III.		* 7-pr. 200 lbs. (steel) IV.		* Howitzers.			
Short.	Long.	Short.	Long.	Short.	Long.	Short.	Long.	Short.	Long.	8-inch of 46cwt.	6·3 in.	6·6 in.	8 inch of 70cwt.

7-inch.		64-Pounder.		
6½ and 7 Tons.	90 Cwt.	64 Cwt. (Wrought iron.)	$\frac{64}{32}$ Converted.	Should be issued with (*See* note p.).
§ 1231. (a) 0° to 5°	§2549. (a) 0° to 5°	§ 1032. (a) 0° to 5°	—	Nil.
D. S. 14 lbs. Yds. 1,200.	—	—	—	
Steel Shell, 22 lbs. Yds. 2,700.	—	—	—	
C. S. 14 lbs. Yds. 2,200.	C. S. Full, 14 lbs. Yds. 2,200.	—	—	
,, Fuze 1·5	,, Fuze 15.	—	—	
D. S. 14 lbs. Fuze 8.	—	—	—	
§§ 1438, 2676. (a) a d 0° to 5°	—	§ 1476. (a) d 0° to 5°	§§ 1476, 2066. 0° to 5°	
D. S. Full, 14 lbs. Yds. 1,200.	—	—	—	
Battg. 22 lbs. Yds. 2,700. *Pal. S. or S. & C. S. Battg. 35 lbs. P. Yds.* 3,000.	—	—	—	
C. S. Full, 14 lbs. Yds. 2,200.	—	Full, 8 lbs. Yds. 2,000.	Full, 8 lbs. Yds. 2,000.	
C. S. Full, 14 lbs. Fuze 15.	—	S. Full, 8 lbs. Fuze 13.	S. Full, 8 lbs. Fuze 13.	
D. S. Full, 14 lbs. Fuze 8.	—	—	—	
§ 2198. (a) b	—	§ 2198. (a) b	§§ 2066, 2198. (a) a	A. B.

a The graduations are carried as far as the shape of the gun will permit, and the bars
The heads are fitted with deflection leaves like those of the tangent sight and is termed the
Has the deep notch being required for S. S.

7-inch.		64-Pounder.		
6½ and 7 Tons.	90-Cwt.	64-Cwt. Wrought Iron.	$\frac{64}{32}$ Converted.	Should be issued with (*See* note p. .
—	—	§ 2903. (8) & 0° to 5°	—	
—	—	Reduced, 8 lbs. Fuze 13.	—	
—	—	,, ,, Yds. 2,000.	do. for this gun	A.B.C. & F.
—	—	Full, 10 lbs. Yds. 2,500.	—	
—	—	,, ,, Fuze 16.	—	
—	—	—	—	
—	—	—	*0° to 12°	
—	—	—	—	
—	—	—	Full, 8 lbs. Fuze 31.	
—	—	—	,, 8 lbs. Yds. 4,000.	
—	—	—	—	
—	—	—	—	

TABLE XXIV.

Wood Scales for R.M.L. Guns.

TABLE XXIV.—Wood Scales for R.M.L. Guns.

Muzzle.

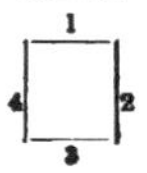

Mark of Scale.	No. of Face, vide Diagram.	† 100 Ton.	† 80 Ton.	† 12"·5 38 Ton.	† 12" 35 Ton.	† 12" 25 Ton.	11" 25 Ton. Mark II	10" 18 Ton.	9" 12 Ton.	8" 9 Ton.	7-Inch. 6½ Tons.	7-Inch. 90 cwt.	64-pr. 64 Cwt. g	64-pr. 71 Cwt., converted.
I.	1	—	—	—	—	—	§ 3153. 6°D to 10°E.	* § 1757. (2) 7°D. to 12°E.	§ 1360 (1) 7°D. to 22°E.	§ 1479 (3) 7°D. to 22°E.	§ 1204 (1) 6° D. to 20° E.	§ 2635 (4) 7° D. to 22 E.	§ 1204 (1) e 6° D. to 12° E.	§ 1752 (6) 6° D. to 12° E.
	2	—	—	—	—	—	C.S. Full 60 lbs. P. Yds. 4,000.	—	S. or S. Full 30 lbs. Yds. 2,000.	S. or S. Full 20 lbs. Yds. 2,000.	This was a side scale.	Shell Full 14 lbs. Yds. 3,000.	This is a side scale.	This is a side scale.
	3	—	—	—	—	—	—	—	—	—	—	—	—	—
	4	—	—	—	—	—	S. or S. Batt. 85 lbs. P. Yds. 4,800.	—	S. or S. Battg. 43 lbs. Yds. 2,000.	S. or S. Battg. 30 lbs. Yds. 2,000.	—	—	—	—
II.	1	—	—	—	—	—	—	(5) e 7°D. to 12°E.	§ 1478 (3) a 7°D. to 22°E.	(3) e 7°D. to 22°E.	1306 (1) b 7° D. to 22° E.	—	(7) f 8° D. to 13° E.	—
	2	—	—	—	—	—	—	S. or S. Full, 44 lbs. P. Yds 2,000.	S. or S. Full 30 lbs. Yds. 2,000	S. or S. Full 20 lbs. P. Yds. 2,000.	S. or S. Full. 14 lbs. Yds. 1,500.	—	Shell 6 lbs. Yds. 3,600.	—
	3	—	—	—	—	—	—	C.S. Battg. 70 lbs. P. Yds 3,000. d	—	—	D.S. Full. 14 lbs. Yds. 1,200.	—	Shell 8 lbs. Yds. 4,000.	—
	4	—	—	—	—	—	—	S. or S. Battg. 70 lbs. P. Yds. 3,000.	S. or S. Battg. 43 lbs. Yds. 2,000.	S. or S. Battg. 35 lbs. P. Yds. 2,500.	S. or S. Battg. 22 lbs. Yds. 1,600.	—	Shell 10 lbs. Yds. 5,000.	—

a Has a sliding gun metal clamp and two small studs. b Square in section and yards graduations added. c Graduation for P. powder, and head of scale painted in different colours. d C.S. is cut on the top of this face. * D. depression, E. elevation. (1) 1866, (2) 1867, (3) 1868, (4) 1869, (5) 1872, (6) 1874, (7) 1876.
e For guns mounted on wooden carriages. f Guns mounted on iron carriages. g 64-pr. fires 10-lb. charges, have a cascable scale, and use wood scale Mark II., vide p. 181.
† Special wood scales for these guns are manufactured as demanded, to suit the different modes in which mounted.

TABLE XXIV.—Wood Scales for R.M.L. Guns—*continued.*

Mark of Scale	No. of Face, vide Diagram.	100 Ton	80 Ton.	12"·5 38 Ton.	12" 35 Ton.	12" 25 Ton.	11" 25 Ton.	10" 18 Ton.	9" 12 Ton.	8" 9 Ton.	7-Inch. 6½ Tons.	7-Inch. 90 cwt.	64 pr. 64 cwt.	64 pr. 71 Cwt.
III.	1	—	—	—	—	—	—	—	([8]) 7° D. to 22° E. c	—	§1587 ([8]) a 7° D. to 22° E.			
	2	—	—	—	—	—	—	—	S. or S. Full. 30 lbs. Yds. 2,000.	—	S. or S. Full 14 lbs. Yds. 1,500.			
	3	—	—	—	—	—	—	—	—	—	D.S. Full 14 lbs. Yds. 1,200.			
	4	—	—	—	—	—	—	—	S. or S. Battg. 50 lbs. P. Yds. 3,000.	—	S. or S. Battg. 22 lbs. Yds. 1,500.			
IV.	1	—	—	—	—	—	—	—	—	—	([9]) 7° D. to 22° E. c			
	2	—	—	—	—	—	—	—	—	—	S. or S. Full 14 lbs. Yds. 1,500.			
	3	—	—	—	—	—	—	—	—	—	D.S. Full 14 lbs. Yds. 1,200.			
	4	—	—	—	—	—	—	—	—	—	S. or S. Battg. 30 lbs. P. Yds. 3,000.			

a. Has a sliding gun metal clamp and two small studs. *c.* Graduation for P. powder, and head of scale painted in different colours. ([8]) 1868, ([9]) 1872.

TABLE XXVA.

LIST OF SIGHTS AND FITTINGS REQUIRED FOR UNDERMENTIONED GUNS.

Mounted in Different Methods.

NATURE.	ORDINARY SERVICE. Sights. Tangent Scale.	Sights. Centre hind.	Sights. Centre fore.	Sights. Trunnion.	Sights. Muzzle.	Pieces, pivot, complete with screws.	Clamps moveable for tangent sight.	Sockets, metal, with set screws for C.H. sight.	Screws, preserving. Plate, elevating.	Screws, preserving. Plate, guide.	Screws, preserving. Pin, friction, tube.	Washers for trunnions.	Studs, trunnions, M.I.	S.S. only. Pin, friction, tube.	S.S. only. Plate, guide.	Supplied for Ordinary Service by Commissary General. Index plates complete, with readers and screws.	Supplied for Ordinary Service by Commissary General. Derrick, complete.
12·5 inch, 38 ton	3	—	1	2	—	2	3	—	12	1	2	2	2	1	1	—	1
10 ,, 18 ,,	2	1	1	2	—	2	2	1	8	1	2	—	—	1	1	2	1
9 ,, 12 ,,	2	1	1	2	—	2	2	1	8	1	2	—	—	1	1	2	1
7 ,, 7 ,,	2	1	1	2	—	2	2	1	8	1	2	—	—	—	—	—	—
64-pr. 64 cwt. Mark III. ...	2	1	1	2	—	1	2	1	4	1	2	—	—	1	1	—	—
64-pr. converted, 58 cwt. ...	—	1	1	—	1	—	—	1	—	1	2	—	—	—	—	—	—
40-pr. Mark II.	2	—	—	2	—	1	2	—	3	—	—	—	—	—	—	—	—
25-pr. 18 cwt. ...	2	—	—	2	—	1	2	—	3	—	—	—	—	—	—	—	—

NATURE.	ISSUED IN ADDITION FOR Moncrieff Carriage. Mark I. Plate, trunnion.	Mark I. Bolt, eye, for trunnion.	Mark I. Sights with screws. Fore scale.	Mark I. Sights with screws. Hind bracket, reflecting sight.	Mark I. Screws, preserving sights.	Mark II. Plate. Trunnion.	Mark II. Plate. Cascable.	Mark II. Bolt, eye, for trunnion.	Mark II. Piece, pivot, complete, with bolt washer and screws complete.	Mark II. Sights, reflecting. Trunnion with bracket and screws complete.	Mark II. Sights, reflecting. With graduated bar for rocker and screws complete.	Mark II. Screw, preserving. Piece, pivot.	Mark II. Screw, preserving. Bolt, eye holes.	Mark II. Screw, preserving. Sights.	Small port carriage. Pieces, pivot, complete.	Small port carriage. Screws, preserving, plate, elevating.	Small port carriage. Trunnion studs, M. II. with gudgeon.	Overbank carriage. Pieces, pivot, complete, with screws. Muzzle.	Overbank carriage. Pieces, pivot, complete, with screws. Breech.	Overbank carriage. Screws, preserving, plate, elevating.
12·5 inch, 38 ton	—	—	—	—	—	—	—	—	—	—	—	—	—	—	1	7	2	—	—	—
10 ,, 18 ,,	—	—	—	—	—	—	—	—	—	—	—	—	—	—	2	10	—	—	—	—
9 ,, 12 ,,	—	—	—	—	—	1	2	2	1	1	1	6	2	10	—	—	—	—	—	—
7 ,, 7 ,,	2	2	1	1	7	1	—	2	1	1	1	6	2	9	—	—	—	—	—	—
64-pr. 64 cwt. Mark III. ...	—	—	—	—	—	—	—	—	—	—	—	—	—	—	—	—	—	1	1	12
64-pr. converted, 58 cwt. ...	—	—	—	—	—	1	—	2	1	1	1	6	2	5	—	—	—	—	—	—
40-pr. Mark II.	—	—	—	—	—	—	—	—	—	—	—	—	—	—	—	—	—	1	1	8
25-pr. 18 cwt. ...	—	—	—	—	—	—	—	—	—	—	—	—	—	—	—	—	—	1	1	8

—	...rs. S.S.	7-prs. 200 lbs. L.S.	7-prs. 200 lbs. S.S.	Howitzers. 8" 46 Cwt. L.S.	Howitzers. 6·3" L.S.	Howitzers. 8" 70 Cwt. L.S.	Howitzers. 6·6" L.S.	Page and Price in Vocabulary of Stores, 1875.	Page in this Book.	Remarks.
Band, bronze, derrick, wi[...] screws and two bolts, comp[...] See Derrick, bronze.	—	—	—	—	—	—	—	—		For guns mounted on barbette—12·5" gun, Mark I., 12" 35 tons I., 11" II., 10" I. and II., 9" I. and 9" II., III., IV., and V.
Bearers, shot or shell ...	—	—	—	1	—	—	—	203		
Bolts, eye (see Band, bronze,										
Bolts, iron, elevating eye	1	—	—	—	—	—	—	209		
Brackets, metal { trunnion	—	—	—	—	—	—	—	213		
Brackets, metal { sight ...	—	—	—	—	—	—	—	213		
Brackets, metal { centre fore	—	—	—	—	—	—	—	—		
Clamps { metal, for wood sc[...]	—	—	—	—	—	—	—	243		
Clamps { moveable for tang[...]	—	—	—	1	1	—	—	243		
Derricks, bronze, muzzle, Band, bronze, 4 fixing scre[...]	—	—	—	—	—	—	—	—		For guns not mounted behind iron shields.
Guard, metal, vent ...	—	—	—	—	—	—	—	257		When mounted on gun boats of the "Blazer" class.
Pieces, pivot (for elevating [...] with plate, elevating, ste[...] pin, and screws.	—	—	—	1	1	—	—	281		
Pins, iron { friction tube	1	—	1	—	—	—	—	282		* When used as siege ordnance.
Pins, iron { keep piece, pivo[...]	—	—	—	1*	1*	—	—	282		† For Mark III. gun.
Pivots, steel, for elevating ar[...]	—	—	—	1*	1*	—	—	284		* When used for siege ordnance. † For Mark III. gun.
Plates { iron, guide ...	1	—	1	—	—	—	—	—		* When used for siege ordnance. † For Mark III. gun.
Plates { metal, elevating { top	—	—	—	—	—	—	—	—		
Plates { metal, elevating { bottom	—	—	—	—	—	—	—	—		
Plates { metal, elevating { right	—	—	—	1*	1*	—	—	285		
Plates { metal, elevating { left	—	—	—	—	—	—	—	285		
Plates { metal, elevating { muzzle	—	—	—	—	—	—	—	—		
Plates { metal, elevating { breech	—	—	—	—	—	—	—	—		† For overbank carriages.
Plates { metal index { right	—	—	—	—	—	—	—	—		
Plates { metal index { left	—	—	—	—	—	—	—	—		
Plates { preserving, muzzle s[...]	1*	—	—	1	1	—	—	—		* 9-pr. 6 cwt. S.S. Mark I.
Prickers, priming steel. { 29-inch	—	—	—	—	—	—	—	289		
Prickers, priming steel. { 23 ,,	—	—	—	—	—	—	—	289		
Prickers, priming steel. { 17 ,,	—	—	—	—	—	—	—	289		
Prickers, priming steel. { 7½ ,,	1	1	1	—	—	—	—	289		
Reader for index plate { right-hand	—	—	—	—	—	—	—	—		
Reader for index plate { left ,,	—	—	—	—	—	—	—	—		
Scales tangent { wood	—	—	—	—	—	—	—	295		
Scales tangent { wood, side ...	—	—	—	—	—	—	—	296		
Screws, copper, set { right-ha[...]	—	—	—	—	—	—	—	—		
Screws, copper, set { left ,,	—	—	—	—	—	—	—	—		
Screws, fixing { Band (see Band, bronze,										
Screws, fixing { bracket, trunnion, sight	—	—	—	—	—	—	—	299		
Screws, fixing { do. centre fore sight	—	—	—	—	—	—	—	—		
Screws, fixing { centre fore muzzle sight	3*	—	—	3	3	—	—	—		* 9-pr. 6 cwt. S.S. Mark I.
Screws, fixing { elevating plate ...	—	—	—	3*	3	—	—	300		* When used as siege ordnance. † For Mark III. gun.
Screws, fixing { guard, vent... ...	—	—	—	—	—	—	—	—		
Screws, fixing { plates, index ...	—	—	—	—	—	—	—	—		
Screws, fixing { reader for index plate	—	—	—	—	—	—	—	—		
Screws, fixing { sights on counterweight carriages, Mark I.	—	—	—	—	—	—	—	315		
	—	—	—	—	—	—	—	315		
Screws, fixing { do. Mark II. { t[...]	—	—	—	—	—	—	—	—		
Screws, fixing { do. Mark II. { g[...]	—	—	—	—	—	—	—	—		* 16-pr. guns and upwards have two of these screws, but being fixed in the gun are considered part of it.
Screws, fixing { sockets for { tangent sigh[...]	—	—	—	—	—	—	—	—		
Screws, fixing { sockets for { centre hind [...]	—	—	—	—	—	—	—	—		
Screws, Preserving. { bracket { trunnion sig[...]	—	—	—	—	—	—	—	300		
Screws, Preserving. { bracket { centre fore s[...]	—	—	—	—	—	—	—	—		
Screws, Preserving. { centre, fore, muzzle sig[...]	2*	—	—	2	2	—	—	—		* 9-pr. 6 cwt. S.S. I., and attaches the preserving patch.
	1*	1	1	—	—	—	—	—		* 9-pr. 8 cwt. S.S.

ıgs, R.M.L. Guns—*continued.*

...verted.						9-prs.		7-prs. 200 lbs.		Howitzers.				Page and Price in Vocabulary of Stores, 1875.	Page in this book.	Remarks.
...pr. 71 Cwt.	64-pr. 58 Cwt.	40-pr. 35 Cwt.	25 pr. 18 Cwt.	16-pr. 12 Cwt.	12-pr. 8 Cwt.					8″	6·3″	8″·70 cwt.	6·6″			
S.S.	L.S.	L.S.	L.S.	L.S.	L.S.	L.S.	S.S.	L.S.	S.S.	L.S.	L.S	L.S.	L.S.			
—	—	3*	3*	—	—	—	—	—	—	3	3	—	—	—		* When used as siege ordnance. † For Mark III. gun.
2	2	2	2	—	—	—	2	2	2	—	—	—	—	—		* When mounted on gun boats of the "Blazer" class.
—	—	—	—	—	—	—	—	—	—	—	—	—	—	—		
1	1	1	1	—	—	—	1	1	1	—	—	—	—	—		
—	—	—	—	—	—	—	—	—	—	—	—	—	—	—		
—	4	—	—	—	—	—	—	—	—	—	—	—	—	—		
—	3	—	—	—	—	—	—	—	—	—	—	—	—	—		
—	3	—	—	—	—	—	—	—	—	—	—	—	—	—		
—	6	—	—	—	—	—	—	—	—	—	—	—	—	—		
—	1	—	—	—	—	—	—	—	—	—	—	—	—	—		
—	—	—	—	—	—	—	—	—	—	1	1	—	—	—		
—	1	—	—	—	—	—	—	—	—	—	—	—	—	—		* Have sliding leaf heads.
—	1	—	—	—	—	1	1	1	1	1	1	—	—	—		
—	1	—	—	—	—	—	—	—	—	—	—	—	—	—		
—	1	—	—	—	—	—	—	—	—	—	—	—	—	—		
—	1	—	—	—	—	—	—	—	—	—	—	—	—	—		
—	1	—	—	—	—	—	—	—	—	—	—	—	—	—		
—	—	—	—	—	—	—	—	—	—	—	—	—	—	—		
—	—	—	—	—	—	—	—	—	—	—	—	—	—	—		
2	—	—	—	—	—	—	—	—	—	1	1	—	—	—		
—	—	—	—	—	—	1*	—	—	—	—	—	—	—	—		* The mark of gun must be stated owing to difference of rad:us.
—	—	—	—	—	—	1*	—	—	—	—	—	—	—	—		* Do. do.
—	—	—	—	—	—	—	1*	1*	1*	—	—	—	—	—		* Do. do.
—	—	—	—	—	—	—	1*	1*	1*	—	—	—	—	—		* Do. do.
—	—	2	2	2	—	—	—	—	—	—	—	—	—	—		* Mark III., steel tubed guns.
2	—	2	2	—	—	—	—	—	—	—	—	—	—	—		
—	—	—	—	2	—	—	—	—	—	—	—	—	—	—		
—	1	—	—	—	—	—	—	—	—	—	—	—	—	—		
—	—	—	—	—	—	—	—	—	—	—	—	—	—	—		* See remarks to screws, fixing for sockets, tangent sight
—	—	—	—	—	—	1	1	1	1	—	—	—	—	—		
—	1	—	—	—	—	1	—	—	—	—	—	—	—	—		
—	—	—	—	1	—	—	—	—	—	—	—	—	—	—		
—	—	—	—	—	—	—	—	—	—	—	—	—	—	—		* Universal pattern.
—	—	—	—	—	—	1	1	*1	1	—	—	—	—	—		
—	—	1	1	1	—	—	—	—	—	—	1	—	1	—		
—	—	—	—	—	—	—	—	—	—	1	—	1	—	—		
—	—	—	—	—	—	—	—	—	—	—	—	—	—	—		
—	—	—	—	—	—	—	—	—	—	—	—	—	—	—		
—	—	1	1	—	—	—	—	—	—	1	1	—	—	—		
—	—	—	—	—	—	—	—	—	—	—	—	—	—	—		
1	1	1	1	—	—	1	1	1	1	—	—	—	—	—		

CHAP. X.

CHAPTER X.

MANUFACTURE OF R.M.L. CONVERTED GUNS; THEIR SIGHTS AND STORES.

Early attempts to convert S.B. into Rifled Guns.—Palliser method preferred and adopted in 1863.—Nature of S.B. pieces converted.—**Mode of conversion,** taking the 8-inch S.B. as an example.—Boring out the S.B. piece.—Manufacture of wrought iron barrel.—A tube, B tube, gas escape, how formed.—Insertion of barrel into casing.—Securing the same in its place.—Completion of gun.—Rifling with plain groove.—**Venting**, different from that of built up guns.—Operation of venting.—Proof and examination.—$\frac{64}{32}$-pr. of 58 cwts. converted.—Conversion of 68-pr. S.B. into $\frac{80}{68}$-pr. R.M.L. of 5 tons.—**Sighting of converted guns.**—The two heavier natures side-sighted.—Wood scales used.—Fittings and small stores.—Brackets for side and centre fore sights.

CONVERSION OF S.B. INTO RIFLED GUNS.

Conversion of S.B. into rifled guns.

When rifled guns came into use there existed in the armament of all nations a number of S.B. cast iron guns, and, for the sake of economy, it was attempted to turn them into rifled pieces. The material, however, was found too weak of itself, and different modes of strengthening the guns were tried.

Early experiments.

As early as 1855 attempts were made to strengthen such guns by encasing them in rear of the trunnions with a wrought iron or other jacket, and betweeh that date and 1863 various plans of this and other descriptions were tried unsuccessfully. In that year Major (now Sir William) Palliser proposed to line cast iron guns with coiled iron barrels, fitting comparatively loosely into the casing until expanded by the heavy proof rounds. This method appeared to be more promising than any previously tried, and was moreover founded upon correct principles, the stronger material being placed next the charge.

Palliser's system.

Several guns converted on the Palliser principle gave very fair results upon trial, and showed themselves more powerful pieces than the smoothbores from which they were made. It was consequently determined to convert a large number of S.B. ordnance in this way, for although such pieces would be much inferior to built up Woolwich ordnance, yet where the range was limited and there were no iron plates to pierce they would prove useful.

Natures of S.B. pieces converted.

The O.S.C. proposed several natures for conversion, but it was finally decided that only the following natures of S.B. guns should be converted. They are, as will be seen, those of the newest and best construction amongst S.B cast iron guns, viz.:

68-pr. of 95 cwts., S.B.,	into 80-pr.	R.M.L., of	5 tons.
8-inch of 65 ,, ,, (throwing 50 lb. shell)	,, 64-pr.	,,	71 cwts.
32-pr. of 58 cwts., S.B.	,, 64-pr.	,,	58 ,,

Mode of conversion.

The mode of conversion consists in boring out the old gun and making a wrought iron tube to fit the casing thus prepared; this tube is slightly smaller than the bore of the casing, and is pushed into it

CHAP. X.

without much force being necessary.* When fitted into its place it is secured there by means of a cast iron collar screwed into the muzzle end of the casing over a shoulder on the end of the tube; a wrought iron plug is also screwed through the casing underneath and into the barrel, preventing any chance of the latter shifting round.

The principles of conversion are identical in all cases, though the dimensions differ somewhat; we will therefore describe in detail the conversion of an 8-inch gun.

Conversion of 8-inch S.B. of 65 cwt. into 64-pr. 71 cwt.

§ 1752.

The cast iron gun.

The cast iron gun is rough, second and finished bored to 10″·5 diameter, gauged, and horseshoe gauges prepared for turning the *A* tube. A variation of ± 0″·1 is allowed in the diameter of the bore of the cast iron-casing, but should there be any difference in the diameter at one part or another it must result in a taper from muzzle to breech. The play between the tube and the casing is not allowed to exceed 0″·007 for a length of 24″ from the breech end, and 0″·015 for the remainder of the length.

The muzzle is recessed and threaded for the cast iron collar (the use of which is to keep the tube in position), and the gas channel is bored through the breech. This is under the cascable in the first 212 guns converted at Elswick, and in all the others is to the right top of the cascable so as to be clear of the breeching rope.

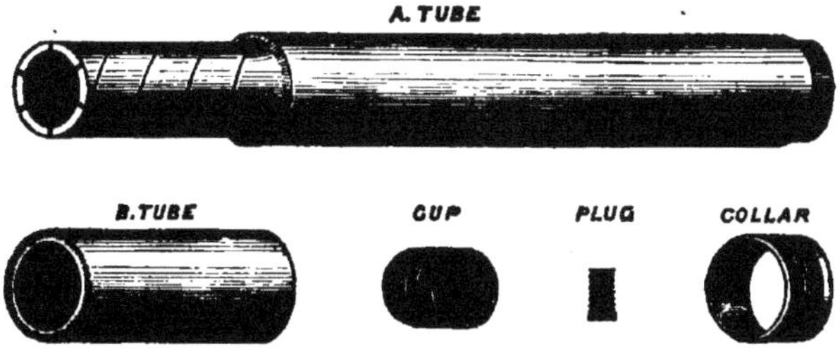

A tube of wrought iron.

The coils for the tube are made entirely of departmental bar iron specially prepared by being put three times through the rolling mills. The tube is formed of five coils united together in the usual manner, and is rough and fine bored to 6″·238 diameter, and the recess in the breech cut and tapped for the wrought iron cup. The cup for closing the breech end of the barrel is forged and stamped into shape under a steam hammer. It is turned inside and out and a screw cut on the exterior with a thread of five to the inch. It is then screwed tightly home.

Gas channel.

The tube in this state is proved with water pressure of 120 lbs. on the square inch to ascertain that the cup fits tightly and that there is no leakage. The breech end of the *A* tube is then turned over a length of 32″ for the *B* tube† previously bored, and a spiral gas channel 0″·05 deep and 0″·1 wide is cut round its exterior communicating with the star grooves cut in the end of the barrel, and the gas escape through the cast-iron breech.

B tube.

The *B* tube consists of two coils united, and being rough turned to 10″·75 and finished bored to 8″, it is shrunk on with 0″·003 shrinkage in the diameter. The tube is made double at this part in order that the

* After proof, however, the barrel being permanently expanded, fits tightly against the interior of the casing. This is clearly shown when it is attempted to force out of its casing a tube condemned at proof, which is done by hydraulic pressure applied from breech end.

† This must not be confounded with the "*B* tube" of a wrought iron gun which forms the chase, while in converted guns it goes over the breech end of the *A* tube.

CHAP. X.

gas may escape through the gas channel without bursting the gun in the event of the inner layer splitting. The whole tube is then rough turned and the bore broached to 6"·29* and examined, after which the exterior is fine turned to fit the cast iron casing with the requisite amount of play.

Putting the gun together.

The tube is now fitted into the casing, the greatest care being taken that the breech end bears fairly against the cast iron; the curved part of the end of the barrel is described with a longer radius than the corresponding curve in the cast iron so as not to be in contact with it at that part. The space thus left between the two prevents the tube acting as a wedge to split open the cast iron. When the tube is properly adjusted a cast iron collar is securely screwed into the muzzle end of the casing and over a shoulder of the tube.

Cast iron collar.

A hole 1"·25 in diameter is drilled through the cast iron, and a short distance into the tube at 29" from the trunnions under the chase, and, being tapped, a wrought iron pin is screwed in to prevent the tube from shifting round.

Wrought iron pin.

The muzzle of the gun is then cut and faced, and the bore lapped and rifled.

$\frac{\text{64-pr.}}{\text{8-inch}}$ *M.L. Converted Gun, of 71 cwts. L.S. and S.S.* § 1752.

Scale $\frac{1}{32}$nd.

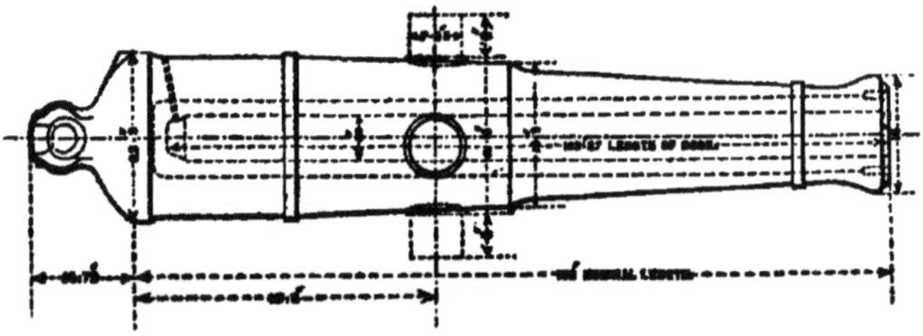

The gun is rifled in the same machine, and in a similar manner to other R.M.L. guns, but the groove is the plain groove of section below, as mentioned at p. 45.

Plain groove.

RIFLING. Scale $\frac{1}{12}$th.

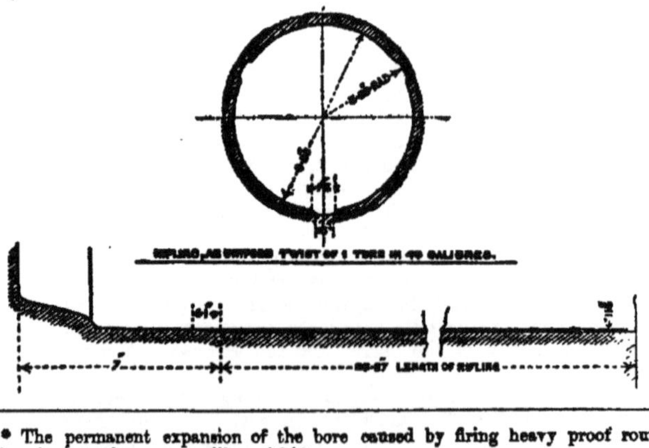

* The permanent expansion of the bore caused by firing heavy proof round brings it up to the proper calibre of 6"·3.

CHAP. X.

Venting.

Venting.

The vent patch is removed,* as it would interfere with the lanyard when used through the guide plate on S.S., and the old vent closed with a wrought iron screw plug, a new vent being drilled a little from the breech end. This is bushed permanently before proof with a "through vent" (seven threads to the inch) of hardened copper screwed through the barrel into the breech cup, and perpendicular to the surface of the cup, *i e.*, at an angle of 12° 25′ to the vertical.† The lower thread of the screw in the gun is cut away and the end of the vent is set up into the recess thus formed.‡

CONVERTED M.L. GUNS, VENTING. Scale ½.

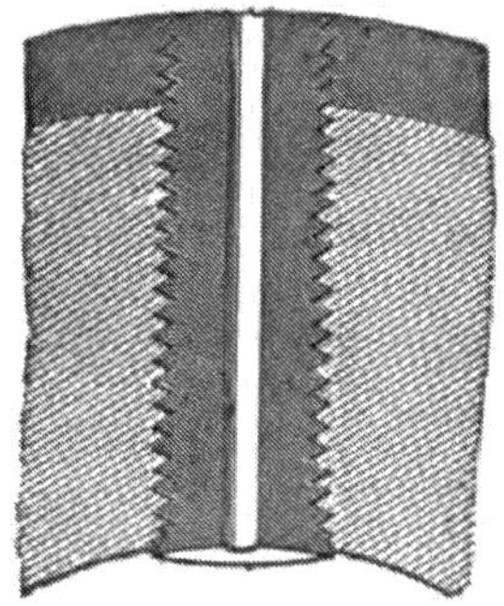

The process of venting is as follows (vide Plate on next page for tools mentioned):—§

The thread of the through bush is turned off for a length of about ¾ inch, and the bush is then screwed in by hand by means of a lever wrench. To insure that only the required amount projects into the bore, a "*block stop for vent*" is passed into the chamber. This is an iron block attached to an iron bar, and having a portion slotted away of the depth sufficient for projecting part of the bush to fit into; the bar is kept in centre of the bore by means of two wood discs.

A block stop for vent.

A¹.

When the end of the bush comes in contact with the bottom of the slot on the "stop block," the operation of screwing is discontinued, and the projecting copper "upset" into the recess as follows.

B¹ and B².

C.

D.

E.

A split head or "*expanding block*" of wrought iron, fitting the shape of the chamber, is pushed up into the latter by means of an iron "*extracting hook;*" into that part of the block immediately under the copper bush a piece of hard steel is dovetailed. An iron *tube* called guide cylinder, the inner end of which fits over the "expanding block" is next passed up the bore, and through this guiding tube is passed a solid iron *wedge*, which being forced into the "split head" or "expanding block" presses out the sides of the latter, and so sets up the copper into the recess in the cup. The wedge is at one end of a stout

---

* $\frac{\text{64-pr.}}{\text{8-inch}}$ only.

† With $\frac{64}{32}$-pr. and $\frac{80}{68}$-pr. the hole used for the old vent bush is used for venting and the vents are therefore inclined at the same angle (from 9½° to 10°) as in the S.B. guns from which they were converted.

‡ The first 207 guns converted at Elswick, from 8″ gun, had a vent bush 1″ diameter, 12 threads to the inch, but when they require re-venting it wi be with the service bush.

§ A list of the special tools required for this operation and that of re-venting is given at p. 322, Chapter XII. At present, such sets of special tools have only been issued to Portsmouth, Malta, Hong Kong, and Esquimault, and also to the Flagship of our Mediterranean Squadron.

SPECIAL TOOLS TO BE EMPLOYED IN RE-VENTING CONVERTED GUNS.

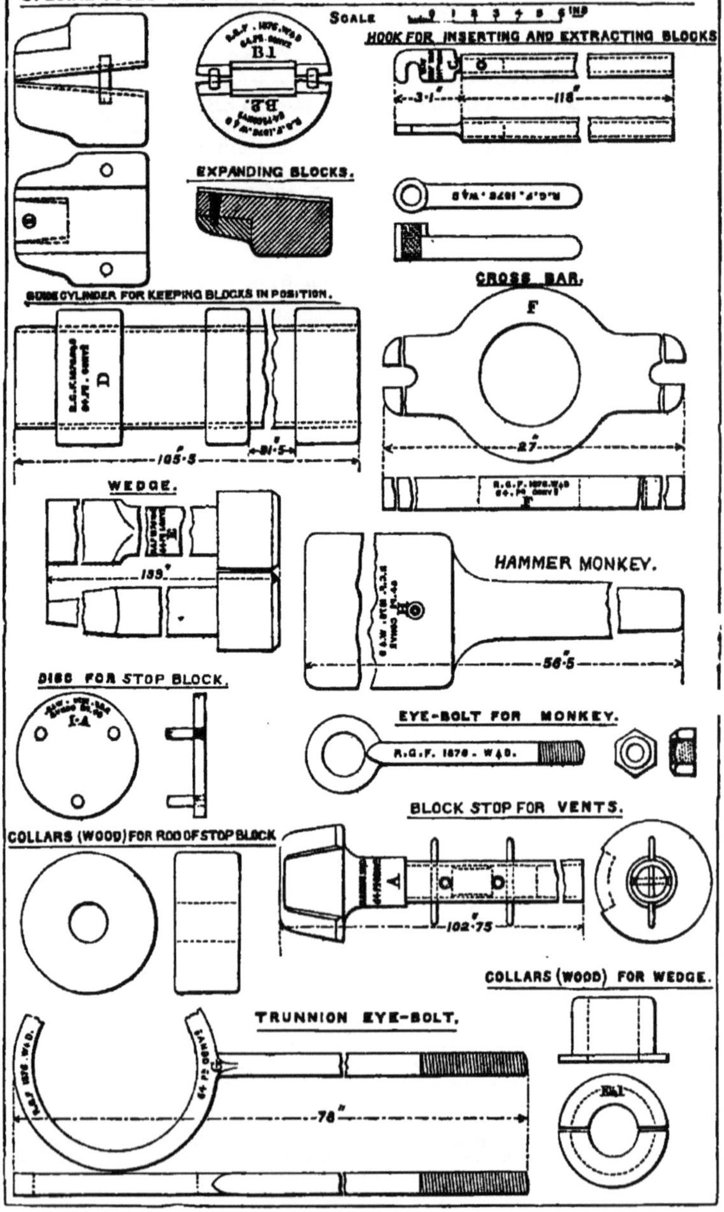

CHAP. X.

H. iron bar, the outer end of which is struck by a "*hammer monkey*" worked by two or three men, as a considerable amount of power is required for the setting up of the hardened copper bush. This hammer is slung from an iron triangle gyn, 10 feet high, or a small service gyn will answer.

E.I. F. G. To keep the bar to which the wedge is attached in the centre of the bore it passes through a "*wood collar*," and also through the centre of a "*cross bar*" in front of the muzzle, which is retained in position by a frame consisting of two rods with loops fitting over the trunnions and secured by nuts and screws to the ends of the collar.

When the copper is sufficiently "set up," the bars are loosed from the collar and the latter removed. The wedge and cylinder guide are then withdrawn and the expanding block removed by the extracting hook after which a gutta-percha impression is taken of the bottom of the vent to ascertain if the operation has been performed completely; in that case the copper should still project into the bore for about 0·05 inch.

Proof and Examination.

Proof and examination. These guns are proved like other R.M.L. ordnance with two rounds, 1¼ service charge and service projectile, and the expansion caused by proof must not exceed certain limits, No part of the bore before proof is allowed to be more than 0″·02 under the gauge (6″·29) nor 0″·04 over that gauge after proof. After proof the guns are gauged, examined with gutta-percha, and again tested with water pressure of 120 lbs. on the square inch to see that the breech is still perfectly tight

Fire-proof.

Waterproof.

Wrought iron cup. Owing to the defects inherent to wrought iron as a material for the inner barrel of a gun, converted guns not unfrequently fail at proof, generally speaking through defects in the wrought iron cup which closes the end of the breech, though sometimes the tube itself splits.

Removing a tube from its casing. Should a tube fail it is forced out of the casing by hydraulic pressure applied to the breech end; the nozzle of the hydraulic apparatus being passed through the gas escape channel, which is enlarged for this purpose.

Lining and sighting. If the gun pass proof, however, it has now to be lined* and sighted. These operations are performed as in the case of other R.M.L. guns (vide pp. 179, 180), except that the sockets for the fore sights are not fitted into holes bored in the metal of the gun but into gun metal brackets which are themselves secured to the piece by two screws.

$\frac{64}{32}$-pr R.M.L. Gun L.S. of 58 *cwts.*

§ 2066. The method of conversion of these guns is identical with that just described. They are, however, only sighted centrally.

Venting. They are vented in the original position, and those guns which had

Scale $\frac{1}{32}$nd.

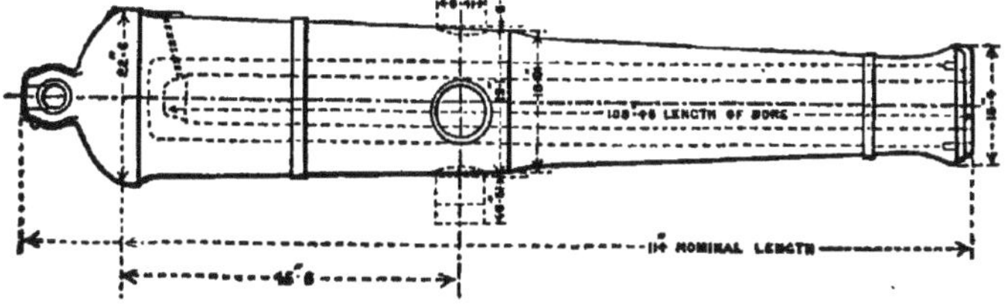

---

* The vertical line on muzzle is extended over the cast iron casing, so as to show any shifting of the tube.

six thread bushes * as S.B. pieces are re-vented with bushes of the same thread, this divergence from the pattern being stamped on the vent field. CHAP. X.

The vent patch is not removed, these guns being intended for L.S. only.

$\frac{80}{68}$-*pr. R.M.L. Gun L.S., of 5 tons.*

The conversion of these guns is identical with that of the 64-pr. of 58 cwts., but the rifling is on the "Woolwich" system, the width of the groove being 1″·3 and depth 0″·145. § 2220.

They are vented the same as the 58-cwt. gun, and guns having six thread bushes are re-bushed with the same. Venting.

They are for L.S. only.

Scale $\frac{1}{32}$nd.

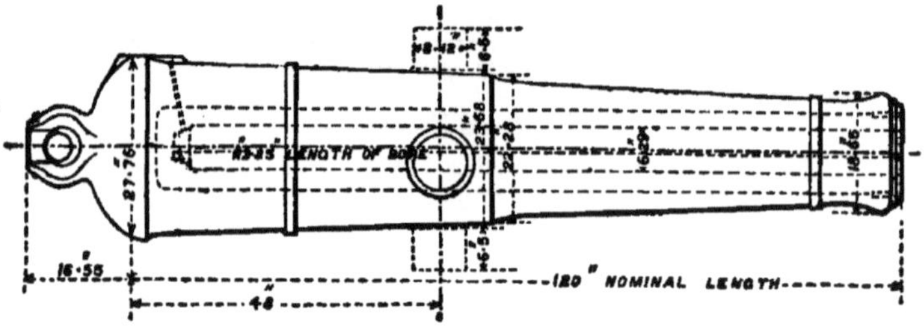

Sights.

The 80-pr. of 5 tons and 64-pr. of 71 cwt. are side sighted and have two tangent scale sights and two trunnion sights as shown below. The tangent scale sights are similar to those described for other R.M.L. guns of 5 tons and upwards (vide p. 183), *i.e.*, are provided with the deep notch, and have no small elevating nut. The trunnion sights are high and of the drop pattern, and are fitted to the gun by gun metal brackets, to avoid boring into the cast iron casing. $\frac{80}{68}$pr. of 5 tons and $\frac{64\text{-pr.}}{8 \text{ inch}}$ of 71 cwts.

Sighting of 64-pr. of 71 cwt. and 80-pr. of 5 tons.

SIGHTING. Scale $\frac{1}{12}$th.

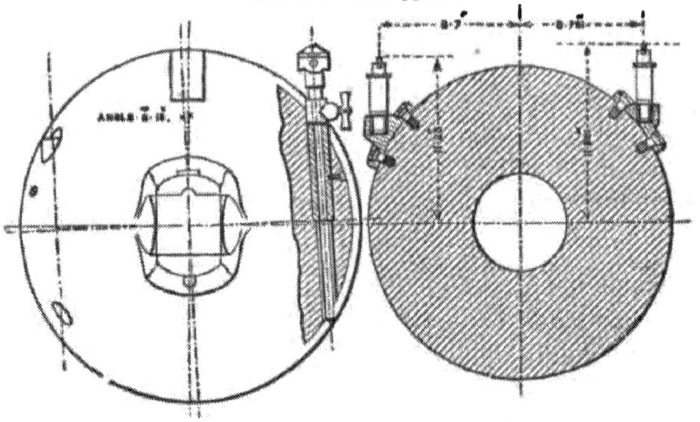

* Converted guns having the 6 thread bush, are stamped on vent patch "vent, 6 threads."

CHAP. X.

$\frac{64}{32}$ pr. of 58 cwt.

64-pr. guns of 58 cwt. converted previous to 1877, were sighted centrally and provided only with two sights, a centre hind sight, (hexagonal and with plain head similar to the built up 64 pr. of 64 cwt.) and a centre fore sight of drop pattern fitting into a gun metal bracket. Since that date and for future conversion, the gun will be provided with three sights. These consist of a long hexagonal centre hind sight graduated up to 12 degrees, and provided with a cross head and sliding leaf for deflection (vide Mark VI. p. 226). The centre fore sight is as before, while the muzzle sight is a small screw sight, fixed in a recess, cut in the swell of the muzzle.

The angle of derivation of sights for all converted guns is 2° 16'. (For details of graduations, &c., see Table p. 226).

Ordnance Cast Iron R.M.L. Converted 64-pr. 58 cwt.

GENERAL ARRANGEMENT OF SIGHTS.

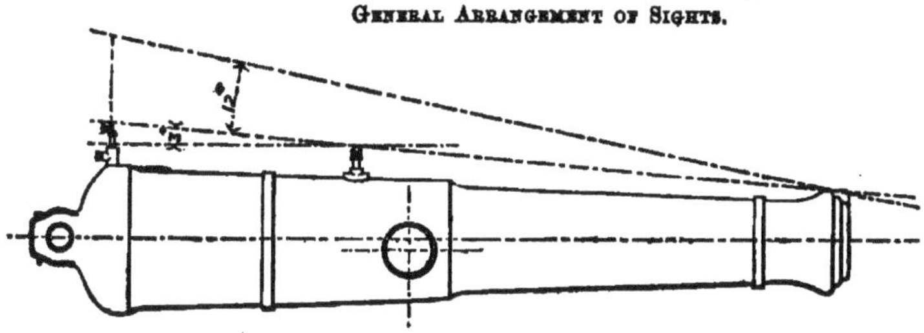

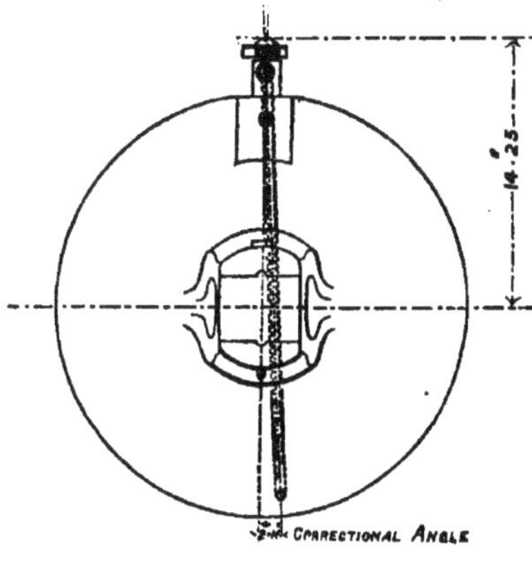

SIGHT, CENTRE HIND.

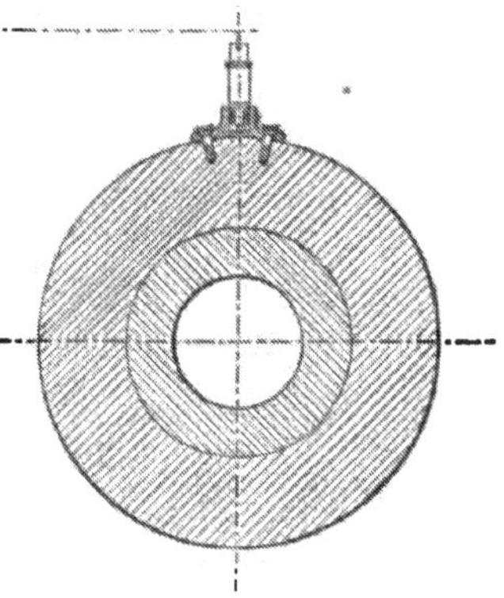

SIGHT, CENTRE FORE.

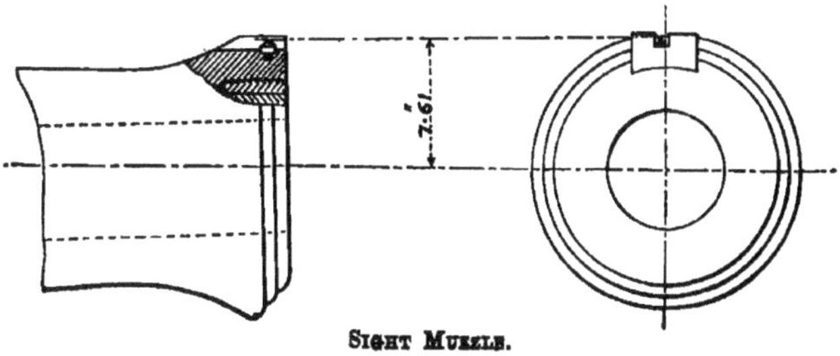

SIGHT MUZZLE.

Besides these sights the 64-pr. of 71 cwts. S.S. is supplied with the wood side scale shown below. CHAP. X.

WOOD SIDE SCALE. Scale ¼th.

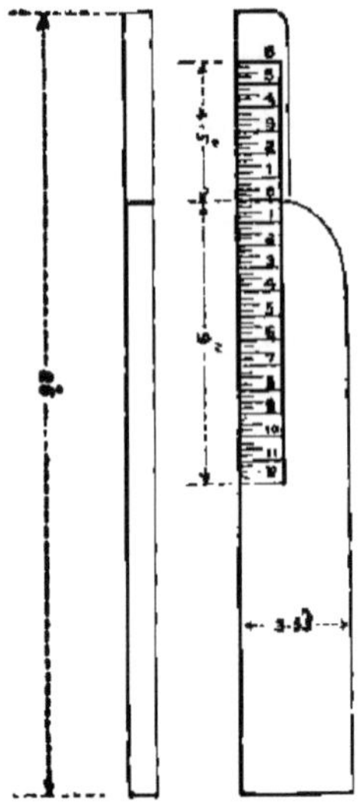

Small stores and fittings.

The following are the stores belonging to these guns as mentioned in Table, p. 231.

Bearers, shot or shell.

For the 80-pr. guns only.

Brackets right and left hand.

These are of gun metal and attached to cast iron casing of the gun by two screws, and contain the socket for the drop trunnion sights. To bore holes in the cast iron casing for these sockets would weaken it too much. They are for the 80-pr. 5 tons and 64-pr. 71 cwts., and must be removed when the gun is required for transport. §§ 1752, 2066 2220.

Bracket, centre fore sight.

This is similar to the above bracket, but only for the 64-pr. 58 cwt. for its centre fore drop sight, and attached to the second reinforce by two screws. It must be removed when the gun is required for transport. § 2066.

Clamps, moveable, for tangent sight.

Are of gun metal and are for the 80-pr. 5 tons, and 64-pr. of 71 cwt., see p. 133. §§ 1144, 1357.

CHAP. X.

Guide plates.

§ 688. Are of steel, attached to the right rear of the vent, and the pattern is universal for all S.S. guns, see p 205. The 80-pr. 5 tons, and 64-prs. of 71 and 58 cwts. are prepared for the reception of this fitting in the event of the guns being required for S.S.

Pins, iron, friction tube.

§ 1141. This is attached to the left front of the vent, and is of wrought iron case hardened, and the pattern is universal for all S.S. guns. For further remarks see heading above.

Pricker, priming iron.

§ 1212. 17 inches long and serve to prick the cartridge. All the converted guns use it.

Screws fixing.

See Brackets and Sockets.

Screws, preserving.

These fill up the holes occupied by the brackets for the sights when the guns are dismounted, and also the holes for the friction tube pin and guide plate when the piece is in use by the L.S.

Sockets, metal.

§ 1481. Centre hind sight with set screw for 64-pr. 58 cwt. and let into a hole being fixed by a screw. It must be removed when the gun is required for transport.

Spikes.

Common and Spring. Vide Table p. 281.

Wrench, pin, friction tube.

§ 1213. This wrench for the converted guns is required to screw or unscrew the friction tube pin, guide plate, and the fixing and preserving screws.

CHAPTER XI.

RIFLED M.L. ORDNANCE IN THE SERVICE, AND REMARKS UPON THE SAME.

CHAP. XI.

Different Classes of R.M.L. Ordnance.—Mountain or Boat, Field, Siege Medium, and Heavy.—General remarks as to employment.—Marks or patterns. —Gas escapes.—Exterior form.—**Mountain or Boat Guns.**—7-pr. of five different patterns.—**Field Guns.**—Four natures of 9-prs., 16-pr. guns.—**Siege Pieces.**—Light and Heavy 6·3-inch and 8-inch howitzers, 25-pr., 40-pr., and 64-pr. guns of latest pattern.—**Medium guns.**—Converted and other 64-pr. and 80-pr. guns.—7-inch of 90 cwts.—**Heavy Guns.**—Different natures of, 7-inch, 8-inch, and 9-inch; two patterns of 10-inch and 11-inch guns.—12-inch guns of 25 and 35 tons.—12·5-inch of 38 tons.—13·05-inch gun practically obsolete.—**Experimental Pieces.**—16-inch gun of 80 tons.—Rifled howitzers.—13-pr. field gun.—Table of R.M.L. Ordnance, with dimensions, rifling, &c.

Classes of R.M.L. Ordnance.

The general mode of construction of the guns now manufactured has been explained in Chapter VIII., and in the following chapter will be given the details of construction of all our existing R.M.L. pieces, including some natures which are still in the service but of which no more will be manufactured. R.M.L. ordnance.

Our R.M.L. ordnance* may be divided into— Classification.

(1.) *Mountain or Boat Guns.*

7-prs. of 150 to 224 lbs. Mountain or boat.

(2.) *Field, Boat, or Field Marine.*

9-prs. of 6 or 8 cwts.
16-pr. of 12 cwts. Field, boat, or field marine.

(3.) *Siege or Position.*

25-pr. of 18 cwts. Siege or position.
40-prs. of 34 or 35 cwts.
64-pr. (Mark III. with steel tube) of 64 cwts.
6·3-inch howitzer of 18 cwts.
8 „ „ 46 „

* Vide table, p. 291, for list with dimensions.

CHAP. XI.

Medium.

(4.) *Medium.*

64-pr. built up (except Mark III. with steel tube) of 64 cwts.
64-pr. converted of 58 cwts. & 71 cwts.
80-pr. " 5 tons.
7-inch of 90 cwts.

Heavy.

(5.) *Heavy.*

7-inch	of	6½ and 7 tons.
8	"	9 tons.
9	"	12 "
10	"	18 "
11	"	25 "
12	"	25 and 35 tons.
12·5	"	38 tons.
16	"	80 "
17·72-inch	of	100 tons.

General Remarks.

Before entering into the details of the various natures it may be well to make some general remarks as to R.M.L. service ordnance.

§§ 899, 1061. (1.) *As to Use.*—For S.S. only we have the 7-inch of 6½ tons and of 90 cwts. as well as the 64-pr. built up guns, except such as are appropriated for siege purposes.

The 7-in. of 7 tons, the $\frac{80}{68}$-pr. 5 tons and $\frac{64}{32}$-pr. 58 cwt. converted guns are meant for L.S. only. All other heavy guns are used both for L.S. and S.S.

Siege ordnance. All the siege pieces or guns of position are L.S. Where guns of that size would be employed for S.S. as deck guns or otherwise 20-pr. and 40-pr. R.B.L. are generally used.

Field, boat or Field marine. For Field service we have the 9prs. L.S. of 6 and 8 cwt. respectively and the 16-pr. For heavy boat service or field marine, the 9-prs. S.S. of 6 and 8 cwt.*

Boat or mountain. For boat or mountain service we have the small 7-pr. guns, which are L.S. and S.S.

(2.) *Marks or Patterns of Guns.*—As explained at pp. 161, 162, all built up Woolwich guns below the 9-inch are now made on the Fraser construction, with one layer of metal over the breech end in the shape of the jacket, the 7″ Mark IV and reduced 7-inch of 90 cwts. excepted (vide p. 169). The different marks of such guns, therefore, only differ in the manner in which the B layer of metal is put on, as in the 40-pr., or in length or other dimensions, as in the 9pr.

Patterns of guns. In the case of the heavier guns, however, where there is more than one construction on which the guns were formerly manufactured, the following table may be useful, as it shows in a condensed form the various patterns of guns in use. All these guns, it will be seen, are now made on the Fraser principle, with either one or two layers of iron over the steel tube.

---

* Most of these latter have been altered from the L.S. Mark I. 9-pr. of 8 cwts.

TABLE XXVI.

TABLE showing the CONSTRUCTION of the VARIOUS MARKS of BUILT UP M.L. GUNS, 64-prs. and upwards.†

Nature.	Original Construction, pp.	Fraser Modification with Forged Breech-piece, pp.	Fraser Modification with One Layer pp.	Fraser Modification with Two Layers, pp.	Remarks.
	Mark.	Mark.	Mark.	Mark.	* These are now the service patterns for future manufacture.
64-pr., 64 cwt, ..	I.§	II.§	III.*§	—	|| Iron tubes.
7-inch, 90 cwt. ..	—	—	I.¶	—	¶ Steel tubes.
7-inch, 6½ or 7 tons ..	I.||	II.§	III.¶	IV.*¶	§ Some with iron, and some with steel tubes.
8 " 9 tons ..	I.§	II.||	III.*¶	—	
9 " 12 " ..	I.§	II.¶	III.¶	IV., V.*¶	
10 " 18 " ..	—	—	I.¶	II.*¶	
11 " 25 " ..	—	—	I.¶	II.*¶	
12 " 25 " ..	I.¶	—	—*	II.*¶	
12-inch, 35 tons ..	—	—	—	I.*¶	
12·5 inch, 38 tons ..	—	—	—	I.*¶	
16 inch, 80 tons ..	—	—	—	I.¶	
17·72 inch, 100 tons ..	—	—	—		

Gas escapes.

(3.) *Gas escapes.*—As a precaution against accident in case the inner tube should split, most R.M.L. guns are furnished with gas escapes, *i.e.*, with a channel through which, should the inner barrel crack, the gas will escape and give warning that firing should be stopped.

All guns having cascable screws, *i.e.*, the R.M.L. howitzers as well as the 40-pr. guns and upwards, have gas escapes, except Mark I. 64-pr. wrought iron guns, which have none, being manufactured before their introduction; but any of these latter guns that may require retubing would of course be furnished with such an escape.

The converted guns are also furnished with gas escapes, as explained in Chapter X., p. 234.

In the 25-pr. and downwards no such escape exists, as in these pieces the steel tube projects beyond the breech end of the jacket.

To form the gas escape in the built up guns, 40-pr. and upwards, a part of the thread of cascable screw is turned off, so that when the cascable is home against the end of steel tube an annular space is left; this is made to communicate with the outside of the gun by means of a channel cut through the cascable itself or through the metal of the gun, so that in the event of the tube splitting in the vicnity of the chamber the gas escaping through the channel would indicate the fact to the detachment at the gun, when firing should immediately cease.

* There are two 12-inch 25-ton guns on this construction similar to Mark I. 11-inch, but they do not constitute a separate pattern, being known by their Nos. (20 and 21).

† Guns marked F. or F.I. are of Mark II. construction.

	"	F.II.	"	III.	"
64-pr. guns	"	B.	"	II.	"
"	"	D.II.	"	III.	"

CHAP. XI.

In heavy guns of original construction.

With Marks I. and II. of 7, 8, and 9-inch guns, and Mark I. 12-inch of 25 tons, which have a forged breech-piece, a hole 0''·3 diameter was drilled through the breech-piece, at right angles to the cascable screw, so as to meet the annular space on the cascable, a groove also being cut along the outer edge of the breech-piece so as to meet this hole. In this pattern the gas escape comes out underneath the cascable, between the breech-piece and the coil overlapping it.

In other marks the channel is cut along the cascable screw and comes out at the right side of the cascable, or under the loop, in guns manufactured before 1869.

Exterior form.

(4.) *Exterior Form.*—It is convenient sometimes to know at a glance the nature and pattern of a gun. As to howitzers and the smaller guns, no remarks are required, but the following may be useful in the case of the heavier natures, of which there are several marks, or patterns:—

§ 1335.

Exterior Form.

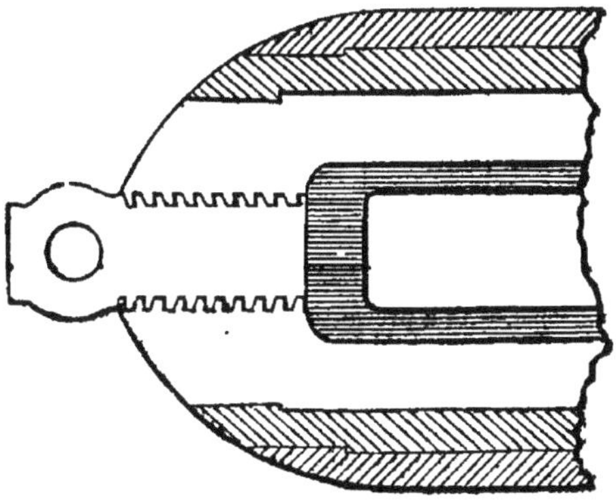

Breech rounded.

Original construction, breech rounded off.

Mark I. 64-pr., 7-inch, 8-inch, and 9-inch.—Are rounded off at the breech, because at the time of their introduction the breeching rope was used through the breeching loop, which was provided with a moveable block and pin, and consequently this shape of breech was adopted to save the rope from wear. They have also several steps in front of the trunnions.

Mark I, 12-inch and 13-inch are stepped at the breech.

Mark II. 64-pr., 7-inch, 8-inch, and 9-inch.—The curve at the breech is broken by the breech-piece which forms a step; the breeching rope being carried through the carriage, the moveable block in the breeching loop was abolished, and the thin edges of the layers of coil caused by the rounded outline was considered objectionable owing to the liability of the edges setting up. This and all future patterns have but two steps in front of the trunnions.

CHAP. XI.

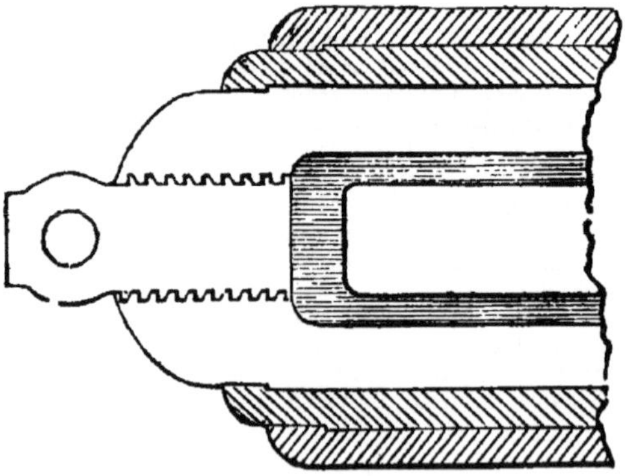

Breech cut in steps.

Mark III. 64-pr., 7-inch, 8-inch, and 9-inch, and Mark I. 10-inch and 11-inch.—Having but one layer at the breech it is rounded off. Fraser modification with one layer.

Mark IV. 7-inch, Marks IV. and V. 9-inch, and Mark II. 10-inch, 11-inch, and 12-inch, Mark I. 12·5-inch, Mark I. 16-inch.—As a rounded outline in this pattern would introduce the objectionable thin edge, the breech is cut in steps, corresponding to the layers of coils. Fraser modification, with two layers.

(5.) *Steel tubes.*—The maximum thickness of these inner barrels varies from about 0·25 of a calibre in the larger guns to 0·33 of a calibre in the smaller pieces. This greatest thickness is naturally given at the breech end over the powder chamber, except in some of the field pieces where the barrel is not supported towards the muzzle by any coil shrunk on, and in which the tube is consequently thicker in the chase than at the breech end. Steel tubes.

The thickness of the tube is not uniform in the heaviest guns of latest pattern; in all pieces above the 7-inch gun (excepting the 10-inch of 18 tons) it decreases towards the muzzle, and in the 35-ton, 38-ton, and 80-ton gun this is readily seen.

A thinner chase is quite strong enough to withstand the strain, which decreases rapidly towards the muzzle.

Mountain or Boat Guns.

7-pr. Rifled M.L. Guns.

There are seven patterns of this gun in the service, three of bronze and four of steel, but the number of any one pattern made is very small, with the exception of Mark III. and IV., steel, of which many have been made. 7-pr. guns.

The rifling of the whole of these guns (both bronze and steel) is identical, and they fire the same ammunition.

CHAP. XI.

7-pr. bronze, Mark I.

Bronze. Mark I. Weight, 224 lbs. Calibre 3″.

Scale 1 inch = 1 foot.

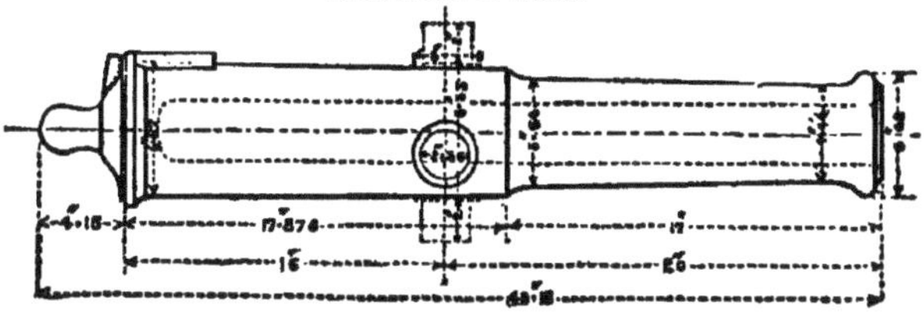

RIFLING. Scale 3 inches = 1 foot.

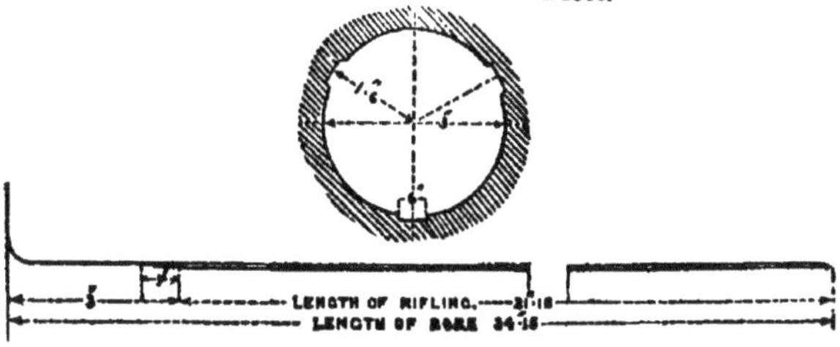

§ 1146. O.S.C. Proceedings, 1865, pp. 212, 303, 304; and 1866, p. 233.

In 1865 some mountain guns were required on an emergency by the Indian Government to accompany the expedition to Bhootan, and 10 steel guns were demanded by the late Ordnance Select Committee, five of 190 lbs. weight and five of 150 lbs. As these guns could not however, be supplied in time for Bhootan, six bronze 3-pr. S.B. guns of 2¼ cwt. were turned down to a weight of 224 lbs., bored to 3″, and rifled on the French system with a twist of one turn in 20 calibres.

This gun was found to be too heavy for the mules in India, and a gun of 200 lbs. weight was asked for.

In order therefore to utilize the existing stock of bronze 3-prs., a gun was reduced to the required weight and adopted in 1866 as Mark II.

7-pr. Bronze. Mark II. Weight, 200 lbs.

Scale 1 inch = 1 foot.

7-pr. bronze, Mark II.

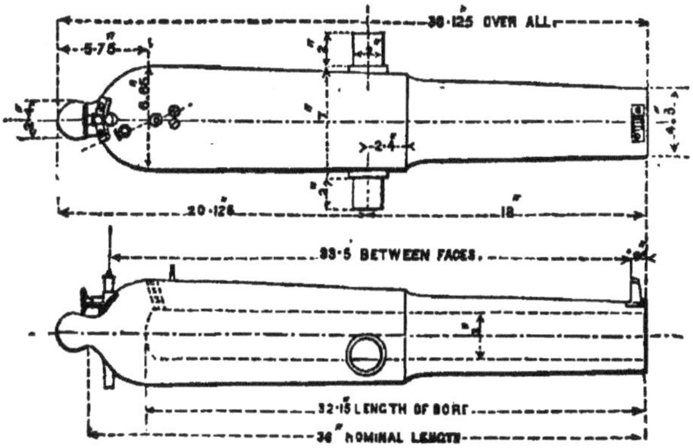

SIGHTING. Scale 1 inch = 1 foot.

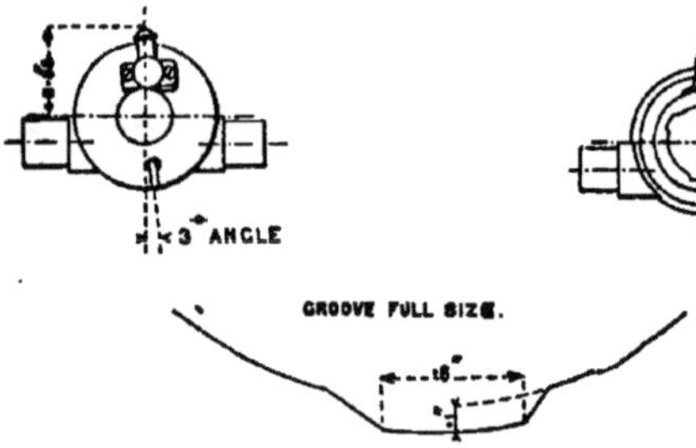

This differs from Mark I. gun in having the exterior turned perfectly plain, and being 2″ shorter in the bore. The swell of the muzzle is removed and a dispart sight screwed on the gun. § 1935. O.S.C. Proceedings, 1866, p. 107.

A pattern of this gun was sent to India in 1866 to govern manufacture, and in 1867 twelve were sent to Ireland during the Fenian disturbances. In 1870 six were sent to Canada for the Red River Expedition, and being also approved of for boat service, it was then definitely adopted into the service.

The form and preponderance of this gun (being only a conversion) are not satisfactory, and experiments were carried out with a view to the adoption of a new bronze gun of a better construction, but intended to fire the same ammunition with a heavier charge. The gun recommended is shown at Plate I.

Mark III., bronze.

On 21/1/73/ it was ordered that no more bronze 7-pr. R.M.L. were to be made. Only two of this Mark III. pattern have been manufactured. 7-pr. bronze, Mark III.

Steel. Mark I. 190 lbs.

This was the first steel gun adopted, and five were made in 1865 and sent to Bhootan. The pattern is obsolete, and no more are made, but those already issued are retained in the service in India. 7-pr. steel. Mark I.

Steel. Mark II. 150 lbs.

This is the piece commonly known as the Abyssinian gun, twelve having been sent with that expedition. They were sighted on the right side only, the tangent sight working in a gun metal socket screwed to the breech, and the dispart sight screwed into the side of the muzzle. § 1506. 7-pr. steel, Mark II.

The cascable had a projection underneath, which fitted into a slot in the elevating screw. A pattern of this gun was never sealed, and the gun was altered on the experience gained in Abyssinia into the present pattern. Thirteen manufactured.

CHAP. XI.

Scale 1¼ inch = 1 foot.

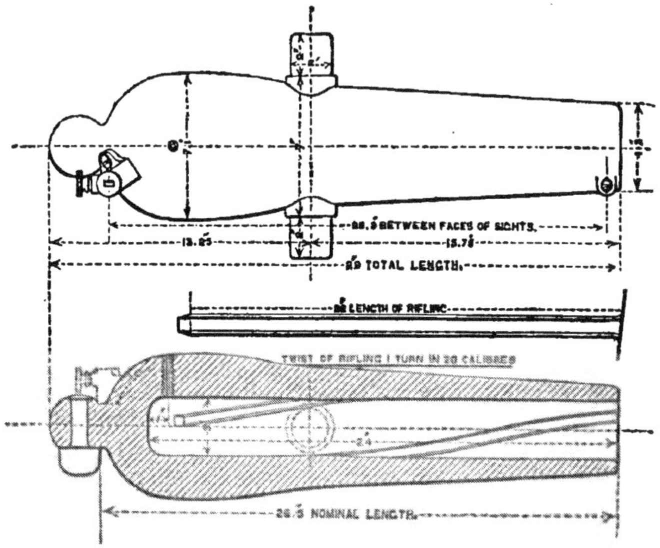

SIGHTING. Scale 1¼ inch = 1 foot.

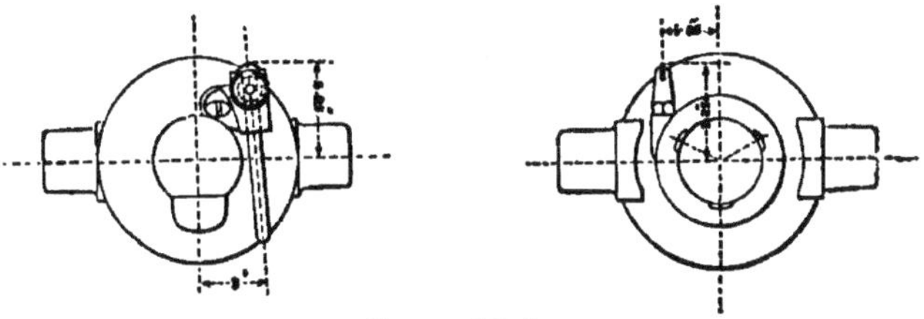

GROOVE, full size.

Steel. Mark III. 150 lbs.

§ 1717. 7-pr. steel, Mark III. This gun differs from Mark II. in being centre sighted (see page 249 as the side sights were found to be liable to injury), and also in having a horizontal hole bored through the cascable, through which a rod can be passed, so as to facilitate loading or unloading, and also to enable the gun to be readily carried by men over country impassable for animals.

Scale 1½ inch = 1 foot.

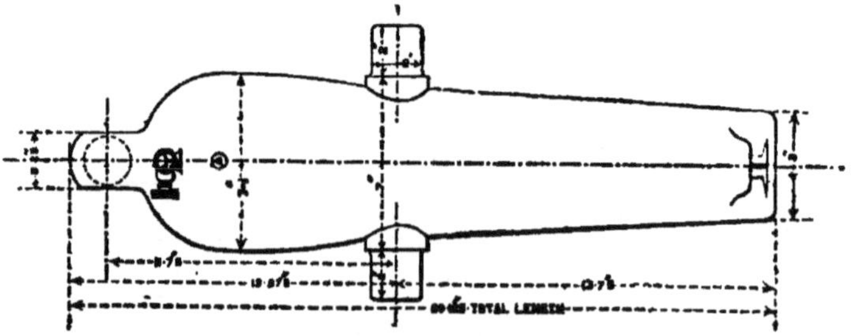

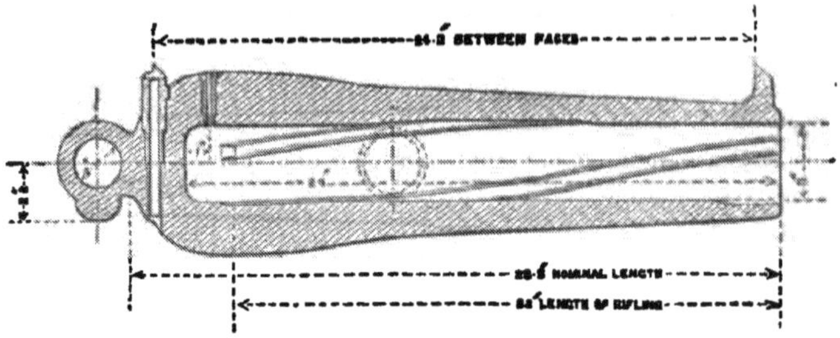

SIGHTING. 1½ inch = 1 foot.

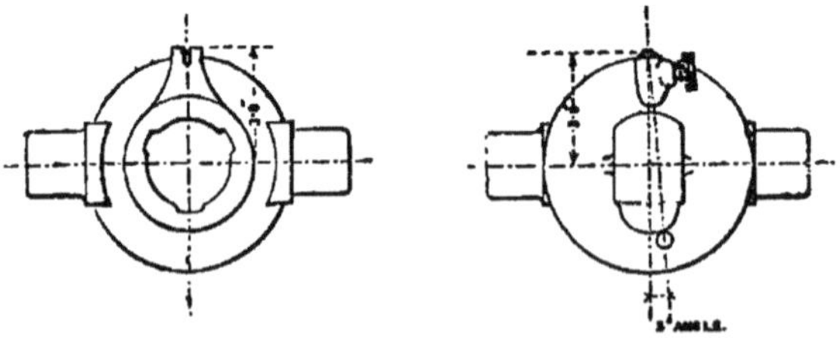

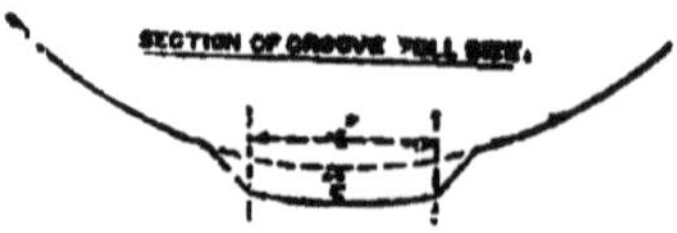

Steel. Mark IV. 200 lbs. L.S. and S.S.

§ 2198. 7-pr. steel, Mark IV.

This gun was introduced into the service in July 1873; it is much longer than Mark III., and externally resembles the bronze, 7-pr. Mark II. It is made out of a solid block of steel, and has no swell at the muzzle, but a small dispart patch into which the fore sight screws.

CHAP. XI.

The hole through the cascable is here small, this piece being intended for mule carriage if employed for L.S.

This piece would be used in the siege train to take the place of the old S.B. bronze mortars.

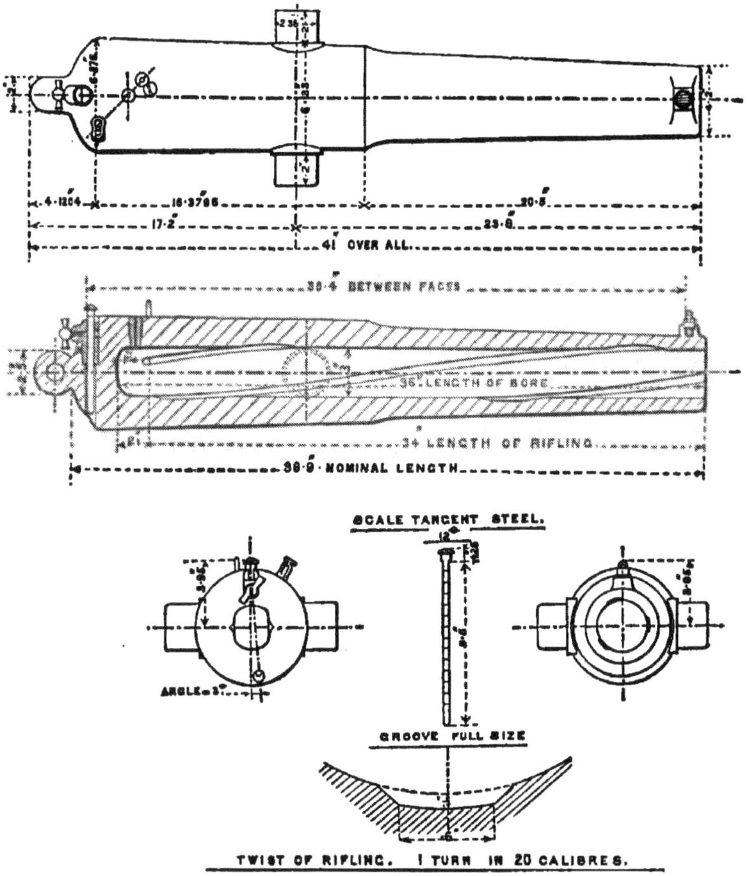

Field Guns.

9-pr. Rifled M.L. Guns.

9-prs. of four natures.

We have four different natures of 9-prs. in the service; two for L.S. and two for S.S. They all have the same calibre and rifling.

The L.S. are 9-prs. of		8*	cwts.	Mark	I.
"	"	6	"	"	II.
Those for S.S.	"	8*	"	"	II.
"	"	6	"	"	I.

It will be seen that the 6-cwt. gun Mark I. is the shortest of all. It is meant for boat service.

The 8-cwt. gun* will be boat and field marine; and the 6-cwt. Mark II., which has the greatest length of bore, will be the Horse Artillery and light field battery gun.†

Bronze 9-pr.

In 1869 a bronze 9-pr. gun of 8 cwts. was introduced for Indian service, with the calibre rifling, &c., of our present 9-prs. Several batteries of these pieces were issued, but as the material of which they were made proved unsatisfactory, their manufacture soon ceased, and they were all withdrawn from the service in 1874.‡

Since that date all our Horse Artillery have been armed with 9-prs. of 6 cwts., and half of the field batteries with 9-prs. of 8 cwts. The latter, however, are being now altered for sea service, and the field batteries having 9-prs. will be equipped entirely with those of 6 cwts.

All 9-prs. consist of:—

- *A* tube, toughened steel.
- Jacket, composed of a single coil, trunnion ring and a coil in front of trunnions welded together.
- All have the French modified rifling, with a uniform twist of 1 in 30 calibres. They have rear vents striking the cartridge near the bottom.

The following diagrams, &c., give details of the 9-prs.

Mark I. 8 cwt. L.S.

Scale ⅛ inch = 1 foot.

§ 2067.

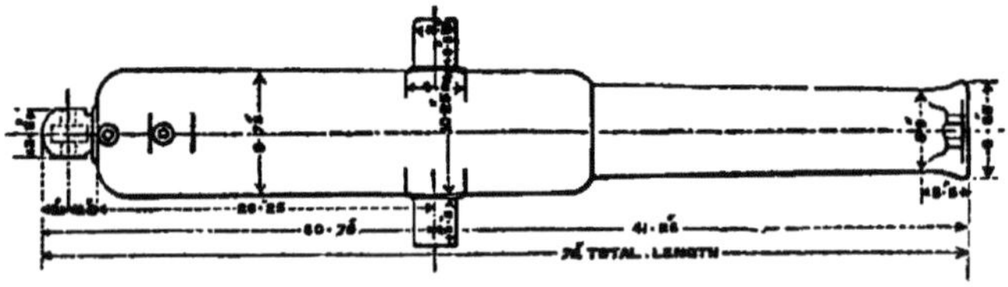

9-pr. Mark I. L.S.

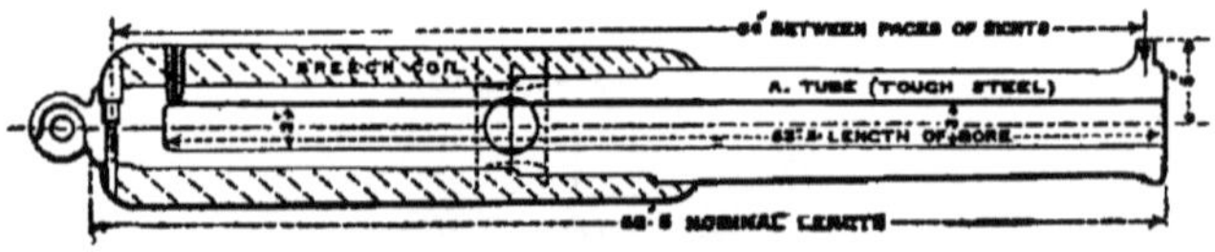

These guns are being altered to S.S. by turning off the projection at muzzle and providing the S.S. sights mentioned at pp. 186 and 189.

---

* L.S. of 8 cwts. will be all altered and become S.S. guns, receiving a new register number, and the Mark being altered from I. to II.

† For light field batteries we shall soon have the 13-pr. of 8 cwt., now being experimented with, vide p. 77.

‡ Vide also p. 73. A number of these bronze guns were rendered unserviceable by ordinary practice firing in peace time.

CHAP. XI.

§ 2674.

Mark I. 6 cwt. S.S.

9-pr. Mark I. S.S.

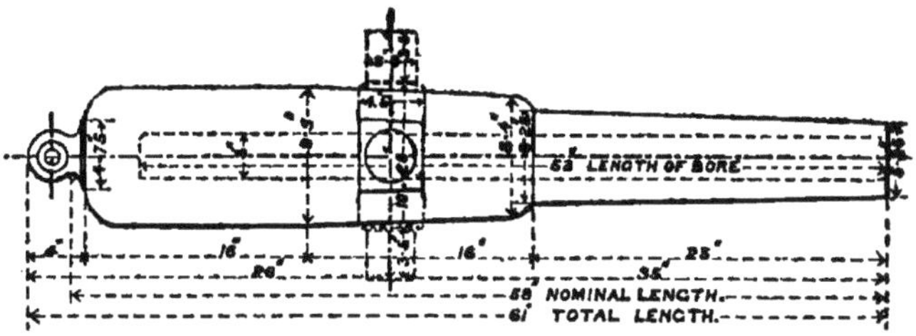

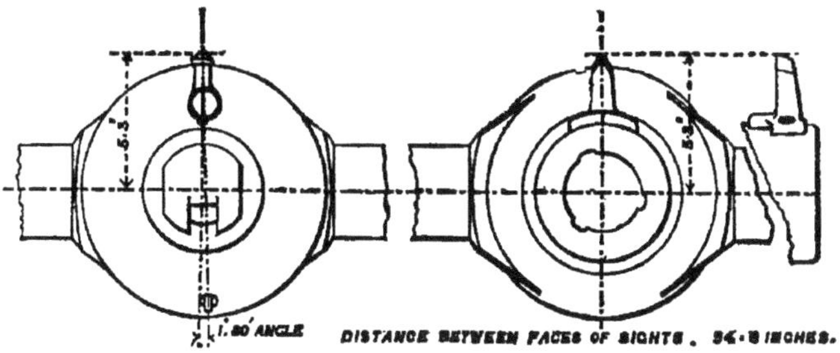

The 6-cwt. gun, Mark I., S.S., is used by the Navy; some of the pattern have been ordered for the Indian Naval Service.

Only a small number (45) of these have been made, as the Navy prefer the 9-pr. of 8 cwt. Mark II.

---

§ 2599.

Mark II. 8 cwt. S.S.

SCALE 1/4 INCH TO 1 FOOT.

9-pr. Mark II. S.S.

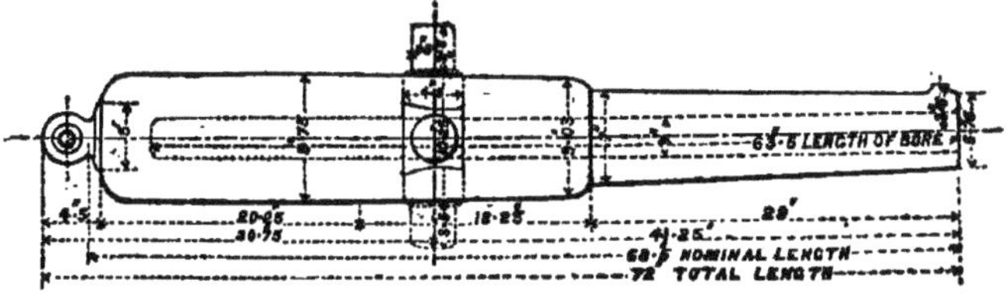

CHAP. XI.

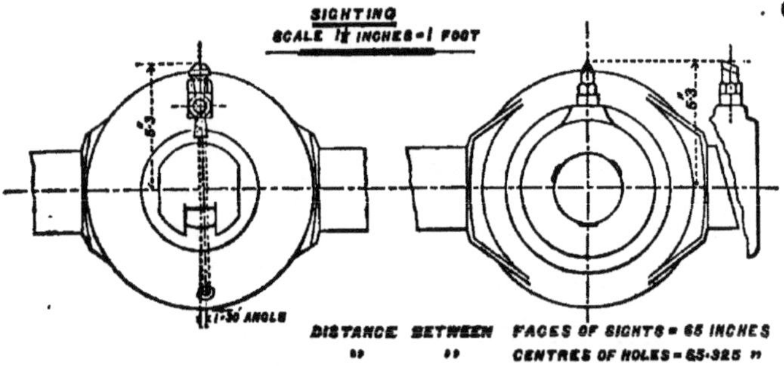

In 1873 a pattern gun of 8 cwt., Mark II., was sealed for S.S. It is the same generally as the L.S. 9-pr. of 8 cwt., and uses the same ammunition. It differs from the latter gun in having no swell on the muzzle.

Mark II. 6 cwt. L.S.

§ 2683.

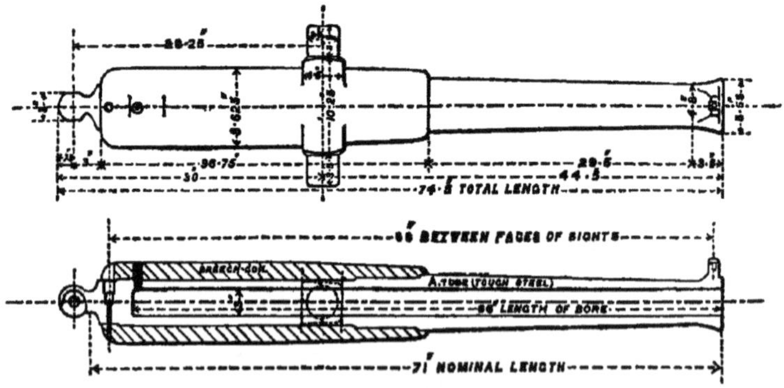

This 9-pr. R.M.L. gun of 6 cwt., L.S., Mark II., was introduced in September 1874. It only differs from the 9-pr. of 8 cwt., L.S., in weight, and dimensions generally.

9-pr. Mark II. L.S.

We now come to the heavy field guns with which the greater portion of our field batteries are armed.

16-*pr. Rifled M.L. Gun of* 12 *cwt. Calibre* 3″·6, *L.S.*

In 1870, a special Committee recommended the adoption of a rifled gun of about 12 cwt., having a calibre between 3″·5 and 3″·7.

Two guns were made, one with 3″·6 calibre and the other 3″·3, and the former was ultimately adopted for the armament of a portion of our field batteries.

Report of Experimental Department of D. of A., 1870, p. 341.

CHAP. XI.

§ 2221.

Mark I. L.S.

Scale ⅛ inch = 1 foot.

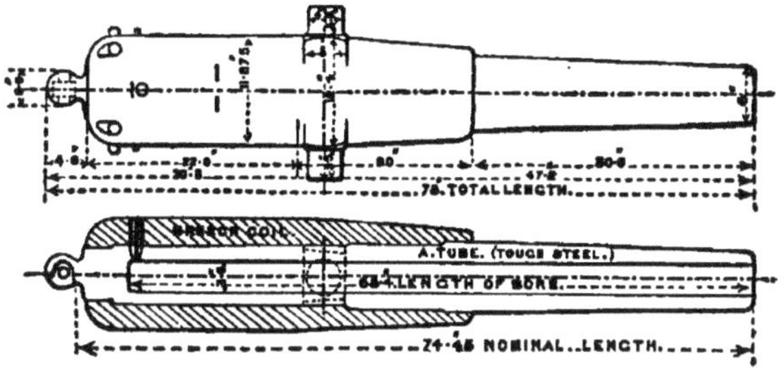

SIGHTING. 1 inch = 1 foot.

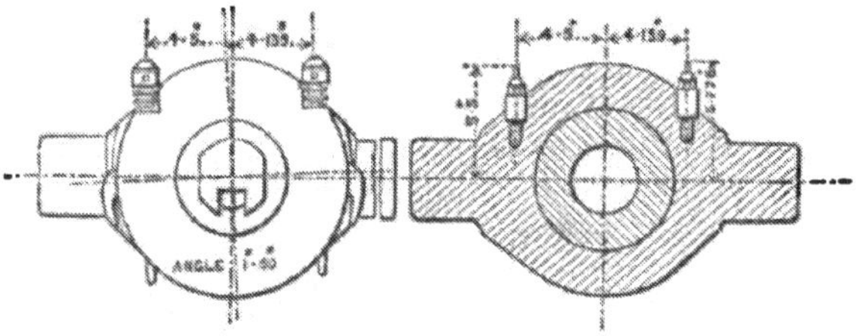

RIFLING. 3 inches = 1 foot.

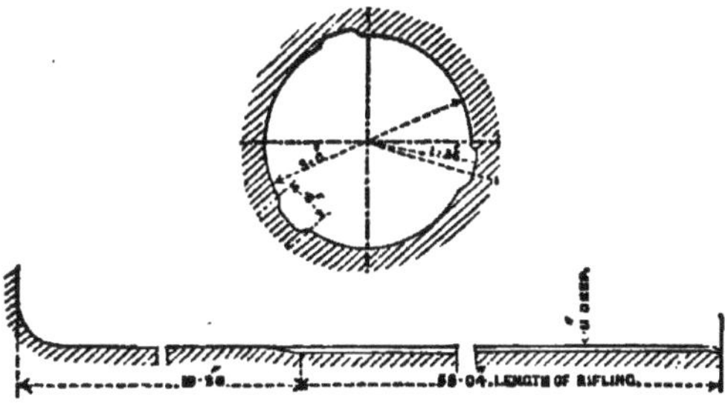

Consists of:—

A tube (toughened steel).
Jacket, composed of a single coil, trunnion ring, and a coil in front of the trunnions welded together.

The form of groove is the "modified French," as in the 9-pr. guns, and the twist 1 in 30 cals.

The siege train will consist of the following proportions to each unit of 30 pieces.* Siege ordnance.

† Light unit	25-prs.	10	Light.
	40-prs., Mark II. ..	10	
	6·3-inch howitzers ..	10	
† Heavy unit	40-prs. Mark II. ..	8	Heavy.
	64-prs., „ III., with steel tube	8	
	8-inch howitzers ..	14	

Each unit is associated with six 7-prs. of 200 lbs. Mark IV.

25-pr. Rifled M.L. Gun of 18 cwt. Mark I. L.S. Calibre 4″. § 2673.

This gun was proposed in January 1871, by the Superintendent, R.G.F., as a light siege gun and gun of position, intermediate between the 16-pr. and 40-pr. 25-prs. of 18 cwt.

Its introduction was decided on in April 1874.

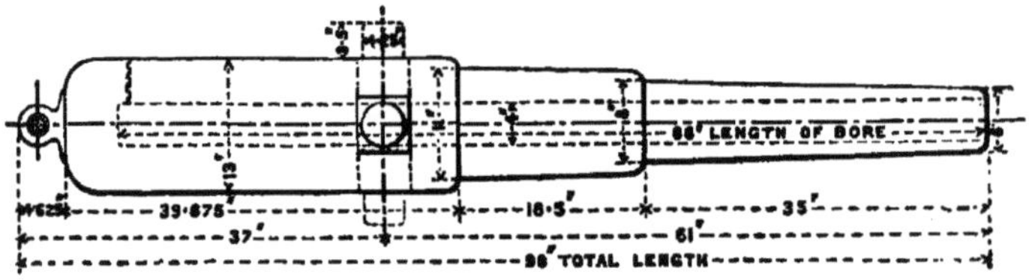

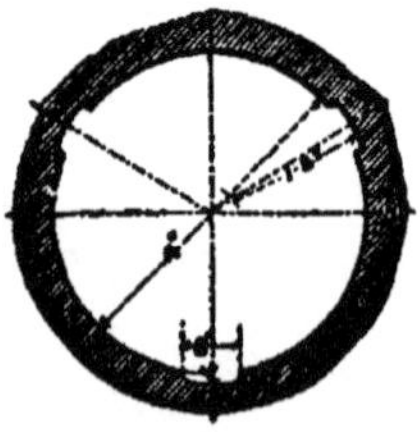

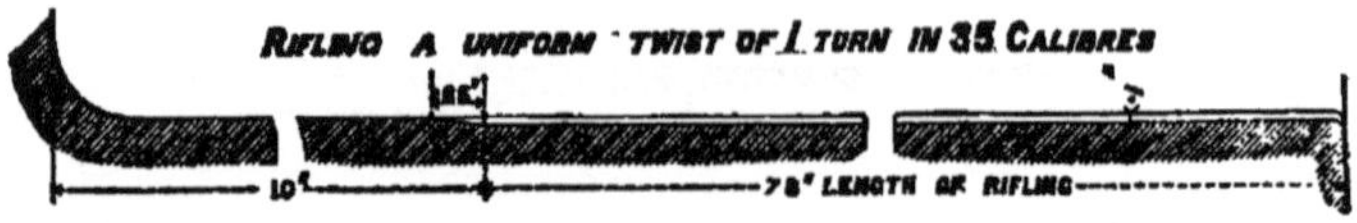

Consists of :—

A tube (toughened steel).
B coil.
Jacket, composed of a single coil, and trunnion ring, welded together.

---

* According to that laid down in Army Circular, February, 1878.
† It is also proposed that a small proportion of 7″ guns of 90 cwt. should form part of a siege train for direct fire.

CHAP. XI.

§ 2478.
40-prs.

40-pr. Rifled M.L. Gun. Calibre 4″·75.

It was at one time proposed to convert the B.L. 40-pr. guns into muzzle-loaders, but it was decided in preference to make new guns, first, because the B.L. guns as they exist are excellent weapons, and the cost of converting and re-tubing them would be little less than that of new guns; secondly, if converted without being re-tubed the guns would have to be weakened by being bored out.

Introduction of Mark I.

In 1871, therefore, a R.M.L. 40-pr. was designed, Mark I.

During further experiments carried on after a few of these guns had been manufactured, some irregularity of velocity, &c. appeared to be due to the incomplete burning of the charge; a longer gun of the same weight and calibre was therefore tried with such satisfactory results that its introduction into the service was approved in March 1874. These guns have both the Woolwich groove, with a twist of 1 in 35.

Mark II.

Only 20 Mark I. have been made, and no more will be manufactured; for the future all will be of Mark II. pattern.

Mark I. has a rear vent, striking the bore near the bottom. Mark II. a forward vent,* striking the powder chamber in the centre, 7 inches from the bottom of bore.

---

§ 2478.

40-pr. Mark I. L.S.

Mark I. of 34 cwt. L.S.

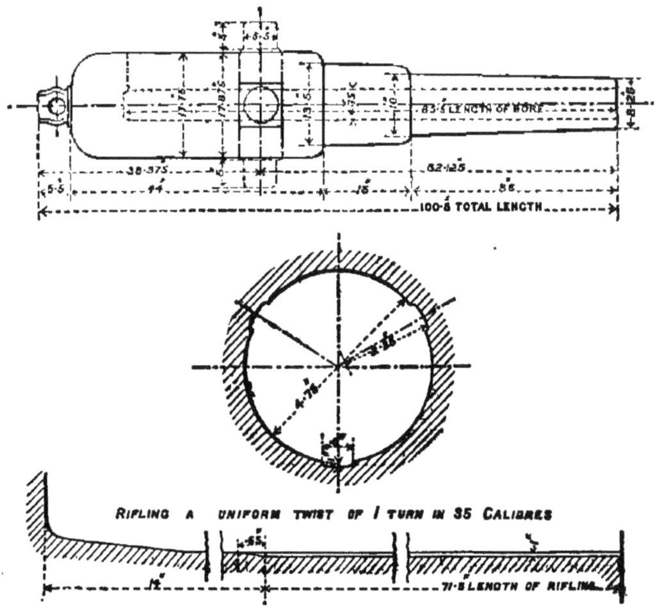

---

* Up to 1878, Mark II. guns were vented as Mark I, but in future all Mark II. guns passing through the department will be vented as above.

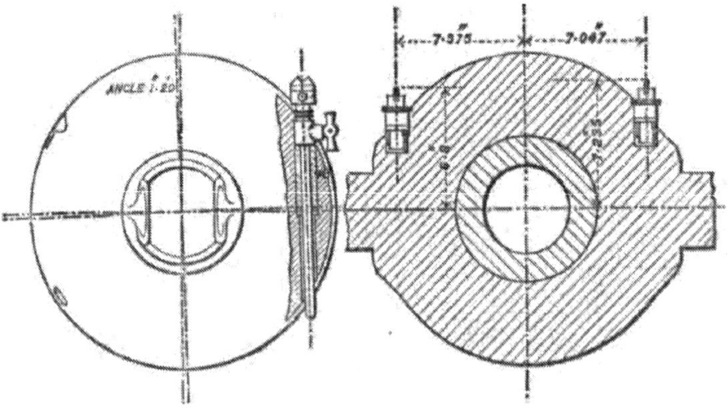

Consists of :—

A tube (toughened steel).

B tube.

Breech coil, composed of a single coil, trunnion ring, and coil in front of trunnions welded together.

Cascable.

Mark II. of 35 cwt. L.S.

§ 2672.

40-pr. Mark II. L.S.

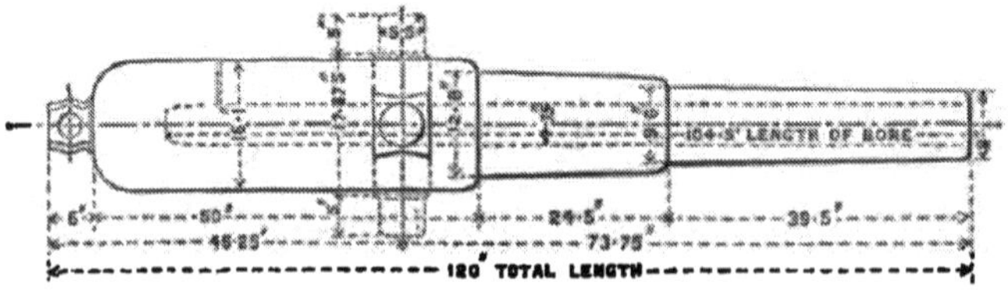

Its construction differs slightly from that of Mark I., and it consists of :—

A tube (toughened steel).

B tube.

1 *B* coil.

Breech coil, composed of a single coil and trunnion ring welded together.

Cascable.

6·3-inch Rifled M.L. Howitzer of 18 cwt.

Having in view the success of the 8-inch howitzer (vide p. 261), and being anxious to try a lighter nature of rifled howitzer to fit the 40-pr. carriage, and to fire a shell of about 60 lbs. weight, the Committee on High Angle Fire proposed in March 1874 that a howitzer should be constructed of 6·3 inch calibre. 6·3-inch howitzer.

CHAP. XI.

§ 3281.

Scale 1 inch = 1 foot.

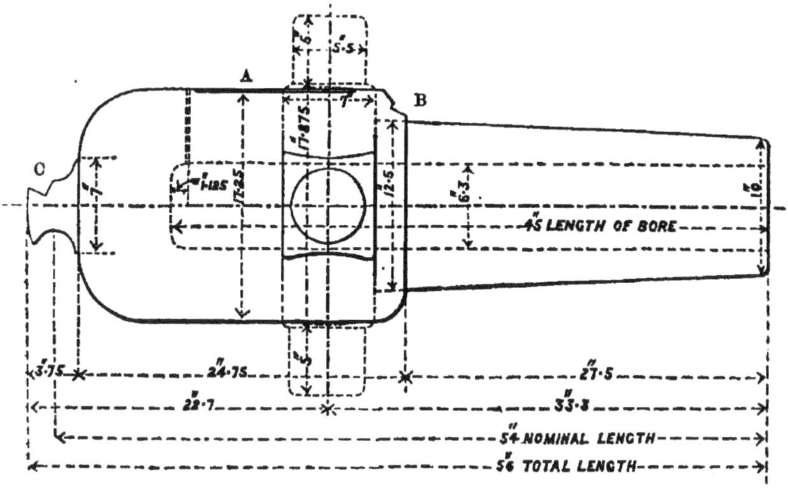

Scale 2 inches = 1 foot.

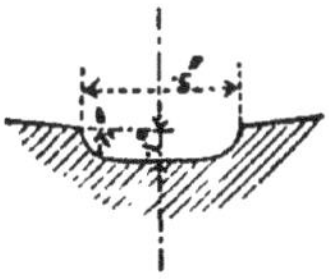

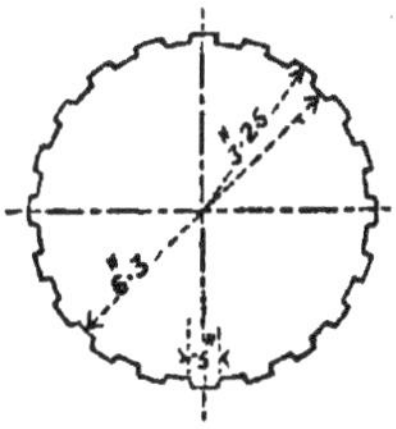

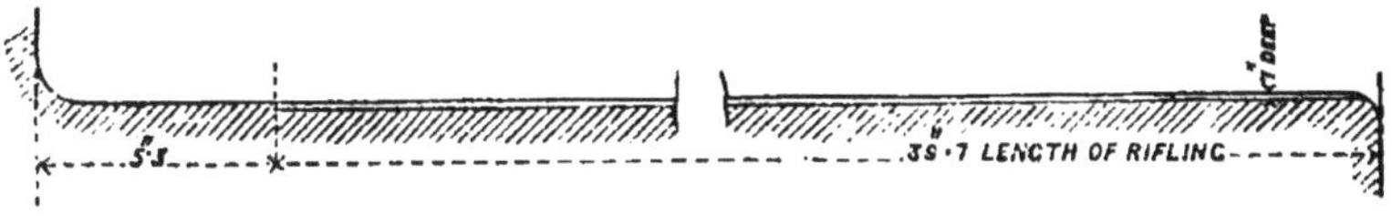

The rifling is polygrooved with a twist increasing from 1 in 100 to 1 in 35; number of grooves 20.

Quadrant planes. There are three planes, A, B, and C on the howitzer, one on the top on which to place the quadrant, one across the trunnion ring, and one across the cascable button for levelling the trunnions.

Venting. The howitzer is vented vertically at a point 1·125 inches from bottom of bore. The end of the bore is slightly rounded.

Stop. In order to ensure an exact space in the bore being left for the charge, the grooves terminate abruptly at a distance of 5·3 inches from the end of the bore, so as to prevent the projectile being driven beyond this point in ramming home. In some of the howitzers already made, however, the rifling extends 2·8 inches beyond this point, but in this portion the grooves are only ·05 inch in depth.

64-pr. Rifled M.L. Gun of 64 cwt. Mark III. (Siege.)

The 64-pr. siege gun is the Mark III. pattern (vide p. 172), but all § 2084.
those employed for this service have steel tubes, and have therefore Siege 64-pr.
been either manufactured as new guns since 1871, or else have been re-tubed with steel.

The several parts are—

A tube (toughened steel).

B tube.

Breech coil or jacket, consisting of breech coil, trunnion ring and muzzle coil welded together.

Cascable.

8-inch Rifled M.L. Howitzer of 46 cwt. L.S. Mark I. § 2507.

The 8-inch howitzer was the first piece of ordnance introduced into our service to supersede S.B. mortars for high angle fire.

Towards the close of 1872 it was ordered that S.B. mortars should no longer constitute a part of the siege train, and that 8-inch howitzers should be substituted in their place.

Several of these howitzers were therefore estimated for in 1873–4, 8-inch.
and up to this date a number have been manufactured.

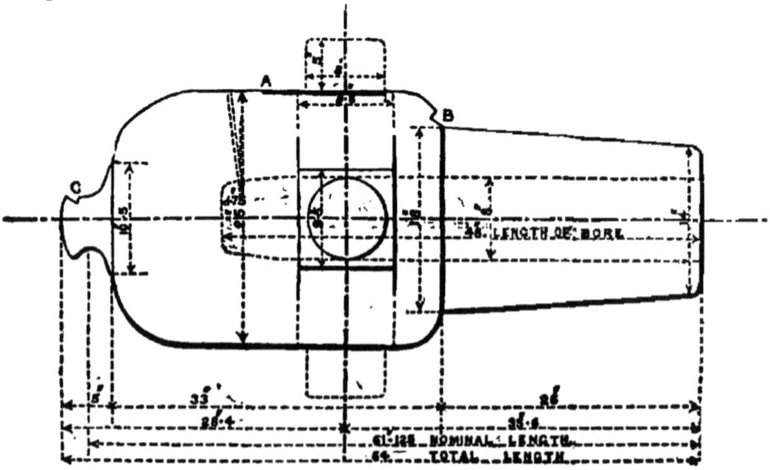

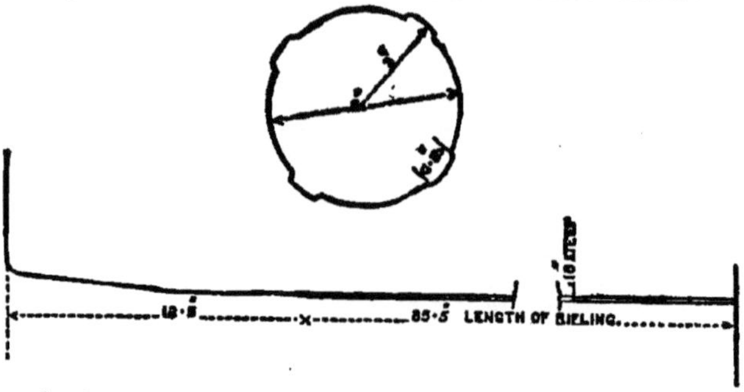

CHAP. XI. This howitzer consists of—

A tube (of toughened steel).
B tube.
Breech coil (or jacket), composed of a single coil and trunnion ring.
Cascable.

It has the Woolwich rifling, with a uniform twist of 1 in 16 calibres with four grooves of the same size as the 8-inch R.M.L. gun.

The vent strikes the chamber near the bottom at right angles.

The quadrant planes are similar to the 6·3 inch howitzer.

Medium Guns.

We may term the 64-prs., other than siege pieces, the three natures of converted guns, and the 7-inch of 90 cwts.,* medium guns.

Medium guns 64-prs.

64-pr. Rifled M.L. Guns of 64 *cwt. L.S. and S.S. Calibre* 6″·3.

There are three patterns of this gun. (See Plate V.)

The 64-pr. was the R.M.L gun first introduced, so that we find a number of these pieces built up on the old method of construction. Indeed the construction and external appearance of Mark I. of these guns is identical with that of the 64-pr. wedge, for which they were originally intended. They were adopted in 1864 for the Navy as a broadside or pivot gun to replace the 64-pr. wedge gun. The calibre of 6″·3 was chosen to permit of firing 32-pr. smoothbore spherical projectiles in cases of emergency.

Those tubed or re-tubed with steel, and also such of Marks I. and II. as have been re-tubed with iron, have the plain groove. The remainder of Mark I. have the older shunt groove. Vide also pp. 29 and 30.

In 1866 it was approved that this nature should replace the 7″ B.L. guns in the Navy, in cases in which the latter were found too heavy.

These guns are used solely as shell guns, not being sufficiently
§ 1113. powerful against iron clad ships.

§ 1062. **Mark I.**

Scale ⅛ inch = 1 foot.

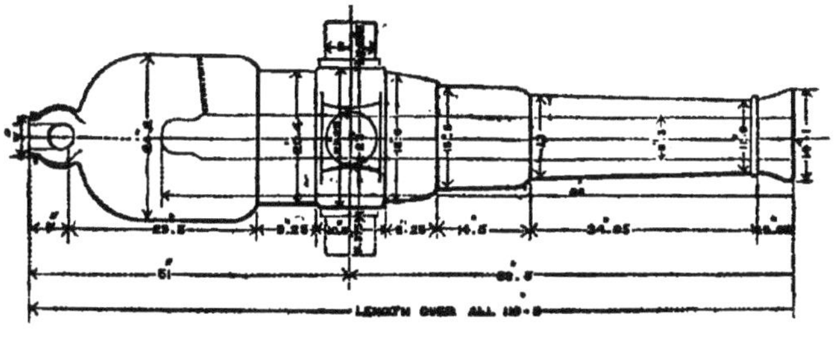

* Vide note p. 244.

The end of the bore is closed by a copper cup, backed up by the cascable. CHAP. XI.

64-pr. Mark I.

Sighting. Scale ¾ inch = 1 foot.

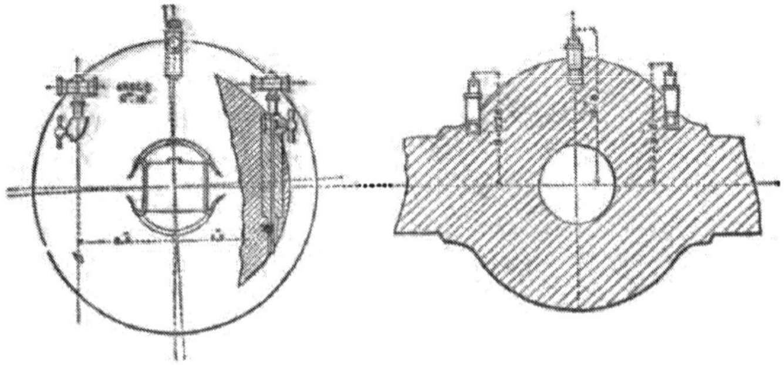

Mark I. consists of:— § 1032.

A tube (coiled iron).	4 coils.
Breech-piece and *B* tube.	Cascable.
Trunnion ring.	

Mark II.

Guns of this nature were recommended for manufacture in 1866 being less expensive than Mark I., and equally efficient. The end of the bore is closed by a wrought iron plug, a copper disc intervening between it and the cascable. O.S.C. Proceedings, 1866, p. 237.

These guns differ entirely in exterior appearance from Mark I., the swell at the muzzle is dispensed with, and the gun is cylindrical from the breech to a little in front of the trunnion, the curve of the breech is also broken so as to form a step.

They are marked B on the left trunnion, that being the designation of this pattern when introduced into the service. 1608.

Scale ¼ inch = 1 foot.

64-pr. Mark II. or B pattern

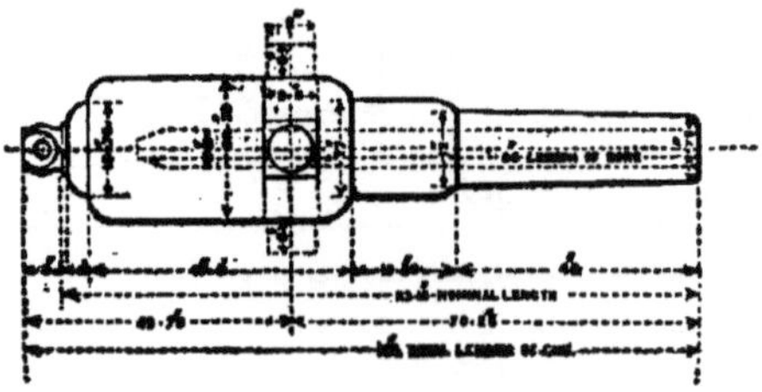

SIGHTING. Scale ⅜ inch = 1 foot.

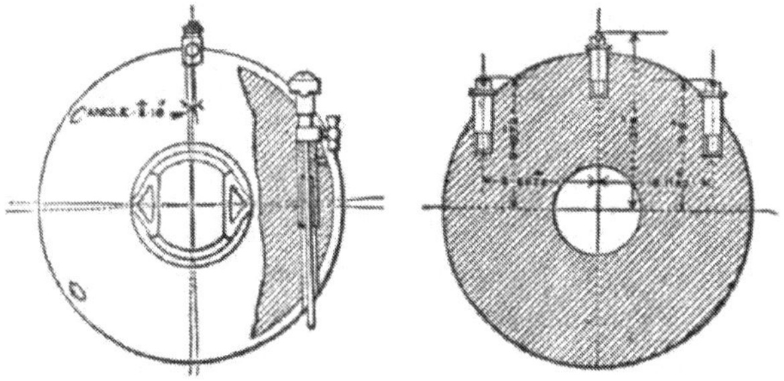

RIFLING. Scale 2 inches = 1 foot.

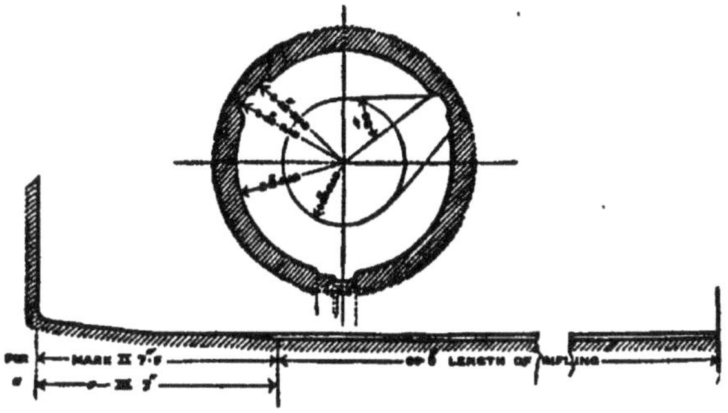

MARK II. consists of :—

§ 1608. *A* tube (coiled iron), double at the chase.
Breech-piece (a solid forging).
Breech coil, composed of a double coil and trunnion ring welded together.
Coil in front of the trunnions or belt.
Cascable.

Rifling.—Three plain grooves, uniform twist of 1 in 40.

Mark III.

64-pr. Mark III. or D pattern. O.S.C. Proceedings, 1867, p. 11. O.S.C. Proceedings, 1866, p. 352. § 1608.

The manufacture of Mark III. was approved in 1867, experiments having proved that this construction is stronger than that in which the solid forged breech-piece is used.

Its external appearance is the same as Mark II., excepting that the breech is rounded off.

Guns of this pattern issued prior to March 1868, have D stamped on the left trunnion ;* that being its designation when introduced.

Those made since April 1871 have solid ended steel tubes, and a *B* tube shrunk over the chase. Most of the guns made with steel tubes

---

* Under the D in such guns, the numeral II was stamped. In all guns passing through R.G.F., this II. is altered to M. III., and the letter D removed.

are for the siege train, as explained at p. 261, the guns having iron tubes are employed as S.S. guns. CHAP. XI.

Scale ¼ inch = 1 foot.

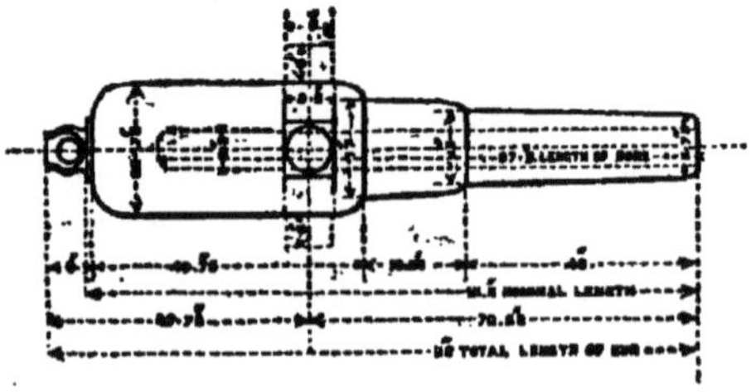

MARK III. consists of :— § 1608.

A tube, coiled iron (double at the chase).
Breech coil, composed of a triple coil, trunnion ring, and a single coil welded together.
Cascable.
Rifling as in Mark II.

Converted Guns.

As described in Chapter X., the manufacture of converted guns on the Palliser principle, where a wrought-iron barrel is inserted into a cast-iron casing, was commenced in 1863. Some of these converted guns were made by contract, but most of them in the R.G.F. Converted guns.

$\frac{64}{32}$-pr. (converted) 58 cwt. L.S.

A number of these guns have been made for sea fronts where range is limited. The construction is described at p. 233.

Scale $\frac{1}{32}$nd.

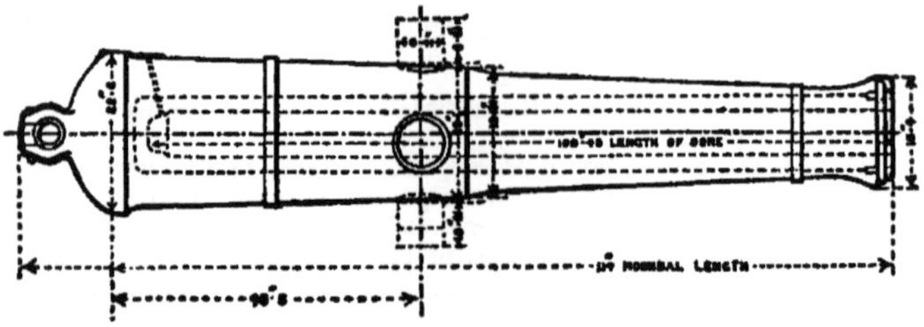

CHAP. XI.

$\frac{64}{32}$-pr. of 58 cwt.

RIFLING. Scale $\frac{1}{8}$th.

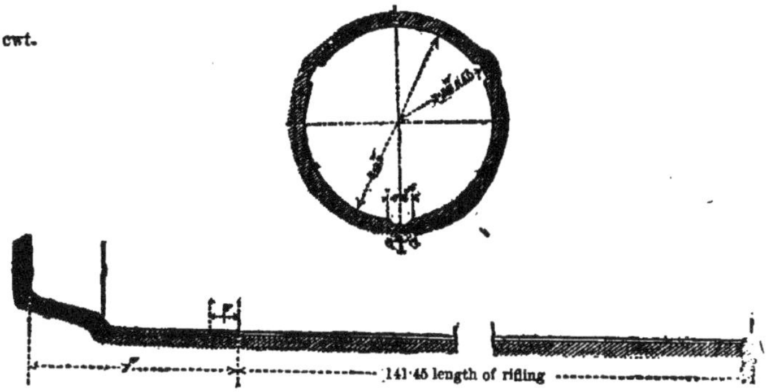

§ 2066. This 64-pr. of 58 cwt. consists of:—

Cast iron casing.
Barrel { *A* tube, coiled iron.
B " " "
Cast iron collar.
Wrought iron screw plug.
Rifling three plain grooves, twist 1 in 40.

$\frac{\text{64-pr.}}{\text{8-in.}}$ R.M.L. (converted) of 71 cwt. L.S. and S.S.

§ 1752. This gun is meant for the armament of unarmoured ships and land fronts of fortifications. Vide, for description, p. 233, where a diagram of the sights, &c. will also be found.

64-pr. / 8-inch of 71 cwt. This 64-pr. of 71 cwts. consists of:—

Cast iron casing.
Barrel { *A* tube, coiled iron.
B " " "
Cast iron collar.
Wrought iron screw plug.
Rifling as in 64-pr. of 58 cwt.

§ 2220

$\frac{80}{68}$-pr. R.M.L. (converted) gun of 5 tons. L.S.

This our heaviest converted gun is a L.S. gun and would be employed for the same purpose as the $\frac{64}{32}$-pr. 58 cwt. where a somewhat more powerful piece is wanted.

$\frac{80}{68}$pr. of 5 tons.

Scale $\frac{1}{32}$nd.

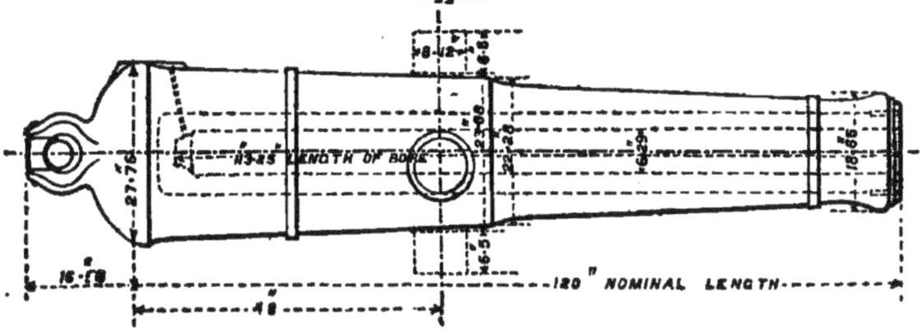

SIGHTING. Scale $\frac{1}{12}$th.

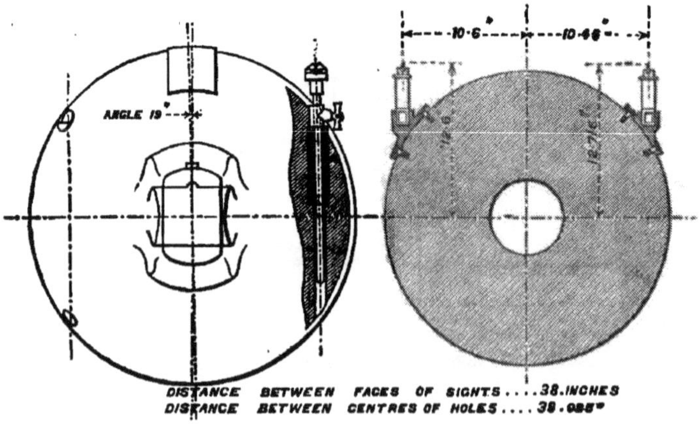

RIFLING. Scale $\frac{1}{6}$th.

Parts the same as for the two preceding guns.
Rifling. Woolwich groove, with a twist of 1 in 40.

7-inch R.M.L. gun of 90 cwts. Mark I. S.S.

At the request of the Admiralty some 7-inch $6\frac{1}{2}$-ton guns under manufacture were completed in 1874 as 90 cwt. guns for the armament of unarmoured vessels. The bore of the piece remains the same, but the exterior is turned down all over and slightly shortened. In order to rectify the preponderance and the position of the trunnions, a large part of the jacket is also turned off and a new jacket shrunk over the breech, where a portion of the old jacket forms a breech-piece. § 2363.

CHAP. XI.

7-inch of 90 cwts. Mark I. reduced from those of 6½ tons.

Many service 6½ ton guns Mark III have since been reduced to 90 cwt. guns as described, and a few new guns have been made of similar construction, the coiled breech-piece in the latter case being made of two coils united, which together resemble the breech coil and 1 B coil of 10″ guns and upwards.

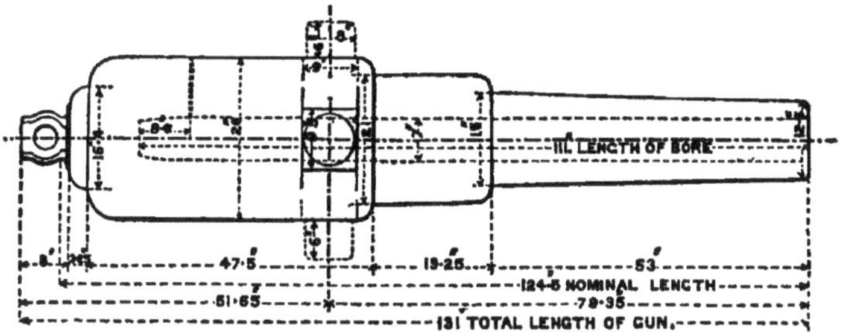

Consists of:—

A tube (toughened steel).
B tube (chase).
Coiled breech-piece.
C coil or jacket:—A single coil and trunnion ring welded together.
Cascable.

Heavy Guns.

Employment of heavy guns.

We now come to guns which might be employed against armour plates with fair effect from ranges of 1,000 up to 3,000 yards. Their powers increase very much as they become larger until in the 80-ton gun we have a piece of ordnance which can pierce with ease the thickest armour at present afloat (24 inches) up to a range of about 1,000 yards.

All these pieces have the Woolwich rifling with increasing twist, except the 7-inch where it is uniform. In all of them also the vent strikes the bore at right angles, and at a distance of $\frac{4}{10}$ the length of the cartridge from the bottom of the bore.*

7-inch Rifled M.L. Guns of 6½ tons, S.S. only.

There are three patterns of this gun.

7-inch guns of 6½ tons, Mark I.

It is built on the original construction, the coils being separate and hooked. It was adopted in 1865 as a broadside or pivot gun for frigates, to replace the 7-inch B.L. and 68-pr. S.B. guns, and is now very extensively used, 331 having been made.

Externally the breech is rounded off, giving the gun somewhat the shape of a soda-water bottle, and it has several steps in front of the trunnions.

O.S.C. Proceedings, 1865 pp. 227, 317.

These guns are in total length 18 inches shorter than the land service 7-inch gun, being a length more suited to the requirements of the Navy.

---

* With the 80-ton gun it is possible that an axial vent may be employed.

Mark I.

Scale ½ inch = 1 foot.

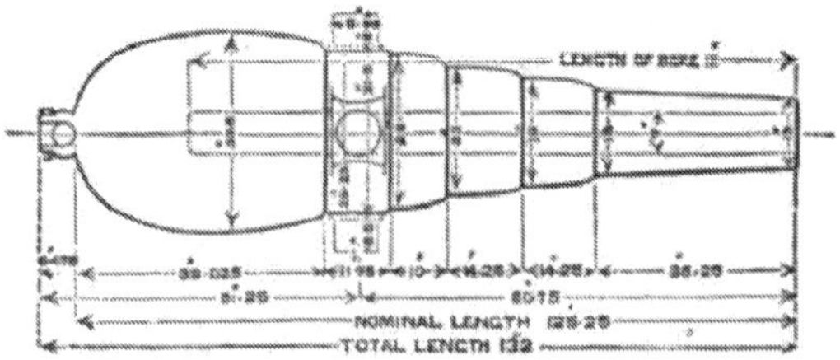

SIGHTING. Scale ½ inch = 1 foot.

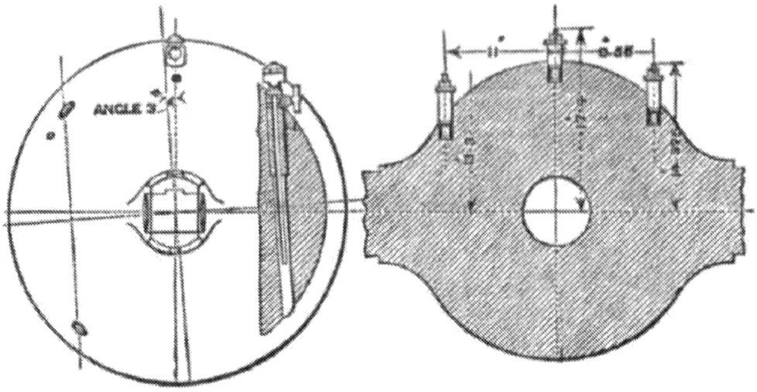

RIFLING. Scale 1½ inch = 1 foot.

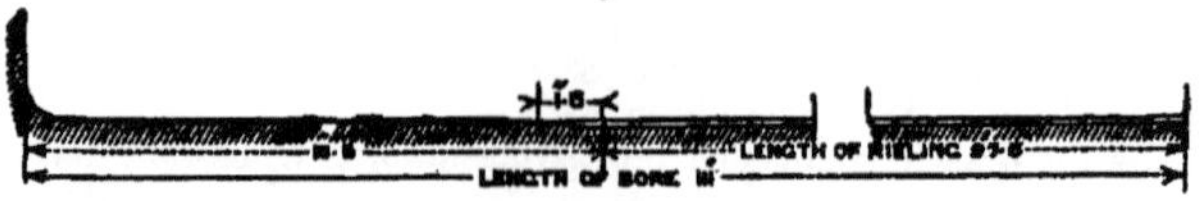

MARK I. consists of :— § 1231.

A tube (toughened steel).
Breech-piece and *B* tube.
Trunnion ring.
Six coils.
Cascable.

CHAP. XI.

Mark II.

7-inch 6½ tons Mark II.

Very few of this pattern were manufactured. They were introduced in 1866 and then designated and marked on the left trunnion F. I. (Fraser construction); guns of this pattern issued prior to 31st March 1868, are thus known.

§§ 1462, 1596, 1644.

They had coiled iron barrels closed by an iron plug, as steel was not then finally adopted as the material for inner barrels. Any of them, however, which have since been re-tubed have solid ended steel barrels.

Scale ¼ inch = 1 foot.

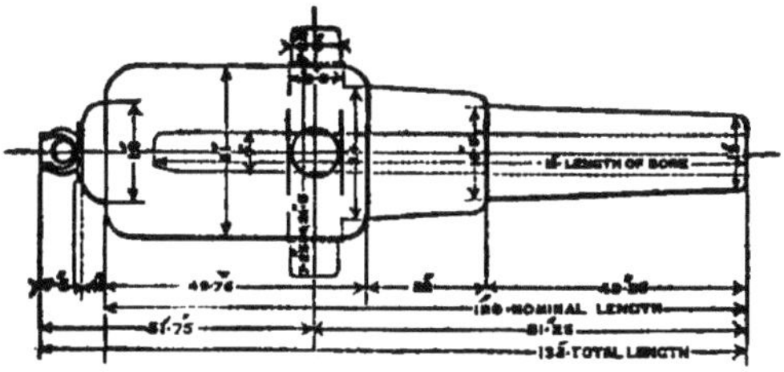

SIGHTING. Scale ¾ inch = 1 foot.

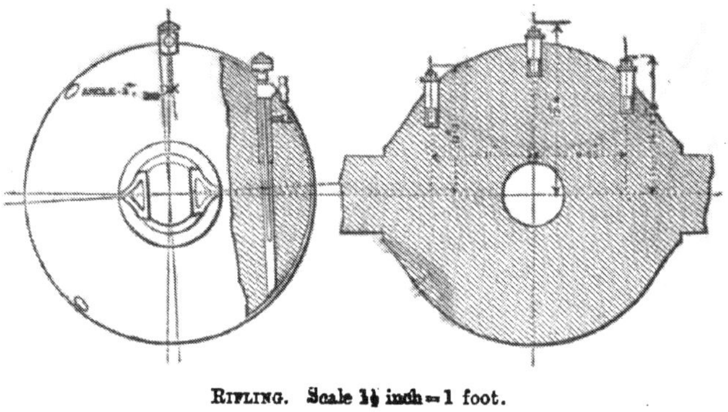

RIFLING. Scale 1½ inch = 1 foot.

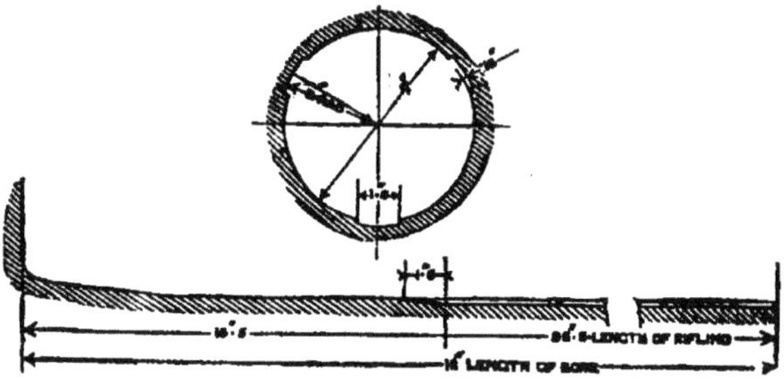

MARK II. consists of:— CHAP. XI.

§ 1644.

A tube (coiled iron).
Breech-piece.
B tube (the coil in front of the trunnions, and chase).
Breech coil:—a double coil and trunnion ring welded together.
Cascable.

Mark III.

§§ 1462–1596. 7-inch 6½ tons Mark III.

This construction adopted in 1866 superseded Mark II., the solid forged breech-piece being abandoned and the metal over the breech being put on in one thickness. Those issued prior to March 1868 had F II.* stamped on the left trunnion.

Scale ¼ inch = 1 foot.

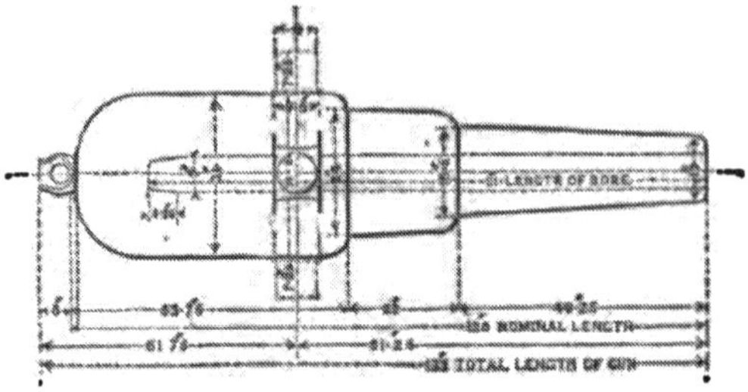

MARK III. consists of:— § 1644.

A tube (toughened steel).
B tube (chase).
Breech coil:—a triple coil, trunnion ring and coil in front of the trunnions welded together.
Cascable.

7-inch R.M.L. Gun of 7 tons. L.S. only.

7-inch 7 tons, Mark I.

There are four patterns of this gun. (See Plates).

This nature is entirely a land service gun, and was introduced in 1865 as a battering gun for coast defence.

Mark I. is built on the original construction. The *B* tube of this pattern is covered by an additional thin coil so as to reduce the preponderance, which was found to be excessive. 51 were manufactured.

Mark I.

O.S.C. Proceedings, 1865, p. 316.

O.S.C. Proceedings, 1866, p. 119.

Scale ¼ inch = 1 foot

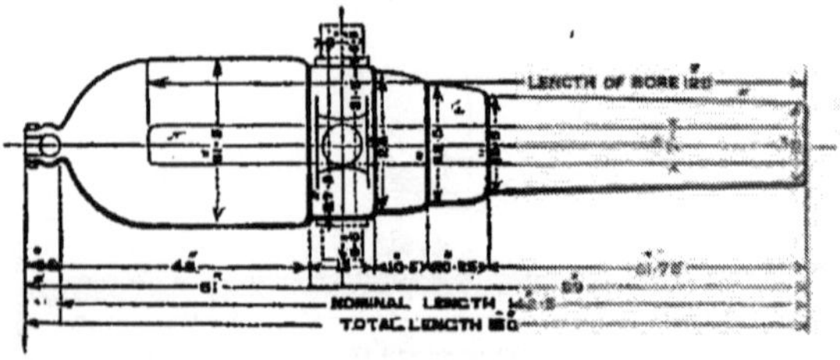

* Such guns passing through the R.G.F. will have letter F removed, and II. altered to III.

CHAP. XI.

SIGHTING. Scale ⅛ inch = 1 foot.

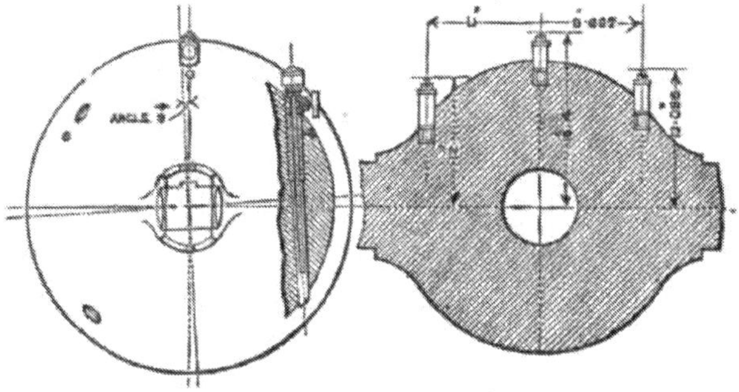

RIFLING. Scale 1½ inch = 1 foot.

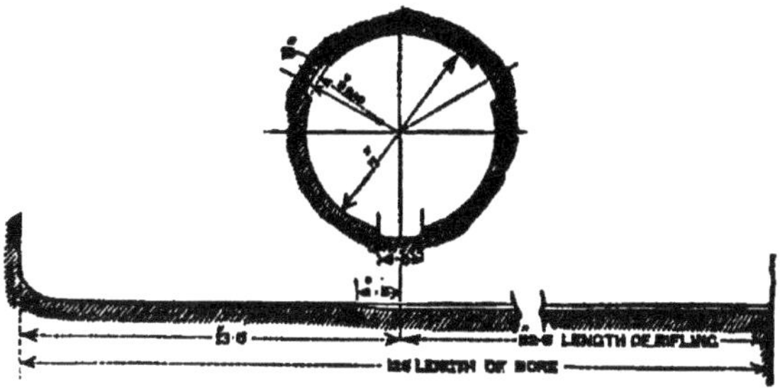

§ 1230. MARK I. consists of:—

- *A* tube.
- Breech-piece and *B* tube.
- Trunnion ring.
- Five coils.
- Cascable.

Mark II.

7-inch 7 tons; Mark II. Introduced in 1866, consists of same parts as Mark II., 7-inch 6½-ton gun, and differs from it only in length. There were only two made, and they are marked F. I. on the left trunnion.

§ 1607. Scale ⅛ inch = 1 foot.

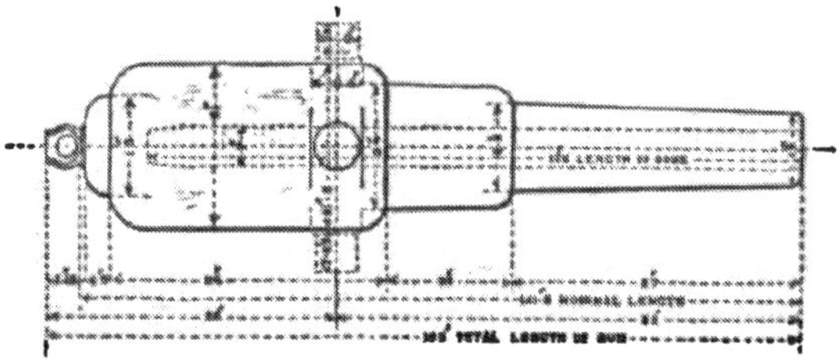

SIGHTING. Scale ⅛ inch = 1 foot.

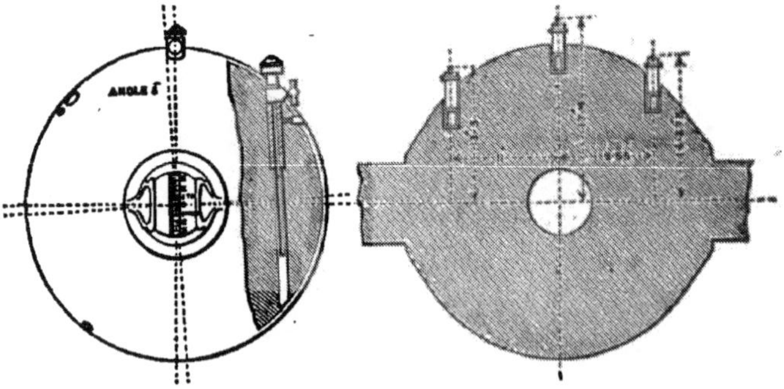

RIFLING. Scale 1½ inch = 1 foot.

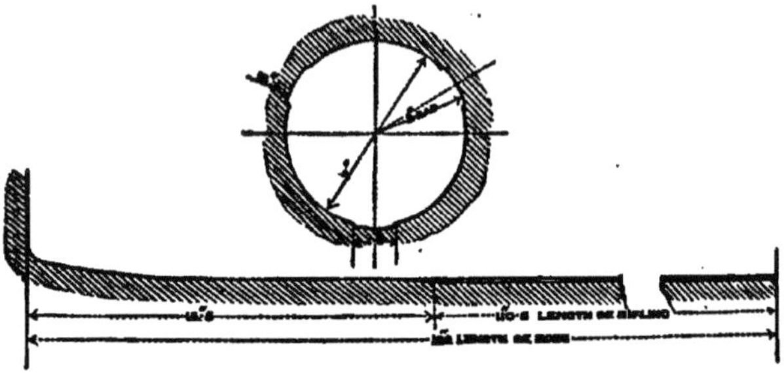

MARK III., introduced in 1866, differs only in length from Mark III., 7-inch 6½-ton gun. Previous to April 1868 these guns were marked F. II.* on the left trunnion. 7-inch of 7 tons. Mark III.

Mark III.

Scale ¼ inch = 1 foot. § 1607.

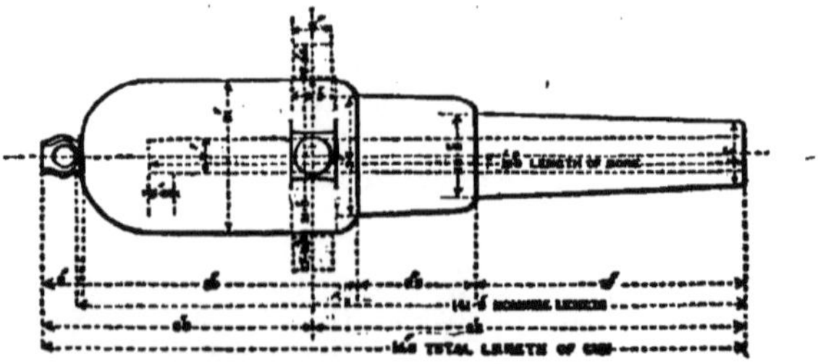

The several parts are:—
A tube (toughened steel).
B tube.

---

* Vide note p. 271.

CHAP. XI.

Jacket, consisting of a breech coil, trunnion ring and muzzle coil.
Cascable.

Mark IV.

Already fully described, see pp. 163, 168, and Plate XI.

8-inch of 9 tons, Mark I.

8-inch Rifled M.L. Gun of 9 tons, L.S. and S.S.

There are three patterns of this gun. (See Plate XII).

These guns were introduced in 1866 for S.S., and a few have since been made for L.S.

The number made of this pattern is 76.

Mark I.

Scale ¼ inch = 1 foot.

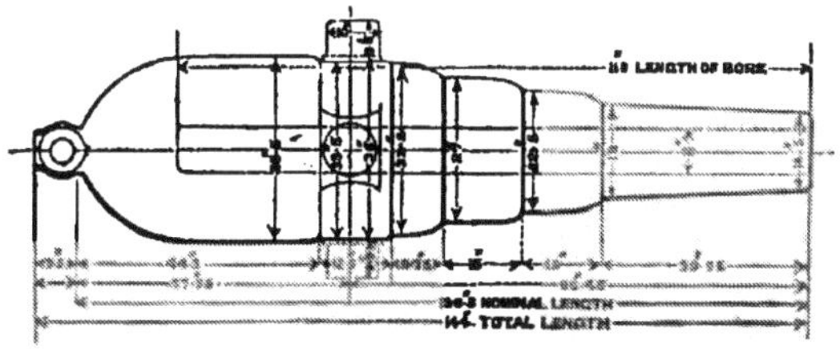

SIGHTING. Scale ¼ inch = 1 foot.

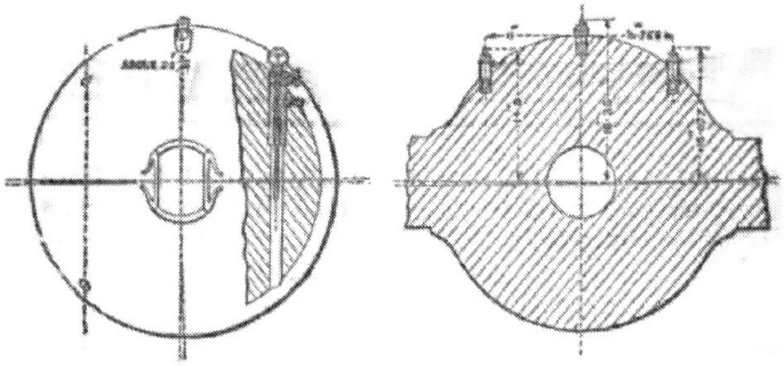

RIFLING. Scale 1½ inch = 1 foot.

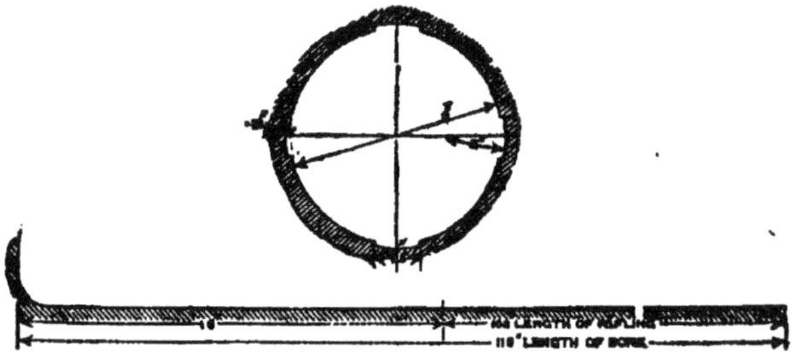

§ 1289.

MARK I. consists of :—

A tube.
Breech-piece and *B* tube.
Trunnion-ring.

Five coils.
Cascable.

8-inch of 9 tons, Mark II.

Mark II.

This gun, introduced in 1866, consists of same parts as Mark II. 7-inch 6½-ton gun. Only six were made, and they are marked F. 1.

Scale ¼ inch = 1 foot.

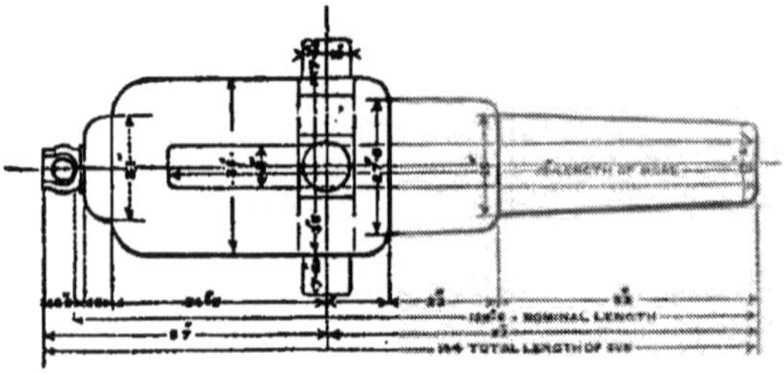

§ 1643.

SIGHTING. Scale ½ inch = 1 foot.

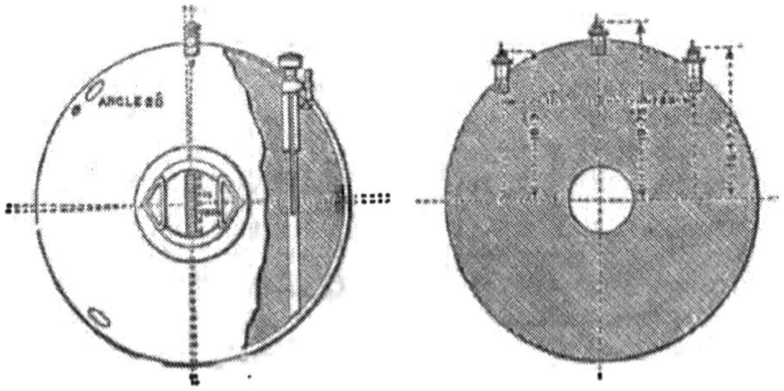

RIFLING. Scale 1½ inch = 1 foot.

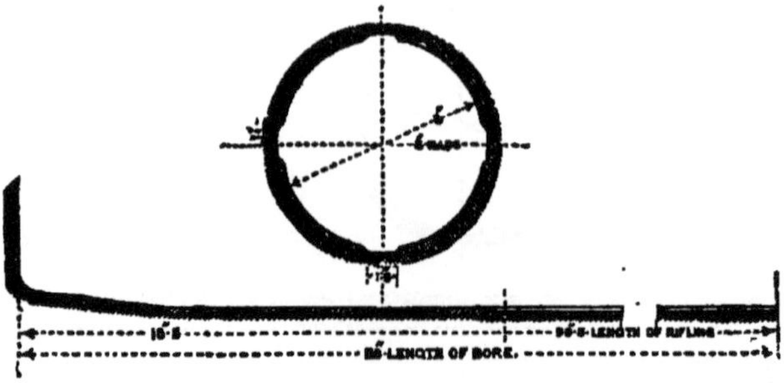

CHAP. XI.

Mark III.

8-inch of 9 tons, Mark III.

Mark III., introduced in 1866, consists of same parts as Mark III. 7-inch 6½-ton guns. Those made previous to April 1868 were marked F. II.*

Scale ¼ inch = 1 foot.

§ 1013.

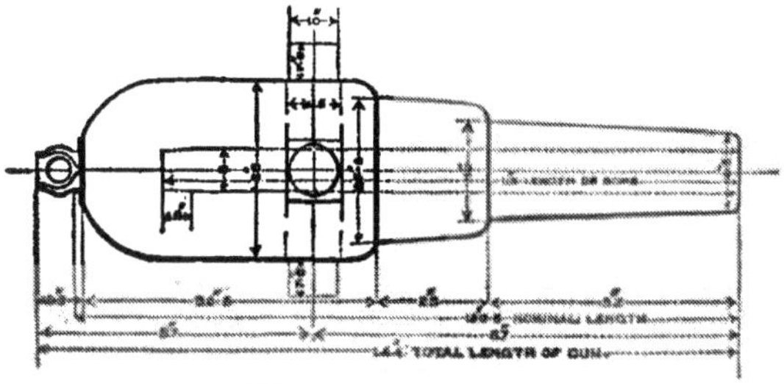

9-inch Rifled M.L. Gun of 12 tons, L.S. and S.S.

There are five patterns of this gun. (See Plate XIII).

Mark I.

9-inch of 12 tons, Mark I.

Introduced in 1865 as a broadside gun for iron clad ships, and also for the defence of harbour and sea fronts. 190 were made.

Scale ¼ inch = 1 foot.

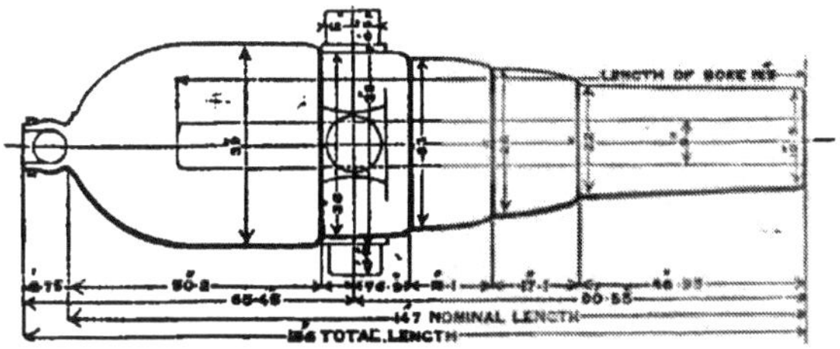

Sighting. Scale ¼ inch = 1 foot.

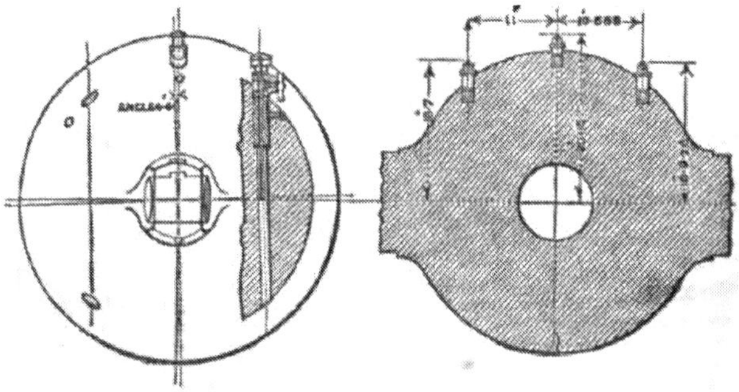

* Vide Note p. 271.

RIFLING. Scale 1½ inch = 1 foot.

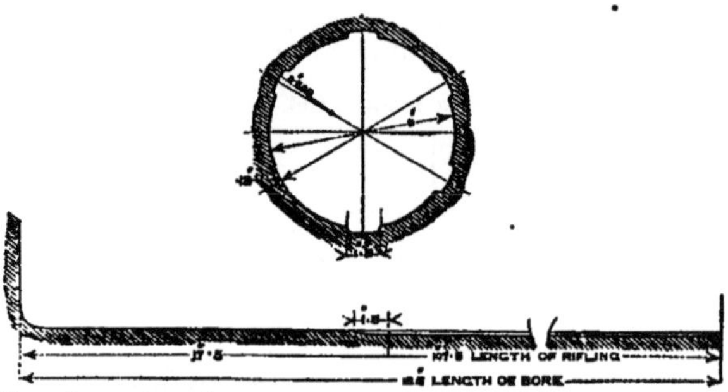

MARK I. consists of:—

A tube (toughened steel).
Breech-piece.
B tube.
Trunnion ring.
Seven coils.
Cascable.

Mark II.

9-inch of 12 tons, Mark II.

Introduced in 1866, only 26 being made. They differ from Mark II. 7-inch and 8-inch in having steel barrels. They are marked F. I. on the left trunnion.*

Scale ¼ inch = 1 foot.

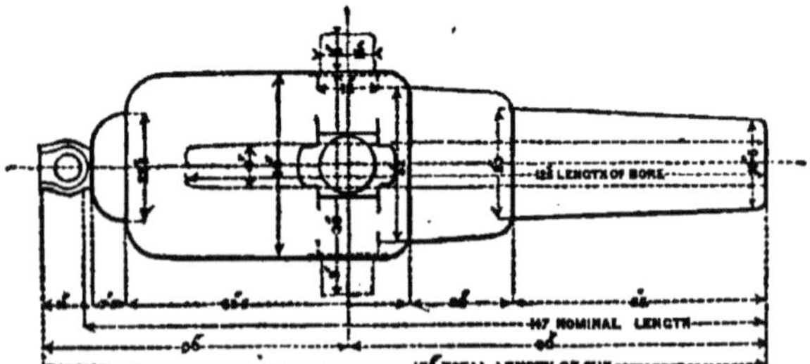

SIGHTING. Scale ½ inch = 1 foot.

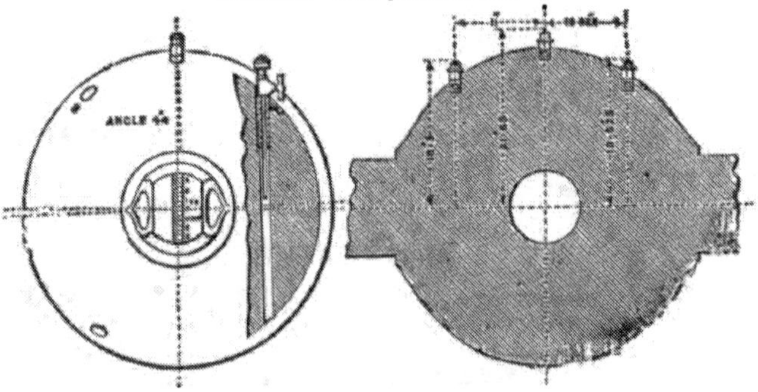

* See Note p. 271.

CHAP. XI.

RIFLING. Scale 1¼ inch = 1 foot.

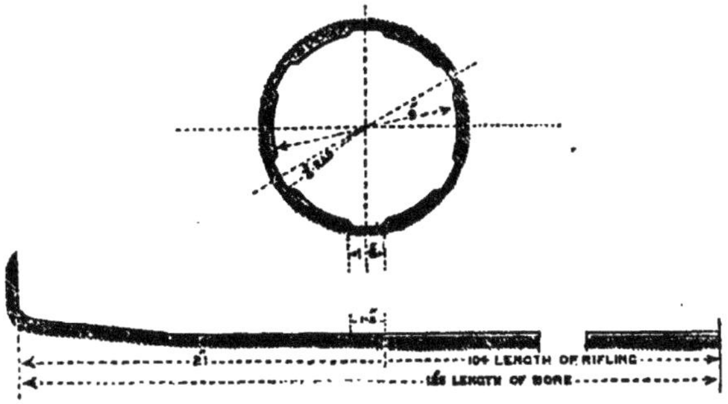

§ 1612. Mark II. consists of :—

A tube (toughened steel.

Breech-piece. (Solid forging).

B tube.

Breech coil :—Double coil, trunnion ring, and a coil in front of the trunnions welded together.

Cascable.

Mark III.

9-inch of 12 tons, Mark III.

Was introduced in 1866, and is similar in construction and exterior form to Mark III. 7-inch. Those manufactured previous to April 1868 are marked F. II.* 136 of this pattern were manufactured.

Scale ⅛ inch—1 foot.

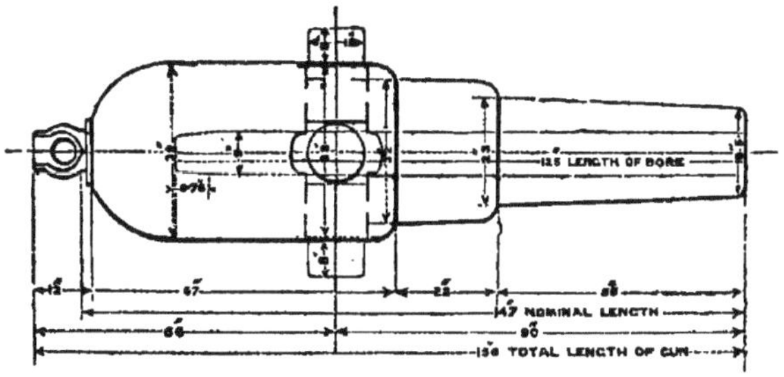

§ 1612. MARK III. consists of :—

A tube (toughened steel).

B tube.

Breech coil :—Triple coil, trunnion ring, and a double coil in front of the trunnions welded together.

Cascable.

Mark IV. and V.

9-inch of 12 tons, Marks, IV and V.

Mark IV. pattern was adopted in 1869. Mark V. differs from IV. in preponderance only, the trunnions being placed ⅜" further to the rear.

---

* Vide Note p. 271.

CHAP. XI.

§ 1906.

O.S.C. Proceedings, 1869, p. 172.

These pieces have two layers over the breech, this construction being considered preferable to the Mark III. for heavy guns, on account of the difficulty of ensuring the soundness of the interior of a large mass of iron.

Scale ¼ inch = 1 foot.

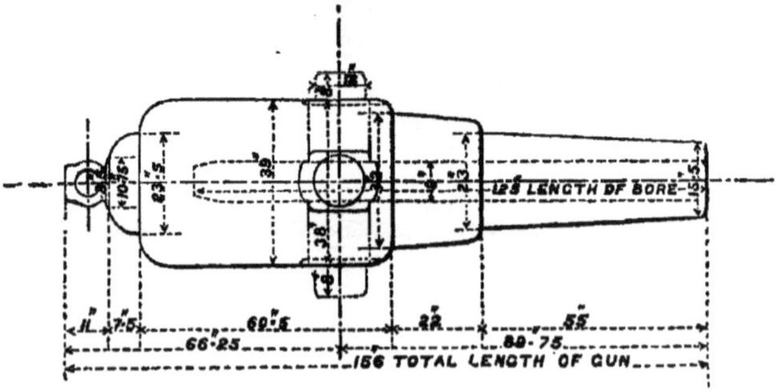

Marks IV. and V. consist of:—

A tube (toughened steel).

B tube.

Coiled breech-piece.

Breech coil:—double coil, trunnion ring, and single coil in front of the trunnions welded together.

Cascable.

10-inch Rifled M.L. Gun of 18 tons. L.S. and S.S.

There are two patterns of this gun. (See Plate XIV).

Mark I.

Mark I was proposed by Commodore Heath, R.N., in 1865, owing to the success of the 9-inch gun, and introduced in 1868 for the Navy as a powerful broadside gun, H.M.S. "Hercules" being armed with it. The 10-inch is now used by the land service for coast defence.

10-inch of 18 tons, Mark I.

There are 18 guns of this pattern.

Scale ¼ inch = 1 foot.

O.S.C. Proceedings, 1865, p. 390.

O.S.C. Proceedings, 1868, p. 86.

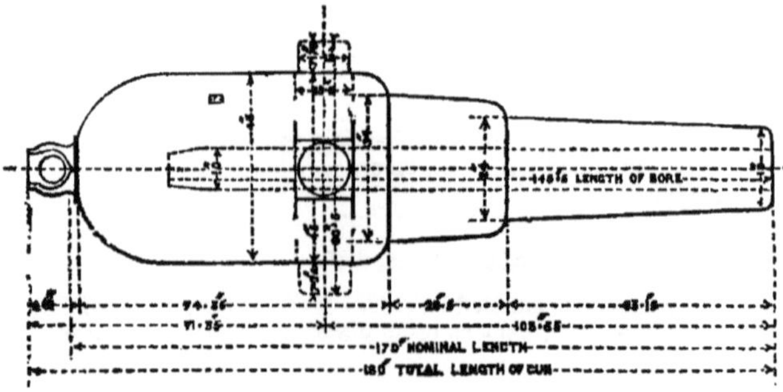

CHAP. XI.

10-in. of 18 tons. Mark I.

SIGHTING. Scale ½ inch = 1 foot

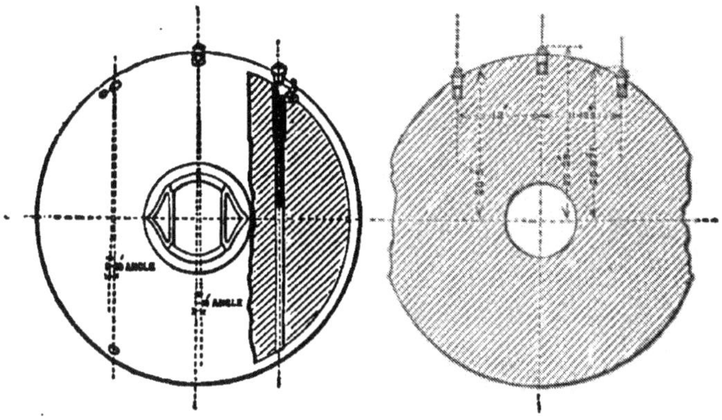

RIFLING. Scale 1 inch = 1 foot.

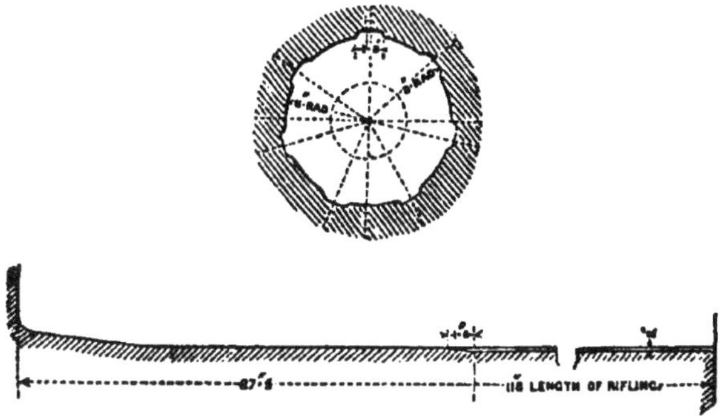

§ 1686. MARK I. consists of:—

A tube (toughened steel).

B tube.

Jacket:—Breech coil, trunnion ring, and a muzzle coil in front of the trunnions welded together.

Cascable.

Mark II.

Adopted in 1869.

10-inch of 18 tons, Mark II.

Scale ¼ inch ± 1 foot.

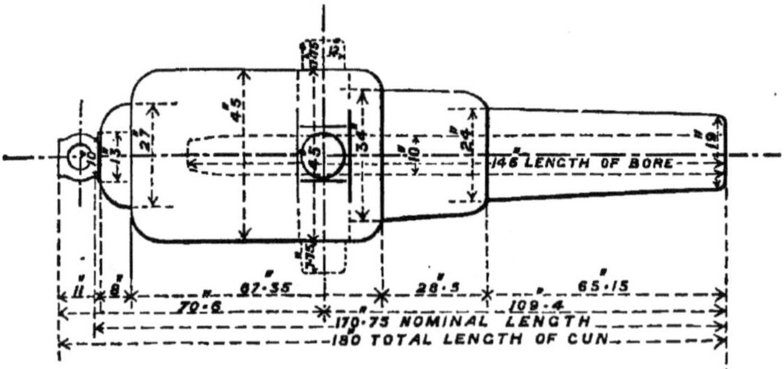

Mark II. consists of:—

O.S.C. Proceedings, 1869, p. 122. § 1905.

- *A* tube (toughened steel).
- *B* tube.
- Coil in front of the trunnions. (Belt).
- Coiled breech-piece.
- Jacket:—Breech coil, trunnion ring, and ring coil welded together.
- Cascable.

11-inch Rifled M.L. Gun of 25 tons, L.S. and S.S.

There are two patterns of this gun. (See Plate XV).
This gun was introduced in 1867.
It is for use in the Navy and by L.S. for coast defence.

11-inch of 25 tons.

Mark I.

Only seven of this pattern have been made.

11-inch of 25 tons, Mark I.

Scale $\frac{1}{48}$.

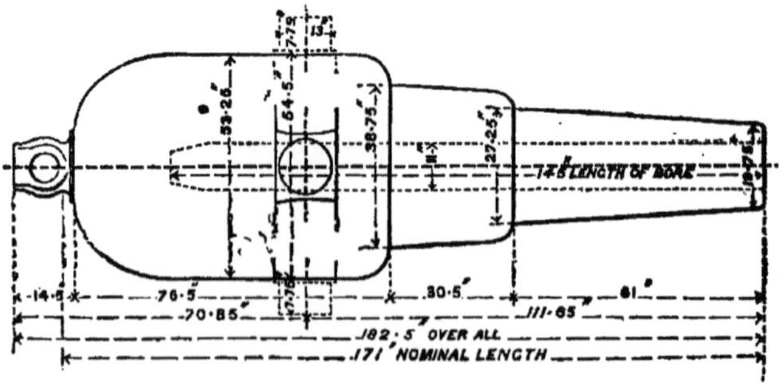

CHAP. XI.

RIFLING. Scale $\frac{1}{12}$.

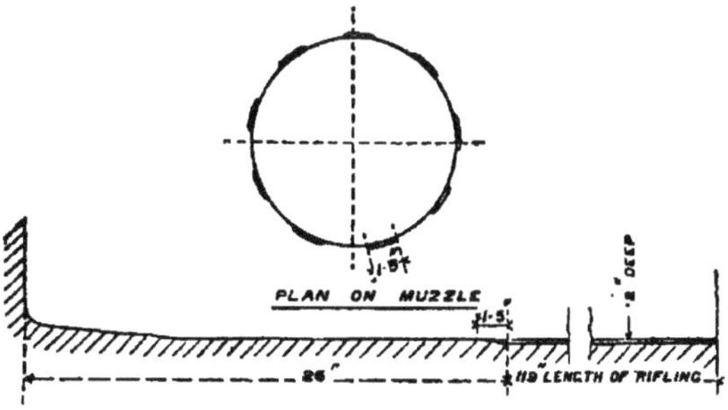

§ 2102. MARK I. consists of:—

A tube (toughened steel).
B tube.
Jacket:—Breech coil, trunnion ring, and muzzle coil welded together.
Cascable.

Mark II.*

11-inch of 25 tons, Mark II. This gun was adopted in 1871 for the same reasons as to manufacture as already mentioned in the case of 9-inch guns Mark IV., vide pp. 278, 279.

Scale $\frac{1}{48}$.

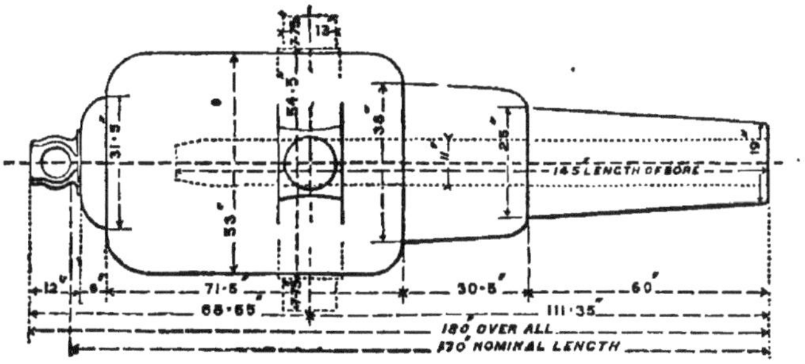

§ 2102. MARK II. consists of.—

A tube (toughened steel).
B tube.
Coil in front of the trunnions (belt).
Coiled breech-piece.
Jacket.—Breech coil, trunnion ring, and ring coil welded together.
Cascable.

---

* One of these guns, No. 68, has a forged breech-piece, being a 12" of 25 tons Mark II., retubed as an 11" gun.

12-inch Rifled M.L. Gun of 25 tons, L.S. and S.S.

There are two patterns of this gun. (See Plate XVI).

Mark I.

This gun was recommended in 1864 on the belief that it would give higher muzzle velocity than the 13-inch gun, but the trial was not carried out till 1866. 12-inch of 25 tons, Mark I.

There are only four of this construction in the service, and their weight is 23½ tons, but by an order of 3rd July 1868, to avoid confusion and the necessity of separate series on account of weight it was approved that they all should bear the same designation, viz., Ordnance, Rifled M.L. 12-inch of 25-tons, this being the weight of the later patterns. O.S.C. Proceedings, 1864, p. 190.

12-inch guns are used for turret ships and coast defence.

Scale $\frac{1}{48}$.

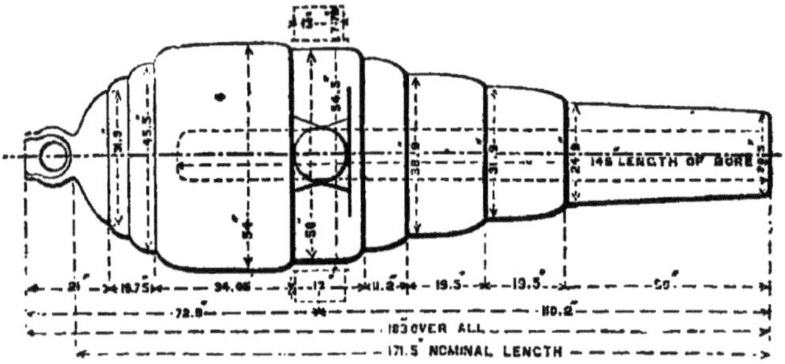

Rifling. Scale $\frac{1}{12}$.

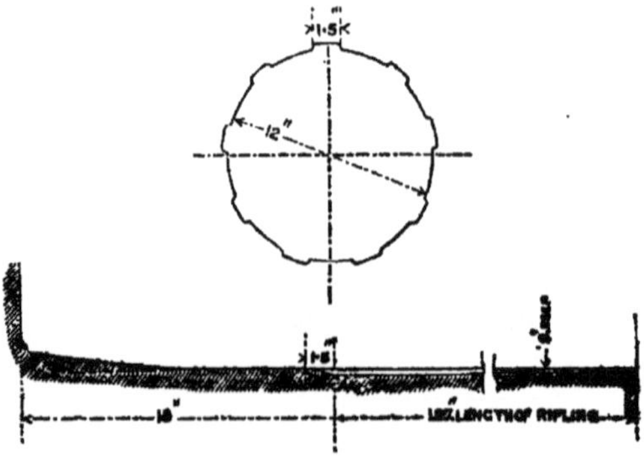

Mark I. consists of :— § 2028.

A tube (toughened steel).

CHAP. XI.

B tube.
Breech-piece.
Trunnion ring.
7 coils.
Cascable.

Mark II.

12-inch of 25 tons, Mark II.

This pattern was introduced in 1866.

The number of 12″ 25-ton guns manufactured up to June 1871 is 15. Of these 4 are on the original or Mark I. construction and 8 on the Mark II. pattern.

O.S.C. Proceedings, 1866, p. 256.

The remaining two guns, Nos. 20 and 21, are of the same construction as the 11″ gun Mark I., but they do not form a distinct pattern, and are known by their numbers. The breech of these two guns is rounded.

Scale $\frac{1}{48}$.

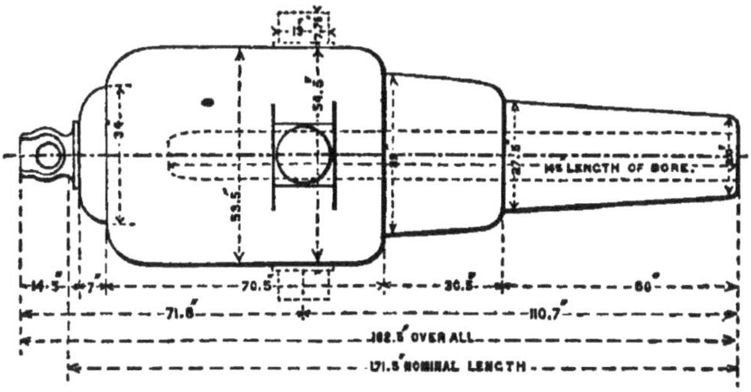

§ 2022.

Mark II. consists of:—

A tube (toughened steel).
B tube.
Triple coil in front of trunnions.
Coiled breech-piece.
Jacket:—Breech-coil, trunnion ring, and ring coil, welded together.
Cascable.

12-inch Rifled M.L. Gun of 35 tons. Mark I. S.S.

There is only one pattern of this gun. (See Plate XVII.)

12-inch of 35 tons, Mark I.

The first of these guns was completed in February 1871 as a 700-pr. of 11″ ·6 calibre, but experiments having proved that this calibre is not suitable for the efficient combustion of 120 lbs. of "P" (Pebble) powder, the proposed battering charge of the gun, it was decided to bore it out experimentally to 12″.

This calibre was finally adopted for guns of this weight, of which only 15 have been made. The cascable terminates in a plain button instead of a loop.

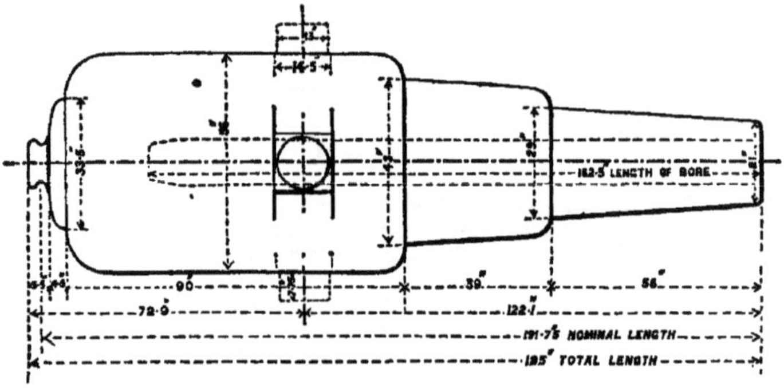

Consists of:— § 2410.

A tube (toughened steel).
B tube.
Triple coil in front of the trunnion.
Coiled breech-piece,
Jacket:—Breech coil, trunnion ring, and ring coil welded together.
Cascable.

12·5-inch Rifled M.L. Gun of 38 tons, Mark I.

See Plate XVIII.

The 12-inch gun of 35-tons being rather short for its other dimensions, a pattern for a 38-ton gun about three feet longer in the bore was provisionally approved of in 1873. 12·5-inch of 38 tons, Mark I.

The calibre was in 1874 approved of being 12.5 inches.* § 2792.

In construction this gun exactly resembles the 35-ton gun, already described, from which it differs in dimensions only.

Scale $\frac{3}{16}$ inch = 1 foot.

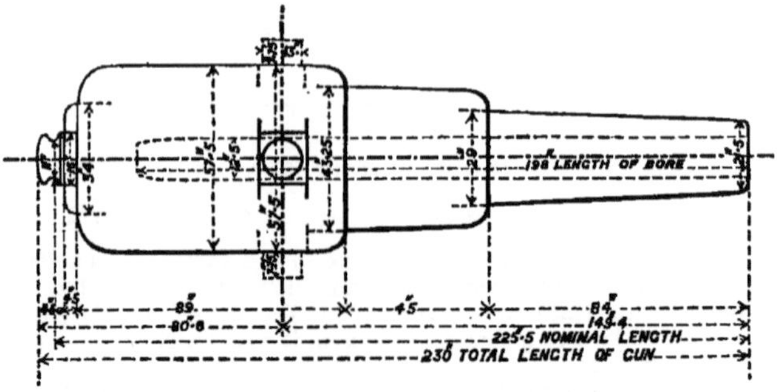

---

* Two 38-ton guns, Nos. 2 and 3, on board H.M.S. "Thunderer" retain the calibre of 12" temporarily.

CHAP. XI.

RIFLING. Scale 1 inch = 1 foot.

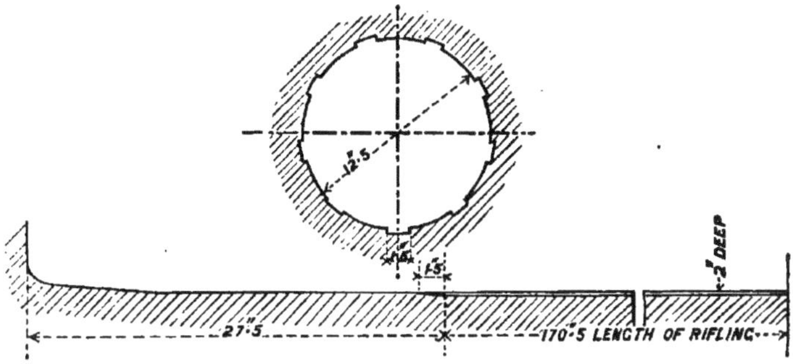

13″·05 Rifled M.L. Gun of 23 tons. Mark I. L.S.

13·05-inch of 23 tons, Mark I.

There is only one pattern of this gun, which was brought forward in 1862 as a 600-pr. to have a calibre of 13·3 inches. In 1864 four of these guns were ordered having a calibre of 13 inches, shunt, rifling, uniform spiral and *A* tubes of untempered steel, closed by the Elswick loose end similar to that in the 64-pr., Mark II. (See Plate V).

We may look upon them as obsolete, for no ammunition is made for these guns, one of which is in the Royal Military Repository, and the other in Woolwich Arsenal.

Scale $\frac{1}{48}$.

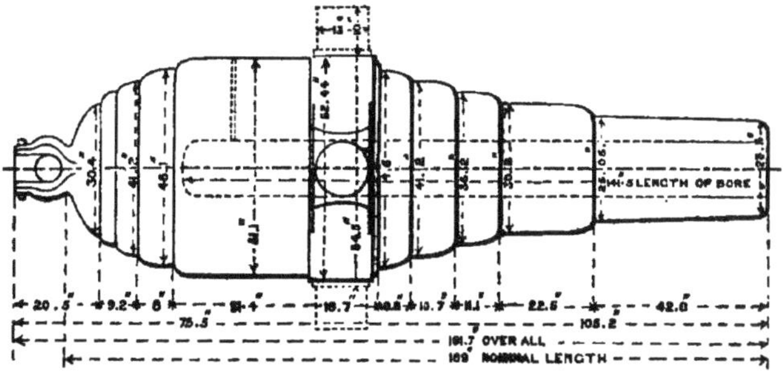

Note.—Of these four guns, three were rendered unserviceable, two by splitting their *A* tube and some of the coils, and the third by the splitting of its outer coil. The latter was repaired, and together with the fourth gun, which passed proof, is now serviceable. The 12″ calibre having been adopted, the rifling of these two guns was converted to the Woolwich groove, the calibre being increased to 13·05 inches.

RIFLING. Scale $\frac{1}{37}$.

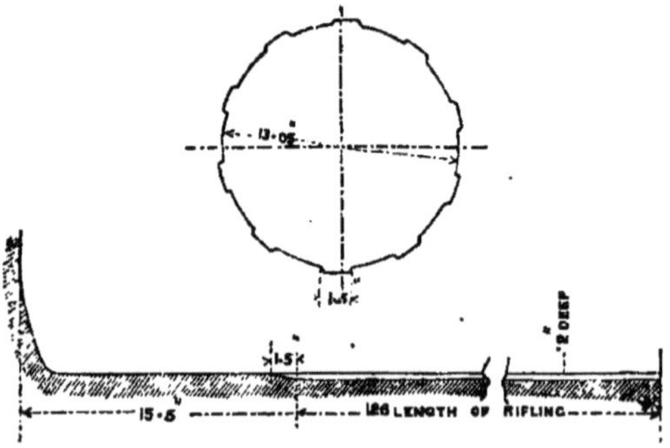

Consists of.—

A tube, untempered steel, with *loose* end.
B tube.
Breech-piece.
Trunnion ring.
15 coils.
Cascable.

16-inch Rifled M.L. Gun, L.S. and S.S., of 80 tons.

See Plate XIX.

In 1873 it was proposed to construct a more powerful gun than any of the service ordnance, and that an experimental gun should be manufactured to weigh about 75 tons. 16-inch guns proposed in 1873.

In 1874 the Secretary of State for War ordered the manufacture of a gun to weigh 80 tons.

It was completed in 1875, being bored up to 14·5 inches, and in September of that year was fired at the Proof Butts, giving very satisfactory results. Experimental calibres of 16-inch gun.

The gun was subsequently bored up to 15 inches.

Since then it has been bored up to its final calibre of 16 inches, and fired with most satisfactory results at Shoeburyness, vide p. 2, 86, 87.

It consists of:—

A tube (toughened steel).
B tube.
1 *B* coil (or belt).
2 *B* coil.
Coiled breech-piece.
C coil (or jacket) made of 4 single coils, a ring coil, and a trunnion ring.
Cascable.

The rifling, sighting, venting, &c. of the piece are not yet finally settled upon. The gun is chambered, the size of chamber being 18

CHAP. XI.

inches in diameter and of a length of 59·6 inches. A stop will prevent the gas check on projectile from entering the chamber.

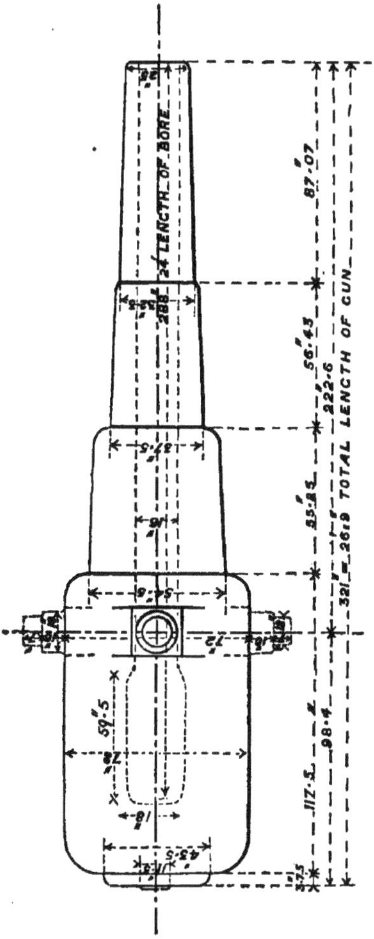

Scale 3/16 inch = 1 foot.

17·72-inch Rifled M.L. Gun L.S. and S.S. of 100 tons.

See Plate XX.

In 1878 four of these guns were purchased from the Elswick Ordnance Company.

ORDNANCE WROUGHT IRON RIFLED M.L. 17·72-INCH, 100 TONS.
Scale 3/16 = 1 foot.

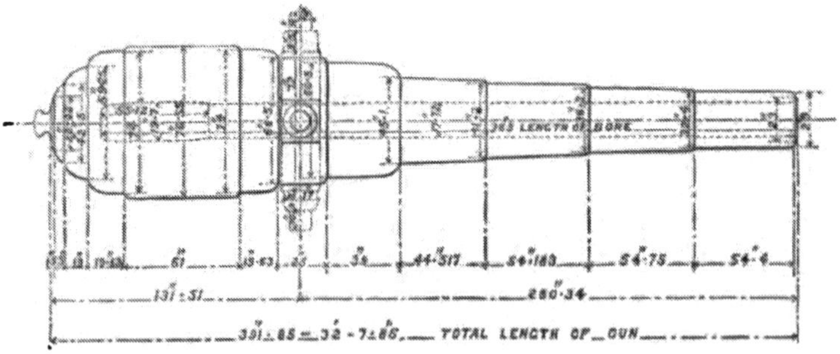

The *A* tube consists of a breech and muzzle portion 1 A, and 1 B, with a ring over the joint.

Over the *A* tube are the following coils of wrought iron 2 A, 2 B, 2 C, 2 D, 2 D′, 2 F, 2 G, and 2 H. The coils 2 E, and 2 E′ come immediately over the 2 D and 2 D′ coils.

The third layer over the breech is formed by the 3 A, 3 B and 3 C coils and a separate forged trunnion ring.

The fourth and outer layer consists of the 4 A′, 4 A and 4 B coils. Portions of the surface of some of the coils are serrated to afford additional grip longitudinally when shrunk one over the other.

Rifling.

These guns are polygrooved, the number of grooves being 28. The spiral is an increasing twist from 1 in 150 calibres at breech to 1 in 50 at 2·88 inches from muzzle remainder uniform. The chamber is in the form of a cylinder. 19·7 inches in diameter terminated at each end by the frustrum of a cone. The distance from where gas check is stopped by the termination of the grooves to end of chamber is 59·72 inches. The venting and sighting are not yet settled.

8-inch R.M.L Howitzer of 70 cwt. L.S.

ORDNANCE WROUGHT IRON R.M.L. HOWITZER, 8-INCH, 70 CWT. MARK I.
Scale ½-in. = 1-foot.

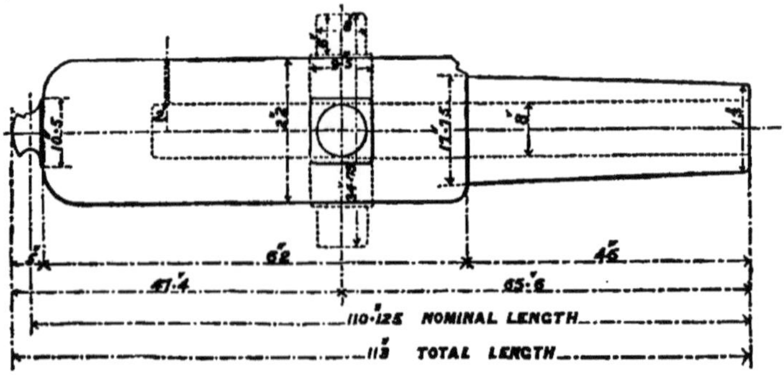

CHAP. XII.

At the end of 1876, a howitzer was proposed to weigh about 70 cwt. to be employed in the siege train.

During the early part of 1878 this piece was tried at Shoeburyness with good results, and the manufacture of these pieces being now proceeded with, they will shortly be introduced into the service.

The rifling is polygrooved, and the number of grooves 24. Spiral increasing from 1 in 100 to 1 in 35.

Three quadrant planes on surface of piece, similar to the light howitzers.

Stop. The gas check on the base of projectile is stopped at a distance of 8 inches from the bottom of the bore, by putting a stop in the grooves at that point. This gives a definite air space for the charge. The piece is vented vertically, the vent striking bore 2″ from bottom.

The method of sighting this howitzer has not yet been definitely settled. (See p. 292 for details of grooves, twist of rifling, &c.)

6·6-inch R.M.L. Howitzer of 36 cwt. L.S.

ORDNANCE WROUGHT IRON R.M.L. HOWITZER, 6·6-INCH, 36 CWT. MARK I.

Scale ⅜-in. = 1-foot.

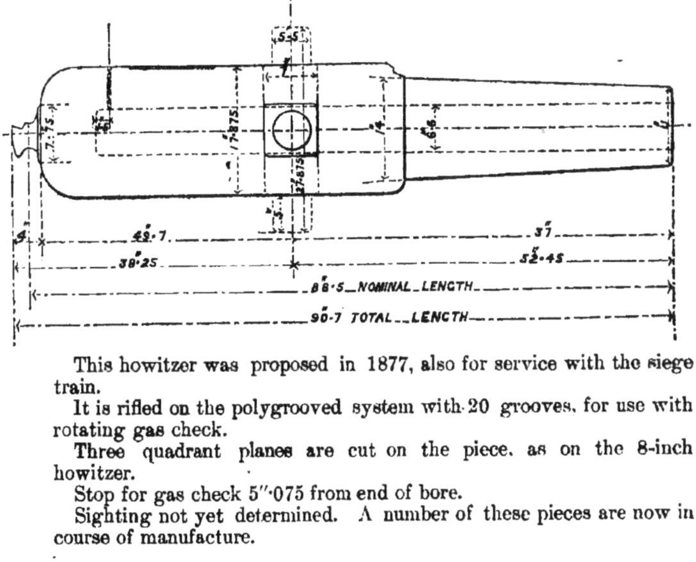

This howitzer was proposed in 1877, also for service with the siege train.

It is rifled on the polygrooved system with 20 grooves, for use with rotating gas check.

Three quadrant planes are cut on the piece, as on the 8-inch howitzer.

Stop for gas check 5″·075 from end of bore.

Sighting not yet determined. A number of these pieces are now in course of manufacture.

13-pr. Rifled M.L. Gun of 8 cwt. L.S.

In 1878 an experimental gun of this nature was completed. It is intended for use as a light field gun, with Horse Artillery and light field batteries.

It consists of :—

- *A* tube (toughened steel).
- Jacket consisting of a breech coil, trunnion ring, and muzzle coil.

TABLE XXVII.—Dimensions, Rifling, &c. of Muzzle-Loading Guns.

Nature, Nominal Weight and Service.	Length. Gun.*	Length. Bore.	Length. In calibres.	Length. Rifling.	Chamber. Length.	Chamber. Diameter.	Chamber. Capacity.	Rifling. System.	Rifling. Spiral.	Grooves. Number.	Grooves. Depth.	Grooves. Width.	Average Preponderance.
	ft. in.	in.		in.	in.	in.	c. in.				in.	in.	cwt.
17″·72 of 100 tons.—M.L., L.S., and S.S.	32 10	363	20	302·88	55·12	19·7	16,637	Polygrooved	Increasing from 1 in 150 to 1 in 50, at 2·88 inches from muzzle, then uniform 1 in 50.	28	·125	1·1	
16″ of 80 tons.—L.S. and S.S., I.	26 9	288	18	228·4	59·6	18·0	14,450	Do.	—	32	·1	·1	
12·5″ of 38 tons.—L.S. and S.S., I.	18 9¼	198	16	170·5				Woolwich.	Increasing, from 0 to 1 in 35.	9	·2	1·5	Preponderance of guns of 12 tons and upwards not to exceed 3 cwt.
12″ of 35 tons.—S.S., I.	15 11¼	162·5	13·5	135					Do.	9	·2	1·5	
13·05″ of 23 tons.—L.S., Mark I.	14 1	141·5	10·8	126	—	Unchambered.	—	Do.	Uniform, 1 turn in 55 calibres of 13.″	10	·2	1·5	
12″ of 25 tons.—L.S. and S.S., I. and II.	14 3½	145	12	127	—	o.	—	Do.	Increasing, from 1 in 100 at breech to 1 in 50 at muzzle.	9	·2	1·5	
11″ of 25 tons L.S. { I.	14 3 }	145	13	119	—	Do.	—	Do.	Increasing, from 0 to 1 in 35.	9		1·5	
{ II.	14 2 }												
10″ of 18 tons L.S. and S.S. { I.	14 2 }	145·5	14·5	118	—	Do.	—	Do.	Increasing, from 1 in 100 to 1 in 40.	7	·2	1·5	
{ II.	14 2⅜ }												
9″ of 12 tons L.S. and S.S. { I.	12 3	125	14	107·5	—	Do.	—	Do.	Increasing, from 0 to 1 in 45 calibres.	6	·18	1·5	Nil.
{ II., III., IV., and V. ...	12 3	125	14	104†	—	Do.	—	Do.	Do.	6	·18	1·5	5
8″ of 9 tons, S.S. { I.	11 4½	118	13	102	—	Do.	—	Do.	Increasing from 0 to 1 in 40 calibres.	4	·18	1·5	Nil.
{ II. and III. ...	11 4½	118	13	99·5†	—	Do.	—	Do.	Do.	4	·18	1·5	4
7″ { of 7 tons, L.S. { I.	11 10·8	126	18	112·5	—	Do.	—	Do.	Uniform, 1 turn in 35 calibres.	3	·18	1·5	4·5
{ II. and III.	11 9½	126	18	110·5	—	Do.	—	Do.	Do.	3	·18	1·5	3
{ of 6½ tons, S.S. { I.	10 5¼	111	16	97·5	—	Do.	—	Do.	Do.	3	·18	1·5	5·5
{ II. and III. ...	10 6	111	16	95·5	—	Do.	—	Do.	Do.	3	·18	1·5	3
7″ of 90 cwt., S.S., I.	10 4½	111	16	95·5	—	Do.	—	Do.	Do.	3	·18	1·5	5
80-pr. 6·3″ } of 5 tons, L.S., I.	10 0	113·25	18	106·25	—	Do.	—	Do.	Uniform, 1 turn in 40 calibres.	3	·145	1·3	9·75

* The length of a Rifled M.L. gun is measured from the face of the muzzle to the smallest diameter of the cascable, excepting those having base rings, from behind which to the muzzle constitutes their nominal length.

† Those made before 1st January 1868, have length of rifling same as Mark I.

Note. For charges, &c. *see* table, p. 428*c*.

TABLE XXVII.—*continued.*

Nature, Nominal Weight, and Service.	Length.				Chamber.			Rifling.		Grooves.			Average Prepon-derance.
	Gun.*	Bore.	In calibres.	Rifling.	Length.	Diameter.	Capacity.	System.	Spiral.	Number.	Depth.	Width	
	ft. in.	in.		in.	in.	in. ‖	c. in.				in.	in.	
64-pr. 6·3" of 64 cwt. S.S. and L.S. I. ...	9 3¼	98	15·5	90·5	—	Unchambered.	—	Shunt‡	Uniform, 1 turn in 40 calibres.	3	·11 & ·08	·6 & ·4	7
64-pr. of 64 cwt. S.S. and L.S. II. ...	9 5	98	Do.	90·5	—	Do.	—	or Plain.	Do.	3	·11 & ·08	·6 & ·4	3
64-pr. of 64 cwt. S.S. and L.S. III. ...	9 3¼	97·5	Do.	90·5				Plain.	Do.	3	·11 & ·08	·6 & ·4	3·75
64-pr. of 71 cwt. L.S. and S.S., I.	9 0	103·27	16·4	96·27	—	Do.	—	Do.	Do.	3	·115	·6	6·375
64-pr. of 58 cwt. L.S., I.	9 6	108·45	17·2	101·45	—	Do.	—	Do.	Do.	3	·115	·6	—
40-pr 4·75" of 34 cwt., L.S., I.	8 0	85·5	18·0	72·5	—	Do.	—	Woolwich.	Uniform, 1 turn in 35 calibres.	3	·1	·8	28 lbs.
40-pr of 35 cwt., L.S., II.... ...	9 6¼	104·5	22·0	90·5	—	Do.	—	Do.	Do.	3	·1	·8	26
25-pr. 4" of 18 cwt., L.S., I.	7 10½	88	22·0	78	—	Do.	—	Do.	Do.	3	·1	·8	14
16-pr. 3·6" of 12 cwt., L.S., I.	6 2.45	68·4	19·0	58·04	—	Do.	—	Modified French.	Uniform, 1 turn in 30 calibres.	3	·11	·8	7·5
13-pr. 3" of 8 cwt., L.S., I.	7 3.96	84·0	28·0	67	14·33	3·15	110·38	Polygrooved.	Increasing, from 0 to 1 in 30.	10	·05	·5	—
9-pr. 3" of 8 cwt. L.S. I....	5 8½	63·5	21	59·8	—	Unchambered.	—	Modified French.	Uniform, 1 turn in 30 calibres.	3	·11	·8	7
9-pr. of 8 cwt. S.S. II.	5 8½	63·5	21	59·8	—	Do.	—	Do.	Do.	3	·11	·8	7
9-pr. of 6 cwt. S.S. I. ...	4 10	53	17·6	49·3	—	Do.	—	Do.	Do.	3	·11	·8	3
9-pr. of 6 cwt. L.S. II.	5 11	66	22	62·3	—	Do.	—	Do.	Do.	3	·11	·8	10
7-pr. 3" of 150 lb. steel (mountain),* I., II., and III.	2 2½	24	8	22	—	Do.	—	French.	Uniform, 1 turn in 20 calibres.	3	·1	·6	3 lbs. III. 5 lbs. I. II.
7-pr. of 200 lb. steel—IV.	3 2 3/16	36	12	34	—	Do.	—	Do.	Do.	3	·1	·6	5
7-pr. of 200 lb. bronze (boat). II.	3 0	32·15	11	29·15	—	Do.	—	Do.	Do.	3	·1	·6	45
7-pr. of 224 lb. bronze, Mark III.	3 0	34	11	32	—	Do.	—	Do.	Do.		·1	·6	5
8" Howitzer of 46 cwt. L.S., I. ...	5 1	48	6	35·5	—	—	—	Woolwich.	Uniform, 1 turn in 16 calibres.	4	·18	1·5	2
6·3" Howitzer of 18 cwt., L.S., I. ...	4 6	45	7	42·5	5·3	6·3	161·26	Polygrooved.	Increasing, from 1 turn in 100 to 1 in 35.	20	·05	·5	—
8" Howitzer of 70 cwt., L.S., I. ...	9 2·125	96	12	93·5	8	8	397	Do.	Increasing, 1 in 100 to 1 in 35,	24	·05 & ·1	·7	—
6·6 Howitzer of 36 cwt., L.S., I. ...	7 4·5	79·2	12	76·7	5·075	6·6	169·5	Do.	Do.	20	·05 & ·1	·7	—

* The length of a rifled M.L. gun is measured from the face of the muzzle to the smallest diameter of the cascable excepting those having base rings, from behind which to the muzzle constitutes their nominal length.

‡ As explained at pp. 45 and 262, all 64-prs. having steel tubes whether originally or re-tubed Mark III. guns, and also all of Marks I. and II. re-tubed with iron have the plain groove.

Note. For charges, &c., *see* table, p. 428*a*.

CHAPTER XII.

EXAMINATION, PRESERVATION, AND REPAIRS OF ORDNANCE AND STORES.

CHAP. XII.

Examination of Ordnance.—Memorandum of examination issued with every gun.—Periods at which different natures of ordnance must be examined.—Mode of examination.—How to examine the bore.—When impressions should be taken.—Mode of taking gutta percha impressions with B.L. and with M.L. guns.—Natures of defects usually found in bore.—How the vent should be examined.—Natures of defects in or near vent.—As to sentencing for re-venting.—**Examination of B.L. fittings.**—Defects on exterior of gun of little importance.—**Repairs.**—What can be performed in the field.—Re-coppering, adjusting new fore-sights or clamping screw for hind sights.—Re-venting.—Repairs at certain stations.—Adjusting Millar's sights to a S.B. gun.—Re-venting heavy guns.—Re-bushing 7-inch R.B.L.—Repairs to be carried out in an arsenal.—**Preservation of Guns and Fittings.**—Painting, browning, lacquering bore.—Vent plugs.—Sights, how preserved.—Blueing.—Bronzing.—Preserving screws should be used.—**Tables.**—Cleaning and examining tools for S.B. Ordnance.—Rifled Ordnance.—Venting tools.

EXAMINATION.*

Every artilleryman should know how to examine the weapons with which he works, and should understand what defects in guns are serious and what defects may be disregarded, while it is of great importance that the examination of both guns and fittings should be very searching and exact, otherwise a small flaw left unnoticed may endanger the life of the gun in future. *Importance of examination.*

Guns must therefore be frequently examined at out-stations,† but as it requires great practice and experience in order to become a competent judge of the various conditions of all the different natures of guns in the service, it has been directed that *final* condemnation shall be pronounced only by the authorities of the Royal Gun Factories. *Final condemnation.*

In all cases, however, where there are sufficiently serious defects (p. 298), or if there be any doubt as to the serviceable state of the gun it must be provisionally condemned and a report made of the same as mentioned in Appendix, p. 372. *Provisional condemnation.*

As described at p. 177, every gun is carefully examined after proof and before being passed into the service, and in the case of R.B.L. guns with reference also to its breech fittings. *Guns examined, &c. before issue.*

* Vide instructions relating to fitting, care, and preservation of ordnance, issued with Army Circular, 1878.

† This will be done by the Inspector of Warlike Stores or other competent person as mentioned Appendix, p. 372.

CHAP. XII.

It is then marked with the broad arrow in front of the vent, and if it be a rifled gun a memorandum of examination is filled in, which is in fact a register sheet for that particular piece.

Memorandum of examination.

This memorandum of examination always accompanies the gun and is in possession of the officer in whose charge the piece may be. In it is given the information required by any one who has to examine a gun, viz., the material of the bore, and, in the case of muzzle-loading ordnance, a short description of the construction, with a woodcut showing the gun in section. The defects in the gun at the time of *its issue*,* the number of rounds it has fired, and the subsequent examinations are also stated. When a gun is returned to the R.G.F. for repairs, a new memorandum of examination is made out after the repairs are completed, and sent out with the gun on re-issue.

Periods of Examination.

Periods of examination.

With regard to the periods of examination the following rules are laid down:

Every gun must be examined after firing a given number of rounds with projectile as under:—

S.B. guns. Rifled guns.

S.B. Cast iron guns	Firing 10 lbs. charges † and upward	100
	Under 10 lbs. charge	200
Rifled guns B.L. and M.L.	9-inch guns and upwards ..	50
	8″, 7″, and 64 prs.	100
	40-prs. and under.	150

If however there should be any appearance of fissures about the vent, or other defects likely to develop, all guns will be examined after every 50 rounds, or more frequently as examining officer may direct.

Mode of Examination.

Preparing a gun for examination.

The bore should be thoroughly cleaned, as it is not possible to detect small defects, which may sometimes be of importance, if the bore be in a rusty or very greasy state. If care has previously been taken in keeping a gun tolerably clean, it will probably be sufficiently prepared for examination by washing, brushing, and drying with tow or a clean sponge head. If, however, there be hard rust which will not yield, or a thick coating of lacquer or grease, the bore may be cleaned either by firing, if circumstances admit, one or two scaling charges, about one-third the full service charge, without projectiles, which will usually loosen the scale, or by the use of hot water and potash, in the following manner:—

Cleaning with potash water.

About a gallon of boiling water is poured on one pound of black American potash, and an old sponge covered with a canvas cap, or some substitute to make it tight to the bore, is dipped into the solution. The bore is then rubbed till the dirt is loose, when the hard brush will remove it; it is then wiped dry with tow, &c., and slightly oiled. The potash water must be used very hot, and the sponge must be very tight, or the process is ineffectual. If the dirt be very thick in the small grooves of the Armstrong B.L. guns, a common pricker with the point filed flat is useful. No sharp edged or pointed scrapers should

---

* *See* p. 297. The position of defects developed on service are noted in a similar manner.

† 68-pr. | 32-pr., 63 cwt.
56-pr. | 32-pr., 58 cwt.
42-pr. | 32-pr., 56 cwt.

CHAP. XII.

be employed for cleaning the bores of rifled guns; they are unnecessary and are liable to injure the rifling.

Description of impressions generally required.

As to taking impressions genenerally speaking:—With S. B. guns it will usually be only necessary to take one of the vent and that portion of the bore immediately adjoining it. With R.B.L. pieces impressions would only be taken when flaws were seen, and as a rule of the breech bush in the gun and of the face of the vent piece. With converted guns impressions of the whole of the powder chamber showing also the cup and the joint where the cup closes the end of the bore should be taken. With R.M.L. steel-lined guns, unless scoring, damage by shell, or incipient cracks are thought possibly to exist, an impression of the vent would very often be sufficient.

Examination of bore (including Chamber).

Bore.

For the purpose of this examination tools are provided for S.B. and Rifled guns respectively as shown in Tables, pp. .

Examination by means of lamp.

The bore, being cleaned, should be examined by aid of a lamp; if the surface is slightly wet the detection of defects by this means is greatly facilitated. A sharp pointed pricker is used to ascertain the extent and position of any flaw, the stave being graduated in inches so that the distance from the muzzle may be readily ascertained. A spring searcher is also used to detect defects, and with B.L. guns in such a manner that each groove shall be traversed in succession by one of the points. A pricker with "blunted point," or rather a flat edge, is also supplied, and is useful in searching the defects in coiled barrels.

Spring searcher.

Blunted pricker.

Defects in S.B. guns.

Should any defects be discovered they need not be taken notice of, in the case of S.B. guns, unless they are 0″·1 deep in rear of, or 0″·2 deep in front of the trunnions, or unless they have jagged edges likely to retain pieces of cartridge; and in the case of rifled guns, unless they are new defects not shown on the memorandum of examination, or old ones which have materially increased. If the defects, however, be of the natures mentioned, impressions must be taken of the bore.

Defects in rifled guns.

Impressions to be taken.

Temporary impression.

For S.B. guns an impression of the defect will be taken by means of a pricker having a copper pan attached to it below. This impression is only a temporary one, taken with a mixture of soft soap, bees-wax, and treacle.*

Instrument for taking impressions of bores.

A special instrument is issued with rifled guns to take an impression of bore (including powder chamber). To use this instrument "for taking impressions of bores," prepared gutta percha (Vide appendix, p. 374), is laid on a gun metal plate secured by screws to a frame. The upper part of the frame itself answers as a plate for some natures. The frame is attached to a long hollow tube, and is so made that the upper portion can be raised or lowered by bell crank levers, worked by a right and left handed screw. The screw forms part of a rod passing through the frame handle, and to the outer end of the bar is attached a cross handle, by working which, the plate carrying the gutta percha can be forced up against the bore or powder chamber when required, or lowered when the gutta percha is sufficiently hardened.

Should this instrument not be available, wood blocks such as those shown below for taking impressions of the whole length of the bore may be used. They can be made by any carpenter,† but it requires some practice to take good impressions with them. It is therefore advisable to take impressions of the powder chamber if possible with the special instrument provided with examining tools.

---

* Should a permanent impression be required it must be taken with gutta-percha with whatever means may be available.

† These blocks are therefore not issued.

CHAP. XII.

Long impressions. § 1625.

In all cases where there is any doubt as to the state of the bore in the case of rifled guns, impressions should be taken of the whole length by means of these blocks.

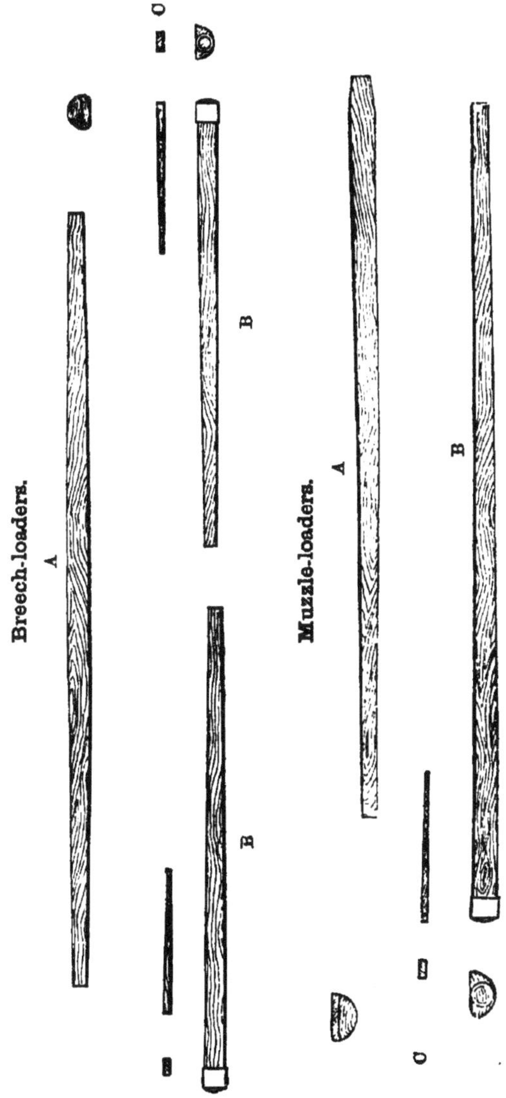

Wood Blocks for taking Impressions of the Bores of Guns- Scale ¼ inch = 1 foot.

Method of using wood

The wood blocks are thus used:—

Blocks A (tapering from the centre for B.L. guns, and from the breech for M.L. guns) with their wedges B should be made to suit the diameter of the bores to be taken, leaving room for about 0·25 inch of gutta percha when the wedge or wedges are driven home, then proceed as follows:—

Preparation of the gutta percha.

A sufficient quantity of gutta-percha prepared as in Appendix, p. 374 is laid along the block A, which has been previously prepared by rubbing it over with a little soft soap. The gun is so placed that the impression required will be taken upwards, the block A is inserted into the bore, and the wedge B (if a B.L. gun by simultaneous blows at both ends) is driven well home with mauls; a small wedge C is then forced between the ends of the blocks A and B.

Method of using the Wood Blocks for taking Gutta-Percha Impressions of the Bores of Guns.

Scale $\frac{1}{48}$ size.

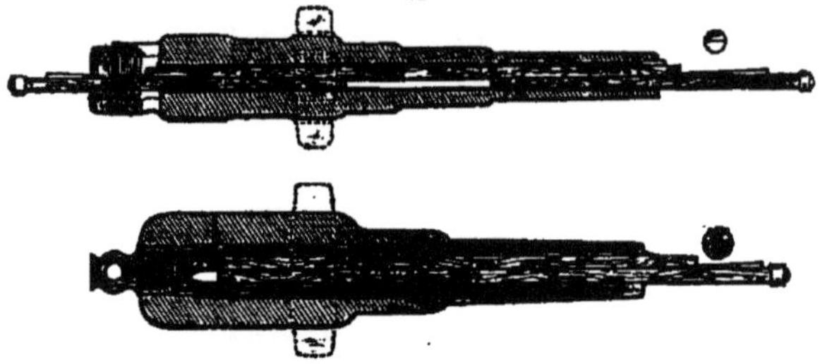

This can easily be withdrawn in about 10 or 15 minutes, according to the temperature, when the impression has become cold, and thus gives slackness to the wedge B and the block A, which are withdrawn in the order named together with the impressions, which can be readily removed from the block, being prevented from sticking by the soft soap.

Before impressions are taken the bore should be quite clean but slightly greasy; if quite dry the gutta-percha will adhere to it, and the impression been damaged in removal.

Position of flaws, &c., how noted returns.

In order to note in returns* the position of any flaw, &c. upon an impression, their distance from the muzzle is measured in inches and noted as up, right of up, &c., as the case may be.

Defects in powder chamber of R.B.L. guns are measured from the breech end, and noted in a similar manner to defects in the bore.

Circumferential defects are noted round the bore in the direction of the movement of the hands of a clock, *e.g.*, R of Up to R to D, L of D to L to L of Up.

IMPRESSIONS ARRANGED FOR RECORDING DEFECTS.

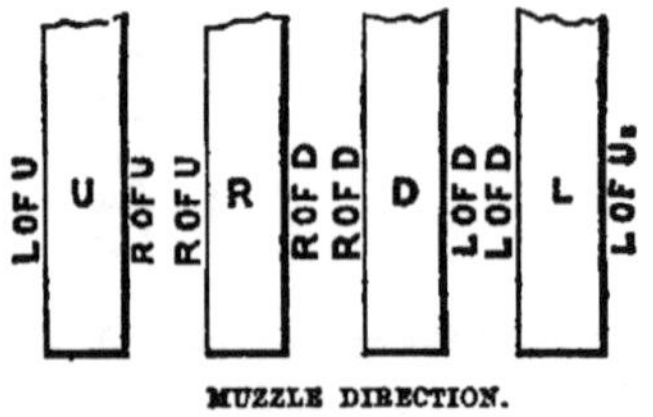

MUZZLE DIRECTION.

* Vide Appendix, pp. 375, 376.

CHAP. XII.

Labelling impressions.

When an impression is forwarded with a return for report or decision, a label will be gummed to the back, showing the name of the station, the date of taking the impression, the direction of the muzzle, the nature and number of the gun, the position of the fissures (if it be in any part but the vent) defined according to directions.

The impressions should be reduced to the smallest dimensions compatible with showing the whole of the defect.

Defects in Bore.

S.B. guns. Corrosion of bore.

In S.B. guns the bore is liable to be much corroded if exposed to the action of salt water, and occasionally a flaw in the metal leaving a hole may also occur. Generally speaking, however, the defects which condemn such pieces will be found at or about the vent, vide p. 300.

Rifled guns.

With rifled guns the following defects are likely to be found.

Tool marks.

(1.) "Tool marks," or slight irregularities and scratches caused in manufacture during the boring and rifling. A chip or grain of sand getting between boring or rifling head will cause such tool marks, while the emery powder used in lapping leaves very fine marks on the hard surface of the steel tubes.

With steel tubes, on account of the hard smooth surface, these are more apparent and have been at times mistaken for fine cracks when running along the bore. To an experienced eye, however, the difference is easily perceptible.

Wearing of the grooves.

(2.) Slight wear of the sides of the grooves from the friction and over-riding of the studs of projectile is also found at times, but this takes place as a rule only in guns having iron tubes, or when with steel tubes a uniform twist is employed with a heavy gun.

Defects in wrought iron barrels.

(3.) With wrought iron "A" tubes* imperfections of weld frequently exist, and scarcely any are quite free from them; they are, however, of little consequence as a rule, unless they are found to increase very much when the gun is used.

The following are the technical names of the several defects likely to be found in wrought iron barrels:—

"Coil marks," where the line of coil running round the barrel is visible, the weld not being quite perfect; but in this case the defect has no appreciable depth; and is of little consequence.

"Defective welds," the same sort of defect as "coil marks," but deeper and more important; they run round the bore in coiled barrels and along it in solid forged barrels.

"Specks," small pots and pinholes in the metal, caused by dirt in iron; blisters, &c. These sometimes occur in clusters.

"Flaws," larger defects of the same nature.

"Longitudinal cracks" are also found in solid forged barrels, caused by the gas eating into the defective welds and splitting the barrel lengthways.

"Scoring" or "Guttering" about the seat of the projectile, caused by the rush of gas through the windage in M.L. guns, occurs in the coiled barrels of heavy guns (few of which have been made), and "longitudinal cracks" are also sometimes developed.

The experience of years has proved that flaws, coil marks, and

---

* The material of which the barrels are made has for some years past been stamped on the muzzle, and is also entered in the "memorandum of examination."

even defective welds are of little importance in guns not exceeding the size and power of the 7-inch B.L. gun. The importance of a defect depends in a great measure on its position in the gun, one in rear of the trunnions, especially if in the powder chamber, being more dangerous than one of the same nature and extent in front of the trunnions, as the powder gas acts much more rapidly upon it, and it is liable moreover in M.L. guns to hold a piece of ignited cartridge. Few instances have occurred in which defects in coiled barrels have caused accidents or have increased in any material degree after issue, unless they exist in the powder chamber, and for this reason no guns have been issued for some years past with any defect except of the most trifling character in that part of the bore. Certain defects of little importance in coiled barrels.

It has been found that in M.L. guns having coiled iron barrels (64-prs. and a few of the heavier natures, see p. 31), the tube is liable to split in the chamber in continuation of the edge of one of the grooves. Longitudinal cracks in barrels.

(4.) Steel barrels are fortunately free from the defects so inherent to wrought iron, and generally show us a beautifully smooth and even surface, very slightly scored by a number of very minute tool marks. Occasionally a small piece is found chipped out as it were, but leaving a smooth hollow behind it, which if small is quite an unimportant defect. Defects in steel barrels.

Longitudinal cracks at times occur which develop into splits. When such are found they are of far more importance than similar cracks in a wrought iron tube. Longitudinal cracks.

A few instances have occurred in which steel tubes have split on service from the bursting of a shell inside the bore; in some of these the tube split at the muzzle. In cases where steel barrels have split it M.L. experimental guns when being tested for endurance, the crack has commenced at the edge of one of the grooves (as in coiled barrels) and extended into the powder chamber; as a crack in the chamber, should it exist, might be so fine as not to be visible from the muzzle, it is essential that impressions in gutta-percha should in all cases be taken of the chamber of rifled M.L. guns, as before stated, and care must be taken to discriminate between a crack and a superficial streak or tool mark. Splitting of steel barrels, how caused.

(5.) Scoring or guttering was caused by the rush of gas round the projectile in R.M.L. guns due to windage, and was common to both iron and steel barrels. This wear is principally over the top of the projectile where windage is greatest owing to projectile resting on its lower studs. In small guns this was very slight, but when very heavy charges were used the surface of the bore was gradually eaten away to a considerable extent, but even in extreme cases scoring has not caused the destruction of a gun, though in some instances acting on the corner of a groove it has tended to split the tube. Where gas checks are employed, this defect exists in a very slight degree. Scoring or guttering. Prevention of scoring by gas checks. Gas checks to prevent scoring.

(6.) Again, we have sometimes "dents" or "abrasions" caused by the bursting of a shell in the bore; these are as a rule of very little importance. In one or two instances muzzles have been blown off from this cause; but generally the only result is to cut up and graze the bore more or less. It is found that such injuries seldom interfere with the efficiency of the piece, and an armourer can generally file down any metal set up in the bore. Dents or abrasions due to shell bursting.

Sentence as to bore.

In sentencing a gun according to the state of the bore, it is essential to discriminate between defects which are characteristic of the Sentence as to bore.

CHAP. XII. material, and cannot wholly be avoided in manufacture, and those which are created or developed on service, such as cracks in a steel tube. In coiled barrels defects are often numerous and generally of little importance, while in steel barrels the case is reversed; defects seldom occur, but when they do they are generally speaking of great importance.

It is almost impossible to lay down any definite rules as regards the extent or depth of a defect which should necessitate the condemnation of a barrel; a great deal must be left to the judgment and experience of the inspector.

Sentence on coiled barrels. Unless there is reason to believe that there has been some material change from the former state of the defects, a gun with a coiled barrel need not be condemned. Speaking generally, the depth of a defect is of more importance than its extent, but, should a defective weld run *right round* the bore, the gun would be liable to part at that point, and must be considered unserviceable. The best method of testing a gun is to take an impression of the defect, then to fire a few rounds and take another impression; if on comparing these impressions the defect does not appear to have increased, the barrel may be considered as serviceable.

Solid forged. In solid forged barrels a flaw running lengthways has a tendency to develop into a crack, especially if it occur in the powder chamber. If the inspector finds a case of this kind he will put the gun aside, but he will endeavour to discriminate between this and a mere streaky line, which is unimportant.

Steel barrel. As to steel barrels, any crack, however fine, would necessitate the provisional condemnation of the gun.

Examination of Vent.

Examination of vent. Besides examining the bore of a gun, it is necessary especially to examine the vent-bush itself, and that part of the chamber or bore of the piece where it enters. For this purpose in S.B. and R.M.L. guns,

Gauging. the vent channel is thoroughly cleared and scraped, and then gauged by means of a set of gauges issued for the purpose.

It is not uncommon to find a "choke" where the metal has been set up near the bottom of the vent; this should be removed by a drill or rimer before the gauge is taken.

Impressions of vents. A clean impression of the bottom of the vent must also be taken, this will be taken with the improved instruments for taking impressions of bores, the plates for which are now fitted with ends to suit the chambers of guns, or they may be taken with the instrument provided among the tools for venting ordnance.

In using the instrument for taking impressions of vents, a small piece of prepared gutta-percha is placed on the pan of the instrument, (which should have been previously fitted with a pad of gutta-percha) and screwed up against the vent or other part of the bore of which an impression is to be taken. It is there left until cold (from ten to twenty minutes according to the weather), and the instrument is then withdrawn. A little soft soap, or if that be not available, common soap and water, oil or grease will prevent the impression sticking to the pad. The bore of the gun should be slightly greased. Too much pressure must not be applied, otherwise the impression will be very thin, and if the defect be deep, it will be difficult to remove the gutta-percha. A good deal of practice is required to get good smooth impressions, and several impressions of the vent have sometimes to be taken before one is obtained which can be relied on to show any hair lines.

R.B.L. guns are vented through the breech block or vent piece, see p. 303, for mode of examining the same.

Defects as to vent, and sentence.

The effect of service on a vent is seen either by a gradual increase to the channel of the vent itself, by an irregular wearing away of the bottom, by the metal of the vent setting up, and the gas forming a hollow ring round it, or by fissures or hair lines radiating into the metal of the bore from the edge of the vent bush. **Effect of firing on vents.**

The following are the existing regulations as to sentencing guns, either for re-venting* or condemnation, according to the state of their vents.

1 (*a*). S.B. and converted guns.—To be revented if 0″·3 gauge passes down the vent. **Re-venting.**

1 (*b*). All other Rifled guns —To be revented if 0″·25 gauge passes down the vent.

2 (*a*). S.B. unbushed† guns.—Should there be a cavity at the bottom, or any hair lines or fissures, the extent of the defects will be measured on the impression, If they be of a length beyond ·2-inch, but not exceeding ·35-inch from the original centre, the gun will be sentenced for cone venting; if ·35-inch, but not ·5-inch, for through venting; if beyond that limit, the gun will be reported provisionally as unserviceable. **Unbushed guns.**

UNBUSHED GUNS.

2 (*b*). S.B. Bushed guns. The defects usually found round the vent of bushed guns, are the giving way of the iron round the bush, from the gas getting in between the two metals (see Fig. 2), and the fissures or hair lines, which radiate in the iron, from the edge of the vent bush (see Fig. 3). **Bushed guns.**

---

* For re-venting, vide p. 314.

† Very few of such guns would now be found, as all guns issued from R.G.F. since 1855 have been bushed. Still S.B. guns might have to be drawn from stores abroad where unbushed guns are occasionally on charge.

CHAP. XII.

Bushed Guns.—Fig. 2.

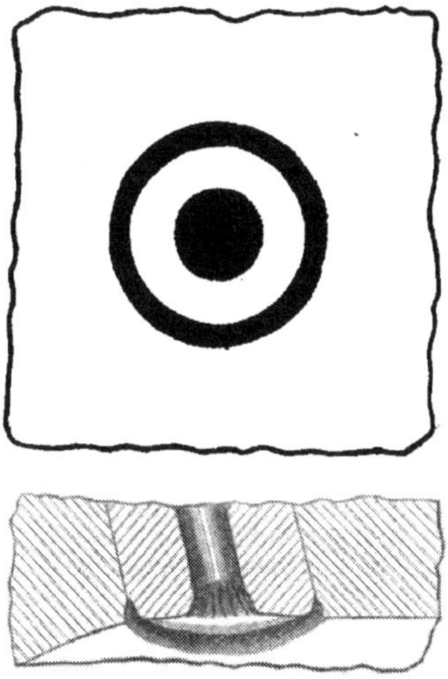

Bushed Guns.—Fig. 3.

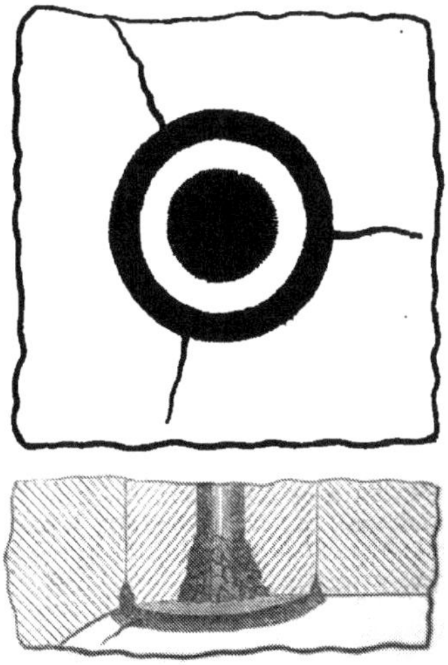

Ring worm round bush.

The metal round a bush begins to give way almost immediately after a gun is vented, forming a hollow ring round it, which gradually increases. So long as this wear is uniform, and the edges are not jagged it is not of importance ; and guns will not be re-vented or condemned for this cause until the ring has become 0·1-inch deep, or 0·1-inch wide. If, however, the edges are ragged, or if one side has given way much more than the other, so as to be likely to hold a piece of unconsumed cartridge, the examiner must use his discretion as to sentencing the gun, it being impossible to lay down fixed rules for all cases.

Fissures, or hair lines. S.B. guns.

Fissures or hair lines, radiating in the iron from the edge of the bush, will be found in bushed guns. There will often be one on either side, and a third to the front. The examiner will carefully trace the fine lines on the gutta-percha impressions, and if they extend in S.B. guns more than one-tenth part of the circumference of the bore in any direction, measured from the original centre, he will provisionally condemn the gun. One-tenth may be taken as—

68-pr.	$2\frac{1}{2}$ inches.
32 „	2 „
24 „	$1\frac{3}{4}$ „

Vents of R.M.L. guns.

Ring round bush.

(3.) R.M.L. Should the ring (Fig. 2) above, worn away round bottom of bush, be 0·″075 wide or 0″·1 deep, or jagged and irregular, so as to be in the opinion of the examiner likely to retain a piece of cartridge, the gun would be condemned for re-bushing and if possible with a cone bush. Much wear of the lower portion of the fire channel in the vent will also necessitate the reventing of the gun. The amount and nature of wear, which should condemn a vent, must be left to the judgment of the examiner. Generally speaking, if any sign of fissures of the copper is apparent, or if the thickness of the wall is much reduced in any part, the vent must be condemned. A reduction to a thickness of 0″·15 at the bottom would usually necessitate condemnation, though no absolute rule can be laid down on the subject.

Hair lines.

In case of hair lines radiating from the edge of the vent in steel lined pieces the gun must be provisionally condemned should the hair lines be 1-inch in length, unless when directly to the front or rear, when the limit of condemnation is smaller, ½-inch, owing to that position being the worst for thedevelopment of a crack. The measurements are taken from the edge of the bush.

Fissures.

In heavy guns with steel barrels a so-called fissure frequently develops in front of the vent (caused apparently by the rush of gas through the vent channel). This fissure partakes at first of the nature of a slight scoring. If not attended to it may increase in depth into a crack: such a fissure should, therefore, be filed out, and an examiner must use his judgment as to when this operation is necessary, but in no case should such scoring be allowed to attain a depth of 0″·01.

Although a properly instructed examiner will generally be able with the aid of the above rules to sentence a vent, either as serviceable for another series of rounds, or for reventing, yet many cases will arise in which it will be advisable for the vent to be sentenced as fit to fire some smaller number of rounds, and then to be further examined. Such cases must be left to the discretion of those whose duty it is to give sentence on the impressions taken.

Examination of R.B.L. fittings.

Examination of breech-screw, lever, tappet, and keep pins.

(1.) The breech-screw will be examined with the straight-edge, in order to ascertain that the face is quite flat and true; it it be not, it will be filed; the thread should be examined by tapping with a wooden mallet, and should not be broken or burred, but a considerable portion may be removed, if injured, without destroying the efficiency of the screw.

The lever and tappet should be sound; the lever handles of naval guns are sometimes broken off, but the lever can still be used in this state, though not so conveniently. The keep pins must be sound.

Vent-piece.

(2.) The vent-piece is the most important fitting, and should be perfectly sound, neither cracked nor bulged. The back and sides when tested by the straight-edge, should be quite flat and true: the fracture of vent-pieces is frequently owing to the back not being true to the face of the screw. The copper ring on the vent-piece, as well as the breech bush at the end of the barrel, must be sufficiently high to prevent the action of the gas on any part of the iron. The angle face of the 7-inch vent-pieces should be flat, and should work truly against the end of the barrel, and the "nose" should fit closely, but not too tightly, into its place.

Vent bush.

The copper bushes in the neck of the vent-piece should be in good order; if they are so much worn that a 0·3-inch gauge can pass through, the friction tube is liable to be pulled out without being fired, and the bushes will be renewed from the spare sets issued for the purpose. A cavity frequently forms at the angle of the vent channel, but this (which should be examined with a probe) does not entail the immediate condemnation of the vent-piece, unless the examiner considers it dangerously large.

Cross head.

The cross-head should not be loose, as instances have occurred of its being broken off whilst firing. In all cases before taking a vent-piece into use for practice, it will be advisable to test the soundness of the crosshead, as well as that of the neck by tapping with a hammer.

Breech bush and vent piece copper ring.

The bush of a R.B.L. gun will be sentenced to be renewed if found to be so much expanded that the gas could escape between it and the tin cup or vent-piece, and when it is not possible to remedy the same by facing. In order to prevent serious damage to gun and vent-piece especial attention should be paid, both before and during practice, to the breech bush in the gun, and the copper ring on the face of the vent-piece. If at all dented they should be carefully faced before use. During practice, the officer in charge should ensure these points being looked to constantly, and stop practice with any gun or vent-piece, in which eating away of the copper by the escape of gas begins to show itself. When this once sets up a few rounds are often sufficient to damage permanently the gun or vent-piece.

Examination of exterior.

Defects on exterior of guns. In examining the exterior of a cast iron gun the points to pay attention to are the soundness of the trunnions and cascable loop.

Very considerable defects may exist on the exterior of a wrought iron gun, without the strength being affected. Occasionally on firing, one of the outer tubes develops a flaw running round the gun, due to the coiled tube having been in an undue state of tension. Such defects are as a rule unimportant, and are easily repaired when the gun passes through the R.G.F.

Defective welds on the exterior near the muzzle are sometimes developed in the *B* tube; these are of no practical importance, and a gun should not be condemned on such grounds alone, though it should be exchanged when an opportunity offers.

If it be found that a shell has burst in the bore, the exterior will be thoroughly scraped with old swords, and cleaned (with potash water, if necessary), in order to ascertain whether it is perfectly sound.

Rifled guns. It also occasionally happens with wrought iron guns that on firing the outer coils shift; if on examination the shifting is found considerable, the gun will be provisionally condemned, but a slight shift, which is sometimes perceptible when the gun is first used, and which has gone no further afterwards, may be disregarded. Unless there be reason to suspect damage on the exterior, it will not be necessary to scrape the paint off whenever a gun is examined. Large defects on the exterior are noted on the "Memorandum of Examination."

Preservation of Guns.

As a general rule all guns forming the armament of a fortress are to be cleaned and painted biennially, but if guns mounted on the sea faces of works are in a bad state from exposure to the spray of the sea, or those in casemates from the damp and dripping, they are to be cleaned and painted every year, and oftener if considered necessary by the Commanding Officer, Royal Artillery.

All guns, whether B.L. or M.L., should be kept in good preservation, the exterior being protected from the effects of the atmosphere by a sufficient coating of paint, and the bore by being lacquered, when not in constant use, or by being well cleaned and oiled during practice. The smaller natures of rifled guns (below 20-pr. B.L.) are not painted however, but are protected by browning the exterior, and keeping the bore cleaned and oiled.

How prepared for painting. The gun is to be scraped on the exterior, until the old paint and all rust which may appear beneath it are entirely removed. The sight notches and all marks are to be completely cleaned out and rendered distinct, and the gun is afterwards to be wiped over with a piece of old canvas or cloth.

The vents are to be cleaned with vent scrapers. The exterior receives two coats of Pulford's magnetic paint. (For the necessary tools for painting see p. 317.

(The operation of browning is described in Appendix II. p. 317.)

The lacquer can be removed from the bores in a few minutes by brushing the bore with hot potash solution (for proportions of lacquer see p. 330). The lacquer is applied to the bore by means of brushes attached to a stave, two staves are used, and are provided with spring sockets (see table p.317). One is used to lacquer the sides, the other the end of bore. In the case of guns having a gas escape channel, the latter should always be kept clear, the outer end being merely stopped with plugs of greased tow, when the guns are not in use.

Preservation of Sights.

When mounted in exposed positions, all the sights should be removed from the guns, and kept in store, the holes in the guns being filled with a plug of greased tow to keep out the rain and dirt. These plugs can be readily removed when it is required to fit the sights to the guns, and particular attention should be paid to the prevention of rust or grit accumulating in the sight recesses.

The set screw for clamping the centre hind sight not being removable from the socket, should be tested to see that it works freely.

The sights themselves should be kept clean, free from grit and oiled; the sliding leaf, as well as the collars of the centre, fore, and trunnion sights should have free play.

The exposed portions of the sights are bronzed if made of gun metal and blued if of steel. This is done to preserve them from corrosion, and on no account are these parts to be burnished or cleaned in such a manner as to remove the bronzing or bluing, more than it is of necessity worn off by fair wear. (For mode of performing these operations see Appendix II. p. 377).

Preservation of Fittings.

The breech screw and bright parts of R.B.L. guns in store, or mounted, where rarely or never used, will be coated with a lubricant consisting of equal parts of powdered chalk and cocoa-nut oil. (See p. 330). *Fittings of R.B.L. guns.*

When R.B.L. guns are not in use, the following parts should be removed and laid up in store. The vent piece and all the B.L. fittings, except the breech-screw itself.

The muzzles of M.L. guns are stopped with tampions (those of B.L. guns left open) and those of mortars covered with wooden caps to keep out moisture, while vent plugs are used with mounted guns. Guns whether mounted or lying on skidding should be depressed at the muzzle to prevent rain or moisture lodging inside.

These plates are removed for transport, and the holes in the guns filled with preserving screws. *Plates metal elevating.*

Guns fitted for L.S. have the friction tube pin holes and the guide plate hole filled by preserving screws, and it is advisable that these screws should be occasionally removed and oiled, to prevent them becoming fixed by rust. *Preserving screws.*

Transport.

In preparing ordnance for transport, the sights, pivot pieces, friction tube pins, sockets for centre hind sights, and in fact, any metal which would project beyond the surface of the gun, and be liable to injury during transport, should be removed, and the holes filled either with preserving screws, or with tow and tallow.

For land transport the smaller natures of guns are packed in jute bags, and for sea transport in boxes. Guns above 16-prs. however, have no such protection.

REPAIRS.

Repairs in the field.

The repairs of guns and fittings may be divided into:—

I. Repairs which can be carried out in the field with the stores issued for the purpose, or such as can be improvised.

Repairs at stations.

II. Such as can be carried out at certain stations provided with special means.

Repairs in an arsenal.

III. Those which must be performed in a large arsenal.

I.—REPAIRS IN THE FIELD.

(*a.*) *R.B.L. Coppering, i.e., replacing or repairing breech bush* copper or vent-piece copper ring, as the case may be.*

Coppering B.L. guns.

Detailed instructions are contained in the boxes of implements† issued for refacing and renewing both the vent-piece copper ring, and the breech bush copper. In the operation of refacing, only just sufficient copper must be removed to render the angle face quite smooth and true.

Inserting a breech bush copper.

A bearing is put into the bore in front of the bush for the boring spindle to work upon, and the face of the breech bush which is to fit against the A tube being red-leaded, the bush is screwed into the gun as a trial, and on being unscrewed, if the red lead shows it does not fit all round, it is scraped or filed down on the high parts. It is important that the face should fit perfectly tight to the barrel, for if the slightest space be left, the powder gas would eat into it. On being screwed in finally it is sent well home by striking the lever with a handspike. The ring is then upset with the upsetting block; it is next bored out, the spindle is introduced through the breech-screw, there being two bearings in the breech-screw, one in front and one behind, and the knife is fixed through the spindle in the vent slot; the spindle is turned by a wrench and fed to its work by means of the

---

* As to 7-inch bush, vide p. 125.

† Each B.L. battery, in addition to spare articles connected with the breech-loading apparatus, is supplied with a box of facing implements, weighing 106 lbs., and containing 25 articles required for refacing the vent-pieces and bush rings, and with a set of special tools in two boxes, together weighing 83 lbs.

breech-screw. After boring, a different tool is fitted through the spindle, and the copper is faced to within ·03 of an inch of the face of the *A* tube, the cone part being left ·15 broad. CHAP. XII.

SCREWING IN BREECH BUSH COPPER.

(Diagram No. 1.)

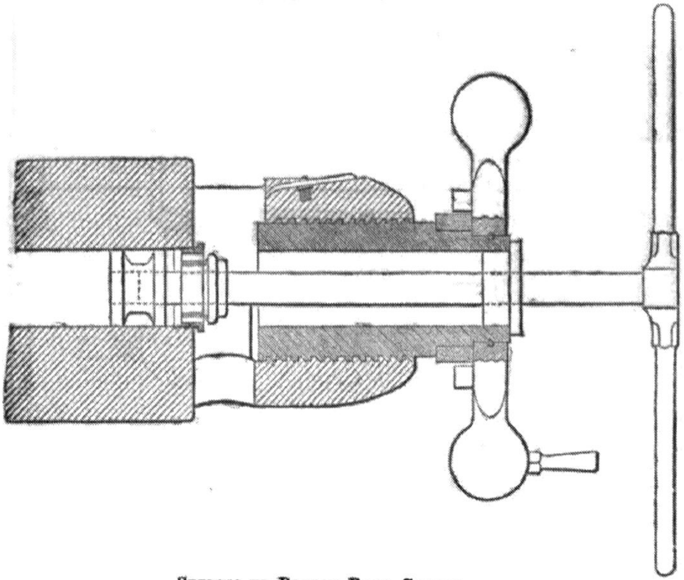

SETTING-UP BREECH BUSH COPPER.

(Diagram No. 2.)

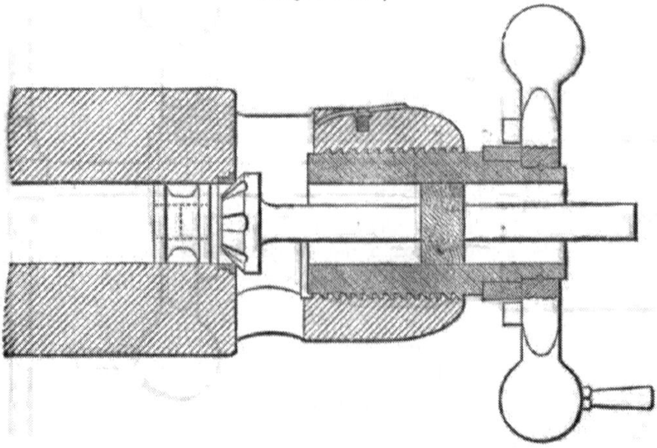

BORING AND FACING BREECH BUSH COPPER.
(Diagram No. 3.)

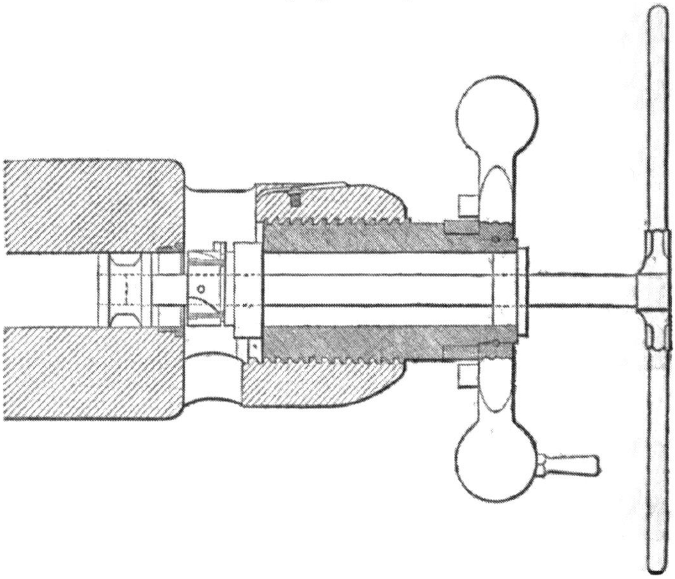

ANGLE-FACING BREECH BUSH COPPER.
(Diagram No. 4.)

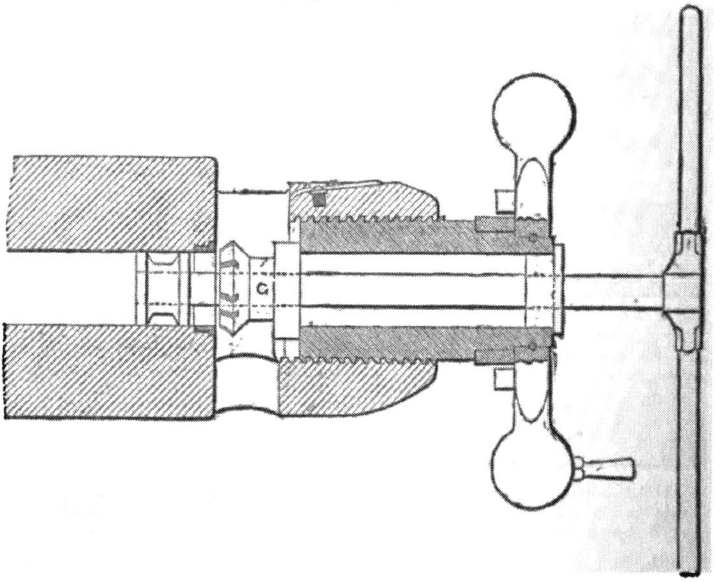

CHAP. XII.

Refacing vent-piece.

The vent-piece copper ring can be repeatedly refaced until the angle face and the back edges meet. After this it can be removed by striking it a few smart blows with a hammer on the cone face, when it is so expanded that it flies off. The new one is put on by hand, and the vent-piece having been placed in the gun front to the rear, the ring is forced on by screwing up the breech-screw. It is well to place one of the guide blocks in the face of the breech-screw to prevent its injuring the copper.

The operation of refacing is shown in the diagram below; the facing tool forming a conical face exactly fitting into the hollow made in breech brush.

ANGLE FACING VENT-PIECE RING. (Diagram No. 5.)

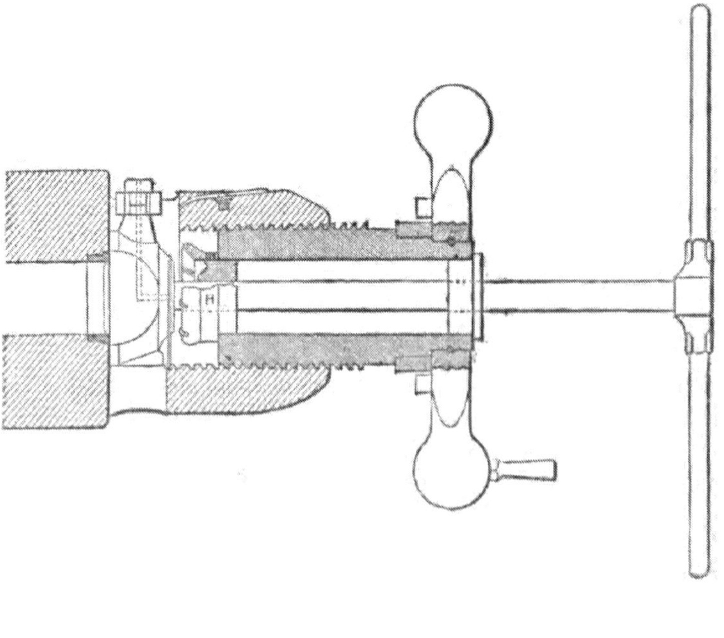

(b.) Re-venting a Vent-piece.

To re-vent R.B.L. in the field.

In boring out the hold bush a bearing can be obtained for the drill by lashing a handspike across the wheels and performing the operation with the vent-pice placed in the gun. Care must be taken to drill right down to the bottom of the copper before removing the screwed piece at the top, otherwise some difficulty will be found in removing the lower piece (or pieces). The bush is to be renewed from the spare ones issued for the purpose in the manner described at page 126.

(c.) *Adjusting New Trunnion Sights.*

Adjusting new trunnion sights.

Fore sights, whether on the trunnion ring or elsewhere, are liable to get damaged on service, and as the spare sights are issued with rough leaves, some simple mode of adjusting a new fore sight should be known.

Re-adjusting Trunnion Sights of B.L. Field Guns on Service without the aid of Sighting Instruments.

As sighting instruments are no longer issued to batteries and only to some stations, it may be found necessary to adjust damaged trunnion sights on service without them, this may be done in the following manner, viz., by making wooden copies of the proper instruments.

Requisite materials.

The requisite materials are :—

Two blocks or discs of wood to fit into breech and muzzle.

Two rectangular pieces of wood long enough to project beyond the tangent sights on each side, and of such width that the upper edge shall be level with the top of the right tangent sight, when the lower edge coincides with the line of horizontal axis marked on the breech and muzzle.

Four screws to fasten these pieces together.

Two silk or fine thread lines.

Operation.

The object of the following operation is to obtain a line on each side of the gun through the tangent sight notch parallel to the axis of the bore.

(1.) Remove the breech-screw and fasten the breech and muzzle discs in the gun securely by wedges.

(2.) Fasten the two rectangular pieces together and plane their lower edges *a b* (see Fig.) level, treating them as one piece. Then square up a line *c d* at the centre of each at right angles to *a b*, and cut away the wood on one side of *c*, in order that *c d* may be brought against the vertical line on the face of the breech or muzzle.

FOR 12-PR., BREECH-SCREW GUN.

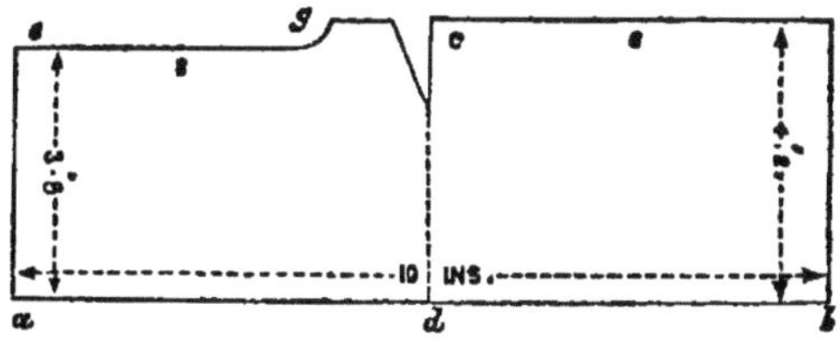

(3.) Now bring *a b* against the horizontal line, and *c d* to the vertical line marked on the breech, and fasten the boards by two screws to the disc in the breech of the gun.

(4.) Cut away the top until it is level with the *bottom* of the notch of (the tangent sight on each side; the left side *e g* will require to be cut away more than the right.

CHAP. XII.

(5.) Then with a rule brought by eye parallel to the axis of the gun, mark lines *s s* on the top of the boards opposite the tangent sight notches on each side; this can be done sufficiently accurately by eye, as any error is corrected in the after operations.

(6.) Remove the boards from the gun and separate them, fastening one at the breech and the other at the muzzle, taking care to adjust them to the horizontal and vertical lines on the gun.

(7.) Stretch silk lines on each side of the gun from the marks *s s* on the breech and muzzle board; they should pass exactly through the bottom of the notch of the tangent sights. If they do not they must be moved to the right or left until they do, but care must be taken always *to move the line the same amount both at breech and muzzle*, so as to keep it parallel to its original position. These lines are parallel to the axis of the gun.

Adjusting new trunnion sights.

(8.) The leaf is filed down to the right height, *i.e.*, the level of the silk line. In doing this it is usual to raise the silk both at breech and muzzle by the thickness of a piece of paper, when it should just clear the leaf, and when the paper is removed it should touch the same.

(9.) The position for the apex on the trunnion sight is then ascertained by the length of radius for each nature of gun (see table), measured from the notch on the tangent sight. The apex is not in the axis line of the sight, except in 6-prs. and 9-prs.

In 12-prs. it is	0″·05 to the rear.
„ 20 prs.	
„ 40-prs,	0″·2 „
„ 7-inch	

(10.) Remove the sight from the gun, place it in a vice, and file down the front and back slopes. The slopes should be sufficient to form a point when the gun is at its highest elevation.

(11.) Replace the sight in the gun and file down the lateral slopes, so that a true edge is obtained under the silk line, care being taken that the top of the leaf is not made too sharp, as it would be liable to injury. In the case of a screw trunnion sight, a curved line is marked by a scriber, to show the position of the metal surface of the trunnion when the sight is screwed home.

(12.) The trunnion sight is again removed and the back slope is roughed, so that in laying a gun there may not be too bright a reflection presented to the eye.

(d.) *R.M.L.—Replacing a broken Clamping Screw, 9-pr. L.S.*

Replacing broken clamping screw.

The mill-headed screw employed to clamp the tangent sight of the 9-pr. R.M.L. guns, Mark I. and II., L.S., is occasionally broken off near the head, and can be replaced as follows:—

Remove the screw (A) which secures the steel projecting patch (B), drive out the latter in the direction of the breech by means of a copper set.

With a piece of hard wood applied through the hole bored in the gun for the reception of the sight, drive out the gun metal sight socket.

Take out the broken clamping screw through the hole in the back of the socket, replace it by a new one, and put the socket back into its place, securing it there by reversing the foregoing operations.

CHAP. XII.

9-PR. WROUGHT IRON R.M.L. GUN OF 8 CWT.

Diagram illustrating the replacing of a broken Clamping Screw.
¼ full Scale.

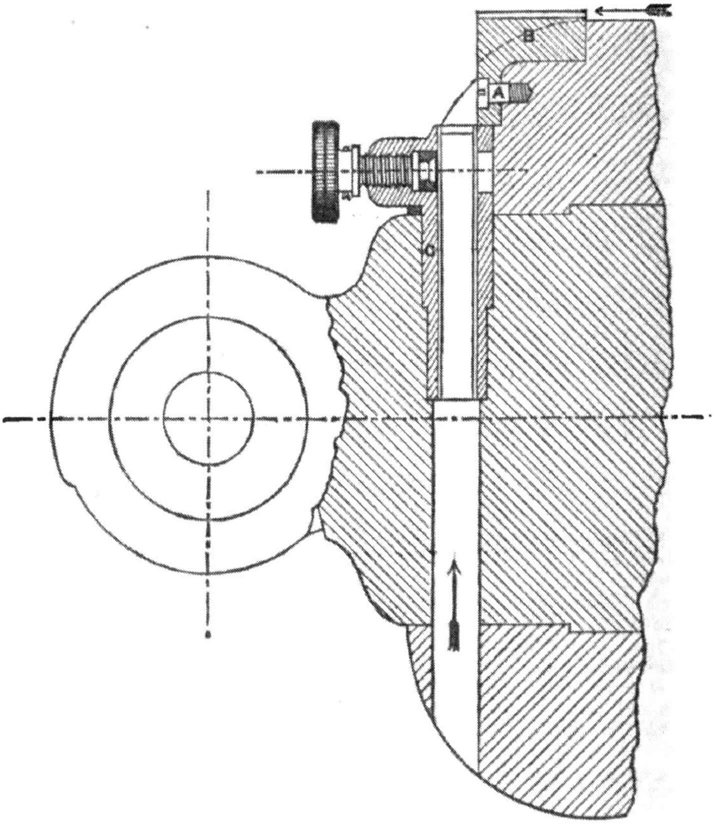

(e.) Adjusting New Foresight.

Adjusting R.M.L. fore-sights on service.

7-pr. or 9-pr.—Spare fore sights are issued rough, in case the sight in the gun should get broken, and a wrench is supplied for removing and replacing it. To fit the new one it will be necessary to level the gun longitudinally along the bore, and to screw the rough sight home into the dispart patch, bringing its leaf parallel to the axis. Then file the top down until it is level with the bottom of the notch on the tangent sight. This can be done by placing a piece of metal the thickness of the depth of notch on the top of the fore sight. Measure the radius also from the back of the tangent sight, and file the back and front slopes. With a rule or straight-edge mark the centre line on the top of the leaf opposite the line of vertical axis engraved on the dispart patch and muzzle of the gun, and file up the front side slopes to this line. Remove the sight from the gun, rough the back slope, clean

CHAP. XII.

and blue it (as described at p. 377), and replace it in the gun in the same position.

Adjusting a new trunnion sight for a 16-pr. R.M.L. gun.

16-pr.—To adjust a new screw trunnion sight for a 16-pr. R.M.L. gun, proceed as follows:—

Screw in the new sight with rough leaf, then place the gun upon level ground and lay with an undamaged sight upon some distant object (not less than 2,000 yards away), the tangent sight being at zero. File down the top of the leaf of the new sight until it is in a line with the bottom of the notch of its tangent sight and the object.

Mark the length of the radius from the back of the tangent sight, mark the muzzle direction on the sight leaf, then remove the sight and file up the front and back slopes.

Replace the sight and file down the side slopes until the gun is accurately laid on the same object by both sights.

Rough the back slope, polish and blue the sight, and so complete it.

N.B.—There is a slight error in this method, for the line of sight will cross at the distance of the point upon which the gun was laid.

The following is another method which may be adopted, but which is less accurate than the foregoing in some points and requires more time and material.

It is less accurate in some cases because the sights are not interchangeable in reality (as mentioned at p. 189), and they are here treated as though they were.

1. Place a block of wood in the muzzle of the gun, and on its face pivot a piece of wood, having an oblong slit cut in it through which a screw passes, forming the pivot, and also serving to tighten it against the disc in the bore.

2. Stretch a silk line from the bottom of the tangent sight notch across the top of the undamaged trunnion sight, and attach it to the piece of wood pivoted on the muzzle disc, which must then be screwed up tightly. The silk line will give us a line parallel to the axis of the piece, and passing just over the centre of the leaf of the fore sight.

3. Remove undamaged sight, and in its place screw that which is to be adjusted (with a rough leaf), file its leaf down until level with the silk line, mark the apex by measuring the length of radius, also muzzle direction of leaf, remove the sight and file down the front and back slopes, then replace it in the gun and file the side slopes, using the silk line as a guide.

4. Take out the sight so adjusted, and screw it into the hole on the other side of the gun, replacing it by the one originally removed.

(6.) *Venting Tools.*

There are two lists of these tools, vide pp. 322, 326.

One list embraces those required for converted guns and all heavier pieces than 64-prs.

The other list contains the required for 64-pr. R.M.L. and smaller guns and Howitzers.

For siege train chests containing the proper tools for venting are supplied for every 30 pieces of ordnance.*

---

* These are carried in three boxes, one of which contains spare vents.

CHAP. XII.

II.—Repairs which can be performed only at certain Stations or on board certain Ships.

(1.) *Adjusting Millar's sights of a S.B Gun.**

Adjusting sights, S.B. guns.

(1.) The gun having been carefully levelled laterally across the trunnions, and longitudinally in the bore, the fore and hind sights are to be adjusted so as to fulfil the following conditions:—

1st. They are to be the exact distance apart according to the short radius given in No. III. Table, p. 65.

2nd. When the scale is at zero, the line joining the top of hind and the top of foresight must be parallel to the axis.

3rd. When the scale is raised to the full elevation marked on it,—that is, the "clearance angle,"—the top of the scale, the apex of the fore sight, and the highest point on the muzzle of the gun, must be in the same line.

4th. The line of sight must be made to coincide exactly with the vertical plane as the line of metal.

An old fore sight is placed loosely on the gun over the second reinforce ring. This and the hind sight are adjusted by hand in the first place until the conditions are fulfilled.

The angular level is used to bring the scale to the angle of 76°. The position of the hind sight screw holes are then marked with a scriber on the gun through the holes previously drilled and punched through sight block and lead packing.

The holes are drilled by aid of the instrument called "machine drilling hind sight," and care must be taken to drill the holes perpendicular to the face of the sighting block, so that the heads of the screws may rest fairly on it; the holes are then tapped.

The hind sight being fixed, the scale is raised to a little more than the clearance angle, and a silk cord stretched from the notch on it to that on the muzzle. The real foresight is now adjusted so as to bring its top under the silk cord at the proper distance from the tangent sight.

The position of the fore sight screw holes being marked, they are drilled and tapped, and the sight screwed on. The head of the sight is then filed down to the proper height, and the position of the ridge being marked, it is unscrewed, and the side slopes filed down. When again screwed on, the sighting is tested to ascertain whether the whole of the conditions are fulfilled.

The sights and lead packing are then marked with the number of the gun to which they have been fitted, and the screws are also marked for their particular holes.

When guns are mounted, the sight screw holes are filled with preserving screws, but these are to be removed when the guns are shifted, as the screw heads are liable to be broken off; the holes are then filled with tallow and white lead composition.

(2.) *Re-venting S.B. and R.M.L. Guns.*

S.B. ordnance.

(*a.*) The operation for S.B. can be carried on, wherever the set of tools at p. 322 exist, and is performed by R.G.F. Artificers at home, and by Armstrong Armourers abroad. Vide Appendix, p. 372, for Regulations.

Heavy R.M.L. guns.

(*b.*) In the case of heavy R.M.L. guns the necessary tools are only supplied to the following stations, Woolwich, Malta, Gibraltar, Ber-

* The necessary implements for the performance of this adjustment are given in table, p. 320.

muda, Devonport,* Hong-Kong, and Esquimault, and also, to some men-of war where qualified engineers are present who have been especially instructed for the purpose.

Special tools for converted guns.

(*c.*) Special tools, p. 286, being required for reventing converted guns, this service can only be carried out at certain stations.

(*a* & *b.*) With regard to re-bushing a S.B. or R.M.L. gun proceed as follows :—

Re-venting with Cone Vent.

Re-venting with cone vent.

Proceed as if venting a gun for the first time, p. 60, using the narrow drill (B); when the hole is some four or five inches deep, turn up the gun and make the hole square with the graver's chisel; hammer into it one of the shortest of the square drifts, and try to unscrew the old bush with a wrench. Sometimes the whole of the copper vent will come out, but if the copper has been much set up in firing it will be much tighter and probably break; the drilling must then be repeated until the whole is removed.

When the old copper vent is taken away, ascertain, either by measuring it or taking an impression of the thread in the gun, whether it is a six or seven thread pitch; then use the corresponding set of taps commencing with the lowest number to clear the thread. If the wrong set of taps be used the damage will be irreparable.

The hole in the gun is next cleaned with tow, and the copper vent, well oiled, is screwed in with as much power as three men can bring to bear upon it. The head should not be wrenched off, as a fracture might occur below the surface of the metal of the gun.

The new bush when properly fixed will project about a quarter of an inch into the bore, and about two inches above the surface of the gun.

Take an impression of the part in the bore with wax composed of

Bees-wax..	2 parts	Boiled together.
Soft soap..	1 „	
Treacle	1 „	

This will show whether the cone is well home, and that that there is no space left between the copper and iron.

If the bush is home proceed to cut off the end in the bore. The instrument consists of a cutting tool supported by a metal head at the end of a long bar; the bar is kept in the axis of the bore by passing through a collar fitting into the muzzle, it is worked from side to side by two levers being fed up by a small screw at the end of the frame; the spiral spring against the muzzle collar makes the knife work regularly. Care must be taken not to cut into the iron of the gun. It is probable that the end of the bush will not be cut off quite flush at first, so another impression is taken, and, if necessary, the knife must be fed out with a small piece of tin, and the process repeated.

It is necessary that the copper and iron in the bore should be perfectly flush with each other, the action of the powder being much greater on anything projecting, however slightly, into the bore.

Saw off the outside end of the vent to within about a quarter of an inch of the patch; chip a little copper away from the mouth of the vent to prevent it becoming choked when hammered, chisel it also at the edges, then hammer it well, next chisel it off flush and open the mouth of the vent, and pass the set of rimers down one after the other, and gauge. File the surface, take another wax impression of the inside, and if all is right the operation is finished.

---

* To other stations at home, qualified artificers with the necessary tools are sent from time to time.

CHAP. XII.

Re-venting with Through Vents.

Re-venting with through vent.

This is precisely the same operation as venting with cone vents, except that the thread is carried right through to the bore. The chief precaution to be taken in venting with a through vent is not to make the hole too wide at the bottom, and not to break away the last thread.

Re-venting Converted Guns.

Re-venting converted guns.

Regarding (c.) for converted guns* proceed as above as to removing old bush.

A hole being drilled through the old bush, and then enlarged for a depth of about 2 inches, the corners are squared, a drift driven in, and the vent screwed out, it will eventually break away from the portion "set up" from the interior of the chamber, which can be driven down into the bore, and so taken out.

The thread having been cleaned out, the new bush is screwed in, set up and completed as described in the case of first venting such a gun, vide p. 236.

In case, however, of re-venting one of the guns mentioned in note, p. 236, which were originally vented with cone vents of exceptional diameter and thread, the old bush is removed as usual, but when that operation is completed the gun must be prepared for a new bush as follows:—

Drill out the hole to proper size for through bush of service pattern and tap the thread for the same as usual.

Then cut out the last thread of female thread in the cup by means of a "rose cutter," take an impression to see if this is properly removed, and then re-vent as already described.

(3.) *Re-bushing 7-inch R.B.L.*

Re-bushing 7" R.B.L. guns.

As the re-bushing of 7-inch R.B.L. guns is a more difficult operation than that of copper bushing the smaller natures, it would be performed only where the necessary tools are provided, which is at all large stations and sub-arsenals at home and abroad.

The set of facing implements is given at p. 321. The tools are of course much stronger than those in the smaller set.

(4.) *Adjustment of Sights.*

Adjusting sights and sight sockets.

Sights, sight sockets, &c. can also be adjusted and repaired at most of our large stations wherever there is a fire-master, as well at some others.

III.—Repairs to be performed in an Arsenal.

Important repairs.

All important repairs, such as re-tubing, re-sighting, &c., must be performed at Woolwich Arsenal.

To re-tube a built up R.M.L. gun, the usual practice is to cut the chase through in front of the trunnion coil or 1 B coil in heavy guns; to bore out the steel tube in breech portion, and to use the latter as a jacket for a new gun which would be otherwise built up as usual.

Re-tubing.

A converted gun can be prepared for re-tubing by removing the cast iron collar and screw plug, stopping up the vent after removing the bush, and then forcing out the tube by hydraulic pressure, the gas escape hole being enlarged for that purpose to admit the nozzle of the tube through which the water is forced.

---

* For tools required, vide table, p. 322.

TABLE XXVIII.

LIST of TOOLS required for the CLEANING and EXAMINATION PAINTING and LACQUERING of the various Natures of CAST IRON and BRONZE ORDNANCE.

—	Calibre. 10-in.	68-pr. and 8-in.	56-pr.	42-pr.	32-pr.	24-pr.	18-pr.	12-pr.	9-pr.	6-pr.	3-pr.
Bit, vent, 12-in.	a	a	a	a	a	a	a	a	a	a	a
Brace, armourer's	a	a	a	a	a	a	a	a	a	a	a
Brushes, hard, round, 10-in.	1	—	—	—	—	—	—	—	—	—	—
Brushes, hard, round, 68-pr. and 8-inch	—	1	—	—	—	—	—	—	—	—	—
Brushes, hard, round, 56 „	—	—	1	—	—	—	—	—	—	—	—
Brushes, hard, round, 42 „	—	—	—	1	—	—	—	—	—	—	—
Brushes, hard, round, 32 „	—	—	—	—	1	—	—	—	—	—	—
Brushes, hard, round, 24 „	—	—	—	—	—	1	—	—	—	—	—
Brushes, hard, round, 18 „	—	—	—	—	—	—	1	—	—	—	—
Brushes, hard, round, 12 „	—	—	—	—	—	—	—	1	—	—	—
Brushes, hard, round, 9 „	—	—	—	—	—	—	—	—	1	—	—
Brushes, hard, round, 6 „	—	—	—	—	—	—	—	—	—	1	nil†
Brushes, Paint, ground	—	—	—	—	—	—	—	—	—	—	—
Brushes, Paint, unground	—	—	—	—	—	—	—	—	—	—	—
Brushes, Soft or Turk's head, large	a	a	a	—	—	—	—	—	—	—	—
Brushes, Soft or Turk's head, small	—	—	—	a	a	a	a	a	a	a	a
Drill, vent, 15-in.	a	a	a	a	a	a	a	a	a	a	a
Gauges, vent, set of 4, short	a	a	a	a	a	a	a	a	a	a	a
Instrument, taking impressions of vents complete with blocks and pans	a	a	a	a	a	a	a	a	a	a	nil†
Knives, cleaning, ordnance, pairs, 10-in.	1	—	—	—	—	—	—	—	—	—	—
Knives, cleaning, ordnance, pairs, 68-pr. and 8-inch	—	1	—	—	—	—	—	—	—	—	—
Knives, cleaning, ordnance, pairs, 56 „	—	—	1	—	—	—	—	—	—	—	—
Knives, cleaning, ordnance, pairs, 42 „	—	—	—	1	—	—	—	—	—	—	—
Knives, cleaning, ordnance, pairs, 32 „	—	—	—	—	1	—	—	—	—	—	—
Knives, cleaning, ordnance, pairs, 24 „	—	—	—	—	—	1	—	—	—	—	—
Knives, cleaning, ordnance, pairs, 18 „	—	—	—	—	—	—	1	—	—	—	—
Knives, cleaning, ordnance, pairs, 12 „	—	—	—	—	—	—	—	1	—	—	—
Knives, cleaning, ordnance, pairs, 9 „	—	—	—	—	—	—	—	—	1	—	—
Knives, cleaning, ordnance, pairs, 6 „	—	—	—	—	—	—	—	—	—	1	—
Knives, cleaning, ordnance, pairs, 3 „	—	—	—	—	—	—	—	—	—	—	1
Lamp, tin, with rod	a	a	a	a	a	a	a	a	a	a	a
Pot, for softening gutta-percha	a	a	a	a	a	a	a	a	a	a	a
Prickers, with staves, common, with pan	a	a	a	a	a	a	a	a	a	a	a
Prickers, with staves, common, without pan	a	a	a	a	a	a	a	a	a	a	a
Rimer, vent, 15-in.	a	a	a	a	a	a	a	a	a	a	a
Scale, diagonal, brass	a	a	a	a	a	a	a	a	a	a	a
Scrapers, balloon, large	1‡	—	—	—	—	—	—	—	—	—	—
Scrapers, balloon, medium	—	1‡	—	—	—	—	—	—	—	—	—
Scrapers, balloon, small	—	—	—	—	1‡	1‡	—	1‡	—	—	—
Scrapers, fore-right	a	a	a	a	a	a	a	a	a	a	a
Scrapers, half round	a	a	a	a	a	a	a	a	a	a	a
Scrapers, vent, half-round	a	a	a	a	a	a	a	a	a	a	a
Scrapers, vent, spring	a	a	a	a	a	a	a	a	a	a	a
Searchers, spring, 8-prong	a	a	a	a	a	—	—	—	—	—	—
Searchers, spring, 6 „	—	—	—	—	—	a	a	—	—	—	—
Searchers, spring, 4 „	—	—	—	—	—	—	—	a	a	a	•
Springs for knives, 10-in., 8-in., and 56-pr.	a	a	a	—	—	—	—	—	—	—	—
Springs for knives, 42 and 32-pr.	—	—	—	a	a	—	—	—	—	—	—
Springs for knives, 24 „ 18 „	—	—	—	—	—	a	a	—	—	—	—
Springs for knives, 12 „ 9 „	—	—	—	—	—	—	—	a	a	—	—
Springs for knives, 6-pr.	—	—	—	—	—	—	—	—	—	1	—
Springs for knives, 3 „	—	—	—	—	—	—	—	—	—	—	1
Swords, old	a	a	a	a	a	a	a	a	a	a	a
Tongs for grinding knives	a	a	a	a	a	a	a	a	a	a	a
Tool for grinding searcher points	a	a	a	a	a	a	a	a	a	a	a
Wire, lacquering vents	a	a	a	a	a	a	a	a	a	a	a
Wrench for knives	a	a	a	a	a	a	a	a	a	a	a
Sockets, spring with staves§, No. 1 for end of bore	a	a	a	a	a	a	a	a	a	a	a
Sockets, spring with staves§, No. 2 „ „	a	a	a	a	a	a	a	a	a	a	a

a Common to these.

† There is no hard brush for the 3-pr., the small Turk's head is used instead; and the instrument for taking impressions is too large for use with this gun, the pricker with pan, or a piece of wood being used; the smallest spring searcher is also to large, the common pricker taking its place.

‡ Only required for shell guns, howitzers, and mortars, having gomer chambers.

Note.—13-inch mortars are cleaned with the same articles as 10-inch, except that the bore is scraped with old swords, instead of with the spring knives.

§ Required only for painting or lacquering.

TABLE XXIX.—List of Tools required for the Cleaning and Examination, Painting and Lacquering, of the various Natures of Rifled Ordnance.

	Muzzle-loading.														Breech-loading.						Remarks.
	16-in.	12·5-in.	12-in.	11-in.	10-in.	9-in.	8-in. and Howitzer	7-in.	80-pr. 64-pr. and 6″·3 Howitzer.	40-pr.	25-pr.	16-pr.	9-pr.	7-pr.	7-in.	40-pr.	20-pr.	12-pr.	9-pr.	6-pr.	
Bits, vent ... Armstrong ...	—	—	—	—	—	—	—	—	—	—	—	—	—	—	a	a	a	a	a	a	
Bits, vent ... 38-in. ...	a	a	a	a	a	a	a	a	—	—	—	—	—	—	—	—	—	—	—	—	
Bits, vent ... 12 ,, ...	—	—	—	—	—	—	—	—	a	a	a	a	a	a	—	—	—	—	—	—	
Brace, armourers' ... 1	—	—	—	—	—	—	—	—	—	—	—	—	—	—	—	—	—	—	—	—	
Brushes, Gun, piassava with staves. 16·0-in. ...	1	—	—	—	—	—	—	—	—	—	—	—	—	—	—	—	—	—	—	—	
12·5-in. ...	—	1	—	—	—	—	—	—	—	—	—	—	—	—	—	—	—	—	—	—	
12 ,, ...	—	—	1	—	—	—	—	—	—	—	—	—	—	—	—	—	—	—	—	—	
11 ,, ...	—	—	—	1	—	—	—	—	—	—	—	—	—	—	—	—	—	—	—	—	
10 ,, ...	—	—	—	—	1	—	—	—	—	—	—	—	—	—	—	—	—	—	—	—	
9 ,, ...	—	—	—	—	—	1	—	—	—	—	—	—	—	—	—	—	—	—	—	—	
8·0 in. and Howitzer ...	—	—	—	—	—	—	1 †	—	—	—	—	—	—	—	—	—	—	—	—	—	
7 ,, ...	—	—	—	—	—	—	—	a	—	—	—	—	—	—	a	—	—	—	—	—	† The Staves for Howitzers are shorter.
80-pr., 64 pr. and 6″·3 Howitzer	—	—	—	—	—	—	—	—	1 †	—	—	—	—	—	—	—	—	—	—	—	
40 ,, ...	—	—	—	—	—	—	—	—	—	a	—	—	—	—	—	a	—	—	—	—	
25 or 20 ...	—	—	—	—	—	—	—	—	—	—	a	—	—	—	—	—	—	—	—	—	
16 ,, ...	—	—	—	—	—	—	—	—	—	—	—	1	—	—	—	—	—	—	—	—	
12, 9, and 7-pr. ...	—	—	—	—	—	—	—	—	—	—	—	—	a	a	—	—	—	a	a	—	
6-pr. ...	—	—	—	—	—	—	—	—	—	—	—	—	—	—	—	—	—	—	—	1	
Brushes, Paint ground 000 ...	—	—	—	—	—	—	—	—	—	—	—	—	—	—	—	—	—	—	—	—	
Brushes, Paint unground 000 ...	—	—	—	—	—	—	—	—	—	—	—	—	—	—	—	—	—	—	—	—	
Compasses 1	—	—	—	—	—	—	—	—	—	—	—	—	—	—	—	—	—	—	—	—	
Gauges, vent, set of 4 ... long ...	a	a	a	a	a	a	a	a	—	—	—	—	—	—	—	—	—	—	—	—	
Gauges, vent, set of 4 ... short ...	—	—	—	—	—	—	—	—	a	a	a	a	a	a	—	—	—	—	—	—	
Instruments, taking impressions of bores No. 1	—	—	—	—	—	—	—	—	—	—	—	—	—	—	—	—	—	—	—	—	For 40-pr. guns and upwards.
Instruments, taking impressions of bores No. 2	—	—	—	—	—	—	—	—	—	—	—	—	—	—	—	—	—	—	—	—	For guns below the 40-pr.
Instruments, taking impressions of vents complete with blocks and pans A large ...	—	—	—	—	—	—	—	—	—	—	—	—	—	—	—	—	—	—	—	—	
B small ...	—	—	—	—	—	—	—	—	—	—	—	—	—	—	—	—	—	—	—	—	
C converted ...	—	—	—	—	—	—	—	—	—	—	—	—	—	—	—	—	—	—	—	—	
Lamp, tin, with rod ... 1	—	—	—	—	—	—	—	—	—	—	—	—	—	—	—	—	—	—	—	—	

a Common to these.

TABLE XXIX.—List of Tools, &c.—*continued.*

—	Muzzle-loading.															Breech-loading.						Remarks.
	16-in.	12·5-in.	12-in.	11-in.	10-in.	9-in.	8-in. and Ho-witzer.	7-in.	80-pr. & 64-pr. convd.	64-pr. and 6″·3 Ho-witzer.	40-pr.	25-pr.	16-pr.	9-pr.	7-pr.	7-in.	40-pr.	20-pr.	12-pr.	9-pr.	6-pr.	
Plates for instruments taking impressions of bores. 16 in.	1	—	—	—	—	—	—	—	—	—	—	—	—	—	—	—	—	—	—	—	—	
12·5-in.	—	1	—	—	—	—	—	—	—	—	—	—	—	—	—	—	—	—	—	—	—	
12 „	—	—	1	—	—	—	—	—	—	—	—	—	—	—	—	—	—	—	—	—	—	
11 „	—	—	—	1	—	—	—	—	—	—	—	—	—	—	—	—	—	—	—	—	—	
10 „	—	—	—	—	1	—	—	—	—	—	—	—	—	—	—	—	—	—	—	—	—	
9 „	—	—	—	—	—	1	—	—	—	—	—	—	—	—	—	—	—	—	—	—	—	
8 „ and Howitzer	—	—	—	—	—	—	1	—	—	—	—	—	—	—	—	—	—	—	—	—	—	
7 „	—	—	—	—	—	—	—	1	—	—	—	—	—	—	—	1	—	—	—	—	—	
80-pr. and 64-pr. convd.	—	—	—	—	—	—	—	—	1	—	—	—	—	—	—	—	—	—	—	—	—	
64-pr. and 6″·3 Howitzer ...	—	—	—	—	—	—	—	—	—	1	—	—	—	—	—	—	—	—	—	—	—	
25, 20, or 16-pr. ...	—	—	—	—	—	—	—	—	—	—	—	*a*	*a*	—	—	—	—	*a*	—	—	—	Plates are not required for the 40-pr. or 6-pr., as the frames answer the purpose.
12, 9, and 7-pr. ...	—	—	—	—	—	—	—	—	—	—	—	—	—	*a*	*a*	—	—	—	*a*	*a*	—	
Pot, softening gutta percha 1	—	—	—	—	—	—	—	—	—	—	—	—	—	—	—	—	—	—	—	—	—	
Pricker, common, with pan 1	—	—	—	—	—	—	—	—	—	—	—	—	—	—	—	—	—	—	—	—	—	
Pricker (with blunted point) 1	—	—	—	—	—	—	—	—	—	—	—	—	—	—	—	—	—	—	—	—	—	
Rimers, vent 22-in. ...	*a*	*a*	*a*	*a*	*a*	*a*	*a*	—	—	—	—	—	—	—	—	—	—	—	—	—	—	
Rimers, vent 15 „ ...	—	—	—	—	—	—		*a*	*a*	*a*	*a*	*a*	*a*	*a*	*a*	*a*	*a*	*a*	*a*	*a*	*a*	
Scales, diagonal, brass, 12-in 1	—	—	—	—	—	—	—	—	—	—	—	—	—	—	—	—	—	—	—	—	—	
Scrapers, vent, half-round... 24-in. 1	—	—	—	—	—	—	—	—	—	—	—	—	—	—	—	—	—	—	—	—	—	
Scrapers, vent, half-round... 14 „ 1	—	—	—	—	—	—	—	—	—	—	—	—	—	—	—	—	—	—	—	—	—	
Socket spring with stave, deal, for holding brushes No. 1 for end of bore	—	—	—	—	—	—	—	—	—	—	—	—	—	—	—	—	—	—	—	—	—	
No. 2 40-pr. to 10-in. R.M.L.	—	—	—	—	—	—	—	—	—	—	—	—	—	—	—	—	—	—	—	—	—	
No. 2A 11-in. R.M.L. and upwards... ...	—	—	—	—	—	—	—	—	—	—	—	—	—	—	—	—	—	—	—	—	—	
No. 3 25-pr. R.M.L. and below	—	—	—	—	—	—	—	—	—	—	—	—	—	—	—	—	—	—	—	—	—	
Straight-edge, testing breech-screws ... 1	—	—	—	—	—	—	—	—	—	—	—	—	—	—	—	*a*	*a*	*a*	*a*	*a*	*a*	According to requirements.
Swords, old	—	—	—	—	—	—	—	—	—	—	—	—	—	—	—	—	—	—	—	—	—	

* For painting and lacquering only; *a*, common to these.

CHAP. XII.

TABLE XXX.

SET of SIGHTING TOOLS for SMOOTH BORE ORDNANCE.

Articles.			No. to a set.	Remarks.
Battens, wood	inside	large	1	
		small	1	
	outside	large	1	
		small	1	
Braces	armourers		1	
	drilling, hand	medium	1	
		small	1	
Cases, wood, spirit level			1	
Chisels	hand, flat		2	
	graver's		6	
Drills	carronade fixing screws		3	3 sizes.
	gun fixing screws		3	Do.
Hammers, hand, small			1	
Levels	spirit	angular	1	
		with bar	1	
	steel for battens		1	
	wood for trunnions	large	1	
		small	1	
Machine, drilling, with chains complete		fore sight	1	
		hind sight	1	
Plummets	brass		2	
	lead		1	
Punches, steel	centre		1	
	lead		1	
Scribers			6	
T-square, steel			1	
Taps	carronade fixing screws		3	3 sizes.
	gun fixing screws		3	Do.
V-set			1	
Wrenches	socket		1	
	tap		1	

TABLE XXXI.

Detailed List of Facing Implements for Rifled B.L. (Breech Screw) Guns.

6-prs.

In a box, with hinges and hasps, and padlock with two keys.

	Letter.	No. of each.
Blocks, breech bush copper, angle facing	L	1
Blocks, breech bush copper, finish boring	K	1
Blocks, breech bush copper, screwing in	H	1
Blocks, breech bush copper, upsetting	I	1
Blocks, vent-piece ring, angle facing	N	1
Guard, wood, for vent-piece	Q	1
Guides, in breech-screws	D	2
Guides, in powder chamber	E	1
Guides, wood (block upsetting), two parts	O	1
Key, for fixing knives	F	1
Knives, breech bush copper, cutting out	G	1
Knives, breech bush copper, facing	M & M1	2
Knives, breech bush copper, rough boring	J	1
Lever	B	1
Punch, for pin in spindle	R	1
Spanner, for stop washers	P	1
Spindle	A	1
Washers, stop	C	2

12 & 9-prs.

In a box, with hinges and hasps, and padlock with two keys.

	Letter.	No. of each.
Blocks, breech bush copper, angle facing	I	1
Blocks, breech bush copper, finish boring, 3·125 diameter	F1	1
Blocks, breech bush copper, finish boring, 3·2 diameter	F2	1
Blocks, breech bush copper, screwing in	D	1
Blocks, breech bush copper, upsetting	C	1
Blocks, vent-piece ring, angle facing	J	1
Guard, wood, for vent piece	S	1
Guides, in breech screw	M	2
Guides, in powder chamber, 3·125 diameter	N1	1
Guides, in powder chamber, 3·2 diameter	N2	1
Guides, wood (block upsetting), two parts	R	1
Key, for fixing knives	O	1
Knives, breech bush copper, cutting out	K	1
Knives, breech bush copper, facing	G & G1	2
Knives, breech bush copper, rough boring	E	1
Lever	B	1
Punch, for pin in spindle	P	1
Spanner, for stop washers	Q	1
Spindle	A	1
Washers, stop	L	2

20-prs.

In a box, with hinges and hasps, and padlock with two keys.

	Letter.	No of each.
Blocks, breech bush copper, angle facing	K	1
Blocks, breech bush copper, finish boring, 3·875 diameter	I 1	1
Blocks, breech bush copper, finish boring, 3·94 diameter	I 2	1
Blocks, breech bush copper, screwing in	G	1
Blocks, breech bush copper, upsetting	M	1
Blocks, vent-piece ring, angle facing	L	1
Guard, wood, for vent-piece	P	1
Guides, in breech-screw	C	2
Guides, in powder chamber, 3·875 diameter	D1	1
Guides, in powder chamber, 3·94 diameter	D2	1
Guides, wood (block upsetting), two parts	Q	1
Knives, breech bush copper, cutting out	F	1
Knives, breech bush copper, facing	J & J1	2
Knives, breech bush copper, rough boring	H	1
Key, for fixing knives	N	1
Lever	B	1
Punch, for pin in spindle	R	1
Spanner, for stop washers	O	1
Spindle	A	1
Washers, stop	E	2

40-prs. (*See* Note.)

In a box, with hinges and hasps, and padlock with two keys.

	Letter.	No. of each.
Blocks, breech bush copper, angle facing	G	1
Blocks, breech bush copper, cutting out	D	1
Blocks, breech bush copper, finish boring, 4·91 diameter	F1	1
Blocks, breech bush copper, finish boring, 4·96 diameter	F2	1
Blocks, breech bush copper, screwing in	E	1
Blocks, breech bush copper, upsetting	M	2
Blocks, vent-piece ring, angle facing	H	1
Collar, for feed motion	B	1
Guides, in breech-screw	A1 A2	1 1*
Guides, expanding, in powder chamber	C	1*
Guides, wood (block upsetting), two parts	N	1
Handle to hold blocks in vent chamber	P	1
Lever	J	1
Punch, for knives in blocks	K	1
Punch, for pin in spindle	L	1
Spanner, for stop washers	R	1*
Spindle	I	1
Washers, feed	O	1*
Washers, stop	L	1*

7-inch.

In a box, with tray hinges and hasps, and padlock with two keys.

	Letter.	No. of each.
Blocks, breech bush iron, cutting out, boring, facing, tapering, and angle facing.	G	1
Blocks, breech bush iron, screwing in, and rough boring.	F1	1
Blocks, vent-piece, angle facing	K	1
Guard, wood, for vent-piece	O	1
Guides, in breech-screw	C	2
Guides, expanding in powder chamber	D	1
Handle to hold blocks in vent chamber	P	1
Knives cutting out, breech bush, thick iron	E2	2
Knives cutting out, breech bush, thin iron	E1	2
Lever	B	1
Punch, for pin in spindle	N	1
Spanner, for stop washers	M	1
Spindle	A	1
Washers, stop	J	2

Note.—This set of 40-pr. Facing Implements is the new pattern; the old pattern, which is not to be considered obsolete, consists of the same articles, with the omission of those marked *, and the insertion instead of the "Guide expanding," Guides in powder chamber {4·91 diameter—C 1—1. / 4.96 diameter—C —1.} See "Changes in Patterns," § 1072, No. 5.

CHAP. XII.

TABLE XXXII.

List of Venting Tools for R.M.L. and S.B. Guns.

Nature of Article.			For 64-pr. W.I. guns and downwards, and 6″·3 and 8″ Howitzers.	For all cast iron ordnance converted guns, 7″ W.I. and upwards.
Bar iron with spring and metal collar		large, Mark I.		1
		small, in 2 pieces Mark II.	1	
Blocks wood for instruments taking impression of vents	guns, cast iron	18-pr.		1
		24 „		1
		32 „		1
		68 „		1
		8″ and mortar		1
		10″ and mortar		1
		13″ mortar		1
	guns, wrought iron	25-pr.	1	
		40 „	1	
		64 „ Mk. I.	1	
		64 „ „ II. and III.	1	
		7″		1
		8″		1
		9″		1
		10″		1
		11″		1
		12″		1
		12″·5		1
	converted	64-pr.		1
		80 „		1
	Howitzers W.I. iron.	6″·3	1	
		8″	1	
Blocks	metal for gun, W.I. 16-pr.		1	
	expanding pairs converted guns	64-pr.		1
		80 „		1
	stop for vent converted guns	64 „		1
		80 „		1
	4″ treble and double with fall converted guns	64 „		1
		80 „		1
Braces	armourers		1	1
	drilling, large			1
	ratchet, 12-inch		1	
Brush, gun, soft, small		with stave in 1 length		1
		„ „ 2 lengths	1	
Burnishers, spare (set of three)				1
Chests	for tools venting siege train ordnance.	No. 1* Mark I.		
		No. 2† Mark II.		
	for vents for do.	No. 3‡ Mark III.		
Chisels	hand, flat, set of five		1	1
	„ gauge		1	1
	„ gravers, set of three		1	1

		cwts.	qrs.					
*	Weight about	3	2	when filled.	Dimensions	5′ 7″	× 1′ 5″	× 1′ 2.″
†	„	2	2	„	„	3′ 8″	× 1′ 7″	× 1′ 3.″
‡	„	1	2	„	„	1′ 5″	× 1′ 1″	× 1′ 2.″

TABLE XXXII.—*continued.*

NATURE OF ARTICLE.	For 64-pr. W.I. guns and downwards, and 6"·3 and 8" Howitzers.	For all cast iron ordnance converted guns, 7" W.I. and upwards.
*Collars, polystepped for muzzles of ordnance — No. 1, 9-in. to 13-in. R.M.L. ..		1
*Collars, polystepped for muzzles of ordnance — No. 2, { 12-pr. S.B. to 68-pr. / 64-pr. to 8-in. R.M.L. }		1
*Collars, polystepped for muzzles of ordnance — No. 3, 7-pr. to 8-in. for M.L. bar ..	1	
Cutters for metal heads { cast iron ordnance } 12-pr..		1
18 „		1
24 „ set of 2		1
32 „ „ 7		1
68 „		1
8" and mortar		1
10" „ „		1
13" mortar		1
Cutters for metal head { wrought iron ordnance } 7-pr. { 224 lb. and 200 lb.	1	
7-pr. 150 lb.	1	
9 „	1	
16 „	1	
25 „	1	
40 „	1	
64 „ and 6"·3 Howitzer	1	
8" Howitzer..	1	
7" gun		1
8" „		1
9" „		1
10" „		1
11" „		1
12" „		1
12"·5 gun		1
Cutter, rose, for 64-pr. and 80-pr. converted guns ..		1
Cylinder, guide for converted guns		1
Drifts for taking out old vents — ¼"	1	
Drifts for taking out old vents — 6", 8¼", 10¼" } set	1	
Drifts for taking out old vents — F, F_1, F_2, F_3, F_4 } set		1
Drills, cone — ¼"	1	
Drills, cone — 1¼"	1	
Drills, cone — G I., G_1 I. } set		1
Drills for venting, drifting — ¼"	1	
Drills for venting, drifting — 4", 10" } set	1	
Drills for venting, drifting — E, E_1, E_2, E_3, E_4 } set		1

* To replace collars, cast-iron, when the latter are worn out.

CHAP. XII

TABLE XXXII.—*continued.*

Nature of Article.	For 64-pr. W.I. guns and downwards, and 6″·3 and 8″ Howitzers.	For all cast iron ordance converted guns, 7″ W.I. and upwards.
Drills for venting, tapping — ⅜″	1	
Drills for venting, tapping — 4″, 10″ set	1	
Drills for venting, tapping — B, B_1, B_2, B_3, B_4 set		
Drills, hollow — ⅜″	1	
Drills, hollow — 1⅛″	1	
Drills, hollow — A, A_1 set		1
File, bastard (with handle) flat, 14″ safe edge	1	1
Frame, iron, complete for cutting off ends of vents with adjusting bars and extenders, long, set		1
Frame, iron, complete for cutting off ends of vents with adjusting bars and no extenders, short, set	1	
Frame, trunnion, with cross head complete for converted guns (4 pieces)		1
Gauge, vent ·22 — short	1	1
Gauge, vent ·22 — long		1
Gyn, iron, triangle, 10 feet high, for converted guns		1
Hammers — hand — large		1
Hammers — hand — small	1	1
Hammers — monkey for converted guns		1
Heads metal for cutters — cast iron guns — 12-pr.		1
18 „		1
24 „		1
32 „		1
68 „		1
8″ and mortar		1
10″ „ „		1
13″ mortar		1
Heads metal for cutters — wrought iron guns — 7 and 9 pr.	1	
16-pr.	1	
25 „	1	
40 „	1	
64 „ and 6″·3 Howitzer	1	
8″ Howitzer		1
7″ gun		1
8″ „		1
9″ „		1
10″ „		1
11″ „		1
12″ „		1
12″·5 gun		1
Hook, extracting, expanding block, converted guns		
Instrument for taking impressions of vent with stave — wood, Mark I.		1
Instrument for taking impressions of vent with stave — iron in 2 lengths, Mark II.	1	
Levers, iron, for working bar — large set of 2		1
Levers, iron, for working bar — small „ „ 2	1	
Machine, drilling, complete with 2 lengths of chain, 7 ft. each length — large		1
Machine, drilling, complete with 2 lengths of chain, 7 ft. each length — small	1	1

TABLE XXXII.—*continued.*

NATURE OF ARTICLE.	For 64-pr. W.I. guns and downwards, and 6″·3 and 8″ Howitzers.	For all cast iron ordnance converted guns, 7″ W.I. and upwards.
Pan instrument for taking impression of vents — A large		1
Pan instrument for taking impression of vents — B small for 64-pr. R.M.L. and downwards	1	
Pan instrument for taking impression of vents — C 64-pr. and 80-pr. converted		1
Piece, lengthening — for bar iron, Mark I, for 9″ to 12″·5 R.M.L. guns		
Piece, lengthening — for instrument taking impressions of vents, Mark I, for 9″ to 12″·5 R.M.L. guns ..		
Rimers — cone — ⅝″	1	
Rimers — cone — 1½″	1	
Rimers — cone — H, H_1 set		1
Rimers — mouth of vent	1	1
Rimers — vent — short and medium	1	
Rimers — vent — medium and long		1
Saw, with frame 14-inch	1	1
Spanner — for nuts on frame	1	
Spanner — for countersink converted guns		1
Stop, iron, for metal head for cutters — 40-pr...		
Stop, iron, for metal head for cutters — 63 „		
Taps — new, thread — short — ⅝″ fine set 1, 2, and 3	1	
Taps — new, thread — short — 1⅛″ set 1, 2, and 3	1	
Taps — new, thread — short — C_1, C_2, C_3, C_4, C_5 set		1
Taps — new, thread — long — C_3, C_4, C_5 set		1
Taps — old thread — D_3, D_4, D_5 set		1
Wedge, expanding blocks converted guns		1
Wire, directing — small	1	
Wire, directing — large		1
Wrenches for — countersink converted guns		1
Wrenches for — frame adjusting		1
Wrenches for — socket — large		1
Wrenches for — socket — small	1	
Wrenches for — taps		1
Wrenches for — taps and vents — large	1	
Wrenches for — taps and vents — small, 7-pr.	1	
Wrenches for — vents		1

Table XXXIII.

Distribution of Venting Tools for R.M.L. and S.B. Guns, showing what are required for each Nature.

	Ordnance Muzzle Loading.																										Remarks.
	Cast Iron.								Wrought Iron.							Converted.		Howitzers.		Wrought iron.							
	10" and 10" Mortar.	8" and 8" Mortar.	68-pr.	32-pr.	24-pr.	18-pr.	12-pr.	13" Mortar.	12·5-inch.	12-inch.	11-inch.	10-inch.	9-inch.	8-inch.	7-inch.	80-pr.	64-pr.	8-inch.	6·3-inch.	64-pr.	40-pr.	25-pr.	16-pr.	9-pr.	7-pr.	
Bar iron with spring and metal collar { large ... Mk. I. ...	*a*	*a*	*a*	*a*	*a*	*a*	*a*	*a*	nil	*a*	*a*	*a*	*a*	*a*	*a*	...	...	...	...	...	...	...	...	...	...	
Bar iron with spring and metal collar { small, in 2 pieces, Mk. II.	...	...	...	...	...	...	...	...	...	...	...	...	...	...	...	...	...	*a*	*a*	*a*	*a*	*a*	*a*	*a*	*a*	
Blocks { wood for instrument, taking impressions of vents	1	1	1	1	1	1	nil	1	1	1	1	1	1	1	1	1	1	1	1	2*	1	1	...	nil	nil	* 1 block for Mk. I. 1 for Mk. II. & III.
Blocks { metal for instrument, taking impressions of vent	...	...	...	...	...	...	...	...	...	...	...	...	...	...	...	...	...	...	...	...	...	...	1	...	...	
Blocks { expanding pairs, converted guns	...	...	...	...	...	...	...	...	...	...	...	...	...	...	...	1	1	...	...	...	...	...	...	...	...	
Blocks { stop for vent, converted guns	...	...	...	...	...	...	...	...	...	...	...	...	...	...	...	1	1	...	...	...	...	...	...	...	...	
Blocks { 4" double and treble with fall	...	...	...	...	...	...	...	...	...	...	...	...	...	...	...	*a*	*a*	...	...	...	...	...	...	...	...	
Braces { armourers 1	*a*	*a*	*a*	*a*	*a*	*a*	*a*	*a*	*a*	*a*	*a*	*a*	*a*	*a*	*a*	*a*	*a*	*a*	*a*	*a*	*a*	*a*	*a*	*a*	*a*	
Braces { drilling, large 1	*a*	*a*	*a*	*a*	*a*	*a*	*a*	*a*	*a*	*a*	*a*	*a*	*a*	*a*	*a*	*a*	*a*	...	...	...	...	...	...	...	...	
Braces { ratchet, 12-inch 1	...	...	...	...	...	...	...	...	...	...	...	...	...	...	...	...	...	*a*	*a*	*a*	*a*	*a*	*a*	*a*	*a*	
Brush gun, soft, small { with stave in 1 length ...	*a*	*a*	*a*	*a*	*a*	*a*	*a*	*a*	*a*	*a*	*a*	*a*	*a*	*a*	*a*	*a*	*a*	...	...	...	...	...	...	...	...	
Brush gun, soft, small { " " 2 lengths ...	...	...	...	...	...	...	...	...	...	...	...	...	...	...	...	...	...	*a*	*a*	*a*	*a*	*a*	*a*	*a*	*a*	
Burnishers, spare... ... ... set of 3	*a*	*a*	*a*	*a*	*a*	*a*	*a*	*a*	...	...	...	...	...	...	...	...	...	...	...	...	...	...	...	...	...	
Chests { for tools, venting No. 1																										
Chests { siege train ordnance No. 2																										
Chests { for vents for " No. 3																										
Chisels { hand, flat, set of five																										
Chisels { hand, gauge...	*a*	*a*	*a*	*a*	*a*	*a*	*a*	*a*	*a*	*a*	*a*	*a*	*a*	*a*	*a*	*a*	*a*	*a*	*a*	*a*	*a*	*a*	*a*	*a*	*a*	
Chisels { hand, gravers, set of three ...																										
Collars, poly-stepped, for muzzles of ordnance { No. 1, 9" to 13".																										
Collars, poly-stepped, for muzzles of ordnance { No. 2 { 12-pr. S.B. to 68 pr. / 64-pr. R.M.L. to 8".																										
Collars, poly-stepped, for muzzles of ordnance { No. 3, 7-pr. to 8-in. R.M.L. (for M. II bar.)																										
Cutters, for metal head	1	1	1	[illegible]	2	1	1	1		1	1	1	1	1	1	nil	nil	1	1*	1*	1	1	1	1	2	* Interchangeable.

Cutters, rose converted gun ...	...	...	...	...	...	...	...	...	...	...	...	...	...	...	...	...	...	a	a	...	...	...	...	...	...	...	...
Cylinder guide, converted guns	...	...	...	...	...	...	...	...	...	...	...	...	...	...	...	...	...	a	a	...	...	...	...	...	...	...	...
Drifts for taking out old vents ¼ ...	...	...	...	...	...	...	...	...	...	...	...	...	...	...	...	...	...	...	...	...	...	...	...	...	...	...	1
Drifts for taking out old vents 6", 8¼", 10½"	set	...	...	...	...	...	...	...	...	...	...	...	...	...	...	...	...	...	a	a	a	a	a	a	a	...
Drifts for taking out old vents F, F1, F2, F3, F4	,,	a	a	a	a	a	a	a	a	a	a	a	a	a	a	a	a	a	...	...	...	...	...	...	...	...
Drills, cone ¼"	...	...	...	...	...	...	...	...	...	...	...	...	...	...	...	...	...	...	...	...	...	...	...	...	...	1
Drills, cone 1¼'	...	...	...	...	...	...	...	...	...	...	...	...	...	...	...	...	...	...	a	a	a	a	a	a	a	a
Drills, cone G, G1	set	a	a	a	a	a	a	a	a	a	a	a	a	a	a	a	...	...	...	...	...	...	...	...	...	...
Drills for venting, drifting ¼"	...	...	...	...	...	...	...	...	...	...	...	...	...	...	...	...	...	...	...	...	...	...	...	...	...	1
Drills for venting, drifting 4", 10"	set	...	...	...	...	...	...	...	...	...	...	...	...	...	...	...	...	...	a	a	a	a	a	a	a	...
Drills for venting, drifting E, E1, E2, E3, E4	,,	a	a	a	a	a	a	a	a	a	a	a	a	a	a	a	a	a	...	...	...	...	...	...	...	...
Drills for venting, tapping ¼" ...	...	...	...	...	...	...	...	...	...	...	...	...	...	...	...	...	...	...	...	...	...	...	...	...	...	1
Drills for venting, tapping 4", 10"	set	...	...	...	...	...	...	...	...	...	...	...	...	...	...	...	...	...	a	a	a	a	a	a	a	...
Drills for venting, tapping B, B1, B2, B3, B4	,,	a	a	a	a	a	a	a	a	a	a	a	a	a	a	a	a	a	...	...	...	...	...	...	...	...
Drills, hollow ¼	...	...	...	...	...	...	...	...	...	...	...	...	...	...	...	...	...	...	...	...	...	...	...	...	...	1
Drills, hollow 1¼ — ...	...	...	...	...	...	...	...	...	...	...	...	...	...	...	...	...	...	...	a	a	a	a	a	a	a	...
Drills, hollow A, A1	set	a	a	a	a	a	a	a	a	a	a	a	a	a	a	a	a	a	...	...	...	...	...	...	...	...
File, bastard (with handle), flat, 14", safe edge	1	a	a	a	a	a	a	a	a	a	a	a	a	a	a	a	a	a	a	a	a	a	a	a	a	a
Frame, iron, complete, for cutting off ends of vents with adjusting bars and extenders, long	set	a	a	a	a	a	a	a	a	a	a	a	a	a	a	a	...	...	...	...	...	...	...	...	...	...
Frame, iron, complete, for cutting off ends of vents with adjusting bars, no extenders, short	set	...	...	...	...	...	...	...	...	...	...	...	...	...	...	...	...	...	a	a	a	a	a	a	a	a
Frame, trunnion, with cross-head complete for converted guns (4 pieces)		...	...	...	...	...	...	...	...	...	...	...	...	...	...	...	a	a	...	...	...	...	...	...	...	...
Gauge, vent, ·22-inch short	1	a	a	a	a	a	a	a	a	...	...	...	...	...	...	...	a	a	a	a	a	a	a	a	a	a
Gauge, vent, ·22-inch long	1	...	...	...	...	...	...	...	...	a	a	a	a	a	a	a	...	...	...	...	...	...	...	...	...	...

a. Common to these.

DISTRIBUTION OF VENTING TOOLS FOR R.M.L. AND S.B. GUNS—*continued.*

		ORDNANCE MUZZLE LOADING.																									
		Cast Iron.								Wrought Iron.							Converted.		Howitzers.		Wrought Iron.						REMARKS.
		10″ and 10″ Mortar.	8″ and 8″ Mortar.	68-pr.	32-pr.	24-pr.	18-pr.	12-pr.	13″ Mortar.	12·5 inch.	12-inch.	11-inch.	10-inch.	9-inch.	8-inch.	7-inch.	80-pr.	64-pr.	8-inch.	6·3-inch.	64-pr.	40-pr.	25-pr.	16-pr.	9-pr.	7-pr.	
Gyn, iron, triangle, 10 feet high, cont. guns	...	...	...	...	...	...	...	...	...	...	...	...	...	...	...	...	a	a	...	...	...	...	...	...	...	...	
Hammers, hand, large	...	a	a	a	a	a	a	a	a	a	a	a	a	a	a	a	a	a	...	...	...	...	...	...	...	...	
Hammers, hand, small	...	a	a	a	a	a	a	a	a	a	a	a	a	a	a	a	a	a	a	a	a	a	a	a	a	a	
Hammers, monkey, converted guns	...	...	...	...	...	...	...	...	...	...	...	...	...	...	...	...	a	a	...	...	...	...	...	...	...		
Heads, metal, for cutters	...	1	1	1	1	1	1	1	1	1	1	1	1	1	1	1	nil	nil	1	1	com. to both	1	1	1	1	com. to both	
Hook, extracting, expanding blocks converted guns		...	...	...	...	...	...	...	...	...	...	...	...	...	...	...	a	a	...	...	...	...	...	...	...	...	
Instrument for taking impressions of vents, with stave, wood, Mk. I.	...	a	a	a	a	a	a	a	a	nil	nil	nil	nil	nil	a	a	a	a	...	...	...	...	...	...	...	...	
Instrument for taking impressions of vents, with stave, iron, in two lengths, Mk. II.*	...	...	...	...	...	...	...	...	...	...	...	...	...	...	...	...	...	...	...	...	a	a	a	a	a	a	
Levers, iron, for working bar, large ...	2	a	a	a	a	a	a	a	a	a	a	a	a	a	a	a	...	...	...	...	...	...	...	...	...	...	
Levers, iron, for working bar, small ...	2	...	...	...	...	...	...	...	...	...	...	...	...	...	...	...	...	...	a	a	a	a	a	a	a	a	
Machine, drilling, complete, with two lengths of chain (7 ft. each length), large	1	a	a	a	a	a	a	a	a	...	...	...	...	...	...	...	a	a	...	...	...	...	...	...	...	...	
Machine, drilling, complete, with two lengths of chain (7 ft. each length), small	1	...	...	...	...	...	...	...	..	a	a	a	a	a	a	a	...	...	a	a	a	a	a	a	a	a	
Pan, instrument, for taking impressions of vents, A large	1	a	a	a	a	a	a	a	a	a	a	a	a	a	a	a	...	...	...	...	...	...	...	...	...	...	
Pan, instrument, for taking impressions of vents, B small	1	...	...	...	...	...	...	...	...	...	...	...	...	...	...	...	...	...	a	a	a	a	a	a	a	a	
Pan, instrument, for taking impressions of vents, C {64-pr. and 80-pr. converted}	1	...	...	...	...	...	...	...	...	...	...	...	...	...	...	...	a	a	...	...	...		...	...	...	...	
Piece, lengthening, for bar iron, Mk. I., for 9-in. to 12″·5 R.M.L. guns		...	...	...	...	...	...	...	...	a	a	a	a	a	...	...	...	...	...	...	...	...	...	...	...	...	
Piece, lengthening, for instruments, taking impressions of vents, Mk. I., 9 in. to 12″·5 R.M.L. guns	...	...	...	...	...	...	...	...	...	a	a	a	a	a	...	...	...	...	...	...	...	...	...	...	...	...	

Rimers, Cone, ¼"	...	...	...	...	...	...	...	...	...	...	...	...	...	...	...	...	...	...	...	...	...	...	...	...	...	1
Rimers, Cone, 1½"	...	...	...	...	...	...	...	...	...	...	...	...	...	...	...	...	...	...	a	a	a	a	a	a	a	...
Rimers, Cone, H, H1 set ...	...	a	a	a	a	a	a	a	a	a	a	a	a	a	a	a	...	...	...	...	...	...	...	...	...	...
Rimers, mouth of vent S.S. guns ...	...	...	...	...	...	...	...	...	...	...	Uni	ver	sal		for		S.S.		guns		...	...	...	...	...	...
Rimers, vent, short and medium ...	...	...	...	...	...	...	...	...	...	...	...	...	...	...	...	...	...	...	a	a	a	a	a	a	a	a
Rimers, vent, medium and long ...	...	a	a	a	a	a	a	a	a	a	a	a	a	a	a	a	a	a	...	...	...	...	...	...	...	...
Saw, with frame, 14-inch ...	1	a	a	a	a	a	a	a	a	a	a	a	a	a	a	a	a	a	a	a	a	a	a	a	a	a
Spanner, for nuts on frame ...	...	...	...	...	...	...	...	...	...	...	...	...	...	...	...	...	...	...	a	a	a	a	a	a	a	a
Spanner, for countersink converted guns ...	...	...	...	...	...	...	...	...	...	...	...	...	...	...	...	...	a	a	...	...	...	...	...	...	...	...
Stops, iron for metal head for cutters, Mk. I. for 64 pr. Mk. I. ...	...	...	...	...	...	...	...	...	...	...	...	...	...	...	...	...	...	...	...	...	1		...	...	...	...
Stops, iron for metal head for cutters, Mk. II. for 64 pr. Mk. II. or III.	...	...	...	...	...	...	...	...	...	...	...	...	...	...	...	...	...	...	...	...						1
Stops, iron for metal head for cutters, Mk. III. for 40 pr. or 64 pr. M. III	...	...	...	...	...	...	...	...	...	...	...	...	...	...	...	...	...	...	...	...		...	...	...	...	...
Taps, ¼" fine, set 1, 2, and 3	...	...	...	...	...	...	...	...	...	...	...	...	...	...	...	...	...	...	...	...	...	...	...	...	...	1
Taps, 1½", set 1, 2, and 3	...	...	...	...	...	...	...	...	...	...	...	...	...	...	...	...			a	a	a	a	a	a	a	...
Taps, new thread, short, C1, C2, C3, C4, C5 set ...	...	a	a	a	a	a	a	a	a	a	a	a	a	a	a	a	a	a	...	...	...	...	...	...	...	...
Taps, new thread, long, C3, C4, C5 set ...	...	...	...	...	...	...	...	...	...	a	a	a	a	a	a	a	...	...	...	...	...	...	...	...	...	...
Taps, old thread, D3, D4, D5 set ...	...	a	a	a	a	a	a	a	a	...	...	...	...	...	...	...	a	a	...	...	...	...	...	...	...	...
Wedge, expanding blocks converted guns ...	...	...	...	...	...	...	...	...	...	...	...	...	...	...	...	...	a	a	...	...	...	...	...	...	...	...
Wire, directing, large ...	1	a	a	a	a	a	a	a	a	a	a	a	a	a	a	a	a	a	...	...	...	...	...	...	...	...
Wire, directing, small ...	1	...	...	...	...	...	...	...	...	...	...	...	...	...	...	...	...	...	a	a	a	a	a	a	a	a
Wrenches for countersink converted guns	1	...	...	...	...	...	...	...	...	...	...	...	...	...	...	...	a	a	...	...	...	...	...	...	...	...
Wrenches for frame adjusting ...	1	a	a	a	a	a	a	a	a	a	a	a	a	a	a	a	...	...	...	...	...	...	...	...	...	...
Wrenches for Socket, large ...	1	...	...	...	...	...	...	...	...	a	a	a	a	a	a	a	...	...	...	...	...	...	...	...	...	...
Wrenches for Socket, small ...	1	...	...	...	...	...	...	...	...	...	...	...	...	...	...	...	...	...	a	a	a	a	a	a	a	a
Wrenches for Taps ...	1	a	a	a	a	a	a	a	a	a	a	a	a	a	a	a	a	a	...	...	...	...	...	...	...	...
Wrenches for Taps and vents, large ...	1	...	...	...	...	...	...	...	...	...	...	...	...	...	...	...	...	...	a	a	a	a	a	a	a	a
Wrenches for Taps and vents, small ...	1	...	...	...	...	...	...	...	...	...	...	...	...	...	...	...	...	...	...	...	...	...	...	...	...	1
Wrenches for vents ...	1	a	a	a	a	a	a	a	a	a	a	a	a	a	a	a	a	a	...	...	...	...	...	...	...	...

a Common to these.

CHAP. XII.

The Quantity of PAINT and LACQUER authorized to be used for the various natures of wrought-iron rifled Guns (one Coat), is given in Table below.

			Quantities of	
			Pulford's Paint for each Coat.	Lacquer for Bores, one Coat.
			Lbs. ozs.	Lbs. ozs.
Ordnance, wrought iron	R.M.L.	12·5-in., 38-ton	4 12	0 14
		12-in., 35 „	4 8	0 13
		12-in., 25 „	3 4	0 12
		11 „ .. 25 „	3 4	0 11
		13 „ .. 23 „	3 8	0 12
		10 „ .. 18 „	2 12	0 9
		9 „ .. 12 „	2 0	0 8
		8 „ .. 9 „	1 8	0 8
		7 „ .. 7 and 6½-ton	1 0	0 6
		7 „ .. 90 cwt.	1 0	0 5
		64-pr. .. 64 „	1 0	0 4
		40 „ .. 35 „	0 14	0 4
		9 „ S.S 8 „	0 8	0 · 2
		9 „ S.S 6 „	0 6	0 2
	R.B.L.	7-in. 82 „	1 8	..
		7-in. 72 „	1 8	..
		64-pr. .. 61 „	1 0	..
		40 „ .. 35 or 32 cwt.	0 8	..
		20 „ .. 16 cwt.	0 6	..

Clause 122, Army Circulars, 1878.

The following are the proportions of materials allowed annually for the care and preservation of guns actually mounted on the defences:—

Articles.	Rifled Guns.	Smooth-bore Ordnance.
Bricks, bath	1 to 5 guns.	1 to 20 pieces.
Cloths, sponge	3 for each gun.	1 to 10 „
Composition, lubricating (cocoa nut oil and chalk in equal parts)	1 lb. for 7-inch and ½ lb. for 40-pr. & 20-prs., and 6 ozs. for smaller natures of B.L. guns when not in use.	Nil.
Hemp, undressed	3 lbs. for each B.L. gun.	Nil.
Linen, old	2 lbs. for every 5 guns.	1 lb. to 10 pieces.
Oakum	1 lb. for every 5 guns.	1 lb. to 20 „
Oil, lucca; Oil, mineral	1 quart for each B.L. gun in use. 1 pint if not in use. 1½ quarts for each M.L. gun in open works. 1 quart for each M.L. gun in casemates.	Nil.
Oil, sweet	Nil.	1 pint to 10 pieces.
Soap, soft	2 lbs. for every 5 guns.	1 lb. to 10 pieces.

CHAP. XII.

Should emery or other articles not mentioned in the Table be exceptionally required in certain cases, they will be supplied to Officers Commanding Royal Artillery on special demand, the purpose for which they are required being clearly stated in the requisition. Woollen rags and dressed flax will not be allowed.

Officers in Command of Royal Artillery Districts may also demand on the receipt from store of rifled M.L. or B.L. guns, requiring to be specially cleaned and lubricated when first mounted on the defences, an additional supply of bath bricks, grease, hemp, and oil, in such quantities as may be absolutely requisite, not exceeding one-half the annual proportion. Less than one bath brick need not be demanded

In certain positions, such as those at the sea batteries at Portsmouth, where the armament is much exposed, the annual proportions of grease and oil may be increased 10 per cent., if necessary.

CHAPTER XIII.

CHAP. XIII.

MODE OF MEASURING AND COMPARING POWERS OF GUNS.

Comparative Power of different Pieces of Ordnance.—How to be valued by useful effect of projectile.—This depends principally upon **Velocity of Translation** at muzzle, and resistance of the air.—How, knowing powder charge and weight of projectile, to find out M.V. theoretically.—Practical methods of arriving at M.V.—The Le Boulengé Chronograph.—How to deduce from M.V. the V. of projectile at any range by Professor Bashforth's tables.—Or by Professor Helié's formula.—Examples of obtaining V., &c., by Bashforth's tables.—**Velocity of Rotation.**—Affected by amount of twist and by velocity of translation.—Effects to be produced by projectile.—From field guns.—With siege pieces.—From heavy guns.—As to piercing of iron plates.—Formula used to arrive at penetrative power. —**Work done upon the Gun.**—On the walls of bore.—On the grooves.—Modes of measuring the powder gas pressure.—Empirically by crusher gauges.—Method of using these.—Theoretically by means of the Noble Chronoscope.—**Pressure on Grooves** already described in Chapter II.—Tables for calculating velocities.—Professor Bashforth's.—Ogival headed and spherical.—Velocities of projectiles from our heavy guns at various ranges.—Energy of the same.—**Table of Work realised** in an experimental gun.—Note A., and Tables C. and D. (Captain Andrew Noble, F.R.S.), showing how to arrive theoretically at M.V.—Note B.—Description of the Le Boulengé Chronograph.

Comparative Power of Guns.

How the comparative value of a gun is obtained.

We have in the preceding Chapters described the materials used for, and the actual manufacture, inspection, and proof of, service ordnance and their stores, and have also discussed generally the theories upon which the mode of manufacture of our guns is based. Let us now consider how we can evaluate these guns when made, and in what manner we can compare one piece of ordnance with another—how, in fact we are to measure the power of any gun.

It may be well to repeat here once again that similar power may be obtained from different systems of construction, or, to quote a foreign artillery authority, who says, with reference to the comparative trial of a 96-pr. R.B.L. Krupp gun and a 9″ R.M.L. Woolwich piece, "It is unnecessary that we should under this head do more than point out that the work impressed on the shot has nothing whatever to do with the mode of construction of the gun; that if from two guns shot of the same form and material and endowed with the same velocity are fired, it is a matter of indifference, as far as regards effect, upon what system the guns are constructed."*

* Letter of President of Prussian Select Committee, quoted by Colonel Reilly, C.B., in Report of 1872.

CHAP. XIII.

There are, it is clear, two points to be principally considered:—(1) the work impressed upon the projectile when the gun is fired, and (2) the work done upon the piece itself.

We know that a gun is but an instrument employed for throwing a projectile, and its power may plainly be measured by the effects it can produce by means of this projectile at a given distance or distances. Of any number of similar pieces that which can produce such effects in a maximum degree and with the greatest certainty and regularity in the shortest time must be judged the best; while on the other hand it is evident that in producing these effects the gun itself must not be unduly strained.

Ballistic power first found.

Before finally introducing any new nature of ordnance into the service we consequently test its ballistic power, upon which, as we shall see, its effective power principally depends, and afterwards try the gun for range, accuracy, and amount of deflection by absolute practice with the proposed service charge and projectile at Shoeburyness.*

Subsequent trials made as to range, &c.

The accuracy and deflection will be influenced by the twist of rifling upon which (taken in conjunction with the velocity of translation as explained at p. 41) the velocity of rotation depends.

While testing the ballistic power of the piece with flat headed projectiles† at the R.G.F. Proof Butts in Woolwich Arsenal, the pressures produced upon the walls of the bore at different points are also determined by the methods described further on.‡

(1.) Useful Effect of Projectile.

Work stored up in, or energy of a projectile.

Prior to inquiring further into the actual effects produced by its projectile it may be pointed out that if $v =$ the velocity, and $W =$ numbers of lbs. in a shot, then the work stored up in it, or the kinetic energy§ of translation $= \frac{W v^2}{2 g.}$, representing a certain number of foot-pounds as the case may be. This is the work stored up to produce effect, and the value of the gun may be theoretically measured according to the amount of energy imparted in a variety of ways, *e.g.*, by the amount of energy imparted per cwt. of gun, per lb. of powder, per ton of mean pressure in the powder chamber, or otherwise; the former method is sometimes employed for field guns where the weight of gun is of importance; the two latter enumerated are employed by the Committee on Explosives as shown in Table, p. 358, at the end of Chapter, columns 11 and 12.

Effects produced by a gun.

The effects we require a piece of ordnance to produce vary much,‖ from the piercing of thick iron plates at the distance of a mile or more to the showering of the greatest possible number of deadly bullets amongst advancing skirmishers at very short ranges,—from the opening of a breech through the logs of a simple wooden pah to the battering down of granite casements upon which all the skill and science of the engineer have been spent. Whatever they may be, however, we

* Vide p. 52 for method employed to determine angle of deflection.

† This allows us to use not quite so thick a butt as would be necessary if pointed projectiles were used: even as it is, 60 feet of sand with a backing of 40 feet of earth constitute the butt for the heaviest natures.

‡ Under heading "Work done on Gun," p. 346.

§ "Kinetic" energy being the energy of motion; any body in motion has kinetic energy which it must communicate to some other body during the process of bringing it to rest.

‖ Vide p. 343 of this chapter.

CHAP. XIII.

Velocity of translation.

Velocity of rotation.

shall find that their value depends principally upon the ballistic power of the gun, which with a given projectile is a function of the *velocity of translation* given at the muzzle, for upon this depends the destructive work stored up in the projectile at the point where it is required to act. To ensure the projectile reaching this required point with certainty and regularity it must also be endowed, upon leaving the muzzle, with sufficient *velocity of rotation* to ensure its flying point foremost and steadily to that point.

Velocity of Translation.

We have, then, in the first place, to determine the velocity of translation, from which that of rotation is readily deduced.

Muzzle Velocity.

Determination of muzzle velocity theoretically.

Various attempts have been made to arrive at the M.V. of a projectile by means of determining by calculation the total amount of work done by the explosion of the powder charge (deducting from that the work done upon the gun), and so finding out the work done upon the projectile itself, from which if known, we could work out what the velocity at the muzzle would be.

Researches and calculations of Captain Noble (late R.A.), F.R.S., and Professor Abel, F.R.S.

None of these theoretical solutions have given such results as to be practically useful, until the recent researches and calculations of Captain Noble (late R.A.), F.R.S., and Professor Abel, F.R.S., provided us with data by means of which we can form a tolerable approximation, which will prove of much service in working out the problem as to how any required gun should be constructed. The mode of arriving at the approximate velocity is shown in the example, at p. 360.

The necessary tables, extracted from the work mentioned, together with a short explanation of the same, are given at the end of this chapter, note I.

Robin's ballistic productions. Navez-Leurs, Le Boulengé Bashforth's, Watkins.

Since 1742, when Robins employed the ballistic pendulum for this purpose, the velocity of projectiles have been determined by absolute experiment; but it is only of late years that the introduction of the improved instruments* of Navez-Leurs, Le Boulengé, Bashforth,† Watkins, &c., have enabled us to make perfectly certain of our results.

Obtaining actual M.V. by experiment.

The velocity of the projectile always has been and still is empirically obtained at a given distance from the muzzle, and then the muzzle velocity deduced from this velocity by calculation.

The Boulengé Chronograph.

Neglecting the old experimental modes of obtaining the velocities, we will now proceed to the description of that at present employed at the R.G.F. Butts, viz., the Boulengé chronograph.

The Le Boulengé instrument.

As in all similar instruments, electricity is the agent employed for measuring the velocity, and the general mode of utilizing it is as follows:—

Electric apparatus employed.

Frames with wire stretched across them in connection with primary circuits are placed at certain intervals in front of the gun so that the shot shall cut the wires successively, thereby interrupting the electric currents which pass through them and the instrument connected with

* A description of these and their results will be found at pp. 24–6 in Professor Bashforth's most useful work already mentioned and also in Table, p. 354.

† For determining the velocity of small-arm bullets the Bashforth chronoscope is employed at Woolwich.

CHAP. XIII.

them. A record is thus obtained in the instrument of the precise instant at which the projectile passes each frame. The time occupied in passing from one frame to the other is therefore known, and by dividing the distance between the frames by this time the velocity of the projectile at a point half-way between the frames is found. For instance, suppose the frames to be 120 feet apart, and the time taken by the shot in passing between them to be 0·1 of a second, then the velocity of the projectile will be $\frac{120}{0 \cdot 1} = 1{,}200$ f. s. As the first frame must be placed some distance from the gun to prevent its being damaged by the discharge, the velocity obtained as above is that of the projectile between 100 and 200 feet from the gun, and in order to find the velocity at the muzzle it is necessary to calculate the loss due to the resistance of the air over that space. This is done by means of a formula similar to that used for calculating the remaining velocity of a projectile at any range, being given the velocity at the muzzle, and is explained further on.

Frames in the circuit.

Electro-magnets.

With the Boulengé chronograph,* the shot is made to cut two currents (by passing through two screens), and thus to demagnetise one after another two electro-magnets which had previously supported two heavy bodies. The fall of these bodies, under the action of gravity, is the measure of the time taken by the shot to pass over a known distance.

In this instrument the weight falls *freely*, and without any disturbing influence, in a vertical direction, and the distance actually described by it is measured; the corresponding time being readily ascertained, and the velocity calculated.

By means of a scale previously prepared, the velocity of the projectile can at once be read off without any calculation, an advantage which this instrument possesses over most others.†

The "Registrar."

Trigger.

"Chronometer."

It is so arranged that when the shot strikes the first screen an electro-magnet supporting a long rod covered from the lower part by a zinc tube (and called the chronometer) is demagnetised, and the rod begins to fall; while it is falling, the shot continues to fly forward in the space between the screens, and soon cuts the wires of the second screen, when a second weight called the "registrar," supported by another electro-magnet, is released. The weight falls upon a trigger, and so realeases a circular knife which (carried forward by a spring) strikes the zinc tube on the "chronometer," which is still falling, and makes a perceptible dent in the metal.‡

A very simple relation exists between the height of this indent and the velocity of the projectile. It is evident that the time which elapses after the fall of the chronometer before the registrar is released is the time taken by the projectile in passing over the distance between the screens; the less, therefore, the velocity of the projectile, the further in advance will the chronometer be, and the higher will be the indent.

---

* For a full description of this instrument see a pamphlet published in 1871 by the War Office, "Description and Use of Le Boulengé Chronograph, &c., &c.," by Lieut. (now Captain) C. Jones, R.A.; or a paper by Captain Jones in Proceedings, R.A.I., Vol. IX., No. 6, from which much of this description is taken verbatim.

† Except that lately perfected by Captain Watkin, R.A., a full account of which is given in Proceedings, R.A.I., Vol. IX., p. 14. Every part of the chronograph is of metal, and it is consequently little influenced by changes of climate—a property which renders it peculiarly suitable for use in a country like India, where vulcanite and other insulating materials rapidly deteriorate.

‡ A detailed description of this instrument is given in Note II., p. 364, at the end of the chapter.

CHAP. XIII.

Plate A.

Fig. 1.—Chronograph in Position for Measuring the Velocity of Projectiles.

Scale $\frac{1}{8}$th inch.

Side view from the left. Front view.

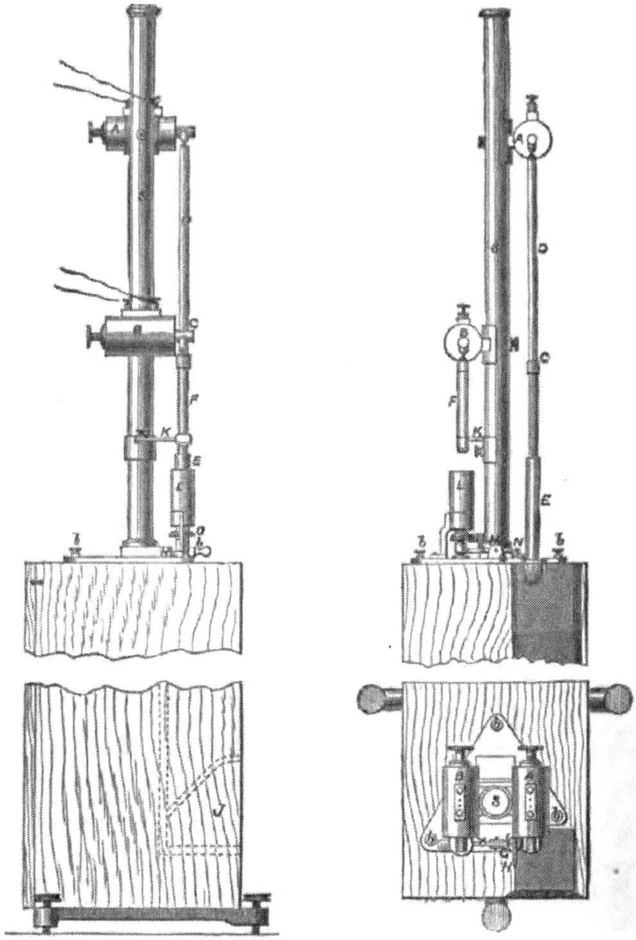

Horizontal projection.

Plate B.

Fig. 1.—The Trigger. Scale ½.

Elevation.

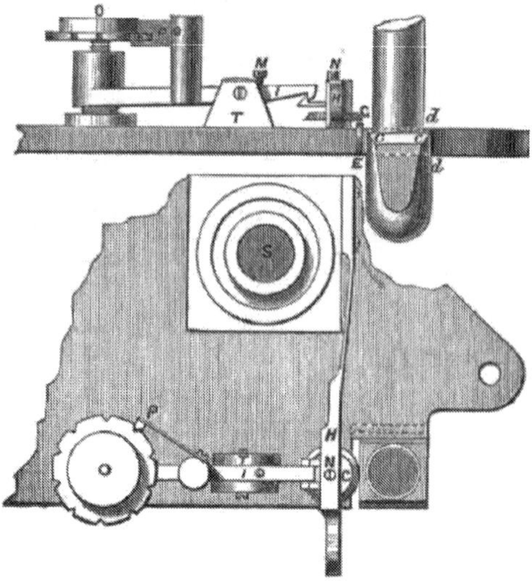

Section on A B.

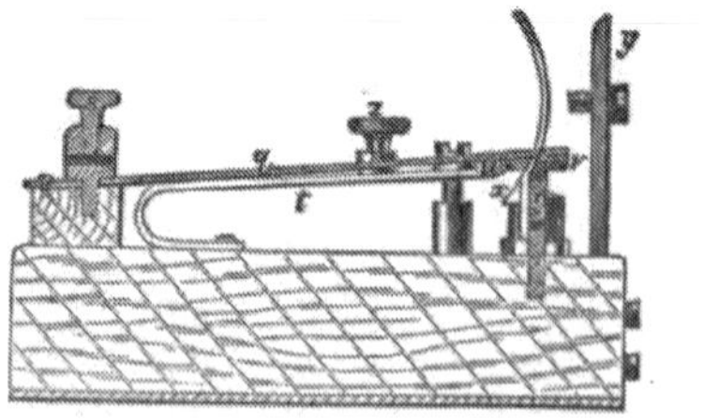

Fig. 2. Disjunctor. Scale ½.

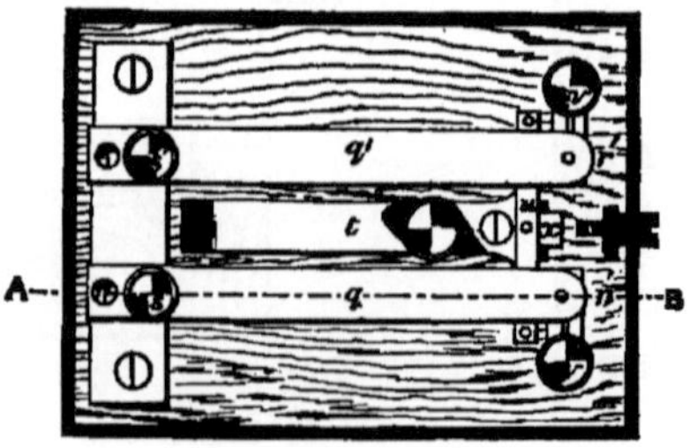

Horizontal projection.

CHAP. XIII.

PLATE C.

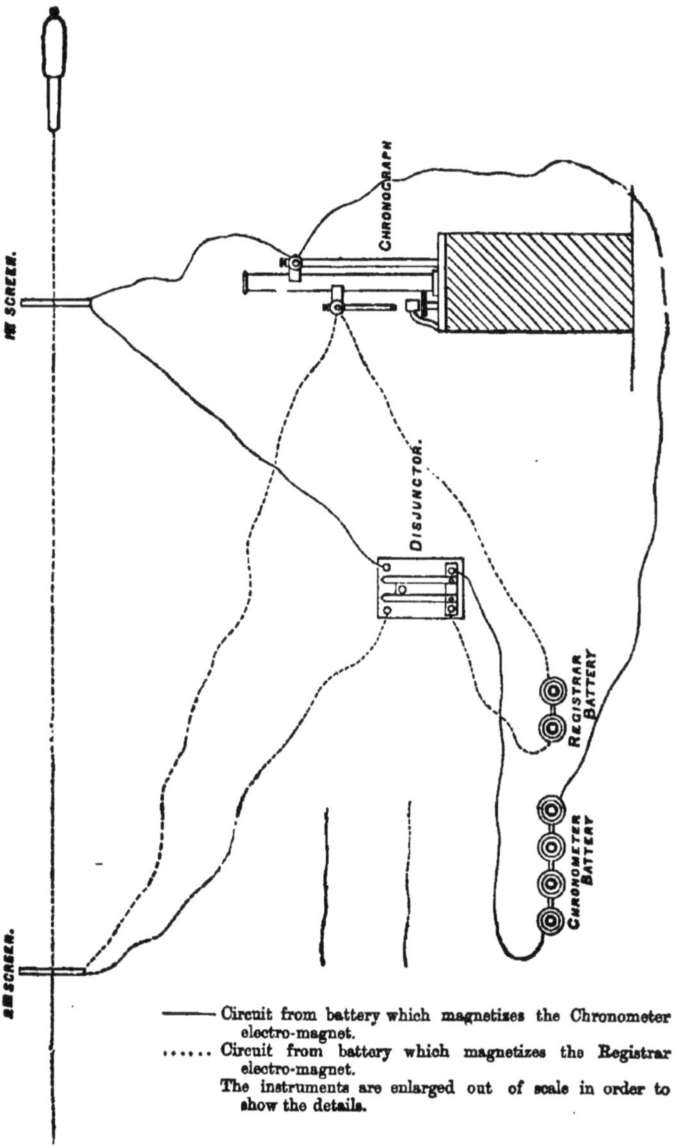

——— Circuit from battery which magnetizes the Chronometer electro-magnet.

...... Circuit from battery which magnetizes the Registrar electro-magnet.

The instruments are enlarged out of scale in order to show the details.

A graduated rule is used for measuring the height of the indent above the zero point. It is of brass, and is graduated on both edges; the upper edge is a scale of equidistant parts, divided into inches and tenths, reading to thousandths with a vernier, and is intended for use in connexion with the tables. CHAP. XIII. Graduated rule.

The lower scale is for reading off the velocity of the projectile without any calculation. It is graduated in feet for a distance between the screens of 120 feet.

Should it be necessary to place the screens nearer to one another, the velocity can be found by multiplying the number read off on the scale by the fraction $\frac{D}{120}$, D being the actual distance between the screens in feet. Employment of the rule.

The zero point on the scale corresponds with the "*origin*," or the point at which the knife marks the zinc if the trigger is set in action when the chronometer is at rest.

The rule is fitted at the zero end with a jointed piece having a slightly conical projection, which enters into a recess in the bob of the chronometer when applied for measuring the marks. Care must be taken not to injure this portion of the scale or the measurement may be rendered inaccurate.

Prior to use, the instrument must be carefully regulated and adjusted, readings being taken until the adjustment is correct. Adjustment of instrument.

To take a velocity, cock the disjunctor by pressing the mill-headed screw (Plate B.) so as to establish the current; suspend the chronometer and then the registrar to their respective electro-magnets; fire the gun through the screens, and then measure by means of the scale the height of the mark made by the knife upon zinc tube of chronometer above the origin; from this calculate the time and corresponding velocity according to the formula in note, p. 364, or else measure the V. at once by means of the scale adapted for the purpose.

The accuracy of this instrument has now been tested by continual use for many years at the R.G.F. Butts, and has proved very satisfactory; its extreme simplicity and ease of manipulation recommend it in preference to other instruments which give an equally accurate result. Accuracy of the instrument.

Deduction of Velocity at any point from M.V.

We see now how we can, at some distance from the muzzle, obtain with great accuracy the velocity of the projectile. Next comes the question, How can we deduce from this velocity that of the projectile at the muzzle or at any point in its trajectory.

In order to obtain the mean velocity we must know what velocity the shot has lost during the time of its flight from the muzzle to the screen, and as it is the resistance of the air which retards the shot and reduces its velocity, we require to know the law of such motion and the mathematical expression for this law. Muzzle velocity. Resistance the air.

This subject has been investigated by mathematicians since the days of Newton, and it is generally acknowledged that the retardation is some function of the velocity. In the formula we employ the retardation of the air is taken (with ogival headed projectiles) to vary as a certain function of the velocity.* On this assumption it is not difficult to Calculation of velocity ac-

* It is, however, with certain velocities supposed to vary as the square or as the sixth power, both Professor Bashforth and General Mayevski observing that the resistance may be said to vary as the sixth power of the velocity, for velocities between about 900 to 1,100 f.s.

CHAP. XIII.

cording to resistance of air.

Bashforth's chronoscope, and his calculations as to resistance of air.

obtain mathematical expressions by means of which we can calculate the velocity of the shot at any point, and also obtain the absolute trajectory* of the projectile or the path it describes in the air.†

In Professor Bashforth's work is given (at pp. 29 to 44) a interesting account of a series of experiments carried out by means of his chronoscope, in the years 1865–70, which satisfactorily established the truth of the theory that for the velocities experimented with (900 to 1,300 f.s.), the resistance does really vary approximately as the cube of the velocity with ogival-headed projectiles. In his work he points out that for practical purposes we may in certain cases neglect the action of gravity, and treat the motion of the projectile as in a straight line when the mean velocity is high and it is desired to find the loss of velocity or the time of flight for a limited space; the less the shot is affected by the resistance of the air the more accurate will these results be, so that the method applies better to pointed and elongated shot than to spherical, and better to solid shot than to shell of the same form.

Professor Bashforth's formula.

‡ Taking the equation of motion as:—

$$\frac{d^2 s}{dt^2}=\frac{dv}{dt}=2bv^3 \text{§}$$

we obtain two equations:—

$$(1)\ \frac{d^2}{w}t=\frac{500}{K}\left\{\left(\frac{1000}{v}\right)^2-\left(\frac{1000}{V}\right)^2\right\}$$

$$(2)\ \frac{d^2}{w}s=\frac{(1000)^2}{K}\left\{\left(\frac{1000}{v}\right)-\left(\frac{1000}{V}\right)\right\}$$

Useful tables by Professor Bashforth.

As, however, the resistance of the air does not exactly vary as the cube of velocity, we must give different numerical values to K, and as this is a troublesome operation the values of the second portion of each of the above equations (1) and (2) have been worked by Professor Bashforth, taking V = to 1700 f. s., and will be found in the tables pp. 366–369, at the end of this chapter. These tables will be found very useful, and enable us to calculate with very little trouble the time

---

* The question of trajectories is shortly alluded to in Appendix, p. 385.

† Professor Helé deduced the law of resistance varying as V^2 even for these low velocities. Professor Helié's formula, employed by Major W. H. Noble, R.A., in calculating many of the range tables given, is:—

$$v=\frac{V}{1+cVx}$$

where c is a constant depending on the form, weight, and velocity of the shot,

x = distance from muzzle in feet,

v = velocity at any point,

V = muzzle velocity,

$c = b\frac{R^2}{W}$,

R = radius of the projectile *in feet*,

W = weight " " in lbs.

b = a variable co-efficient depending on the form of the shot and the velocity of the projectile.

For higher velocities General Mayevski concludes that it varies as the square, or V^2, while Professor Bashforth observes that between 1,100 and 1,350 it varies as V^3, and for higher velocities as V^2.

‡ Bashforth's Motion of Projectiles. Asher and Co., London.

§ Where $b=\frac{Kd^2}{2.w}\cdot\frac{1}{(1000)^2}$

and K is a co-efficient dependent upon the velocity and the form of projectile.

d = diameter of shot in inches,

w = weight of shot in pounds.

CHAP. XIII.

required to describe a given space, or the space described in a given time, if we know the velocity. Equation (1) becomes—

$$\frac{d^2}{w}t = \frac{500}{K}\left\{\left(\frac{1000}{v}\right)^2 - \left(\frac{1000}{1700}\right)^2\right\} - \frac{500}{K}\left\{\left(\frac{1000}{V}\right)^2 - \left(\frac{1000}{1700}\right)^2\right\}$$

$= T_v - T_V$ where T_v and T_V represent the tabular reduced times for velocities v and V respectively with reference to a standard velocity of 1700 f. s.

Similarly $\frac{d^2}{w}$. $S = S_v - S_V$ when S_v and S_V represent the tabular reduced ranges for velocities v and V respectively.

Muzzle velocity, 80-ton gun calculated from experimental data.

* To give a simple instance of the way in which the muzzle velocity can be obtained when we have found by the Le Boulengé instrument or otherwise the velocity, say at 120 feet from the muzzle :—

Suppose a 1,700 lbs. shot be fired from the 80-ton gun, its diameter being 15·98 inches, and that the velocity at 120 feet from the muzzle be shown by the instruments to be 1,500 f. s., then—

$$\frac{d^2}{w} S = S_v - S_V \therefore \frac{(15{\cdot}98)^2}{1700} \times 120 = S_{1500} - S_V.$$

From table $S_{1500} = 876$.

$\therefore S_V = 876 - 18{\cdot}03 = 857{\cdot}97$ or $V = 1504$ from same table.*

This then will be the muzzle velocity of the projectile in question; of course with lighter projectiles and lower velocities the difference would be much more marked.

Examples of use of Bashforth's tables.

The following examples of application of Professor Bashforth's tables are taken from Principles of Gunnery, by Captain Sladen, R.A.

Example (1).—In what range would the velocity of a projectile fired from the 9-in. M.L. gun be reduced from 1420 f. s. to 1240 f. s.?

For the 9-in. shell, $d = 8{\cdot}92$ ins., $w = 250$ lbs; also $V = 1420$, $v = 1240$. Substituting in equation (I.),

$$\frac{d^2}{w} S = S_v - S_V$$

or $\frac{(8{\cdot}92)^2}{250} S = S_{1240} - S_{1420} = 2208{\cdot}6 - 1251{\cdot}1$, by Table I.;

or $.3183 S = 957{\cdot}5$ ft.;

so that $S = 3000$ ft. nearly.

i.e., the velocity would be reduced from 1420 f.s. to 1240 f.s. in about 3000 ft.

Example (2).—In what *time* would the velocity of the same projectile be reduced from 1420 f.s. to 1240 f.s.?

---

* In all these cases—
d = diameter of projectiles in inches,
w = weight of shot in lbs.,
S = distance.
T = time.

Substituting in equation (II.),

$$\frac{d^2}{w}T = T_v - T_r;$$

$$\text{or } \frac{(8{\cdot}92)^2}{250}T = T_{1420} - T_{1240} = 1{\cdot}5285 - {\cdot}8057;$$

$$\text{or } {\cdot}3183T = {\cdot}7228 \text{ sec.};$$

$$\text{so that } T = 2{\cdot}27 \text{ secs.}$$

i.e., the velocity would be reduced from 1420 f.s. to 1240 f.s. in about 2·27 secs.

Example (3).—Find the remaining velocity of the 12·5-in. M.L. gun at 3000 ft.

In this case, $w=800$ lbs., $d=12{\cdot}42$ ins., muzzle velocity $(V)=1420$ f.s. $S=3000$ ft.

Substituting in equation (I.),

$$\frac{d^2}{w}S = S_v - S_r;$$

$$\text{or } \frac{(12{\cdot}42)^2}{800} \times 3000 = S_v - S_{1420}.$$

Transposing, $$S_v = \frac{(12{\cdot}42)^2}{8} \times 30 + S_{1420}$$

$$= 578{\cdot}4 + 1251{\cdot}1 \text{ (since } S_{1420} = 1251{\cdot}1 \text{ by Table I.)};$$

$$\text{or } S_v = 1829{\cdot}5 \text{ ft.};$$

then from Table I. $v=1306{\cdot}6$ f.s.

i.e., the remaining velocity at the distance of 3000 ft. from the muzzle is 1306·6 f.s.

Example (4).—A shell fired from the 64-pr. M.L. gun with a charge of 10 lbs. of powder was observed to strike the crest of a parapet in 3 secs.; the muzzle velocity is known to be 1383 f.s. Find the striking velocity and the range.

In this case, $w=64$ lbs., $d=6{\cdot}22$ ins., $V=1383$ f.s., $T=3$ secs.

Substituting in equation (II.),

$$\frac{d^2}{w}T = T_v - T_r;$$

$$\text{or } \frac{(6{\cdot}22)^2}{64} \times 3 = T_v - T_{1383};$$

$$\text{or } T_v = 1{\cdot}814 + {\cdot}9351 \text{ (since } T_{1383} = {\cdot}9351 \text{ sec. by Table II.)};$$

$$= 2{\cdot}7491 \text{ secs.};$$

then from Table II. $v=1049$ f.s.

i.e., the striking velocity at the end of 3 secs. from firing $=1049$ f.s.

Next to find the range:—

Substituting in equation (I.),

$$\frac{d^2}{w}S = S_v - S_r;$$

$$\text{or } \frac{(6{\cdot}22)^2}{64}S = S_{1049} - S_{1383};$$

$$\text{or } {\cdot}6045S = 3595{\cdot}6 - 1432{\cdot}4;$$

$$\text{or } S = \frac{2163{\cdot}2}{{\cdot}6045} = 3578 \text{ ft.}$$

i.e., the distance of the crest of the parapet from the gun is 3578 ft.

We now see that we can readily find the M.V. given by a gun to its projectile by the use of the Le Boulengé or other instrument which gives us the shot's velocity at some distance from the muzzle, and so enables us by the application of Professor Bashforth's formula to find out the muzzle velocity. Knowing the latter, again, we can within certain limits find out the velocity of the shot at any time of its flight.*

Velocity of Rotation.

Velocity of rotation.

With regard to the velocity of rotation, we must impart to the shot a sufficient amount of kinetic energy of rotation (or work stored up, due to rotation) to keep it steady up to the farthest range required.

As the shot leaves the muzzle, its kinetic energy of rotation is expressed by the formula, $\frac{\varpi'^2}{2}.\ M\ k^2$.†

Where, ϖ'=angular velocity at muzzle$=\left(\frac{V}{n\,d}\right)2\pi$. (vide p. 41).

M=mass of projectile$=\frac{W}{g}$

W=weight of do. in lbs.

k^2=(radius of gyration)$^2=\frac{d^2}{8}$‡

$2\pi=\begin{cases}\text{Four right angles (in circular measure 6·28318), or}\\ \text{one complete turn, as we choose to take it.}\end{cases}$

How to find out amount of rotation required.

We do not know, however, the rate at which rotation is retarded,§ and therefore can only determine experimentally the amount of energy of rotation to be imparted, in order that the shot shall be stable at its maximun range.

The twist necessary (with a given velocity) in a particular gun to determine the shot leaving the muzzle with sufficient energy of rotation being known, we can find out approximately the twist required for another piece firing the same projectile but having a different M.V.,‖

---

* In this manner have been calculated the velocities at the muzzle and at various ranges of projectiles for service guns in table pp. 354–357.

† To give an easy example: Taking the 10-inch service gun—

Where M.V.=1364 f.s. $M=\frac{W}{g}=\frac{400}{2240\cdot g}$ tons.

$n=40$

and $d=\frac{10}{12}$ ft. $k^2=\frac{d^2}{8}$

$$\frac{\varpi'^2}{2}\cdot(M\ k^2)=\tfrac{1}{2}\left\{\frac{1364^2}{n^2\cdot d^2}\ (2\pi^2)\right\}\times\left\{\frac{400}{2240\cdot g}\ \ \frac{d^2}{8}\right\}$$

which when worked out, gives us about 16 foot tons of kinetic energy of rotation.

‡ Looking at the projectile as a cylinder of mass M and radius $\frac{d}{2}$.

§ Not knowing what is the co-efficient of friction of the air.

‖ *e.g.*, Let us take the case of an 8-inch gun—

Where M.V.=1400 f.s. about.

And suppose that we want to know what twist we should give to an 8-inch howitzer, which is to throw the same projectile, but the M.V. of which we require to be low, as it is intended for high-angle fire.

We want, say, a M.V. of 550 f.s.; supposing the time of flight to be about equal, then the D or angular V must be the same in each case, and we have

$$D=\frac{2\pi}{n\,d}V=\frac{2\pi}{40\cdot d}\times 1400 \text{ for 8'' gun,}$$

$$D=\frac{2\pi}{n'd}V'=\frac{2\pi}{n'd}550 \text{ for 8'' howitzer;}$$

but as these are equal—

$$\frac{2}{n}=\frac{55}{7}\qquad\text{or } n'=\frac{110}{7}=15\cdot7$$

or the twist should be about 1 in 16 calibres.

CHAP. XIII. for the kinetic energy of rotation varies with the angular velocity, and that again with the M.V. and length of twist.

Effects produced by a Projectile depend on its Velocity.

Effects to be produced by projectiles. Having ascertained the mode of arriving at the velocity both of translation and rotation, the former at any point of the trajectory and the latter at the muzzle, let us see how we apply this knowledge to ascertain the effects the gun can produce by means of its projectile.

We have the three general cases of—

(1.) Field Ordnance.
(2.) Siege „
(3.) Heavy „

By field guns. (1.) The power of a field gun must be measured by its shell* power which again is proportionate to the range at which it can produce a sufficiently murderous effect and to the effect itself. As to the latter the tendency now is to evaluate it by the number of *effective* bullets or splinters which a shrapnel shell gives over a stated area; the greater the maximum range at which this effect can be produced, *cæteris paribus*, the more powerful is the piece, and as this range depends upon velocity† we must decide upon what velocity is necessary at that maximum range to give the bullets or splinters sufficient energy to disable a man.‡

Again, the manner and direction in which the bullets and splinters spread, and so the amount of destructive space covered by them, whether the shell bursts in the air or upon impact, will depend for any given range upon the flatness or tension of the trajectory, and this again is consequent upon the calibre and M.V. with equal weight of projectile.

We have but few experimental data as yet upon this point and so are tied down to measure the effects of a gun as valued by the effect of its shrapnel shell, by finding out through what extent of range the remaining velocity of the shell will be sufficient to impart to its bullets § the necessary energy for the destruction of life; and in conjunction with this, taking into consideration the number of bullets the shell contains and their individual weight.

Comparing these with the actual effective hits made upon targets by the bullets of loaded shell, and with the deviations, vertical and lateral, of similar shells fired without fuzes, we can arrive at a very tolerable idea of the effects the gun can produce, an idea sufficient for purposes of comparison.

Siege guns. (2.) The effects required to be produced by siege guns are so different from those we wish field guns to afford that we must use a different measure for their value.

A siege piece is either required for direct fire to strike masonry or earthwork with a projectile having sufficient energy to pierce it and afterwards burst, or else to fire at high angles a heavy projectile having a very curved trajectory, also to pierce masonry or earthwork, and to

* In the case of common shell this would of course be much more dependent upon the bursting charge of the shell itself than when shrapnel are employed as the measure of effect.

† With similar calibres and weights of projectile.

‡ According to Lieut.-Colonel Maitland, R.A., this should be about 800 f.s., such with bullets as we employ in our service M.L. shrapnel, vide an interesting paper published in proceedings R.A.I., vol. IX., p. 384.

§ It is perhaps more advisable in the case of shrapnel to leave out altogether the question of splinters, and to compare the available amount of space in the shell for bullets.

CHAP. XIII.

have as heavy a bursting charge as possible, in order to tear through the roofs of magazines, casemates, &c. The weight of the shell as compared with that of the piece and the amount of bursting charge it can contain will therefore determine to a certain extent the power of a siege piece.

Effects produced by heavy guns.

(3.) The effects produced by the projectile of a heavy gun may be measured by its power of piercing iron plates. This power is found (with projectiles of the same form, material,* and weight) to vary approximately with the kinetic energy of a projectile, and inversely as its diameter or circumference, or some function of the same.

The experiments carried on against armour plates have not been sufficiently numerous nor exact to allow of formula being deduced therefrom which we can rely upon as being under all circumstances absolutely accurate. Such formula as those given below prove, however, very useful, not only for the purpose of comparing the penetrating or perforating power (at any particular range) possessed by the projectiles of different pieces, but also as allowing us to calculate approximately what this power will be.

Let W=weight of projectile,
r=radius of „
v=velocity of same at any point,
E=kinetic energy „
e=kinetic energy per inch of shot's circumference.

$t=\left\{\begin{array}{l}\text{thickness of iron plate the projectile}\\ \text{can pierce} \quad .. \quad .. \quad .. \quad ..\end{array}\right\}=\text{perforation.}$

$t'=\left\{\begin{array}{l}\text{thickness of iron it can penetrate with-}\\ \text{out perforation} \quad .. \quad .. \quad ..\end{array}\right\}=\text{penetration.}$

Then $E=\dfrac{W \cdot v^2}{2g}$ and $e=\dfrac{E}{2\pi r}$

The formula (α) of Captain A. Noble, F.R.S., as well as the modification of the same (β) (from the result of further experience) by Major W. H. Noble, R.A., are of the form $t=(b\,e)^c$, where (b) and (c) are constants depending upon the nature of the projectile and the material of the plates to be pierced: with ogival-headed chilled projectiles against wrought iron plates these constants are as shown below.

$$(\alpha) \quad t=\left\{\frac{e}{3 \cdot 18}\right\}^{\frac{1}{1 \cdot 5}}, \text{ or } e=3 \cdot 18\, t^{1 \cdot 5}$$

$$(\beta)\dagger \quad t=\left\{\frac{e}{2 \cdot 52}\right\}^{\frac{1}{1 \cdot 6}}, \text{ or } e=2 \cdot 52\, t^{1 \cdot 6}$$

Captain English, R.E., gives formulæ (γ) for finding t or t', but they are too long to be given here *in extenso*. They will be found in R.E. Corps Papers, 1871, Vol. XIX.

---

* The material is naturally of importance, and here again we find a defect in the lead-coated projectiles used with the heavy Krupp R.B.L. guns, with which, as stated in Captain v. Doppelmair's pamphlet, "we shall not be much in error when calculating the momentum (energy ?) of a lead-coated shot if we only take into account the weights of the cast iron or steel core."

This will account for the want of material difference in the penetrative power as found by experiment of the two guns mentioned in note p. 332, whose respective energies (owing to the greater weight, length, and charge of the R.B.L.) were as 3,817 foot-tons to 3,018 foot-tons with the M.L., which would by the ordinary formula give not equal powers of penetration as observed, but powers in the ratio of 118·5 : 100.

† This formula is that hitherto generally used by us in calculating such tables as that at pp. 354, 355, which give the kinetic energy of shot fired from our service ordnance at various ranges.

CHAP. XIII.

Lieut.-Colonel Maitland, R.A., gives the following (δ) for perforation of iron plates with chilled ogival-headed projectiles.

$$(\delta) \quad t = \frac{E}{(4 \cdot 733) \pi r^2} = \frac{E}{(14 \cdot 87) r^2}$$

In this formula it will be seen that Lieut.-Colonel Maitland has considered the action of the ogival-headed projectile as a wedging rather than a punching action, and takes the sectional area instead of the circumference as used by Captain Noble and Major Noble.*

2. Work done upon the Gun.

Pressure on walls of bore and upon grooves.

Having given a slight sketch of some of the methods by which we arrive at the comparative power of a piece of ordnance, as measured by the effect which its projectile produces, we will proceed to explain in what manner we find out by actual experiment whether, in giving the velocity to the shell which is necessary for the production of such effects, we strain the walls of our gun too much or otherwise; also whether we exert too great a stress or pressure upon the coating or projections and the grooves employed to give rotation.†

The first point mentioned is of very great importance, and, as already mentioned, the labours of the Committee on Explosives, under the presidency of Major-General Younghusband, F.R.S., from 1869 to the present date, have given us invaluable information and a mass of most useful data, and may be said indeed to have revolutionised the manufacture of our powder, and to have elucidated many a problem as to the action of the products of the charge upon both gun and projectile, which before had been insoluble.

Rodman's method of measuring.

Attempts had previously been made to measure experimentally the pressure exerted by the powder gas on the walls of the gun, and the American Ordnance officer Captain Rodman ‡ invented a pressure gauge in 1861, upon which that now employed by us was founded.

Means of determining Pressure in Gun.

Modes used by us to ascertain pressure in bore.

We employ two distinct methods§ of determining the pressure on the walls of the bore, viz.:—

Crusher gauges.

(I.) By means of "crusher gauges," which register the maximum pressure at various parts along the bore.

Chronoscope.

(II.) By the use of the chronoscope, which enables us to measure the velocity of the projectile at several successive points of its passage through the bore, and thence to calculate the mean pressure exerted by the gas at these points.

---

* Many of the experiments on which their formula are based were carried out with flat-headed projectiles, or with projectiles which upon striking the plate became such, practically.

† Vide Chapter II.

‡ Vide "Report of Experiments on the Properties of Metals for Cannon, and the "Qualities of Cannon Powder."—by Captain Rodman, Ordnance Department, U.S.A., 1861.

§ I have taken much of the following description out of a most useful paper on "Instruments for taking Velocities, &c.," by Captain Jones, R.A., in Proceedings, R.A.I., Vol. IX., No. 6.

‖ The idea of ascertaining the pressure by the amount of "crushing" or compression of small cylinders of copper, originated in the Royal Gun Factories, in which department these gauges were first manufactured.

Measuring Pressures by Crusher Gauges.

Description of Crusher gauge.

The crusher gauge ‖ is shown in the diagram on the next page. It consists of a steel bush (Fig. 1), of the same dimensions as the vent-bush used for all our service guns, provided with a moveable base or nozzle screwed on to the bush. The interior of the nozzle is hollowed out so as to form a chamber, in which is placed a steel anvil, with grooves along its surface communicating with a channel corresponding with the vent-channel in a vent bush. Thus any gas that may, by accident, find its way into the crusher chamber has a free escape into the air (this channel has been omitted in the crushers for the 80-ton gun). In the chamber is placed a small cylinder of copper,* *A* (Figs. 1 and 2), half-an-inch in length and $\frac{1}{11}$th of a square inch in sectional area. This cylinder is held lightly (but not prevented from expanding laterally) in the centre of the chamber, by a small watch spring. One end rests against the anvil, while the other end is acted on by a piston, *C* (Fig. 1), which is moveable in the nozzle, and is pressed against the

THE CRUSHER GAUGE.

Fig. 1.

Fig 2.

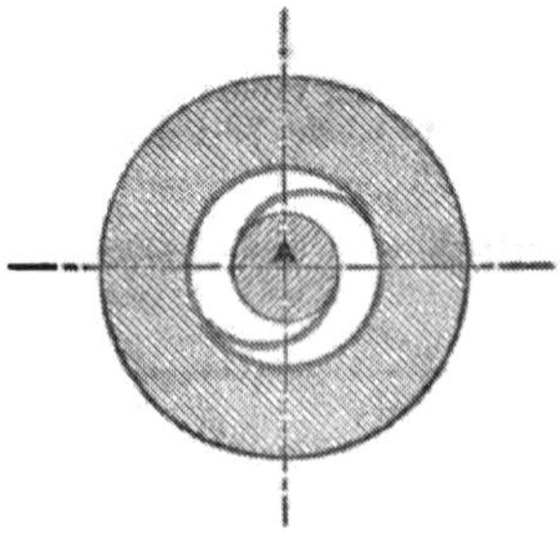

* Where extreme accuracy is required—as in the late experiments with the 80-ton gun—each copper cylinder is tested previous to being inserted in the gauge. This is done by crushing them all up to a given pressure (say 18 tons on the square inch)—rather less than the pressure which the crusher gauge is expected to record—which method of procedure has the additional advantage of eliminating some irregularities to which the crusher guage has been found, under certain circumstances, to be liable.

CHAP. XIII. copper cylinder by a small spring. The lower end of this piston is $\frac{1}{8}$th inch in area (under ordinary circumstances) and is open to the pressure of the powder gas; a small brass cup, or gas check, *D*, being inserted to prevent the gas from penetrating past the piston.

Copper crushers used.

The gauge is used as follows:—A hole is bored in the gun in the required position (see plate D.), into which the gauge is screwed until the end of the plug is flush with the surface of the bore. On the explosion of the charge, the copper cylinder is crushed between the piston and the anvil, and the amount of compression is an indication of the pressure exerted by the gas at that point. The actual pressure corresponding to any given compression of the crusher cylinder is arrived at by compressing a series of similar copper cylinders in the statical testing machine in the Royal Gun Factories, and tabulating the results.

The compression of the cylinder in the crusher gauge is ascertained by removing the plug from the gun, taking out the cylinder, and measuring its length by means of a micrometer. A reference to the tabulated results above mentioned, gives the pressure in tons on the square inch.

When it is not considered advisable to bore holes for the gauges in service guns, the pressure has been obtained at three points—viz. (1) by a crusher in the base of the projectile, (2) by a copper cup, in which a gauge is fastened, at the end of the bore, (3) by inserting a gauge in the vent bush hole, and firing the gun by electricity, wires being passed down the bore from the muzzle to the cartridge. By this means the pressure is approximately arrived at; but since the copper cup is liable to a certain amount of motion, the record given by the gauge inserted in it is not always reliable. The pressure recorded by the gauge in the base of the projectile, when charges of powder suitable to the calibre of the gun are employed, is always somewhat less than the maximum pressure in the powder chamber as shown by the other gauges.*

Regularity of pressures according to law of expansion of gases.

To give some idea of the pressures recorded by the gauges in a large gun, the following table is extracted from the Report of the Committee on Explosives on the 38-ton 12·5-in. gun.

TABLE showing the PRESSURES recorded by the CRUSHER GAUGES in the Bore of the 12·5-in. (38-ton) Gun.

No. of Round.	Charge and Projectile.	Pressure in tons per square inch at																	
		A	1	2	3	4	5	6	7	8	9	10	11	12	13	14	15	16	On base of shot.
72	130 lbs. of 1·5-in. powder, 800-lb. cylinder.	21·5	22·6	22·8	19·2	13·2	9·5	6·3	4·0	2·8	2·9	2·4	2·1	1·4	1·3	1·1	1·0	—	—
138	130 lbs. of 1·7-in. 800 lb. cylinder, with gas check.	27·3	26·5	24·5	—	19·3	13·9	8·9	4·8	3·8	3·2	3·1	2·4	1·6	1·4	1·3	1·2	0·7	—
139	130 lbs. of 1·5-in. 800 lb. cylinder, with gas check.	26·1	27·7	19·7	—	22·4	11·9	9·4	5·1	4·0	3·1	3·0	2·3	1·7	1·6	1·3	1·3	1·0	21·2
40	130 lbs. of 2-in, 800 lb. cylinder, with gas check.	28·7	24·5	22·8	—	18·3	12·4	9·5	4·6	3·2	3·1	2·6	2·3	1·7	1·6	1·3	1·1	0·8	19.3

* This is probably due to the fact that the inertia of the shot is overcome, and it therefore begins to move, before the maximum pressure is reached, and, being in motion, the gauge is not influenced to the same extent as those that are fixed in the walls of the gun.

*Measuring Pressure by use of the Chronoscope.**

Measuring pressure by means of chronoscope.

The instrument (vide plate p. 344) used for measuring the velocity at very small intervals along the bore is the Noble chronoscope.† It enables us to measure the time taken by the shot in passing over certain small successive spaces in the bore of the gun, and calculating therefrom the velocity of the shot in passing over each small space, and ultimately the mean pressure acting on it through this space.

The principle of action consists in registering, by means of electric currents upon a recording surface travelling at a uniform and very high speed, the precise instant at which a shot passes certain defined points in the bore.

It consists of two portions—firstly, the *mechanical arrangement* for obtaining the necessary speed and keeping that speed uniform; secondly, the *electrical recording arrangement.*

The first part of the instrument consists of a series of thin metal discs, *AA* (in fig. 1), each 36 ins. in circumference, fixed at intervals upon a horizontal shaft, *SS*, which is driven at a high speed by a heavy descending weight acting on a chain, *B*, arranged according to a plan originally proposed by Huyghens, through a train of gearing, *F*, mutiplying 200 times.

The *modus operandi* is as follows:—When the requisite speed of rotation has been attained, the stop clock is connected with the shaft *E*, and the time occupied by the wheel, *F*, in making five revolutions—that is, the time occupied by the discs, *A*, in making 1000 revolutions—is recorded.

After the wheel, *F*, has made one revolution, unconnected with the clock, the time of making 1,000 revolutions is again observed, and in the middle of this observation the gun is fired.

After the wheel, *F*, has again made one revolution, the time of 1,000 revolutions is once more recorded, and the instrument is stopped.

The arrangements for obtaining the *electrical records* are as follow:—The edges of the revolving discs are covered with smoke by means of an oil lamp, and are connected with one of the secondary wires, *G* (Fig. 1) of an induction coil. The other secondary wire, *H*, carefully insulated, is brought to a discharger, *Y*, opposite the edge of its corresponding disc, and is fixed so as to be just clear of the latter.

When a spark passes from the discharger to the disc, a minute portion of the lamp-black is burnt away upon that part of the disc which was opposite to the discharge at the instant of the passage of the spark. A distinct spot is left on the blackened disc, the lamp-black at that point having been burnt away by the spark, so that the vessel is shown beneath.

---

* The following description is taken from the first Progress Report of the Committee on Explosives, dated February 1870, as modified by Captain Jones, R.A.

† The invention of Captain A. Noble, F.R.S., late R.A.

The precise rate of the discs is ascertained by means of the stop clock, *D*, which can, at pleasure, be connected or disconnected with the revolving shaft, *E* (Fig. 1), and the time of making any number of revolutions of this shaft can be recorded with accuracy to the one-tenth part of a second.

The speed usually attained in working the instrument is about 1,100 ins. per second linear velocity at the circumference of the revolving discs, so that each inch travelled at that speed represents somewhat less than the one-thousandth part of a second; and as the inch is subdivided by the vernier, *V*, into a thousand parts, a linear representation, at the circumference, is thus obtained of intervals of time as minute as the one-millionth part of a second.

CHAP. XIII.

PLATE D.

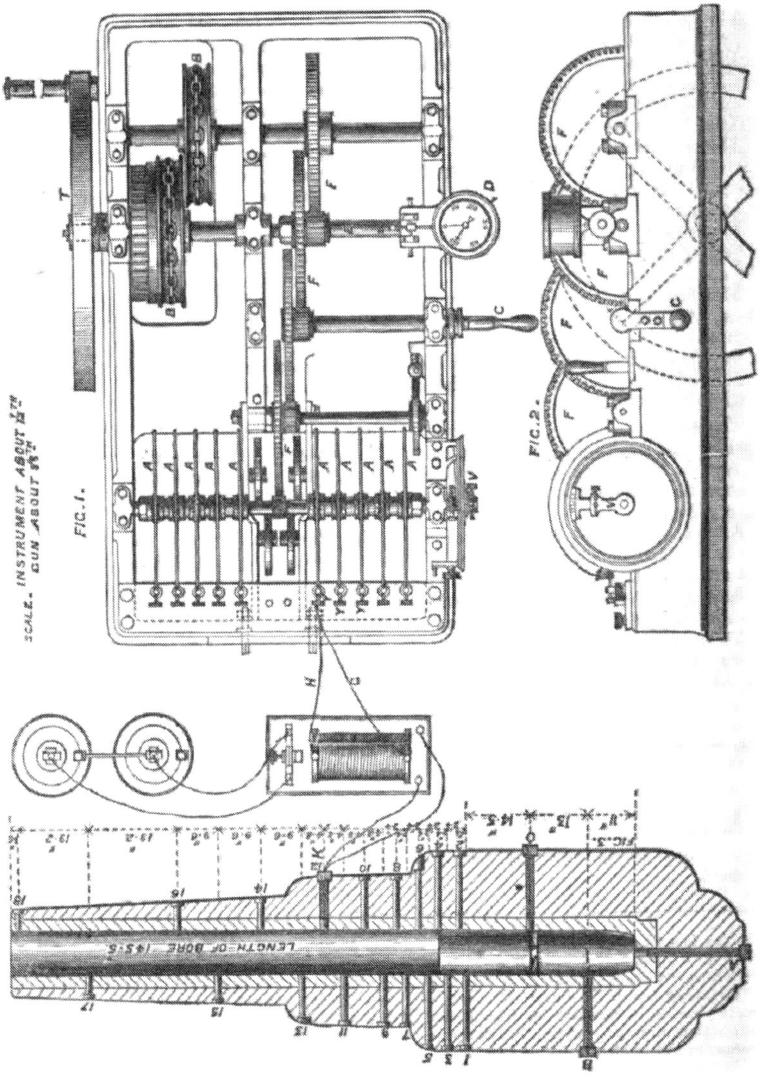

NOTE.—To avoid complication only one induction coil and cell is shown, a cell and coil being required for each disc.
The handwheel shown in Figs. 1 and 2 has been replaced by a lever.

CHAP. XIII.

The mode of connecting the primary wires of the induction coils with the bore of the gun in such a manner that the shot, in passing a defined point, shall sever the primary current, and thereby produce a spark from the secondary, is shown in the diagram above, which represent a longitudinal and a transverse section of the bore, *B*, of the gun, along which the shot *A*, is moving.

Cutting plugs.

A hollow plug, C, is screwed in to the gun, carrying at the end next the bore a cutter, *D*, which projects slightly into the bore. The cutter is held in this position by the primary wire, *e*, which passes in at one

Cutter Plugs.

Scale ½.

Fig. 2. Fig. 1.

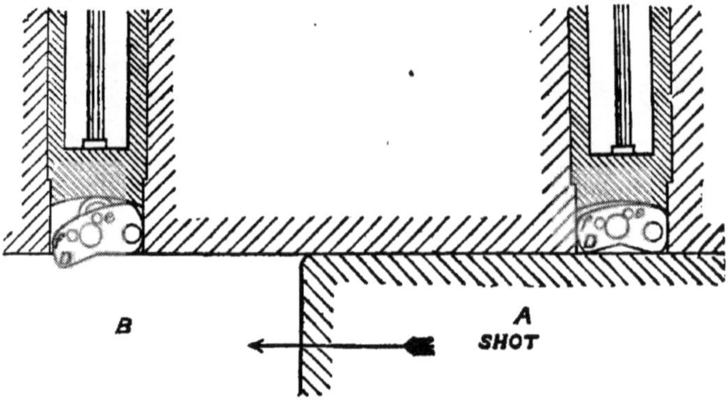

Fig. 3.

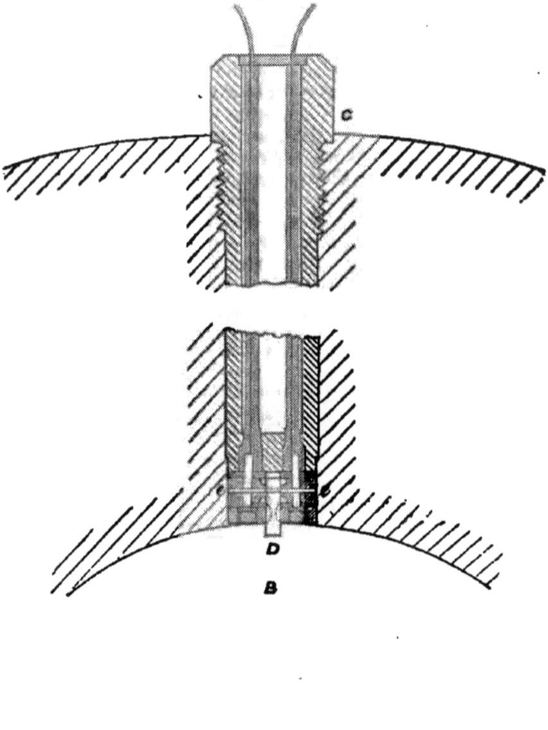

CHAP. XIII.

side of the plug, *C*, then through a hole in the cutter, *D*, and out again at the other side of the plug. The two ends of this wire are connected with the main wires leading to the instrument when the plugs have been fixed in this gun.

When the shot, *A*, is fired, it presses the cutter into the second position as shown in the figure, thereby severing the primary wire, and causing the induced spark instantaneously to pass from the discharger to the disc, its passage being marked by the spot left upon the edge as already described. To prevent a possibility of the cutter being forced down by any gas that may escape past the projectile, a safety pin, *f* secures the cutter firmly in its place. This safety pin is cut simultaneously with the primary wire.

The mode of reading the observations is as follows:—After an experiment has been made, and the series of sparks has been found on the discs, the mark on No. 1 disc is, by means of a micrometer screw brought precisely opposite to the point of the discharger: the vernier *V*, is then attached to the extremity of the shaft, *S*, is firmly fixed there, and is set at zero.

The spark of No. 2 disc is then brought precisely opposite to its discharger, the vernier reading is recorded, and so with the other discs in succession.

From these data are deduced, by correction and interpolation, the times occupied by the shot in passing over each two hundredth ($\frac{1}{200}$) of a foot. From the differences of these times are calculated the mean velocities over each of these minute spaces, and from the velocities the mean pressures necessary to produce them are obtained. These calculations are extremely laborious.

Experimental Gun prepared for Crusher and Cutter Plugs.

The gun shown in plate D. is an 18-ton 10-in. rifled M.L. gun, having a length of bore of 145·5 ins.

Guns as prepared for gauges.

It was tapped at three places in the powder chamber to receive crusher gauges. These holes could be closed with solid plugs when not required. The cutting plugs belonging to the chronoscope were 18 in number, and were fitted into holes at various intervals along the bore, as shown in the plate. When not required for the chronoscope these holes could be used for crusher gauges, or were closed with solid plugs. The projectiles used in these and similar experiments are cast iron cylinders, having a windage of only 0·01 in.

After the discharge, the coppers are taken from the crushers, stamped with the number of the round, and the measurements recorded.

We can employ with great advantage the data obtained as above, with regard to the pressure at different points of the bore, when working out the details of construction of a new gun as to the necessary thickness of metal in the several parts—For this purpose the Table p. is given—This table extracted from Captain Noble's work has been compiled from such results, and enables us at once to see what pressure the walls of the bore will have to withstand at different points along its length, when we know the proposed charge—Given the charge we can readily determine its density at any point, and thence by means of the table the corresponding pressure—We so arrive at an approximation as to the thickness of metal to be given.

Pressure on Grooves.

As regards the pressure upon the grooves of the gun or projectile, it has been already shown, p. 42, that, with our system of rifling, this will vary directly with the pressure on the walls of the bore if the twist be uniform, and if the twist be increasing it will also be in a certain direct relation to the pressure; it is therefore a matter of simple calculation to find what this pressure on the grooves is when we have, by the methods described, found the amount of pressure in the bore of the gun itself.

CHAP. XIII.

TABLE XXXIV.

TABLE of VELOCITIES of PROJECTILES fired with BATTERING CHARGES of P. POWDER from HEAVY RIFLED M.L. GUNS.*

Range.	-inch R.M.L. of tons. Charge, lbs. Projectile, lbs.	-inch R.M.L. of tons. Charge, lbs. Projectile, lbs.	17·72-inch R.M.L. of 100 tons. Charge, lbs. Projectile, lbs.	16-inch R.M.L. of 80 tons. Charge, 425 lbs. P². Projectile, 1,700 lbs.	12·5-inch R.M.L. of 38 tons. Charge, 130 lbs. P². Projectile, 800 lbs.	12-inch R.M.L. of 35 tons. Charge, 110 lbs. P². Projectile, 680 lbs.	Range.
	Velocity.	Velocity.	Velocity.	Velocity.	Velocity.	Velocity.	
yards.	feet.	feet.	feet.	feet.	feet.	feet.	yards.
0				1,520	1,425	1,300	0
100				1,510	1,413	1,289	100
200				1,500	1,401	1,278	200
300				1,490	1,389	1,267	300
400				1,480	1,377	1,256	400
500				1,470	1,365	1,246	500
600				1,460	1,354	1,236	600
700				1,450	1,343	1,226	700
800				1,440	1,332	1,216	800
900				1,431	1,321	1,206	900
1,000				1,432	1,310	1,196	1,000
1,100				1,413	1,299	1,186	1,100
1,200				1,404	1,288	1,175	1,200
1,300				1,395	1,277	1,167	1,300
1,400				1,386	1,267	1,158	1,400
1,500				1,377	1,257	1,149	1,500
1,600				1,368	1,247	1,140	1,600
1,700				1,359	1,237	1,132	1,700
1,800				1,350	1,227	1,124	1,800
1,900				1,341	1,218	1,116	1,900
2,000				1,332	1,209	1,108	2,000
2,100				1,323	1,200	1,100	2,100
2,200				1,315	1,191	1,092	2,200
2,300				1,307	1,182	1,084	2,300
2,400				1,299	1,173	1,077	2,400
2,500				1,291	1,164	1,070	2,500
2,600				1,283	1,155	1,063	2,600
2,700				1,275	1,147	1,056	2,700
2,800				1,267	1,139	1,049	2,800
2,900				1,259	1,131	1,043	2,900
3,000				1,251	1,123	1,037	3,000
3,100				1,243	1,115	1,031	3,100
3,200				1,235	1,107	1,025	3,200
3,300				1,228	1,099	1,020	3,300
3,400				1,221	1,092	1,015	3,400
3,500				1,214	1,085	1,010	3,500
3,600				1,207	1,078	1,005	3,600
3,700				1,200	1,071	1,000	3,700
3,800				1,193	1,064	995	3,800
3,900				1,186	1,058	991	3,900
4,000				1,179	1,052	987	4,000
4,100				1,172	1,046	983	4,100
4,200				1,165	1,040	979	4,200
4,300				1,158	1,034	975	4,300
4,400				1,151	1,029	971	4,400
4,500				1,145	1,024	967	4,500
4,600				1,139	1,019	963	4,600
4,700				1,133	1,014	959	4,700
4,800				1,127	1,009	955	4,800

* Calculated by Professor Bashforth's formula.

TABLE XXXIV.—VELOCITIES of PROJECTILES—*continued.* CHAP. XIII.

Range.	12-inch R.M.L. of 25 tons. Charge, 85 lbs. P. Projectile, 600 lbs.	11-inch R.M.L. of 25 tons. Charge, 85 lbs. P. Projectile, 535 lbs.	10-inch R.M.L. of 18 tons. Charge, 70 lbs. P. Projectile, 400 lbs.	9-inch R.M.L. of 12 tons. Charge, 50 lbs. P. Projectile, 250 lbs.	8-inch R.M.L. of 9 tons. Charge, 35 lbs. P. Projectile, 180 lbs.	7-inch R.M.L of 6½ tons, Charge, 30 lbs. P. Projectile, 115 lbs.	Range.
	Velocity.	Velocity.	Velocity.	Velocity.	Velocity.	Velocity.	
yards.	feet.	feet.	feet.	feet.	feet.	feet.	yards.
0	1,300	1,315	1,364	1,420	1,413	1,525	0
100	1,286	1,302	1,350	1,399	1,391	1,496	100
200	1,273	1,289	1,336	1,379	1,369	1,467	200
300	1,260	1,277	1,322	1,360	1,348	1,438	300
400	1,248	1,265	1,308	1,341	1,327	1,409	400
500	1,236	1,253	1,294	1,322	1,306	1,382	500
600	1,224	1,242	1,280	1,304	1,286	1,356	600
700	1,212	1,231	1,267	1,277	1,267	1,331	700
800	1,201	1,220	1,254	1,270	1,248	1,317	800
900	1,190	1,209	1,241	1,253	1,230	1,284	900
1,000	1,179	1,199	1,228	1,236	1,213	1,261	1,000
1,100	1,168	1,189	1,216	1,220	1,196	1,239	1,100
1,200	1,157	1,179	1,204	1,204	1,180	1,217	1,200
1,300	1,147	1,169	1,192	1,189	1,165	1,197	1,300
1,400	1,137	1,159	1,181	1,174	1,150	1,177	1,400
1,500	1,127	1,149	1,170	1,160	1,136	1,158	1,500
1,600	1,118	1,139	1,159	1,147	1,122	1,141	1,600
1,700	1,109	1,130	1,148	1,134	1,109	1,124	1,700
1,800	1,101	1,122	1,138	1,121	1,097	1,108	1,800
1,900	1,093	1,114	1,128	1,109	1,085	1,093	1,900
2,000	1,085	1,106	1,118	1,097	1,074	1,078	2,000
2,100	1,077	1,098	1,108	1,086	1,063	1,065	2,100
2,200	1,069	1,090	1,099	1,075	1,052	1,053	2,200
2,300	1,062	1,082	1,090	1,065	1,042	1,041	2,300
2,400	1,055	1,075	1,082	1,055	1,032	1,030	2,400
2,500	1,048	1,068	1,074	1,046	1,023	1,019	2,500
2,600	1,041	1,061	1,066	1,037	1,014	1,009	2,600
2,700	1,034	1,054	1,059	1,028	1,005	999	2,700
2,800	1,028	1,048	1,052	1,020	996	990	2,800
2,900	1,022	1,042	1,045	1,012	988	982	2,900
3,000	1,016	1,036	1,038	1,004	980	974	3,000
3,100	1,010	1,030	1,031	997	973	967	3,100
3,200	1,004	1,024	1,025	990	965	960	3,200
3,300	998	1,018	1,019	984	957	953	3,300
3,400	993	1,012	1,013	978	949	946	3,400
3,500	988	1,007	1,007	972	942	939	3,500
3,600	984	1,002	1,002	966	935	932	3,600
3,700	980	998	997	960	928	925	3,700
3,800	976	994	992	955	921	919	3,800
3,900	972	990	987	950	914	913	3,900
4,000	968	986	982	945	907	907	4,000
4,100	964	982	977	940	900	901	4,100
4,200	960	978	972	935	894	895	4,200
4,300	956	974	967	930	888	889	4,300
4,400	952	970	963	925	882	884	4,400
4,500	948	966	958	920	876	879	4,500
4,600	944	962	954	915	870	874	4,600
4,700	940	958	950	910	864	869	4,700
4,800	936	955	946	906	858	864	4,800

TABLE XXXV.

TABLE showing the ENERGY of PROJECTILES fired with BATTERING CHARGES of P. POWDER from HEAVY RIFLED M.L. GUNS.

Range.	-inch R.M.L. of tons. Charge, lbs. Projectile, lbs.			-inch R.M.L. of tons. Charge, lbs. Projectile, lbs.			17·72 inch R.M.L. of 100 tons. Charge, lbs. Projectile, lbs.			16-inch R.M.L. of 80 tons. Charge, 425 lbs. P². Projectile, 1,700 lbs.			12·5-inch R.M.L. of 38 tons. Charge 130 lbs. P². Projectile, 800 lbs.			12-inch R.M.L. of 35 tons. Charge, 110 lbs. P. Projectile, 700 lbs.			Range.
	v.	Total Energy.	Energy per Inch of Shot's Circumference.	v.	Total Energy.	Energy per Inch of Shot's Circumference.	v.	Total Energy.	Energy per Inch of Shot's Circumference.	v.	Total Energy.	Energy per Inch of Shot's Circumference.	v.	Total Energy.	Energy per Inch of Shot's Circumference.	v.	Total Energy.	Energy per Inch of Shot's Circumference.	
yards.	feet.	foot tons.	foot tons.	feet.	foot tons.	foot tons.	feet.	foot tons.	foot tons.	feet.	foot tons.	foot tons.	feet.	foot tons.	foot tons.	feet.	foot tons.	foot tons.	yards.
0										1,520	27,220	542	1,425	11,268	287	1,300	8,200	217	0
200										1,500	26,520	528	1,401	10,892	277	1,278	7,929	210	200
400										1,480	25,820	514	1,377	10,520	268	1,256	7,655	203	400
600										1,460	25,125	500	1,354	10,170	259	1,236	7,413	196	600
800										1,440	24,440	486	1,332	9,840	250	1,216	7,175	190	800
1,000										1,422	23,825	474	1,310	9,520	242	1,196	6,943	184	1,000
1,200										1,404	23,220	462	1,288	9,204	234	1,176	6,714	178	1,200
14,00										1,386	22,640	450	1.267	8,904	226	1,152	6,507	172	1,400
1,600										1,368	22,055	438	1,247	8,621	219	1,140	6,307	167	1,600
1,800										1,350	21,475	427	1,227	8,347	212	1,124	6,129	162	1,800
2,000										1,332	20,910	416	1,209	8,101	206	1,108	5,954	158	2,000
2,200										1,315	20,375	406	1,191	7,867	200	1,092	5,786	154	2,200
2,400										1,299	19,890	396	1,173	7,631	194	1,077	5,631	150	2,400
2,600										1,283	19,390	386	1,155	7,400	188	1,063	5,482	146	2,600
2,800										1,267	18,915	376	1,139	7,195	183	1,049	5,339	142	2,800
3,000										1,251	18,440	367	1,123	6,991	178	1,037	5,217	138	3,000
3,200										1,235	17,965	358	1,107	6,791	173	1,025	5,098	135	3,200
3,400										1,221	17,570	349	1,092	6,611	168	1,015	4,998	132	3,400
3,600										1,207	17,160	341	1,078	6,447	164	1,005	4,900	129	3,600
3,800										1,193	16,764	333	1,064	6,280	160	995	4,804	127	3,800
4,000										1,179	16,389	326	1,062	6,137	156	987	4,728	125	4,000

TABLE XXXV., showing the ENERGY of PROJECTILES—*continued.*

Range.	12-inch R.M.L. of 25 tons. Charge 85 lbs. P. Projectile, 600 lbs.			11-inch R.M.L. of 25 tons. Charge, 85 lbs. P. Projectile, 535 lbs.			10-inch R.M.L. of 18 tons. Charge, 70 lbs. P. Projectile, 400 lbs.			9-inch R.M.L. of 12 tons. Charge, 50 lbs. P. Projectile, 250 lbs.			8-inch R.M.L. of 9 tons. Charge, 35 lbs P. Projectile, 180 lbs.			7-inch R.M.L. of 6½ tons. Charge, 30 lbs. P. Projectile, 115 lbs.		
	v.	Total Energy.	Energy per Inch of Shot's Circumference.	v.	Total Energy.	Energy per Inch of Shot's Circumference.	v.	Total Energy.	Energy per Inch of Shot's Circumference.	v.	Total Energy.	Energy per Inch of Shot's Circumference.	v.	Total Energy.	Energy per Inch of Shot's Circumference.	v.	Total Energy.	Energy per Inch of Shot's Circumference.
yards.	feet.	foot-tons.	foot-tons.	feet.	foot-tons.	foot-tons.	feet.	foot-tons.	foot-tons.	feet.	foot-tons.	foot-tons.	feet.	foot-tons.	foot. tons.	feet.	foot-tons.	foot-tons.
0	1,300	7,030	188	1,315	6,415	187	1,364	5,160	165·6	1,420	3,496	124·7	1,413	2,492	100·2	1,525	1,855	85·3
200	1,273	6,750	181	1,289	6,165	180	1,336	4,950	158·9	1,379	3,297	117·6	1,369	2,339	94·0	1,467	1,716	78·9
400	1,248	6,480	174	1,265	5,935	173	1,308	4,745	152·3	1,341	3,117	111·2	1,327	2,198	88·3	1,409	1,583	72·8
600	1,224	6,230	167	1,242	5,720	167	1,280	4,575	146·8	1,304	2,948	105·2	1,286	2,064	83·0	1,356	1,466	67·5
800	1,201	6,000	161	1,220	5,520	161	1,254	4,360	140·0	1,270	2,796	99·8	1,246	1,944	78·1	1,317	1,383	63·6
1,000	1,179	5,780	155	1,199	5,335	156	1,228	4,185	134·2	1,236	2,648	94·5	1,213	1,837	73·8	1,261	1,268	58·3
1,200	1,157	5,570	149	1,179	5,155	150	1,204	4,020	129·0	1,204	2,513	89·7	1,180	1,738	69·9	1,217	1,181	54·3
1,400	1,137	5,380	144	1,159	4,985	145	1,181	3,870	124·1	1,174	2,389	85·3	1,150	1,651	66·3	1,177	1,105	50·8
1,600	1,118	5,200	139	1,139	4,815	140	1,159	3,725	120·0	1,147	2,281	81·4	1,122	1,571	63·2	1,141	1,038	47·8
1,800	1,101	5,040	135	1,122	4,670	136	1,138	3,590	115·3	1,121	2,178	77·7	1,097	1,502	60·4	1,108	979	45·0
2,000	1,085	4,890	131	1,106	4,540	132	1,118	3,470	111·2	1,097	2,086	74·4	1,074	1,440	57·9	1,078	927	42·6
2,200	1,069	4,750	127	1,090	4,410	129	1,099	3,350	107·5	1,075	2,003	71·5	1,052	1,381	55·5	1,053	884	40·7
2,400	1,055	4,630	124	1,075	4,285	125	1,082	3,245	104·2	1.055	1,930	68·9	1,032	1,329	53·4	1,030	846	38·9
2,600	1,041	4,510	120	1,061	4,175	122	1,066	3,150	101·1	1,037	1,864	66·5	1,014	1,283	51·6	1,009	812	37·3
2,800	1,028	4,400	117	1,048	4,075	119	1,052	3,070	98·5	1,020	1,804	64·4	996	1,238	49·8	990	782	36·0
3,000	1,016	4,300	115	1,036	3,980	116	1,038	2,990	95·9	1,004	1,747	62·4	980	1,199	48·2	973	755	34·7
3,200	1,004	4,200	112	1,024	3,890	113	1,025	2,915	93·5	990	1,699	60·6	965	1,162	46·7	958	732	33·7
3,400	993	4,100	110	1,012	3,800	111	1,013	2,845	91·3	978	1,658	59·2	952	1,131	45·5	945	712	32·8
3,600	984	4,030	108	1,002	3,725	109	1,002	2,785	89·4	966	1,618	57·7	940	1.103	44·3	934	696	32·0
3,800	976	3,960	106	994	3,665	107	992	2,730	87·6	955	1,581	56·4	930	1,080	43·4	924	681	31·3
4,000	968	3,900	104	986	3,605	105	982	2,675	85·8	945	1,548	55·2	920	1,056	42·5	916	669	30·8

NOTE.—The 35-ton 12-inch gun fired with a charge of 120 lbs. of P. powder gives the 700 lbs. projectile an initial velocity of about 1,315 feet and total energy of 8,404 foot-tons, or about 232 foot-tons per inch of shot's circumference.

TABLE XXXVI.*

TABLE showing Mode of Comparing results when firing an Experimental Gun with varying Charges of Powder and different Weights of Projectiles

EXPERIMENTS with 80-ton GUN, Expl. No., Calibre 16 inches.

WORK REALISED.

1	2	3	4	5		6	7	8	9	10	11	12	13	14	15	16
				Velocity.				Energy.					Energy.			
								Total.					Total.		Per Inch of Shot's Circumference.	
No. of Round.	Powder.	Charge.	Projectile.	Muzzle.	At 1,000 Yds.	Maximum Pressure.	Mean pressure on Powder-chamber (A, 1, 2, and 3 gauges).	Theoretic.	Realised.	Per Pound of Powder.	Per Ton of Mean pressure on Powder-Chamber.	Factor of Effect.	At the Muzzle.	At 1,000 Yds.	At the Muzzle.	At 1,000 Yds.
		lbs.	lbs.	ft.	ft.	tons.	tons.	ft.-tons.	ft.-tons.	ft.-tons.	ft.-tons.		ft.-tons.	ft.-tons.	ft.-tons.	ft.-tons.
6	1·5 inch of 22· 3·75	240	1259	1550	1442	25·9	—†	23,040	20,951	87·3	—†	90·9	20,951	18,149	459	358
10	1·7 ,, ,, 13· 4·74	240	1259	1470	1366	20·9	19·5	23,040	18,860	78·6	967·2	81·9	18,860	16,285	414	357
24	2 ,, ,, 1·11·75	220	1259	1493	1388	24·8	22·2	21,762	19,455	88·4	876·2	89·4	19,465	16,814	427	369
10	2 ,, ,, do.	220	1465	1366	1206	27·3	24.4	21,762	18,950	86·	776·6	87·1	18,950	16,795	416	369

* Of this table, Columns 1 to 7 require no explanation; Column 8 gives the theoretic work as calculated from Captain Noble's tables (vide note I, p. , and Column Energy calculated maximum in Table XXXVII.); Column 9 gives actual work realized upon the projectile $\left(\frac{W.V^2}{2g}\right.$ reduced to foot tons$\left.\right)$; Column 12 gives the ratio between 8 and 9, or the percentage of theoretic work actually realized; Column 10 gives the amount of work impressed on the projectile per pound of powder.

Other things being equal it is evident that the smaller the weight of the shot as compared with that of the cartridge the greater the amount of work realized; Column 11 shows the ratio of mean pressure in the powder chamber to the work realized. As the velocity increases a greater increase of pressure will of course be required to produce a constant increment of velocity and so of work. The modes at which we arrive at the contents of Columns 13 to 15 are explained at pp. and .

† With this round the mean pressure was not taken.

TABLE XXXVII.

Table of Pressures; actually found to exist in the Bores of Guns with Pebble or P². Powder, in terms of density.

Mean density of products of combustion.	Pressures observed—Tons per square inch.	Difference in tons — ·01 density.	Mean density of products of combustion.	Pressures observed—Tons per square inch.	Difference in tons — ·01 density.
·90	20·35		·60	11·33	
		·35			·26
·88	19·65		·58	10·81	
		·34			·25
·86	18·87		·56	10·31	
		·34			·24
·85	18·63		·55	10·07	
		·33			·24
·84	17·97		·54	9·59	
		·32			·24
·82	17·33		·52	9·11	
		·32			·24
·80	17·01		·50	8·87	
		·31			·23
·78	16·39		·48	8·41	
		·31			·23
·76	15·78		·46	7·95	
		·30			·22
·75	15·48		·45	7·73	
		·30			·22
·74	14·88		·44	7·29	
		·29			·22
·72	14·31		·42	7·86	
		·28			·21
·70	14·03		·40	6·65	
		·28			·21
·68	13·47		·38	6·21	
		·28			·21
·66	12·92		·35	5·63	
		·27			·19
·65	12·65		·30	4·67	
		·27			·18
·64	12·11		·25	3·77	
		·26			
·62	11·59				
		·26			

Extracted from work by Captain Noble, F.R.S. and Mr. F. Abel, C.B., F.R.S. on Fired Gunpowder.

TABLE XXXVII*a*.

Table of Densities and Volumes of Charges referred to cubic inches of space per pound of powder.

Cubic ins. per lb.	Densities.	Volumes.	Cubic ins. per lb.	Densities.	Volumes.	Cubic ins. per lb.	Densities.	Volumes.
20	1·386	·721	32	·866	1·155	45	·616	1·623
21	1·320	·757	33	·840	1·190	46	·603	1·658
22	1·260	·794	34	·816	1·225	47	·590	1·695
23	1·206	·829	35	·792	1·263	48	·578	1·730
24	1·155	·866	36	·770	1·299	49	·566	1·767
25	1·109	·902	37	·749	1·335	50	·554	1·805
26	1·066	·938	38	·730	1·370	51	·544	1·838
27	1·027	·974	39	·711	1·406	52	·533	1·876
27·73	1·000	1.000	40	·693	1·443	53	·523	1·912
28	·990	1.010	41	·676	1·479	54	·513	1·949
29	·956	1·046	42	·660	1·515	55·46	·500	2·000
30	·924	1·082	43	·645	1·550			
31	·894	1·118	44	·630	1·587			

CHAPTER XIII.—NOTE I.

Theoretical determination of M.V. &c., By Captain A. Noble's Tables.

"From Table XXXVIII we can at once ascertain what is the maximum work to be obtained with any given gun and charge of powder of the same description as our service powders.

Calculation of theoretical work.

"To make use of the table we have only to find the volume occupied by the charge (gravimetric density = 1),* and the number of times this volume is contained in the bore of the gun. The total work per pound which the powder is capable of performing during the given expansion is then taken from the table; and this work being multiplied by the number of pounds in the charge gives the total maximum work. Thus, for example, an 18-ton 10-inch gun, a charge of 70 lbs. pebble powder is fired, and we wish to know what is the maximum work that the charge is capable of performing. We readily find that the length of the gun is such that volume of bore = v = 5·867 vols.;† and from the table we find that 89·4 foot-tons is the maximum work per pound; multiplying by the number of pounds we find that 6258 foot-tons is the maximum work which the whole charge is capable of performing.

"As a matter of course, this maximum effect is only approximated to, not attained; and for actual use it would be necessary to multiply the work so calculated by a factor dependent upon the nature of the powder, the mode of firing it, the weight of the shot, &c.; but in service powders fired under the same circumstances the factor will not vary much. In the experimental powders used by the Committee on Explosives there were, it is true, very considerable differences, the work realized in the same gun varying from 56 foot-tons to 86 foot-tons per pound of powder; but with service powders fired under like conditions this great difference does not exist."

Definition of Factor of effect.

Table XXXIX has been calculated (by Captain Noble) from the above data, and shows:—First, the total work realized per pound of powder burned for every gun, charge, and description of powder in the English service. Secondly, the maximum theoretic work per pound of powder it would be possible to realize with each gun and charge; and thirdly, the factor of effect with each gun and charge, that is, the per-centage of the maximum effect actually realized.

F.E. increases from small to large guns.

If the factors of effect be examined it will be observed how, in spite of the use of slow burning and therefore uneconomical powders in the large guns, the per-centages realized gradually increase as we pass from the smallest to the largest gun in our table; the highest factor being 93 per cent. in the case of a 38-ton gun, the lowest being 50·5 per cent. in the case of the little Abyssinian gun: this difference in effect is no doubt in great measure due to the communication of heat to the bore of the gun.

Examples of use of tables.

As an example to show how these tables can be utilized, we will take the case of a 10-inch R.M.L. gun and compare its actual M.V. as determined by experiment with the results given by the tables.

(1.) To determine M.V.

This experimental 10-inch gun had a length of bore of 145″ and its charge was 70 lbs. of P. powder, while the shot weighed 300 lbs. and the M.V. obtained was 1527 f.s.

We shall find that the number of volumes of expansion in the bore of the charge (gravimetric density = 1) will be 5·867, and looking at the table we see that the corresponding total work is 6258 foot-tons; this has to be multiplied by the factor of effect (depending on the nature of powder and weight of shot), which in this case is 0·7796. And 6258 × 0·7796 gives us W = total work realized

= 4880 foot-tons;

---

* *i.e.*, When a pound of powder is taken as occupying a space of 27·7 cubic inches.

† For length of bore = 145 inches:

$$\because \text{ If } v = \text{volume of bore} = 145 \times 25\pi$$

$$v^1 = \text{ ,, charge (70 lbs.)} = 70 \times 27{\cdot}7$$

$$\text{Then } v = 5{\cdot}867\ v^1$$

substitute this in equation

$$W = \frac{wV^2}{2g} \text{ where } w = \text{projectile, in lbs.}$$

$$V = \text{M.V. in f.s.}$$

$$V = \sqrt{\frac{2.g}{w} W} = \sqrt{\frac{64 \cdot 4 \times 4880 \times 2240}{300}}$$

or V = 1532, which only differs by 5 feet from the ascertained velocity 1527 f.s.

Example (2), To determine length of bore

What should be the length of bore of a 10-inch gun, so that it should have a M.V. of 1650 f.s. and fire a 400 lb. projectile with a charge of 100 lbs. P. powder—

$$\text{Work to be realized} = \frac{400 \times 1650^2}{2240 \times 2g} = 7551 \text{ f.t.}$$

$$\text{Theoretical work, the factor of effect being } \cdot 826 = \frac{7551}{\cdot 826} = 9141 \text{ f.t.}$$

Theoretical work per lb. of powder = 91·41 f.t. = 6·2500 expansions nearly, by Table—∴ vol. of bore = 100 × 27·7 × 6·25 = 17·312 cubic inches

$$\text{and length of bore} = \frac{17312}{25\pi} = 220 \text{ inches.}$$

TABLE XXXVIII.

Giving Total Work in Foot-Tons, per lb. of Powder burnt, which Gunpowder is capable of performing in the bore of a gun, in terms of the Number of Expansions of the Charge.*†

Per lb. burned in foot-tons.	Number of vol. of expansions.	Difference for *one* foot-tons.	Per lb. burned in foot-tons.	Number of vol. of expansions.	Difference for *one* foot-ton.	Per lb. burned in foot-tons.	Number of vol. of expansions.	Difference for *one* foot-ton.
75	3·9652	·1120	95	6·9335	·2202	114	12·5703	·4300
76	4·0772	·1130	96	7·1537	·2209	115	13·0008	·4400
77	4·1902	·1160	97	7·3746	·2222	116	13·4403	·4482
78	4·3062	.1191	98	7·5968	·2455	117	13·8865	·4692
79	4·4253	·1225	99	7·8423	·2455	118	14·3577	·4900
80	4·5478	·1245	100	8·0378	·2455	119	14·8477	·5100
81	4·6723	·1275	101	8·3333	·2686	120	15·3577	·5290
82	4·7998	·1300	102	8·6019	·2686	121	15·8867	·5450
83	4·9298	·1325	103	8·8705	·2893	122	16·4317	·5714
84	5·0623	·1388	104	9·1598	·2916	123	17·0031	·5900
85	5·2011	·1448	105	9·4514	·2950	124	17·5931	·6050
86	5·3459	·1515	106	9.7464	·2970	125	18·1981	·6300
87	5·4974	·1580	107	10.0434	·3300	126	18·8281	·6460
88	5·6554	·1665	108	19·3734	·3300	127	19·4741	·6645
89	5·8219	·1731	109	10·7034	·3750	128	20·1386	·6880
90	5·9950	·1744	110	11·0580	·3750	129	20·8266	·7100
91	6·1690	·1858	111	11·4280	·3763	130	21·5366	·7450
92	6·3542	·1902	112	11·8043	·3630	131	22·2816	·7900
93	6·5444	·1911	113	12·1873	·3830	132	23·0716	
94	6·7355	·1980						

* This Table is extracted from the work by Captain Noble, F.R.S., and F. A. Abel, Esq., C.B., F.R.S., on Fired Gunpowder, by permission of the Authors.

† Assuming the maximum prepare in a closed vessel, at the time of explosion to be 41·5 tons per square inch, when the gravimetric density of the charge is unity.

CAPTAIN NOBLE AND MR. F. A. ABEL ON FIRED GUNPOWDER.*

TABLE XXXIX.—Giving, with the data necessary for Calculation, the Work per lb. of Powder realised, the Total maximum Theoretic Work, and the Factor of Effect for every Gun and Charge in the British Service.

Nature of Gun.	Bore.		Charge.		Projectile.		Gas.		Energy of Powder.			
	Diameter.	Length.	Nature.	Weight.	Weight.	Velocity.	Total Volumes in Bore.	Final Density.	Total.	Realised per lb. of Powder.	Calculated Maximum.	Factor of Effect.
	inches.	calibres.		lbs. ozs.	lbs.	ft. per sec.			foot-tons.	foot-tons.	foot-tons.	
38 tons, M.L.	12·5	16·5	P.	110 0	700	1430	7·342	·1362	9932	90·3	97·0	93·1
35 tons	12	13·5	P.	110 0	700	1300	6·007	·1665	8209	74·6	90·2	82·7
25 tons	12	12·0	P.	85 0	600	1300	6·910	·1447	7036	82·8	94·9	87·3
			P.	85 0	495	1358	6·910	·1447	6334	74·5	94·9	78·6
			P.	55 0	495	1142	10·679	·0936	4479	81·4	108·9	74·8
			R.L.G.	67 0	600	1180	8·765	·1141	5797	86·4	102·8	84·1
			R.L.G.	67 0	495	1271	8·765	·1141	5549	82·8	102·8	80·6
			R.L.G.	50 0	495	1140	11·750	·0851	4464	89·3	111·8	80·0
25 tons	11	13·2	P.	85 0	535	1315	5·585	·1708	6419	75·5	89·2	84·7
			P.	85 0	535	1315	5·855	·1700	6419	75·5	89·2	84·7
			R.L.G.	70 0	535	1217	7·109	·1407	5498	78·6	95·8	82·1
			R.L.G.	70 0	535	1217	7·109	·1407	5498	78·6	95·8	82·1
18 tons	10	14·5	P.	70 0	400	1364	5·867	·1704	5164	73·8	89·4	82·6
			P.	70 0	400	1340	5·867	·1704	4984	71·2	89·4	79·7
			P.	44 0	400	1125	9·334	·1071	3513	79·8	104·7	76·3
			R.L.G.	60 0	400	1298	6·844	·1461	4676	77·9	94·5	82·4
			R.L.G.	40 0	400	1117	10·269	·0974	3463	86·6	107·9	80·3
12 tons	9	13·9	P.	50 0	250	1420	5·742	·1742	3498	70·0	88·6	79·1
			P.	50 0	250	1420	5·742	·1742	3498	70·0	88·6	79·1
			R.L.G.	43 0	250	1336	6·683	·1496	3096	72·0	93·6	77·1
			R.L.G.	43 0	250	1336	6·683	·1496	3096	72·0	93·6	77·1
			R.L.G.	30 0	250	1192	9·566	·1045	2465	82·2	105·2	78·2
9 tons	8	14·8	P.	35 0	180	1413	6·136	·1630	2493	71·3	90·9	78·4
			P.	35 0	180	1413	6·136	·1630	2493	71·3	90·9	78·4
			R.L.G.	30 0	180	1330	7·154	·1398	2209	73·7	96·0	76·8
			R.L.G.	30 0	180	1330	7·154	·1398	2209	73·7	96·0	76·8
			R.L.G.	20 0	180	1163	10·724	·0932	1689	84·5	109·1	77·6
7 tons	7	18·0	P.	30 0	115	1561	5·827	·1716	1945	64·8	89·0	72·9
			P.	30 0	115	1561	5·827	·1716	1945	64·8	89.0	72·9
			R.L.G.	22 0	115	1468	7·948	·1258	1696	77·1	99·4	77·6
			R.L.G.	22 0	115	1468	7·948	·1258	1696	71·1	94·4	77·6
			R.L.G.	14 0	115	1258	12·495	·0800	1263	90·2	113·3	79·7

* Table used by permission of the authors.

TABLE XXXIX.—*continued.*

Nature of Gun.	Bore.		Charge.		Projectile.		Gas.		Energy of Powder.			
	Diameter.	Length.	Nature.	Weight.	Weight.	Velocity.	Total Volumes in Bore.	Final Density.	Total.	Realized per lb. of Powder.	Calculated maximum.	Factor of Effect.
	inches.	calibres.		lbs. oz.	lbs.	ft. per sec.			foot-tons.	foot-tons.	foot-tons.	
			B.	30 0	115	1525	5·148	·1943	1856	61·9	84·6	73·2
			P.	30 0	115	1525	5·148	·1943	1856	61·9	84·6	73·2
6¼ tons	7	15·9	R.L.G.	22 0	115	1430	7·021	·1424	1632	74·2	95·5	77·7
			R.L.G.	22 0	115	1430	7·021	·1424	1632	74·2	95·5	77·7
			R.L.G.	14 0	115	1230	11·039	·0906	1207	86·2	110·0	78·4
80-pr. of 5 tons	6·3	18·0	L.G.	10 0	80	1240	12·748	·0784	835·5	85·4	114·1	74·9
64-pr. of 64 cwt. wrt.-iron ...	6·3	15·5	R.L.G.	8 0	64	1253	13·715	·0729	696·1	87·0	116·0	75·1
			L.G.	8 0	64	1229	13·715	·0729	670·8	83·8	116·0	72·3
64-pr. of 58 cwt.	6·3	17·2	R.L.G.	8 0	64	1245	15·234	·0656	688·3	86·0	118·7	72·5
64-pr. of 71 cwt.	6·3	16·4	R.L.G.	8 0	64	1230	14·518	·0689	671·9	84·0	117·3	71·6
			R.L.G.	8 0	40	1357	6·830	·1464	511·1	63·9	94·6	67·6
40-pr. of 35 cwt.	4·75	18·0	R.L.G.	7 0	40	1336	7·805	·1281	495·4	70·8	99·1	71·5
			R.L.G.	6 0	40	1305	9·105	·1098	472·7	78·8	103·8	76·0
			R.L.G.	5 0	25	1355	6·518	·1534	318·5	63·7	92·8	68·7
25-pr. of 18 cwt.	4·0	18·0	R.L.G.	4 8	25	1320	7·244	·1380	302·3	67·2	96·4	69·8
			R.L.G.	4 0	25	1278	8·151	·1227	283·3	70·8	100·4	70·5
			R.L.G.	3 0	16	1352	8·365	·1195	202·9	67·6	101·0	67·9
16-pr. of 12 cwt.	3·6	19·0	R.L.G.	2 8	16	1273	10·043	·0996	179·9	72·0	106·8	67·5
			R.L.G.	2 0	16	1167	12·541	·0797	151·2	75·6	113·4	66·6
			R.L.G.	1 12	9	1381	9·320	·1073	119·1	68·0	104·6	65·1
9-pr. of 8 cwt.	3·0	21·3	R.L.G.	1 8	9	1325	10·865	·0920	109·6	73·1	109·5	66·9
			R.L.G.	1 4	9	1203	13·026	·0768	90·38	72·3	114·5	63·2
9-pr. of 6 cwt.	3·0	17·5	R.L.G.	1 12	9	1262	7·649	·1307	99·46	56·8	98·1	57·9
			R.L.G.	1 8	9	1234	8·918	·1121	95·10	63·4	103·5	61·3
7-pr. of 224 lb. (bronze)	3·0	11·3	F.G.	0 12	7·25	955	11·538	·0867	45.88	61·2	111·0	55·2
			F.G.	0 10	7·25	854	13·873	·0721	36·69	58·7	116·0	50·6
7-pr. of 150 lb. (steel)	3·0	8·0	F.G.	0 6	7·25	673	16·346	·0612	22·79	60·8	121·0	50·5
7-in. B.L. of 82 cwt.	7·0	14·2	R.L.G.	10 0	110	1013	13·794	·0725	783·2	78·3	116·0	67·5
			R.L.G.	11 0	90	1105	12·541	·0797	847·6	77·1	113·0	68·0
64-pr. B.L. of 61 cwt.	6·4	10·9	R.L.G.	9 0	64	1200	8·982	·1113	639·5	71·1	103·5	68·8
40-pr. B.L. of 35 cwt.	4·75	22·4	R.L.G.	5 0	41	1180	13·590	·0736	396·1	79·2	115·6	68·6
20-pr. B.L. of 16 cwt. L.S. ...	3·75	22·4	R.L.G.	2 8	21	1130	13·377	·0748	186·1	74·4	115·3	64·5
20-pr. B.L. of 13 cwt. S.S. ...	3·75	14·5	R.L.G.	2 8	21	1000	8·672	·1153	145·7	58·3	102·4	57·0
12-pr. B.L. of 8 cwt.	3·0	20·5	R.L.G.	1 8	11·75	1150	10·457	·0956	107·8	71·9	109·0	67·9
9-pr. B.L. of 6 cwt.	3·0	17·7	B.L.G.	1 2	9·25	1057	12·019	·0832	71·71	63·7	112·1	56·9
6-pr. B.L. of 3 cwt.	2·5	21·2	R.L.G.	0 12	6·6	1046	12·500	·8000	50·11	66·8	113·4	59·0

* Table extracted by permission.

CHAPTER XIII.—NOTE II.

Le Boulengé Chronograph.

(Vide Plates, pp. 336–338.)

The Le Boulengé chronograph consists of a hollow brass column *S*, which supports two electro-magnets, *A*, *B*, and a small bracket, *K*. The column stands on a triangular base, upon which is fixed the "*trigger*," *T*. (Fig. 1, p. .)

The *electro-magnet A*, supports a long cylindrical rod, *C* (p. , Fig. 1), suspended vertically, and called the "*chronometer*." This rod is partially covered with two zinc tubes, *D*, *E*, called "*registers*."

The *electro-magnet*, *B*, sustains a shorter rod, *F* (p.), named the "*registrar*."

The "*trigger*" (p.), consists of a circular steel knife, *G*, fixed in a recess of the spring, *H*, by means of the screw, *N*, which forms an axle upon which it can be turned so as to bring a fresh portion of the edge opposite the chronometer.

The spring, *H*, can be "cocked," or restrained, by means of the catch on one end of the lever, *I*.

The other end of this lever carries a disc, *O*, fixed to a screw, by means of which it can be raised or lowered as required.

This disc is vertically below the registrar when suspended to its electro-magnet; consequently, when the current through the second screen is broken, the registrar falls on the disc, and releases the spring, *H*.

The tube, *L* (p.), retains the registrar after its fall.

If it be required to alter the time taken by the registrar to release the knife, it is done by raising or lowering the disc of the trigger by turning it in the direction *with the sun* to *increase* the time, and *against the sun to reduce* it.

The screw has a pitch of one millimètre, and the circumference of the disc is divided by notches into 10 equal parts, in which the paul, *P*, works. By this arrangement the disc can be moved any required number of tenths of a millimètre (within certain limits), and is retained in the required position by the paul.

The screw, *M*, passes through the lever, and acts against the fulcrum supporting it. It is intended for regulating the hold of the catch of the lever on the spring, which should always be as light as possible.

This is regulated once for all; but should the spring at any time show a tendency to escape of itself, this defect can be remedied by slightly withdrawing the screw, *M*.

The *disjunctor* (p. 337, Fig. 2, is composed of a mainspring, *t*, carrying a cross-piece, *u*, covered with insulating material, and passing under the two steel plates, *q q'*. By pressing the milled headed screw, *z*, the spring is compressed and held by the catch, *x*, allowing the plates, *q q'*, to come into contact with the metal pins, *r r'*, and thus complete the circuits by bringing the screws, *s v* and *s' v'*, into connexion with one another. When the catch, *x*, is pressed, the mainspring being released, its cross-piece strikes the two plates exactly at the same instant, raises them from the screws, and thus breaks both currents identically at the same time.

The electro-magnet, A, is magnetised by the current passing through the first screen; consequently, when the shot cuts this screen the *chronometer* is released and falls freely in a vertical direction.

The other electro-magnet is in the circuit through the second screen, so that the *registrar* falls when this screen is cut, and striking the disc on the free end of the lever of the *trigger*, liberates the spring, which carries forward the knife until it strikes the chronometer in its fall and makes an indent in the upper zinc tube.

As stated above, if the trigger be set in action when the chronometer is at rest. a mark will be made by the knife on the zinc, which point we will call the "*origin*," as it is the zero point from which the height of fall of the chronometer must be calculated.

Let *H* be the height above the origin of the mark obtained by firing a projectile through the screens. Since the chronometer follows the law of the fall of heavy bodies.

$$T'_{,} = \sqrt{\frac{2H}{g}}$$

will be the time it was in motion before receiving the impression. Now T'' would be the time required by the projectile to traverse the distance between the screens, supposing that the chronometer commences to fall the instant the projectile passes through the first screen, and further, supposing that it is struck by the knife at the precise instant the shot cuts the second screen. But this is not the case. In fact, after the rupture of the first screen, a certain time, θ, elapses before the electro-magnet is demagnetised sufficiently to free the chronometer; the movement of the chronometer will therefore be delayed, and the observed time consequently diminished, by the quantity θ.

Again, some time elapses between the cutting of the second screen and the moment when the knife reaches the chronometer—viz., the time required for the following operations:—

θ' for the demagnetisation of the electro-magnet supporting the registrar.
t' for the fall of the registrar to the disc of the trigger.
t'' for the disengagement of the catch.
t''' for the knife to pass over the horizontal distance which separates it from the chronometer.

Now, it is evident that the chronometer, before it is struck by the knife, will have been in motion during the sum of the above time in addition to the time taken by the shot in passing over the distance between the screens. Consequently, the observed time T' is too great by the sum of $(\theta' + t' + t'' + t''')$. We have also shown above that T' is too small by the quantity θ—the time required to demagnetise the chronometer electro-magnet. Therefore, to ascertain the true time T, we must deduct from T' the quantity $(\theta' + t' + t'' + t''' - \theta)$, which we will call t.

We have then $T = T' - t$.

Now, suppose $T = 0$, or in other words, suppose the shot to cut both screens simultaneously, then we should have $T' = t$. From which it appears that t would be the time recorded on the chronometer if both currents were cut identically at the same instant. This we can do by using the disjunctor, and we thus obtain a mark, on the lower zinc tube, at a height above the origin equal to the space passed over in the time t, which we call the *disjunctor reading*. The time corresponding to this reading must be deducted from the whole time recorded on the chronometer, to arrive at the time taken by the shot to traverse the distance between the screens. As before stated, the disc of the trigger can be raised or lowered so that the disjunctor reading can be altered (if required) within certain limits, and we can thus regulate the instrument so that the time t shall have a constant value. The value of t for which the velocity scale has been calculated is 0·15 of a second, and the height of the corresponding mark above the origin is 4·345 ins. (110·370 mill.). Starting with this assumption, a scale has been calculated for a distance between the screens of 120 ft., by means of which the velocity of the projectile can be at once determined without the aid of any calculation.

The method of calculating this scale is as follows:—

Suppose the shot to have a velocity of 1,200 ft. a second, it would take $\frac{120}{1200} = 0''\cdot1$ to traverse the distance between the screens.

The instrument will, therefore, mark $0''\cdot15$ (disjunctor reading) + $0''\cdot1$, or $0''\cdot25$, and the corresponding height of fall from the origin will be

$$H = \tfrac{1}{2} g T'^2 = \frac{g \times \overline{0\cdot25}^2}{2} = 12\cdot075 \text{ ins.}$$

Conversely, if the mark on the chronometer is 12·075 ins. above the origin, we know that the velocity of the projectile is 120 ft. a second; the disjunctor reading being at a height corresponding to $0''\cdot15$, and the screens 120 ft. apart.

This calculation has been made for a series of velocities increasing from foot to foot for velocities from 850 to 1,500 feet a second, and increasing by 5 ft. from 1,500 to 1,800 feet a second, and the corresponding heights engraved on the scale supplied with the instrument.

* TABLE XL.

A GENERAL TABLE of VALUES of $\frac{d^2}{w}s$ for OGIVAL-HEADED SHOT.

†V.	0.	1.	2.	3.	4.	5.	6.	7.	8.	9.
f.s.	ft.	ft.	ft.	ft.	ft.	ft.	ft.	ft.	ft.	ft.
50	19604·0	19542·0	19480·2	19418·6	19357·2	19296·1	19235·2	19174·6	19114·3	19054·3
51	18994·6	18935·1	18875·8	18816·7	18757·8	18699·1	18640·7	18582·6	18524·6	18466·8
52	18409·2	18351·9	18294 9	18238·0	18181·3	18124·8	18068·5	18012·5	17956·7	17901·1
53	17845·7	17790·6	17735·6	17680·8	17626·2	17571·9	17517·8	17463·9	17410·1	17356·5
54	17303·2	17250·1	17197·2	17144·4	17091·8	17039·4	16987·1	16935·0	16883·1	16831·5
55	16780·2	16729·1	16678·1	16627·2	16576·5	16526·0	16475 7	16422·6	16375·6	16325·8
56	16276·2	16226·9	16177·7	16128·6	16079·6	16030·8	15982·2	15933·8	15885·6	15837·6
57	15789·8	15742·1	15694·6	15647·2	15600·0	15552·9	15506·0	15459·3	15412·7	15366·3
58	15320·1	15274 0	15228·1	15182·3	15136·7	15091·2	15045·9	15000·7	14955·7	14910·9
59	14866·2	14821·7	14777·3	14733·1	14689·0	14645·1	14601·3	14557·7	14514·2	14470·8
60	14427·7	14384·8	14341·9	14299·1	14256·4	14213·8	14171·3	14129·0	14086·9	14045·0
61	14003·3	13961·7	13920·2	13878·9	13837·6	13796·4	13755·3	13714·4	13673·7	13633·2
62	13592·8	13552·4	13512·2	13472·1	13432·2	13392·4	13352·7	13313·1	13273·7	13234·4
63	13195·2	13156·1	13117·3	13078 5	13039·8	13001·2	12962·8	12924·5	12886·3	12848·2
64	12810·2	12772·3	12734·6	12697·0	12695·5	12622·1	12584·8	12547·6	12510·6	12473·9
65	12436·9	12400·2	12363·7	12327·2	12290·8	12254·5	12218·3	12182·3	12146·4	12110·6
66	12074·9	12039·3	12003·9	11968·5	11933·2	11898·0	11863·0	11828·1	11793·2	11758·4
67	11723·7	11689·1	11654·7	11620·4	11586·2	11552·1	11518·1	11484·2	11450·4	11416·7
68	11383·0	11349·4	11316·0	11282·7	11249·5	11216·3	11183·3	11150·4	11117·5	11084·7
69	11052·0	11019·4	10987·0	10954·6	10922·3	10890·1	10857·9	10825·9	10794·0	10762·2
70	10730·5	10698·9	10667·4	10635·9	10604·5	10573·2	10542·0	10510·9	10479·9	10448·9
71	10418·0	10387·2	10356·5	10326·0	10295·5	10265·1	10234·8	10204·6	10174·4	10144·3
72	10114·3	10084·4	10054·6	10024·9	9995·2	9965·6	9936·1	9906·7	9877·4	9848·1
73	9818·9	9789·8	9760·8	9731·9	9703·0	9674·2	9645·5	9616·9	9588·3	9559·8
74	9531·4	9503·1	9474·9	9446·7	9418·6	9390·6	9362·6	9334·7	9306·9	9279·2
75	9251·6	9224·1	9196·6	9169·2	9141·8	9114·5	9087·3	9060·2	9033·1	9006·1
76	8979·2	8952·4	8925·7	8899·0	8872·3	8845·7	8819·2	8792·8	8766·4	8740·1
77	8713·9	8687·7	8661·6	8635·6	8609·7	8583·8	8558·0	8532·3	8506·0	8480·9
78	8455·3	8429·8	8404·4	8379·0	8353·7	8328·5	8303·3	8278·2	8253·2	8228·2
79	8203·3	8178·5	8153·7	8129·0	8104·3	8079·7	8055·2	8030·8	8006·4	7982·0
80	7957·7	7933·5	7909·3	7885·2	7861·1	7837·1	7813·2	7789·3	7765·5	7741·7
81	7718·0	7694·4	7670·8	7647·3	7623·8	7600·4	7576·0	7553·7	7530·5	7507·3
82	7484·2	7461·1	7438·1	7415·1	7392·2	7369·4	7346·6	7323·9	7301·2	7278·6
83	7256·0	7233·5	7211·1	7188·7	7166·3	7144·0	7121·8	7099·6	7077·5	7055·4
84	7033·4	7011·4	6989·5	6967·6	6945·8	6924·0	6902·2	6880·5	6858·9	6837·3
85	6815·8	6794·4	6773·0	6751·6	6730·3	6709 0	6687·8	6666·6	6645·5	6624·5
86	6603·5	6582·5	6561·6	6540·7	6519·9	6499·1	6478·4	6457·7	6437·1	6416·5
87	6395·9	6375·4	6355·0	6334·6	6314·2	6293·9	6273·7	6253·5	6233·3	6213·2
88	6193·1	6173·1	6153·1	6133·2	6113·3	6093·4	6073·6	6053·8	6034·1	6014·4
89	5994·8	5975·2	5955·7	5936·2	5916·7	5897·3	5877·9	5858·6	5839·3	5820·0
90	5800·8	5781·6	5762·5	5743·5	5724·5	5705·6	5686·7	5667·9	5649·2	5643·5
91	5611·8	5593·2	5574·7	5556·2	5537·7	5519·3	5500·9	5482·6	5464·4	5446·2
92	5428·1	5410·0	5392.0	5374·1	5356·2	5338·3	5320·5	5302·8	5285·1	5267·5
93	5249·9	5232·4	5215·0	5197·6	5180·2	5162·9	5145·7	5128·5	5111·4	5094·3
94	5077·3	5060·4	5043·5	5026·6	5009·8	4993·0	4976·3	4959·6	4943·0	4926·5
95	4910·0	4893·6	4877·3	4861·0	4844·7	4828·5	4812·4	4796·3	4780·3	4764·4
96	4748·5	4732·7	4716·9	4701·2	4685·5	4669·9	4654·4	4638·9	4623·5	4608·1
97	4592·8	4577·6	4562·4	4547·3	4532·2	4517·2	4502·3	4487·4	4472·6	4457·8
98	4443·1	4428·5	4413·9	4399·4	4384·9	4370·5	4356·2	4342·0	4327·8	4313·6
99	4299·5	4285·4	4271·4	4257·5	4243·7	4230·0	4216·4	4202·8	4189·3	4175·8
100	4162·4	4149·1	4135·9	4122·8	4109·7	4096·7	4083·8	4070·9	4058·1	4045·3
101	4032·6	4020·0	4007·5	3995·0	3982·5	3970·1	3957·8	3945·6	3933·5	3921·4
102	3909·5	3897·7	3886·0	3874·3	3862·7	3851·2	3839·8	3828·5	3817·3	3806·9
103	3795·0	3784·0	3773·0	3762·1	3751·2	3740·4	3729·6	3718·9	3708·3	3697·7
104	3687·2	3676·8	3666·4	3656·1	3645·8	3635·6	3625·5	3615·5	3605·5	3595·6
105	3585·8	3576·1	3566·5	3556 9	3547·3	3537·8	3528·4	3519·1	3509·8	3500·6
106	3491·4	3482·3	3473·2	3464·2	3455·2	3446·3	3437·5	3428·7	3419·9	3411·1
107	3402·4	3393·8	3385·2	3376·7	3368·1	3359·6	3351·2	3342·8	3334·4	3326·0
108	3317·7	3309·5	3301·3	3293·1	3284·9	3276·7	3268·6	3260·5	3252·4	3244·3
109	3236·3	3228·4	3220·4	3212·5	3204·5	3196·6	3188·7	3180·9	3173·1	3165·3

* Taken from "Bashforth's Motion of Projectiles," by permission.
V = velocity; f.s. = feet a second.

TABLE XL.—*continued.*

V.	0.	1.	2.	3.	4.	5.	6.	7.	8.	9.
f.s.	ft.	ft.	ft.	ft.	ft.	ft.	ft.	ft.	ft.	ft.
110	3157·5	3149·7	3141·9	3134·2	3126·5	3118·7	3111·0	3103·3	3095·6	3087·9
111	3080·3	3072·8	3065·2	3057·6	3050·1	3042·5	3035·0	3027·5	3020·0	3012·6
112	3005·1	2997·7	2990·3	2982·9	2975·5	2968·1	2960 8	2953·4	2946·1	2938·7
113	2931·4	2924·1	2916·8	2909·6	2902·4	2895·2	2888·1	2880·9	2873·7	2866·6
114	2859·4	2852·3	2845·2	2838·1	2831·0	2823·9	2816·9	2809·8	2802·8	2795·8
115	2788·8	2781·8	2774·9	2767·9	2761·0	2754·0	2747·1	2740·2	2733·3	2726·5
116	2719·6	2712·8	2705·9	2699·1	2692·3	2685·5	2678·7	2672·0	2665·2	2658·5
117	2651·7	2645·0	2638·3	2631·6	2625·0	2618·3	2611·6	2605·0	2598·4	2591·8
118	2585·2	2578·6	2572·1	2565·5	2558·9	2552·4	2545·8	2539·3	2532·8	2526·3
119	2519·8	2513·4	2506·9	2500·4	2404·0	2487·5	2481·1	2474·7	2468·3	2461·9
120	2455·5	2449·1	2442·8	2436·4	2130·1	2423·8	2417·4	2411·1	2404·8	2398·5
121	2392·2	2386·0	2379·7	2373·5	2367·3	2361·0	2354·8	2348·6	2342·4	2336·2
122	2330·0	2323·9	2317·7	2311·6	2305·5	2299·3	2293·2	2287·1	2281·0	2274·9
123	2268·8	2262·7	2256·7	2250·7	2244·6	2238·6	2232·6	2226·6	2200·6	2214·6
124	2208·6	2202·6	2196·7	2190·7	2184·8	2178·8	2172·9	2166 9	2161·0	2155·1
125	2149·2	2143·3	2137·5	2131·6	2125·8	2119·9	2114·1	2108·2	2102·4	2096·6
126	2090·8	2085·0	2079·2	2073·4	2067·6	2061·9	2056·1	2050·4	2044·7	2038·9
127	2033·2	2027·5	2021·8	2016·1	2010·5	2004·8	1999·1	1993·5	1987·8	1982·2
128	1976·5	1970·8	1965·2	1959·6	1954·0	1948·4	1942·8	1937·2	1931·7	1926·1
129	1920·5	1915 0	1909·4	1903·9	1898·4	1892·8	1887·3	1881·8	1876·3	1870·8
130	1865·3	1859·8	1854·4	1848·9	1843·5	1838·0	1832·6	1827·2	1821·7	1816·3
131	1810·9	1805·5	1800·1	1794·7	1789·3	1783·9	1778·5	1773·1	1767·8	1762·4
132	1757·0	1751·7	1746·3	1741 0	1735·7	1730·4	1725·1	1719·8	1714·5	1709·2
133	1703·9	1698·6	1693·4	1688·1	1682·9	1677·6	1672·4	1667·1	1661·9	1656·7
134	1651·4	1646·2	1641·0	1635·8	1630·6	1625·4	1620·2	1615·0	1609·9	1604·7
135	1599·5	1594·4	1589·2	1584·1	1578·9	1573·8	1568·7	1563·5	1558·4	1553·3
136	1548·2	1543·1	1538·0	1533·0	1527·9	1522·8	1517·8	1512·7	1507·7	1502·6
137	1497·6	1492·5	1487·5	1482·5	1477·4	1472·4	1467·4	1462·3	1457·3	1452·3
138	1447·3	1442·3	1437·3	1432·4	1427·4	1422 4	1417·5	1412·5	1407·6	1402·6
139	1397·7	1392·7	1387·8	1382·9	1377·9	1373·0	1368·1	1363·1	1358·2	1353·3
140	1348·4	1343·5	1338·6	1333·7	1328·8	1323·9	1319·0	1314·1	1309·3	1304·4
141	1299·5	1294·7	1289·8	1284·9	1280·1	1275·2	1270·4	1265·6	1260·8	1255·9
142	1251·1	1246·3	1241·5	1236·6	1231·8	1227·0	1222·2	1217·4	1212·6	1207·8
143	1203·0	1198·2	1193·4	1188·7	1183·9	1179·1	1174·4	1169·6	1164·9	1160·2
144	1155·4	1150·7	1145·9	1141·2	1136·4	1131·7	1126·9	1122·2	1117·5	1112·7
145	1108·0	1103·3	1098·6	1093·9	1089·2	1084·5	1079·8	1075·1	1070·4	1065·8
146	1061·1	1056·4	1051·7	1047·1	1042·4	1037·7	1033·1	1028·4	1023·7	1019·1
147	1014·4	1009·8	1005·1	1000·5	995 8	991·2	986·6	981·9	977·3	972·7
148	968·0	963·4	958·8	954·1	949·5	944·9	940·3	935·7	931·1	926·5
149	921·9	917·3	912·7	908·1	903·5	898·9	894·3	889·7	885·1	880·6
150	876·0	871·4	866·9	862·3	857·8	853·2	848·6	844·1	839·5	835·0
151	830 4	825·9	821·3	816·8	812·2	807·7	803·2	798·6	794·1	789·5
152	785·0	780·5	775·9	771·4	766·9	762·3	757·8	753·3	748·7	744·2
153	739·7	735·2	730·7	726·2	721·7	717·2	712·7	708·2	703·7	699·2
154	694·7	690·2	685·7	681·2	676·8	672·3	667·8	663·3	658·8	654·4
155	649·9	645·4	640·9	636·5	632·0	627·5	623·1	618·6	614·2	609·7
156	605·3	600·8	596·4	591·9	587·5	583·0	578·6	574·1	569·7	565·2
157	560·8	556·4	551·9	547·5	543·0	538·6	534·2	529·7	525·3	520·9
158	516·4	512·0	507·6	503·1	498·7	494·3	489·9	485·5	481·1	476·7
159	472·3	467·9	463·5	459·1	454·7	450·3	445·9	441·5	437·1	432·7
160	428·3	423·9	419·5	415·1	410·7	406·4	402·0	397·6	393·2	388·9
161	384·5	380·1	375·8	371·4	367·1	362·7	358·3	354·0	349·6	345·3
162	340·9	336·5	332·2	327·8	323·5	319·1	314·8	310·4	306·1	301·7
163	297·4	293·1	288·7	284·4	280·0	275·7	271·4	267·0	262·7	258·4
164	254·0	249·7	245·4	241·0	236·7	232·4	228·1	223·8	219·5	215·2
165	210·9	206·6	202·3	198·0	193·7	189·4	185·1	180·8	176·5	172·3
166	168·0	163·7	159·5	155·2	150·9	146·7	142·4	138·2	134·0	129·7
167	125·5	121·3	117·0	112·8	108·6	104·3	100·1	95·9	91·7	87·5
168	83·3	79·1	74·9	70·7	66·5	62·3	58·1	53·9	49·8	45·6
169	41·4	37·2	33·1	28·9	24·8	20·7	16·5	12·3	8·2	4·1
170	0·0									

CHAP. XIII.

* TABLE XLI.

A General Table of Values of $\frac{d^2}{w}t$ for Ogival-headed Shot.

†V.	0.	1.	2.	3.	4.	5.	6.	7.	8.	9.
f.s.	secs.	secs.	secs.	secs.	secs.	secs.	secs.	secs.	secs.	secs.
50	26·5083	26·3844	26·2612	26·1388	26·0172	25·8963	25·7762	25.6567	25·5379	25·4198
51	25·3024	25·1857	25·0697	24·9543	24·8396	24·7255	24·6121	24·4994	24·3873	24·2759
52	24·1652	24·0552	23 9457	23·8368	23·7285	23·6209	23·5137	23·4075	23·3017	23·1965
53	23·0919	22·9879	22·8845	22·7817	22·6795	22·5778	22·4766	22·3760	22·2760	22·1766
54	22·0778	21·9795	21·8817	21·7844	21·6877	21·5915	21·4958	21·4006	21·3060	21·2119
55	21·1184	21·0254	20·9328	20·8407	20·7491	20·6580	20·5674	20·4773	20·3877	20·2986
56	20·2100	20·1218	20·0341	19·9468	19·8600	19·7737	19·6879	19·6025	19·5175	19·4330
57	19·3489	19·2653	19·1821	19·0994	19·0171	18·9352	18·8537	18·7726	18·6920	18·6118
58	18·5320	18·4527	18·3738	18·2952	18·2170	18·1392	18·0619	17·9849	17·9083	17·8321
59	17·7563	17·6809	17·6059	17·5312	17·4569	17·3830	17·3094	17·2362	17·1634	17·0910
60	17·0190	16·9474	16·8761	16·8051	16·7344	17·6640	16·9540	16·5244	16·4551	16·3862
61	16·3176	16·2494	16·1815	16·1139	16·0466	15·9797	15 9131	15·8468	15·7809	15·7153
62	15·6500	15·5851	15·5204	15·4560	15·3919	15·3281	15·2646	15·2014	15·1386	15·0761
63	15·0139	14·9520	14·8903	14·8289	14·7678	14·7070	15·6465	14 5863	14·5264	14·4668
64	14·4074	14·3483	14·2895	14·2309	14·1726	14·1146	14·0569	13·9994	13·9422	13·8853
65	13·8286	13·7722	13·7160	13·6601	13·6045	13·5491	13·4940	13·4391	13·3845	13·3301
66	13·2759	13·2220	13·1684	13·1150	13·0618	13·0089	12·9562	12·9038	12·8516	12·7996
67	12·7479	12·6964	12·6451	12·5940	12·5432	12·4926	12·4422	12·3920	12·3421	12·2924
68	12·2429	12·1937	12·1447	12·0958	12·0471	11·9986	11·9504	11·9025	11·8547	11·8071
69	11·7597	11·7125	11·6655	11·6188	11·5723	11·5260	11·4799	11·4340	11·3882	11·3426
70	11·2972	11·2520	11·2070	11·1622	11·1176	11·0732	11·0290	10·9850	10·9412	10·8975
71	10·8540	10·8107	10·7675	10·7246	10·6819	10·6394	10·5970	10·5548	10·5127	10·4708
72	10·4291	10·3875	10·3461	10·3049	10·2639	10·2232	10·1824	10·1421	10·1018	10·0616
73	10·0216	9·9816	9·9420	9·9025	9 8632	9·8241	9·7851	9·7462	9·7075	9·6689
74	9·6305	9·5922	9·5541	9·5162	9·4784	9·4408	9·4034	9·3661	9·3289	9·2919
75	9·2550	9·2183	9·1817	9·1453	9·1090	9·0728	9·0368	9 0009	8 9652	8·9296
76	8·8941	8·8588	8·8236	8·7886	8·7537	8·7189	8·6843	8·6498	8·6155	8·5813
77	8·5472	8·5133	8·4795	8·4458	8·4122	8·3788	8·3455	8.3123	8·2793	8·2464
78	8·2136	8·1809	8·1484	8·1160	8·0837	8·0515	8·0195	7·9876	7·9558	7·9241
79	7·8926	7·8612	7·8299	7·7987	7 7676	7·7366	7·7058	7·6751	7·6445	7·6139
80	7·5835	7·5532	7·5231	7·4931	7·4631	7·4333	7·4036	7·3741	7·3446	7·3152
81	7·2859	7·2567	7·2276	7·1086	7·1698	7·1411	7·1125	7·0840	7·0556	7·0272
82	6·9990	6·9709	6·9429	6·9150	6·8871	6·8594	6.8318	6·8043	6·7769	6·7496
83	6·7224	6·6953	6·6683	6·6414	6·6146	6·5878	6·5612	6·5347	6·5083	6·4819
84	6·4557	6·4296	6·4036	6·3776	6·3517	6·3259	6·3002	6·2746	6·2491	6·2236
85	6·1983	6·1731	6·1480	6·1229	6·0979	6·0730	6·0482	6·0236	5·9990	5·9744
86	5·9499	5·9255	5·9012	5·8770	5·8529	5·8289	5·8049	5·7810	5·7572	5·7335
87	5·7099	5·6864	5·6629	5·6395	5·6162	5·5930	5·5699	5·5468	5·5238	5·5009
88	5·4781	5·4553	5·4326	5·4100	5·3875	5·3651	5·3428	5·3205	5·2983	5·2762
89	5·2541	5·2321	5·2102	5 1884	5·1666	5·1449	5·1233	5·1018	5·0803	5·0589
90	5·0375	5·0163	4·9952	4 9741	4·9531	4·9321	4·9112	4·8904	4·8697	4·8491
91	4 8286	4·8082	4·7878	4·7675	4·7473	4·7272	4·7072	4·6873	4.6674	4·6476
92	4·6279	4·6083	4·5888	4·5693	4·5499	4·5306	4·5114	4·4923	4·4732	4·4542
93	4·4353	4·4165	4·3978	4·3791	4·3605	4·3421	4·3236	4·3053	4·2870	4·2688
94	4·2506	4·2326	4·2147	4·1968	4·1790	4·1612	4·1436	4·1260	4 1085	4·0911
95	4·0738	4·0565	4·0393	4·0222	4·0052	3·9883	3·9714	3·9546	3·9379	3·9213
96	3·9047	3·8882	3·8718	3·8555	3·8393	3·8231	3·8070	3·7910	3·7751	3·7592
97	3·7434	3·7277	3·7121	3·6966	3·6811	3·6657	3·6504	3·6352	3·6200	3·6049
98	3·5899	3·5750	3·5602	3·5455	3·5308	3·5162	3·5017	3·4873	3·4729	3·4586
99	3·4444	3·4303	3·4163	3·4024	3·3885	3·3747	3·3610	3·3474	3·3339	3·3204
100	3·3070	3·2937	3·2805	3·2674	3·2544	3·2414	3·2285	3·2157	3·2030	3·1904
101	3·1778	3·1653	3·1529	3·1406	3·1284	3·1163	3·1042	3·0922	3·0803	3·0685
102	3·0567	3·0451	3·0335	3·0220	3·0106	2·9992	2·9879	2·9767	2·9656	2 9546
103	2·9436	2·9328	2·9220	2·9113	2·9007	2·8902	2·8798	2·8694	2·8591	2·8489
104	2·8388	2·8288	2·8188	2·8089	2·7991	2·7894	2·7798	2·7703	2·7608	2·7514
105	2·7420	2·7328	2·7236	2·7144	2·7053	2·6963	2·6874	2·6785	2·6697	2·6610
106	2·6523	2·6437	2·6351	2·6266	2·6182	2·6098	2·6015	2·5932	2 5850	2·5768
107	2·5687	2·5606	2·5526	2·5446	2·5366	2·5287	2·5208	2·5130	2·5052	2·4975
108	2·4898	2·4821	2·4745	2·4669	2·4593	2·4518	2·4443	2·4369	2·4295	2·4221
109	2·0147	2·4074	2·4001	2·3928	2·3855	2·3783	2·3711	2·3640	2·3568	2·3497

* Taken from "Bashforth's Motion of Projectiles," by permission.
† V = velocity; f.s. = feet a second.

TABLE XLI.—*continued.* CHAP. XIII.

V.	0.	1.	2.	3.	4.	5.	6.	7.	8.	9.
f.s.	secs.	secs.	secs.	secs.	secs.	secs.	secs.	secs.	secs.	secs.
110	2·3426	2·3355	2·3285	2·3214	2·3144	2·3074	2·3005	2·2936	2·2867	2·2798
111	2·2729	2·2661	2·2592	2·2524	2·2456	2·2388	2·2321	2·2254	2·2187	2·2120
112	2·2053	2·1987	2·1921	2·1855	2·1789	2·1723	2·1658	2·1593	2·1528	2·1464
113	2·1399	2·1335	2·1271	2·1207	2·1134	2·1079	2·1016	2·0953	2·0890	2·0827
114	2·0764	2·0702	2·0639	2·0577	2·0515	2·0453	2·0392	2·0030	2·0269	2·0208
115	2·0147	2·0087	2·0026	1·9966	1·9905	1·9845	1·9785	1·9726	1·9667	1·9607
116	1·9548	1·9489	1·9431	1·9372	1·9314	1·9255	1·9197	1·9139	1·9081	1·9024
117	1·8966	1·8909	1·8852	1·8795	1·8738	1·8681	1·8625	1·8568	1·8512	1·8455
118	1·8399	1·8343	1·8287	1·8232	1·8176	1·8121	1·8066	1·8011	1·7956	1·7901
119	1·7847	1·7792	1·7738	1·7684	1·7630	1·7576	1·7523	1·7469	1·7416	1·7362
120	1·7309	1·7256	1·7203	1·7150	1·7098	1·7045	1·6992	1·6940	1·6888	1·6836
121	1·6784	1·6732	1·6681	1·6629	1·6578	1·6526	1·6475	1·6424	1·6373	1·6323
122	1·6272	1·6222	1·6172	1·6121	1·6071	1·6021	1·5971	1·5922	1·5872	1·5823
123	1·5773	1·5724	1·5675	1·5626	1·5577	1·5528	1·5479	1·5431	1·5382	1·5334
124	1·5285	1·5237	1·5189	1·5141	1·5093	1 5045	1·4997	1·4950	1·4903	1·4855
125	1·4808	1·4761	1·4714	1·4667	1·4620	1·4574	1·4527	1·4481	1·4435	1·4388
126	1·4342	1·4296	1·4250	1·4204	1·4158	1·4113	1·4068	1·4022	1·3977	1·3932
127	1·3887	1·3842	1·3797	1·3752	1·3706	1·3663	1·3619	1·3574	1·3530	1·3486
128	1·3442	1·3398	1·3354	1·3310	1·3267	1·3223	1·3180	1·3137	1·3093	1·3050
129	1·3007	1·2964	1·2921	1·2878	1·2835	1·2792	1·2749	1·2707	1·2665	1·2622
130	1·2580	1·2538	1.2496	1·2454	1·2412	1·2370	1·2329	1·2287	1·2246	1·2204
131	1·2163	1·2122	1·2080	1·2039	1·1998	1·1957	1·1916	1·1876	1·1835	1·1794
132	1·1754	1·1713	1·1673	1·1633	1·1592	1·1552	1·1512	1·1472	1·1432	1·1393
133	1·1353	1·1313	1·1274	1·1234	1·1195	1·1156	1·1116	1·1077	1·1038	1·0999
134	1·0960	1·0922	1·0883	1·0844	1·0806	1·0767	1·0729	1·0690	1·0652	1·0614
135	1·0575	1·0537	1·0499	1·0461	1·0423	1·0385	1·0347	1·0309	1·0272	1·0234
136	1·0196	1·0159	1·0121	1·0084	1·0047	1·0009	·9972	·9935	·9898	·9861
137	·9824	·9787	·9751	·9714	·9678	·9641	·9605	·9568	·9532	·9496
138	·9459	·9423	·9387	·9351	·9315	.9279	·9243	·9207	·9172	·9136
139	·9100	·9065	·9029	·8994	·8959	·8923	·8888	·8853	·8818	·8782
140	·8747	·8712	·8677	·8642	·8607	·8572	·8538	·8503	·8468	·8434
141	·8399	·8364	·8330	·8296	·8261	·8227	·8193	·8159	·8125	·8091
142	·8057	·8023	·7989	·7956	·7922	·7888	·7854	·7821	·7787	·7754
143	·7720	·7687	·7654	·7620	·7587	·7554	·7521	·7488	·7454	·7421
144	·7388	·7355	·7322	·7289	·7256	·7223	·7191	·7158	·7125	·7093
145	·7060	·7028	·6996	·6963	·6931	·6899	·6867	·6834	·6802	·6770
146	·6738	·6706	·6674	·6642	·6610	·6578	·6546	·6514	·6483	·6451
147	·6419	·6387	·6356	·6324	·6293	·6261	·6230	·6198	·6167	·6136
148	·6104	·6073	·6042	·6011	·5980	·5949	·5918	·5887	·5856	·5825
149	·5794	·5763	·5732	·5701	·5671	·5640	·5609	·5579	·5548	·5518
150	·5487	·5457	·5426	·5396	·5366	·5335	·5305	·5275	·5244	·5214
151	·5184	·5154	·5124	·5094	·5064	·5034	·5004	·4974	·4944	·4914
152	·4884	·4854	·4824	·4795	·4765	·4735	·4705	·4676	·4646	·4617
153	·4587	·4	·4528	·4499	·4470	·4140	·4411	·4382	·4353	·4323
154	·4294	·4265	·4236	·4207	·4178	·4149	·4120	·1091	·4062	·4033
155	·4004	·3975	·3946	·3918	·3889	·3860	·3831	·3803	·3774	·3745
156	·3717	·3688	·3660	·3632	·3603	·3575	·3547	·3518	·3490	·3462
157	·3433	·3405	·3377	·3349	·3320	·3292	·3264	·3236	·3207	·3179
158	·3151	·3123	·3095	·3067	·3039	·3012	·2984	·2956	·2928	·2901
159	·2873	·2845	·2818	·2790	·2763	·2735	·2707	·2680	·2652	·2625
160	·2597	·2570	·2542	·2515	·2488	·2460	·2433	·2406	·2378	·2351
161	·2324	·2297	·2270	·2243	·2316	·2180	·2162	·2135	·2108	·2081
162	·2054	·2027	·2000	·1973	·1247	·1920	·1893	·1866	·1830	·1813
163	.1786	·1759	·1733	·1706	·1680	·1653	·1627	·1600	·1574	·1548
164	·1521	·1495	·1469	·1442	·1416	·1390	·1364	·1338	·1311	·1285
165	·1259	·1233	·1201	·1181	·1155	·1129	·1103	·1077	·1051	·1026
166	·1000	·0974	·0949	·0923	·0897	·0872	·0846	·0821	·0795	·0770
167	·0745	·0720	·0695	·0669	·0644	·0619	·0594	·0569	·0543	·0518
68	·0493	·0468	·0443	·0418	·0393	·0368	·0343	·0318	·0294	·0269
169	·0244	·0220	·0195	·0171	·0146	·0122	·0098	·0073	·0049	·0024
170	·0000									

APPENDIX.

APPENDIX I.

Returns and Examination of Ordnance.

(1.) The official orders upon this subject are contained in "Instructions relating to the care, preservation, inspection, and fitting of iron ordnance, &c., &c.," issued with Clause 191, Army Circulars, 1878, but the following notes are inserted here for the information of officers concerned.

(1.) Rules for Furnishing Annual Returns.

Returns according to the accompanying forms, p. 375, for S.B., W.O. 1475, and for rifled guns W.O. 1476, will be sent yearly, by officers of Artillery in command of districts,* except in North America, to the Director of Artillery and Stores, on the 1st of June. Those from the North American Artillery stations will be furnished on the 1st November. Officers commanding vessels of war of every description having guns on board, and likewise the Royal Marines, Royal Naval Reserve, and Coast Guard, having ordnance in their charge, will furnish returns on 1st January, through the Admiralty, to the Secretary of State for War. These annual returns are forwarded to Superintendent R. G. F. for report and record. Annual return. Date for furnishing.

It is necessary for the identification of guns that the descriptive marks should be accurately entered in the return.

The weight of the gun is marked on the top of the gun in front of the vent.

The initial of the factory† will be found on the left trunnion in all cases,§ and on rifled guns the numeral of pattern or mark is also on that trunnion.

The date of proof is marked on the reinforce in S.B. guns, and on the left trunnion of rifled guns.§

The register number and years of proof‡ will be found on the reinforce in S.B. cast iron guns, and on the left trunnion in the case of rifled ordnance.§

Under the head of "Nature," the proper name of the gun, with its nominal weight, will be entered; as for instance, "32-pr. of 56 cwts," "9-inch R.M.L. of 12 tons," "40-pr. R.B.L. of 35 cwt." The correct designation of S.B. guns is given in Lists A and B § 1140 List of changes in Artillery material &c., and that of all rifled guns is also given from time to time in the same List of changes, in which a description of every gun appears when finally approved.

The column headed "Date of last examination" will be filled in from the date of the last inspection, made by an Inspector of Warlike Stores, or other qualified person. Date of last examination.

---

* Converted guns and 7-pr. bronze R.M.L. will be included in W.O. Form 1476. A blank return, on W.O. Form 1475, is not required when there are no S.B. guns on charge.

† With S.B. ordnance this would of course denote the foundry where the guns were cast.

‡ On all guns proved since September 1857.

§ Except in the 7-pr. R.M.L., where this information is found on the right trunnion.

Condition of bore and Sentence.

The "Condition of bore" and "Sentence" will be taken from the last report made by the Inspector of Warlike Stores, or other examiner conducting the periodical or special examinations ordered; but if the gun has not been examined, owing to only a few rounds having been fired from it, these columns may be left blank, unless the commanding officer should see cause to call special attention to the gun.

Number of rounds fired.

The "Number of rounds fired" at the date of making the return, will be very carefully entered under the several headings; it is exceedingly important, for sake of the record, that the number should be given correctly. The number of rounds fired with projectiles since previous examination should also be given.

The number of rounds fired with projectiles at the time of making the annual returns will be very carefully recorded from year to year. All the older cast iron guns which were in the service previous to records being kept, have had a number of rounds "assumed" from the size of the vent, in accordance with the instructions of previous circulars. The number of "assumed rounds" will be entered every year in red ink, and the actual number of rounds in black ink. In the column in middle of return should be inserted the number of rounds fired with projectiles since last examination.

Condition of fittings, B.L.

The columns regarding the condition of the fittings of the breech-loading guns, and the vents of muzzle-loading guns, will be filled in from the reports of the Inspector of Warlike Stores, or other examiner.

Column "Remarks."

Particulars of any special defect on the exterior or other part of the gun will be noted in the column of "Remarks," if not entered in any other part of the return; as also, any peculiar circumstance, such as the re-venting of a muzzle-loading gun, the bursting of a shell in the bore, the fracture of fittings, &c. Reference will be made, when necessary, to explanatory documents.

Guns which have not been fired since the previous return will be entered in the return; but the columns headed "Condition of bore," "Sentence," "No. of rounds fired," and condition of fittings, need not be filled in, the remark "Not used since 18 ." being entered against them.

(2.) Rules as to Examination of Guns.

Home Stations, Royal Artillery.

Examination of guns in Royal Artillery charge at home.

For home service the following arrangements will be carried out for guns in Royal Artillery charge:—

As soon as a gun has fired the number of rounds, since previous examination laid down in the regulations, a notification of the circumstance (see W.O. Form, 1473, p. 376) will be sent in the case of Rifled guns to the Director of Artillery, or in the case of S.B. guns, direct to the Superintendent, R.G.F., by the officer commanding the district, and practice from the gun will cease until its condition has been reported upon.

Gutta-pertha impression of vent should accompany the return, as the Director of Artillery, after its inspection, may be able to allow of continuance of practice from the gun without waiting for the quarterly inspection.

At the end of every quarter an examiner and an artificer from the Royal Gun Factories will visit each district, and examine those guns which have fired the prescribed number of rounds and perform such repairs as may be required.

A copy of the Examiner's report will be furnished by him to the officer commanding the R.A. District, or where the guns are mounted, to enable this officer to complete his Annual Return as prescribed.

Should there be no guns in a district requiring such examination or repair, the district will not be visited in that quarter.

Foreign Stations, Royal Artillery.

Examination of guns in Royal Artillery charge abroad.

At all foreign stations the examination will be made under the direction of the officer commanding the Royal Artillery in the district, by an Inspector of Warlike Stores, should there be one at the station, or, if not, by some other competent person. A report of the examination will be made on W.O. Form 1475 or 1476, and forwarded through the same channels as laid down for the annual return.

In such case the word "Special" will be substituted for "Annual" in the heading of the form.

Impressions will not be sent with the report unless there is any doubt as to the serviceability of the gun or guns; but should any gun appear to be in an unserviceable state, or to require re-venting or other repairs beyond what can be effected on the spot, impressions will be forwarded with the report, for the information of the Director of Artillery and Stores, who will give such directions as he may think desirable.

If guns found unserviceable, or requiring re-venting or repair, as above, be mounted in an important position, or if local circumstances render their immediate exchange necessary, they will at once be exchanged, if practicable, by requisition on the Commissary of Ordnance Stores, approved by the General Officer commanding. In such cases the requisition, after being complied with, will be forwarded by the Commissary to the Director of Artillery and Stores.

Note.—Revised instructions with regard to the examination, &c., of Guns in charge of the Royal Navy, Royal Marine Artillery, Royal Marines, Her Majesty's Coast Guard, and the Royal Naval Reserve, are under consideration, and will shortly be officially promulgated.

General Remarks.

When any accident occurs, either at home or abroad, such as the bursting of a shell in the bore, the splitting of a B.L. vent-piece, &c. immediate inquiry should be made into the circumstance, and the gun. If the Commanding Officer considers the damage to be of importance, he will send without delay, a report of the circumstances, through the same channel as his annual return forwarding, if necessary for the illustration of his report, gutta-percha impressions of the damage done to the gun. A similar course will be pursued in regard to naval guns—The "Memorandum of Examination" W.O. Form 1340, of every rifled gun, must be be in possession of the officer in charge of that gun; and when a gun is returned into or issued from store this memorandum must accompany the transfer vouchers.

The number of rounds, that have been fired will be accurately entered in the memorandum, which will be carefully preserved, as containing important information concerning the gun.

(4.) Method of taking Impressions in Gutta-percha and Wax.

Gutta-percha.

The common kind of gutta-percha used for the soles of boots is suitable for the purpose. It can be used over and over again, and need never be thrown away if a little fresh material be added from time to time, to prevent it from becoming brittle. It should not be allowed to become mixed with dirt or grit, and it should be kept in water when not in use.

How prepared.

The method of applying the gutta-percha is as follows: A sufficient quantity having been softened by being put into boiling water, is worked and kneaded on a smooth board, until the air and water are expelled, and a smooth surface obtained. A small lump is then placed on the pan of the instrument (which should have been previously fitted with a pad of gutta-percha) and screwed up against the vent, or other part of the bore, of which an impression is to be taken. It is there left till cold, about 10 or 20 minutes, according to the weather; then the instrument is withdrawn.

How used

A little soft soap, or, if that be not available, common soap and water, oil or grease, will prevent the impression sticking to the pad. The bore of the gun should be slightly greased. Too much pressure must not be applied, otherwise the impression will be very thin, and if the defect be deep it will be difficult to remove the gutta-percha. A good deal of practice is required to get good, smooth impressions; and several impressions of the vent have sometimes to be taken before one is obtained which can be relied on to show any hair lines.

Wax composition.

Wax composition, which may be used instead of gutta-percha, for taking temporary impressions for examination on the spot, is made of beeswax two parts, treacle one part, soft soap one part. The wax should be melted over a slow fire in an iron pot, the treacle being first added and mixed well by stirring; and lastly the soft soap, a little at a time. The mixture must be kept in motion, and when thoroughly mixed poured out, and made into balls when cool. This composition being soft, is always ready for use, but is easily destroyed by handling. The cushion or pad of gutta-percha will be removed from the pan of the instrument before the wax composition is applied.

W.O. Form 1475.

An Annual Return of the State of Cast Iron and Bronze Ordnance, showing the Condition of the Vent and Bore, and the Number of Rounds fired from each Gun.

____________________ 18____

Descriptive Marks.					Actual Weight.			Length.				Condition of								Number of Rounds fired with Projectiles.				Received from		Returned or issued to		
Foundry.												Vent.				Bore.		Sentence.										
													Diameter.															
Number on Trunnion.	Letters.	Date.	Register Number on Reinforce.	Cypher or Badge, if any.	Cwt.	Qrs.	Lbs.	Feet.	Inches.	Nature.	Date of last Examination.	Iron, Copper, or not bushed.	Through gauge.	At bottom.	Extent of Fissures in the Cast Iron round the Vent, measured from the original centre of the Vent.	Examiner's Remarks.	No. of Rounds with projectiles fired since last Examination.	↑ = Serviceable. C> = For cone-venting. T> = Through-venting. + = Unserviceable.	Previously reported.	Since last Report.	Total.	In Store, or whether mounted, and where placed, in Ship or Fortress.	Station, &c.	Date.	Station, &c.	Date.	Remarks.†	
												*																

* The letters (if any) on the cascable or loop to be inserted in this column.
† If any gun has been re-vented at the station since the last return, the date of venting and description of vent will be inserted in this column.

W.O. Form 1476.

An Annual Return of the State of Rifled Ordnance, and of the Number of Rounds fired from each Gun.

_______________ 18____.

Description.			Weight.*				Nature. (See para. 87, p. 24 of Instructions for inspection of Ordnance.	Date of last Examination.	Condition of Bore, specifying any important defects not noted in Memo. of Examination.	Size of Vent.		General state of Vent, and of adjoining part of Bore.	Sentence.			Number of Rounds fired at Date of Return. With Projectile. Battering.			Full.			Reduced.			Number of Rounds with Projectiles fired since last Examination.	In Store, or whether mounted, and where placed, in Ship or Fortress.	Received from		Returned or issued to		Condition of		Remarks.†
Register Number.	Mark.	Initials of Factory, and date.	Tons.	Cwt.	Qrs.	Lbs.				Through Gauge.	Size at Bottom.		↑ Serviceable.	R ‡Repairable.	+ Unserviceable.	Previously reported.	Since.	Total.	Previously reported.	Since.	Total.	Previously reported.	Since.	Total.			Station, &c.	Date.	Station, &c.	Date.	Breech-screw.	Vent-piece.	

N.B.—Duplicate copy of this return not required. * Under 100 cwts. give the full weight in cwts.; 5 tons and over in tons. See § 1081, Changes in Ordnance, &c.

† If any gun has been re-vented at the station since the last return, the date of venting and description of vent should be inserted in this column.

‡ If repairable, give nature of repair required.

W.O. Form 1478.

Royal Artillery_______________District_______________Station_______________, 1878.

Return of Ordnance which have fired the prescribed Number of Rounds since previous Examination.

Nature.	Descriptive Marks.			Weight.				Number of Rounds fired.		Name of Battery where mounted.	When next required for practice.	Remarks.*
	Number on Trunnion.	Letter.	Date.	Tons.	Cwt.	Qrs.	Lbs.	Since last Examination.	Total.			

* State if any impression is sent, or if there is no person at hand to take one.

† In the case of smooth-bore guns this return should be sent direct to the Superintendent, Royal Gun Factories.

† *The Director of Artillery, War Office, Pall Mall, S.W.*

_______________ *In Charge of Guns at Battery.*

_______________ *Commanding the District.*

APPENDIX II.

INSTRUCTIONS FOR BRONZING AND BLUEING SIGHTS AND FITTINGS, BROWNING GUNS, DEEPENING CENTRE HIND SIGHT HOLES, FIXING DERRICKS, &C.

Blueing.

(1.) The steel tangent bar, screw trunnion sight, and all trunnion sight leaves are blued.

Blueing consists simply in covering the surface with a thin film of oxide sufficient to give the article a deep blue colour, and to prevent further oxidation from exposure to the atmosphere. This is easily effected by polishing well the surface of the article, and heating it to about 550° until it assumes a blue colour and then allowing it to cool gradually. A sand bath is generally used in order to obtain a uniform heat, and the bar, &c. is taken out from time to time to watch the change of colour and to prevent its going too far. The temperature may be judged by the colours, which successively appear on the surface of the steel at various low temperatures, viz. :—

At 450° F. the steel becomes a straw colour, at 475° an orange or gold, 500° brown, 530° purple, 550° violet, 580° blue, 610° white, and at 625° red.

Bronzing.

(2.) The exposed gun-metal portions of all the tangent and drop sights are protected from the influence of the atmosphere by "bronzing as follows :—

1st. Polish the parts well and heat them over a spirit lamp or gas.

2nd. Polish with a brush and black lead, to remove all grease, &c.

3rd. The bronzing mixture is then applied to the heated metal. It consists of—

Bichloride of platinum	2 parts.
Corrosive sublimate	1 "
Vinegar	1 "

4th. The parts are next dipped into boxwood sawdust to dry them, and then again polished with black lead to give a body to the colour. The figures, which are left bright, are rubbed with emery cloth, and the whole is finally varnished with shellac and methylated spirits.

Browning.

(3.) 9 and 16-pr. guns are browned, the operation is as follows :—

1. Steam the gun for 10 hours, then wash with a lye of potash (1 lb. black potash to 1 gallon of water) ; repeat until the grease is throughly eradicated.

(A.) If there be no convenience on service for the performance of the steaming process, simple washing may have to be employed instead. The whole object is to get rid of oil which may remain on the surface of the iron, and hence the water should be as hot as can be borne by the operator, who will rub vigorously all over the exterior surface with a clean hard brush ; a little hard soap should be used, and the water should be frequently changed so as to ensure its perfect cleanness. This washing and scrubbing with soap and hot water must be repeated at least three times ; then wash the gun with the lye of potash as aforesaid. (B.) Repeat the process A, consisting of three washings and scrubbings with soap and water, and of one washing with the lye of potash several times ; to obtain perfect cleanness may require many repetitions of the whole process, and care must be taken not to touch the gun with any fatty matter, or even with the hand, as it may take hours of washing to wholly remove the effect.

2. Wash with hydrochloric acid and water (equal parts) to remove oxide, then wash with clean water and wipe dry. 3. Apply browning mixture with a sponge, and let stand for 12 hours in a temperature not less than 60° or more than 100°; then rub off rust with scratch card and brush. The browning mixture is composed of the following:—

Tincture of steel	2 parts.	Spirits of nitre	1½ parts.
Nitric acid	1 "	Spirits of wine	1½ "
Blue vitriol	1 "	Soft water	32 "

4. Apply mixture, let it stand six hours, rub off rust. 5. Repeat No. 4. 6. Apply mixture, let it stand six hours, then boil five minutes in a lye of potash (1 lb. potash to 2 gallons of water), then rub off rust. 7. When cold, repeat No. 4. 8. Repeat No. 4. 9. Apply mixture, stand six hours, then boil as in No. 6 operation, rub off rust, then coat with olive oil.

Care to be taken to well sponge and dry the bore and chambers after each operation of washing, steaming, or boiling.

Stores required for browning a battery. The following is a detail of the stores and quantities of ingredients allowed per battery biennially required for browning a battery of six 9-pr. guns, or six 16-prs.,* viz.:—

Tincture of steel	4 ozs.	To be mixed in 2 quarts of soft water.
Nitric	2 "	
Blue vitriol	2 "	
Spirits of nitre	2 "	
Spirits of wine	3 "	

Hydrochloric acid	6 lbs.
Earthenware pan to hold 6 quarts, for hydrochloric acid	1
American potash	6 lbs.
Wooden pail for do.	1
Oil	½ gill.
Sponge cloth for do.	1
Sponge to apply the browning mixture	1
Flat brush to apply the hydrochloric acid	1
Scratch card to rub the surface of the gun between the coatings	6 ins.
Coals	335 lbs.
Brush, hard	1

Preparing Gun for lengthened hind sight.

(4.) As mentioned at p. 187, lengthened centre hind sights are to be supplied eventually for 9-inch guns and upwards, and for that purpose the socket hole has to be deepened.

The tools required are mentioned below, as well as the necessary instructions regarding their use.

Tools for deepening holes for lengthened Centre Hind Sights, 9-inch R.M.L. Guns and upwards.

Brace, ratchet, 20"		1
Drill	16"	1
	22½"	1
Guide, steel		1
Stamp, D.		1

Instructions for deepening the centre hind sight holes for 9", 10", 11", and 12" guns:—

Take out the metal sight socket and put the steel guide in its place. Erect the sighting machine drill frame (marked L in the tools for sighting ordnance) placing the drill in the steel guide, and bringing the feed screw exactly to it,

* Half as much again for 16-prs. as to proportion of stores.

The drill and feed screw will not be quite perpendicular, but will be inclined at the correctional angle for the gun's deflection. Fasten the frame securely by a strong chain or rope to the gun, placing a block of hard wood under the tail.

The holes in the 12 and 18-ton guns can be sufficiently deepened with the shorter drill alone, but for those in the 25-ton guns it will be necessary to extend the hole farther, and the longer drill must be substituted, when the shorter one has drilled as deeply as the feed screw will drive it. It will be found convenient to mark the proper depth on the drill.

To use the machine, insert the drill, attach the ratchet brace, gently tighten up the feed screw, lubricate with oil and turn the handle.

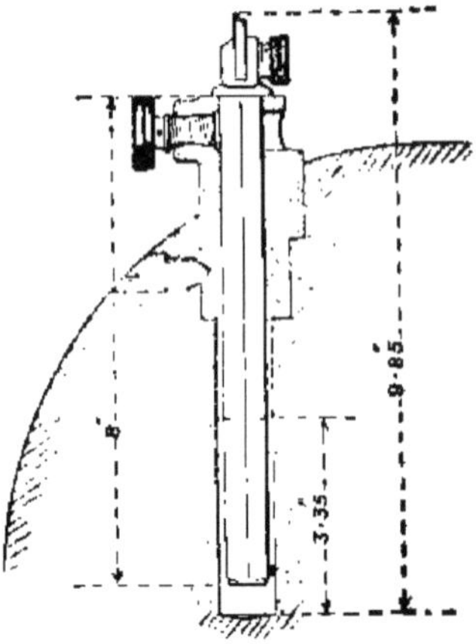

When the proper depth has been attained, remove the apparatus and steel guide, thoroughly clean out the hole and replace the socket; a small D will then be stamped upon the gun in front of the socket hole.

(5.) As mentioned at p. 205, the following instructions apply to the fixing on the guns of the muzzle derricks described at p. 203, Chapter IX. Fixing derricks.

Instructions to be observed in fixing bronze derricks to muzzles of heavy rifled guns :—

1st. Scribe a line upon the top of the chase from the vertical axis line on muzzle for a distance of about 12 inches towards the fore sight.

2nd. Remove the fixing screws and then try on the band ; there may be a slight variation in the diameter of the muzzles of guns of the same nature, but if they are correct the bands would be seated as follows, viz. :—

Distance from face of muzzle to front edge of band	for 9″ M.L. guns $5\frac{1}{2}$″ ; for 10″, 11″, 12″, and $12\frac{1}{2}$″ guns, 6″.

Should the band not reach its seat, it will be necessary to ease it inside with a half-round file until it attains the required position. If, on the other hand, the muzzle of the gun be small, the band must be pushed on as far as it will go.

3rd. When the band has been placed roughly in its position, turn it round until the vertical lines cut on the front and rear faces agree with the scribed line on muzzle mentioned above. When properly adjusted, give the front edge a few gentle taps all round with a piece of wood to drive it on the chase, and thus fix it temporarily in position, but the relations of the lines must not be disturbed in so doing.

4th. Now mark off upon the chase the positions of the holes by means of a steel scriber carefully guided around the interior of the screw holes, and then remove the band. Special attention must be paid to this operation.

5th. Dot round with a centre punch the circles just described, and centre each as nearly as possible ready for drilling.

6th. Erect the drilling apparatus and drill very carefully four holes, each $\frac{5}{16}$ in diameter and $\frac{1}{2}$ deep, to correspond with the plain points of the fixing screws.

Judgment must be exercised in drilling the holes if required, so that they may be perfectly concentric with the dotted circles previously marked off.

At the chief stations, where artificers are at hand, drills can readily be prepared on the spot for this purpose, but in localities where these conveniences do not exist, the drills can be supplied on demand.

7th. After the holes are completed remove the drilling tackle, clean the surface of the chase, take off all "burrs" from the holes, and place the band in position; then insert the screws and send them home firmly and securely.

It will be necessary to adjust the bridge piece which supports the derrick when erected, so that the latter may be brought forward in order to maintain the required relation with the muzzle of the piece. It will be requisite first to level the gun and then to drop a plumb-line from the centre of the loop or eye at the top of the derrick, and the distance measured from the face of the muzzle to this line should be—

		"	
For 9" M.L. gun	....	7·4	Limits of error, one inch minus or plus.
„ 10" „	....	9·4	
„ 11" „	....	10·3	
„ 12" 25-ton „	....	10·45	
„ 12" 35 „ „	....	11·5	
„ 12½" „	....	12·35	

Where the measurement does not comply with the above dimensions, the lower side of bridge piece must be cut away and relieved until the proper overhang has been obtained.

APPENDIX III.

MITRAILLEUR OR MACHINE GUNS.

Mitrailleurs or machine guns of some description now exist in limited numbers in the armaments of most great powers.

With regard to the raison d'être of these weapons, it appeared on the introduction of rifled field guns that the smaller charges used with and motion imparted to the projectile somewhat decreased the effect of case shot, while shell fire (until fuzes are much improved) at short ranges must always be more or less uncertain. Mitrailleurs were therefore made with a view to their affording a fire like that of case (*mitraille*), or a hail of bullets, for ranges up to 1,000 or 1,200 yards. **Use of mitrailleurs.**

In August, 1870, a Special Committee carried out trials which resulted in the purchase of a small number of Gatling guns. **Special Committee, 1870.**

Their report was made in November, 1870, at which date a full knowledge had not been obtained of the effect of the mitrailleurs used in the Franco-German war of that year. Twelve Gatling guns, however, of small calibre, for land service, and 24 of medium, together with 12 of small size for sea service, were ordered as a tentative measure, until further experience was gained. **Gatling guns introduced 1870.**

In November, 1871, the Special Committee above mentioned having prosecuted further inquiry as to the intentions of foreign governments regarding machine guns, and having examined a number of officers who were present with French or German armies during the war, made a second report in which they adhered to their former opinion, recommending the adoption of the larger Gatling gun of 0·65-in. bore for coast defences and naval service, and the smaller of 0·45 in calibre for field purposes. **Adoption recommended of the 0″·65 and 0″·45 Gatling.**

As already mentioned, a small number of these guns had been ordered in 1870, and it was thought advisable, before manufacturing any more, that these should be thoroughly tried in the service. A number of 0·45-in. calibre and of 0·65-in. calibre, have been made for S.S., and lately some smaller 0·45-in. Gatlings have been ordered for Indian service. These Gatling guns were finally approved of as service pieces in 1874.* **Finally approved of in 1874 by § 2647.**

Construction of Service Gatling.

The following are the details of construction of our service Gatling, which belongs to the description of mitrailleurs where the barrels revolve, and are charged simultaneously. Each barrel has its special lock, which accompanies it during revolution, and is also capable of motion backwards and forwards. It pushes the cartridge into its barrel, then serves as a breech, and afterwards extracts the empty cartridge. Each lock has a striker with spiral spring, and an extractor. **§ 2642. § 2647.**

The system is caused to revolve by means of a crank fixed on the right side of the piece, and an automatic "scattering" arrangement has been added, which can be put in gear or not, as required, and which is worked by the crank which moves the system. Each lock can be taken out separately, and replaced by a new one. **Scattering arrangement.**

---

* "It is proposed to keep in store in reserve a few Gatling guns (made by contract), in case they should be required, and also to issue a limited number to ships of war.

"As these weapons, however, are considered complicated and of limited power, it is not contemplated that they will be much employed on active service, and no troops are armed with them.

300 to 400 rounds per minute can be fired from this mitrailleur, and two men are sufficient to serve one in action.

We have three natures in our service, the 0·65-in. and two of 0·45-in. bore, of 3 cwts. and 200 lbs. weight respectively.

	cwt.	qrs.	lbs.
Weight of gun	3	3	24
" carriage and limber (empty)	12	2	13

Barrels. The gun has 10 steel barrels, rifled on the Henry principle, fixed in a circle round a centre shaft of steel. To this shaft are keyed two gun metal discs, through one of which the muzzle ends of barrels pass, while their breech ends are screwed into the other. Vide Plate, page .

Shaft. Breech casing. Cascable plate. The shaft itself is fixed in a "gun frame" (*aa*, Fig. 1) of wrought iron, made of two bars connected in front of muzzles by a curved cross-piece (*n*, Fig. 1). The rear ends of this gun frame are connected by screws to a cast iron box, or "breech casing" (*C*, Fig. 1), which contains the mechanism. In this casing is a vertical diaphragm, through which the shaft passes towards the breech, and the breech end of the casing is closed by a "cascable plate" (*D*, Fig. 1) of cast iron.

Inside the casing, upon the rear end of the shaft, is a small (pinion) worm-wheel (*W*, Fig. 2), which gears into a worm (*f*, Fig. 2) on a crank shaft or spindle (*gg*, Fig. 2), which passes into the breech casing on the right side, and at right angles to the main shaft. By turning a crank handle secured to this spindle, the main shaft and barrels are caused to revolve. When not in use this handle is pushed in out of the way.

Fastened by screws to the gun frame, is a "pivot block" of gun metal (*p*, Fig. 3); a pivot (*P*, Fig. 3) passes through this, and into an iron trunnion plate (*ee*, Fig. 3), and upon it the system turns when lateral spread of bullets is required.

The trunnion plate has projections, or trunnions, on which the system revolves for elevation, and is secured at the rear end by a bolt and nut to a locking bolt plate which fits into an undercut slot in bottom of breech casing.

Scattering fire. Concentrated fire. Locking bolt. When a scattering fire is required, the frame, barrels, &c., turn on this "block" through the required arc by means of an automatic arrangement (*AF*, Fig. 2) worked by the crank handle before mentioned. When such is not required, the fire is concentrated by putting this arrangement out of gear, and preventing any transverse movement by means of a "locking bolt" (*l*, Fig. 2) let down into a slot in trunnion plate at the rear.

Cartridge carrier. Hopper. Feed drum. On the main shaft in rear of the barrels a cast iron cylinder, or "cartridge carrier" (*M*, Fig. 4) is fixed. This has 10 longitudinal grooves, corresponding with the 10 barrels. A gun metal cover, or "hopper" (*B*, Fig. 4), hinged at one side, drops over it. The cover has a longitudinal slot, corresponding to the opening in the "feed drum," this drum is of metal, fits on a pin (*b*, Fig. 4) in centre of hopper, and contains 240 cartridges, in 16 perpendicular columns or channels. It weighs, when full, 50 lbs. (*T*, Fig. 1), which rests upon the upper surface of the "hopper." As each column is exhausted, the drum is turned round by hand until the next one corresponds with the opening in the hopper. Through this slot the cartridges drop (as the shaft revolves) into the several grooves, ready to be pushed by the lock plungers into the barrels corresponding.

Lock chamber. In rear of this, and inside the breech casing, is placed the "lock chamber," which is keyed to and revolves with the main shaft. It is a cylinder of cast-iron, having longitudinal channels through which the "locks" pass.

Locks. Rear guide nuts. Upon the main shaft, again, and against the back of the "lock chamber" is secured a cast-iron "rear guide nut," which keeps the parts firmly together. The locks rest partly upon the outer circumference of this nut; and in the grooves on which they fit, as well as in the channels in the lock chamber, are small slots, in which run studs on the locks, in order to prevent the latter revolving save with the shaft.

Cam. Inside the "casing" is a curved gun metal plate, or cam, by means of which, as the shaft and lock chamber revolve, the locks themselves are pushed forward or back. A piece of steel is let into the front of this cam, against which the butt of each lock bears at the moment the barrel is fired.

Cocking ring. There is also a steel cam, called a cocking ring, which, as the lock chamber

SERVICE GATLING GUN, 0·45 BORE.

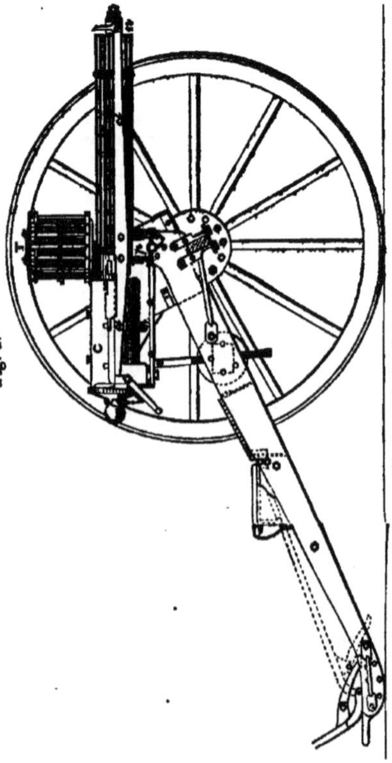

Fig. 1.

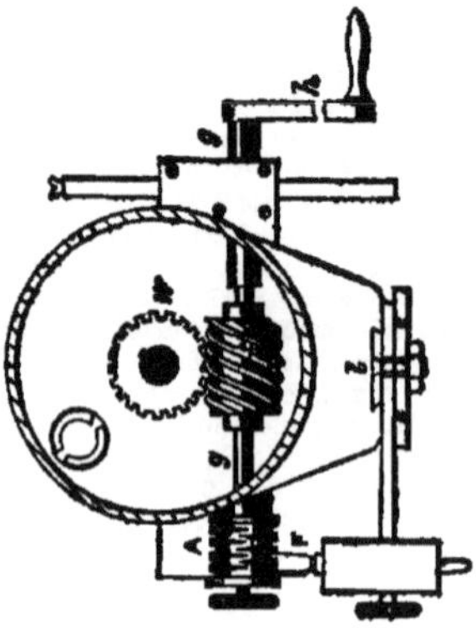

Fig. 2.—Elevation of Breech End with Cascable Plate removed.

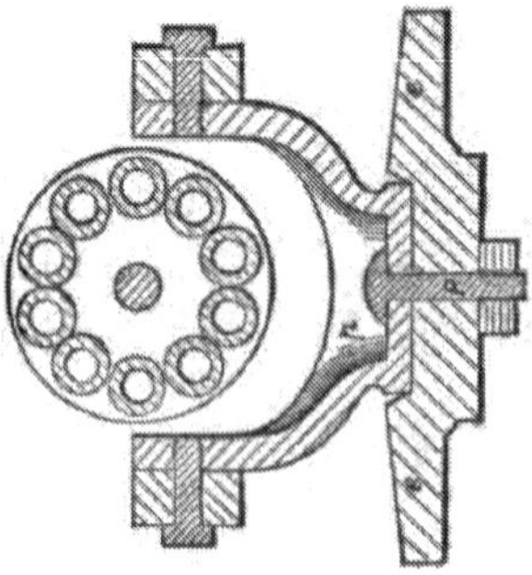

Fig. 3.—Section through Trunnion-Plate, Trunnions, and Pivot Block.

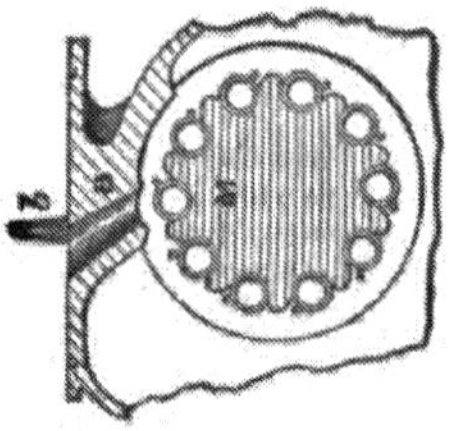

Fig. 4.—Section through Hopper and Cartridge Carrier

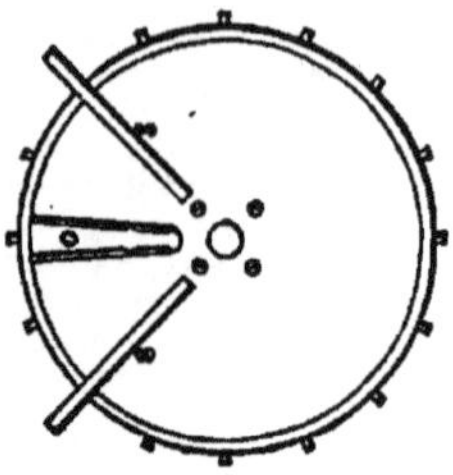

Fig. 5. Plan, Bottom of Drum.

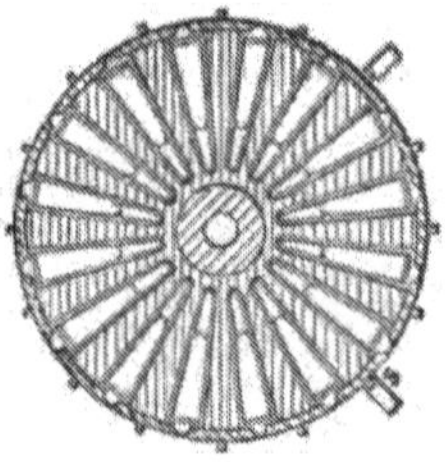

Fig. 6. Horizontal Section through Drum.

Firing pin. revolves, draws back and then releases a spiral spring acting on the "firing pin" or needle of each lock.

The Lock.

Lock. The lock consists of a steel tube or "plunger," about 11½ ins. long, the front end of which, for about 4 ins., is smaller in diameter, and has only a pin hole running through it.

Butt. The remainder is hollow, and slotted out on one side. Its breech end is closed by a steel plug or "butt," screwed in.

Inside is a steel bolt or "hammer," having a projection at the side which passes through the slot in the tube, while to the front part of it is attached a firing pin or "striker," of steel.

A spiral spring is placed over the hammer, being retained by the "butt."

Extractor. To the outside of the tube or lock is fixed a steel extractor, having a hook, which seizes the rim of the cartridge and draws it out as the lock is being withdrawn.

Sighting.

Sights. There are two sights, a fore sight and tangent sight. The former is a plain steel sight attached to the gun frame on the right side at the muzzle, and the latter is a plain steel bar graduated in degrees and yards, working in a socket on the right rear side of the breech casing, and clamped by means of a milled head thumb screw.

The 0·45-in. gun is sighted with a tangent and fore sight up to 2,400 yards (8° 23′ elevation).

Action.

Action. When the gun is in action five cartridges are always in process of loading, and five are in different stages of extraction. Thus, as the system revolves, cartridges drop from the feed drum through slot in the hopper, successively, on the 10 grooves in cartridge carrier; as each lock comes in contact with the cocking ring, the hammer is drawn back and spring compressed; further rotation brings the lock against the gun metal cam, which pushes it forward, driving before it a cartridge from the carrier into its particular barrel. The breech is thus closed, and as the butt comes opposite the steel plate in the cam, the cocking ring releases the spring, and the needle fires the cartridge. The system continues to revolve, and the lock now being drawn back within the chamber, extracts the empty cartridge case while retreating, and the latter falls to the ground.

APPENDIX IV.

RANGE TABLES AND TRAJECTORIES.

If, knowing the M.V. given by a gun, we had any means of ascertaining with certainty from theoretical considerations what would be the exact path described by its projectile through the air, there would be no necessity for obtaining any data for range tables as we now do with each nature of gun, by firing a number of rounds at Shoeburyness.

Range tables.

We cannot, however, as yet determine by mere calculation with sufficient exactness what the trajectory of a given piece will be, although when the angle of elevation is small, we can approximate to it very nearly.

This subject of trajectories, however, is one of theoretical gunnery, and will be found treated of elsewhere.*

In order to make quite certain that our range tables are made out from data obtained under conditions analogous to those which prevail on service, the following method is pursued :—

How obtained.

A specimen gun of the new nature to be introduced † is fired at Shoeburyness with the charge and projectile as settled by previous experimental firing at the R.G.F. Butts and elsewhere, a number of rounds are fired at a given angle, and the mean range carefully measured ; the angle of elevation is then changed and a new series of rounds fired, and so on ; from the data so obtained a curve is then constructed as shown in Plate E. at end of the book, where the practice curve of a 38-ton gun with a charge of 130 lbs. of powder and 800 lbs. shot is given, the ordinates representing the angles of elevation, and the abscissæ the corresponding ranges ; by means of this curve we can, taking any angle of elevation, at once obtain the range corresponding to it for that particular nature of gun, or *vice versâ*.

In a similar manner are obtained the times corresponding to any given range for that gun,‡ and from such curves the range tables given at pp. 386, 427 are constructed by interpolation.

For direct fire these tables give us most of the information we require,§ but for indirect and for curved fire we want more detailed data, including especially the angle of descent, with different charges and elevation, in order to know the facilities given by our guns for breaching revetments, &c., &c., when covered by glacis or otherwise, or for throwing our shells just over the crest of a work.

From theoretical considerations we can find the angle of descent from the formula of Professor Hélie,‖ given in note below, or by the mode employed by Professor Bashforth, vide p. 63 of his work. Vide also Captain Sladen's work already mentioned.

---

* Principles of Gunnery Rifled Ordnance, by Captain Sladen, R.A., Professor of Artillery, R.M. Academy. Vide also a paper by Captain Kensington, R.A., Proceedings R.A.I., Vol. IX., Nos. 3 and 4.

† Or of an existing service nature fired under new conditions as to powder, projectiles, &c.

‡ A stop clock being employed for ascertaining the time of flight for every round fired.

§ And, as explained at p. 340, we can from these data work backwards, and obtain the M.V. if necessary.

‖ Tan. θ = tangt. (angle of descent) $= \frac{gX}{V \cos^2\alpha} + \frac{3gKX}{2\cos^2\alpha} - \tan d.$

Where α = angle of elevation. $K = Z\frac{c}{V}$

X = range.

$Z = \varphi\ (v) = \left.\begin{matrix} 0{\cdot}025 \text{ for 1,200 f.s. and above} \\ 0{\cdot}020 \text{ for 1,200 f.s. and below} \end{matrix}\right\}$ both elongated projectiles.

$C = b\frac{R^2}{W}$. $b = 0{\cdot}00181$ for elongated projectiles.

R = radius of projectile. W = weight of projectile.

APPENDIX V.

Range Tables.*

Range Table for 7-Pr. R.M.L. Gun of 150 lbs. (*Steel.*)

Charge 4 ozs. F.G. powder.

Projectile, double shell.

Range.	Elevation.	Fuze Scale.	Range.	Elevation.	Fuze Scale.
Yards.	° ′		Yards.	° ′	
80	1 7	1	1,080	19 22	16·5
130	1 37	1·5	1,100	19 50	17
160	2 7	2	1,130	20 35	17·5
200	2 37	2·5	1,150	21 4	18
230	3 7	3	1,180	21 52	18·5
270	3 44	3·5	1,200	22 23	19
310	4 21	4	1,220	22 55	19·5
340	4 46	4·5	1,250	23 45	20
380	5 23	5	1,270	24 17	20·5
420	6 0	5·5	1,290	24 49	21
450	6 31	6	1,310	25 21	21·5
480	7 2	6·5	1,330	25 53	22
510	7 33	7	1,360	26 45	22·5
550	8 15	7·5	1,380	27 21	23
580	8 44	8	1,400	27 57	23·5
610	9 22	8·5	1,420	28 33	24
650	10 7	9	1,440	29 9	24·5
680	10 43	9·5	1,460	29 45	25
710	11 19	10	1,480	30 21	25·5
740	11 55	10·5	1,500	30 57	26
770	12 32	11	1,520	31 33	26·5
800	13 9	11·5	1,540	32 9	27
830	13 46	12	1,560	32 46	27·5
860	14 23	12·5	1,580	33 23	28
890	15 0	13	1,600	34 0	28·5
910	15 25	13·5	1,610	34 22	29
940	16 6	14	1,630	35 3	29·5
970	16 48	14·5	1.650	35 45	30
1,000	17 30	15			
1,020	17 56	15·5			
1,050	18 38	16			

* Range Tables for projectiles with gas checks which are not given in these Tables, will shortly be issued, and should be interleaved in the book.

Range Table for 7-Pr. M.L. Steel Gun of 150 lb.

Charge, 6 oz. F.G. powder.

Projectile, common shell.

Range.	Elevation.	Fuze Scale.	Range.	Elevation.	Fuze Scale.
Yards.	° ′		Yards.	° ′	
130	0 30	1	1,440	9 43	13·5
190	0 49	1·5	1,480	10 16	14
250	1 8	2	1,520	10 30	14·5
310	1 27	2·5	1.570	11 0	15
370	1 46	3	1,610	11 27	15·5
430	2 5	3·5	1,650	11 54	16
480	2 21	4	1,690	12 21	16·5
540	2 43	4·5	1.730	12 48	17
600	3 5	5	1,770	13 15	17·5
650	3 25	5·5	1,800	13 37	18
700	3 45	6	1,840	14 4	18·5
750	4 5	6·5	1,870	14 28	19
810	4 30	7	1,910	14 55	19·5
870	4 55	7·5	1,950	15 22	20
920	5 17	8	1,980	15 47	20·5
970	5 39	8·5	2,010	16 12	21
1,020	6 2	9	2,050	16 45	21·5
1,070	6 25	9·5	2,080	17 15	22
1,120	6 49	10			
1,170	7 13	10·5			
1,220	7 37	11			
1,260	7 58	11·5			
1,310	8 28	12			
1,350	8 50	12·5			
1,400	9 20	13			

Range Table for 7-Pr. R.M.L. Gun of 200 lbs. (*Steel.*)

Charge, 4 oz. F.G. powder.

Projectile, double shell.

Range.	Elevation.	Fuze Scale.	Range.	Elevation.	Fuze Scale.
Yards.	° ′		Yards.	° ′	
700	8 12	8·5	1,420	21 23	20
730	8 44	9	1,450	21 58	20·5
760	9 16	9·5	1,470	22 33	21
800	9 48	10	1,500	23 10	21·5
830	10 20	10·5	1,520	23 49	22
860	10 52	11	1,550	24 28	22·5
900	11 24	11·5	1,570	25 7	23
930	11 56	12	1,600	25 48	23·5
960	12 28	12·5	1,620	26 33	24
1,000	13 0	13	1,650	27 18	24·5
1,030	13 36	13·5	1,670	28 3	25
1,060	14 12	14	1,700	28 50	25·5
1,100	14 48	14·5	1,720	29 33	26
1,130	15 24	15	1,740	30 16	26·5
1,160	16 0	15·5	1,760	30 59	27
1,200	16 36	16	1,780	31 42	27·5
1,230	17 16	16·5	1,800	32 46	28
1,260	17 56	17	1,820	33 6	28·5
1,300	18 38	17·5	1,840	33 46	29
1,320	19 10	18	1,860	34 26	29·5
1,350	19 42	18·5	1,880	35 6	30
1,370	20 14	19			
1,400	20 48	19·5			

Range Table for 7-Pr. R.M.L. Gun of 200 lbs. *(Steel.)*

Charge, 12 oz. F.G. powder.

Projectile, common shell.

Range.	Elevation.	Fuze Scale.	Range.	Elevation.	Fuze Scale.
Yards.	° ′		Yards.	° ′	
160	0 30	1	2,060	8 3	16
240	0 44	1·5	2,110	8 20	16·5
320	0 58	2	2,160	8 37	17
400	1 12	2·5	2,210	8 54	17·5
470	1 26	3	2,260	9 12	18
550	1 40	3·5	2,300	9 30	18·5
620	1 54	4	2,350	9 48	19
700	2 9	4·5	2,390	10 6	19·5
770	2 22	5	2,440	10 24	20
840	2 36	5·5	2,480	10 42	20·5
910	2 51	6	2,530	11 2	21
970	3 6	6·5	2,570	11 20	21·5
1,040	3 21	7	2,610	11 38	22
1,100	3 36	7·5	2,650	11 56	22·5
1,170	3 51	8	2,690	12 14	23
1,230	4 6	8·5	2,730	12 31	23·5
1,290	4 21	9	2,770	12 48	24
1,350	4 36	9·5	2,810	13 5	24·5
1,410	4 51	10	2,850	13 22	25
1,470	5 6	10·5	2,890	13 39	25·5
1,530	5 21	11	2,920	13 56	26
1,580	5 37	11·5	2,960	14 13	26·5
1,640	5 53	12	2,990	14 30	27
1,690	6 9	12·5	3,030	14 47	27·5
1,750	6 25	13	3,070	15 4	28
1,800	6 41	13·5	3,100	15 21	28·5
1,860	6 57	14	3,130	15 38	29
1,910	7 13	14·5	3,170	15 55	29·5
1,960	29	15	3,200	16 12	30
2,010	7 46	15·5			

RANGE TABLE for 9-PR. R.M.L. GUN of 8 and 6 cwt.

Charge, 1 lb. 12 oz. R.L.G. powder.
Projectile, shrapnel shell.

Range.	Elevation.	Fuze Scale.	Range.	Elevation.	Fuze Scale.
Yards.	° ′		Yards.	° ′	
200	0 6	1	2,600	6 59	16
320	0 15	1·5	2,660	7 16	16·5
420	0 28	2	2,730	7 33	17
520	0 41	2·5	2,790	7 50	17·5
620	0 54	3	2,860	8 7	18
720	1 7	3·5	2,920	8 24	18·5
820	1 20	4	2,980	8 42	19
910	1 32	4·5	3,040	9 0	19·5
1,000	1 44	5	3,100	9 18	20
1,090	1 56	5·5	3,160	9 36	20·5
1,180	2 9	6	3,220	9 55	21
1,260	2 22	6·5	3,280	10 14	21·5
1,340	2 35	7	3,340	10 33	22
1,420	2 47	7·5	3,400	10 53	22·5
1,490	3 1	8	3,450	11 13	23
1,570	3 15	8·5	3,510	11 33	23·5
1,640	3 29	9	3,560	11 53	24
1,710	3 43	9·5	3,620	12 12	24·5
1,780	3 57	10	3,670	12 31	25
1,850	4 11	10·5	3,730	12 50	25·5
1,920	4 25	11	3,780	13 9	26
1,990	4 39	11·5	3,840	13 28	26·5
2,060	4 54	12	3,900	13 47	27
2,130	5 9	12·5	3,950	14 6	27·5
2,200	5 24	13	4,000	14 24	28
2,270	5 39	13·5	4,050	14 42	28·5
2,340	5 54	14	4,100	15 0	29
2,400	6 10	14·5	4,150	15 17	29·5
2,470	6 26	15	4,200	15 33	30
2,530	6 42	15·5			

N.B.—Common shell being lighter than shrapnel, ranges about 100 yards farther with the same elevation.

Range Table for 9-Pr. R.M.L. Gun of 6 cwt. S.S.

Charge, 1 lb. 8 oz.

Projectile, common shell.

Range.	Elevation.	Fuze Scale.	Range.	Elevation.	Fuze Scale.
Yards.	° ′		Yards.	° ′	
190	0 18	1·	2,520	7 53	16
280	0 30	1·5	2,580	8 10	16·5
380	0 44	2·	2,630	8 26	17·
480	0 58	2·5	2,690	8 43	17·5
570	1 12	3·	2,740	9 0	18·
650	1 25	3·5	2,800	9 18	18·5
740	1 39	4·0	2,850	9 36	19·
830	1 53	4·5	2,900	9 54	19·5
920	2 7	5·	2,950	10 12	20·
1,010	2 21	5·5	3,000	10 30	20·5
1,090	2 35	6·	3,050	10 48	21·
1,170	2 49	6·5	3,100	11 7	21·5
1,250	3 3	7·	3,150	11 27	22·
1,330	3 18	7·5	3,200	11 47	22·5
1,410	3 33	8·	3,240	12 7	23·
1,490	3 48	8·5	3,280	12 28	23·5
1,570	4 3	9·	3,330	12 49	24·
1,650	4 19	9·5	3,370	13 10	24·5
1,730	4 35	10·	3,410	13 32	25·
1,800	4 51	10·5	3,450	13 55	25·5
1,870	5 7	11·	3,490	14 18	26·
1,940	5 23	11·5	3,530	14 42	26·5
2,010	5 39	12·	3,560	15 6	27·
2,080	5 56	12·5	3,600	15 31	27·5
2,150	6 13	13·	3,630	15 57	28·
2,210	6 30	13·5	3,660	16 24	28·5
2,280	6 47	14·	3,700	16 52	29·
2,340	7 3	14·5	3,730	17 22	29·5
2,400	7 20	15·	3,760	17 53	30·
2,460	7 37	15·5			

RANGE TABLE for 16-pr. R.M.L. GUN of 12 cwt.

Charge, 3 lbs. R.L.G. powder.
Projectile, common and shrapnel shell.

Range.	Elevation.	Fuze Scale.	Range.	Elevation.	Fuze Scale.
Yards.	° ′		Yards.	° ′	
200	0 0	1	2,580	6 6	16
290	0 6	1·5	2,650	6 20	16·5
390	0 16	2	2,710	6 37	17
480	0 28	2·5	2,780	6 50	17·5
570	0 40	3	2,840	7 3	18
660	0 52	3·5	2,910	7 16	18·5
750	1 4	4	2,970	7 29	19
840	1 16	4·5	3,040	7 42	19·5
920	1 28	5	3,100	7 55	20
1,010	1 40	5·5	3,160	8 8	20·5
1,090	1 52	6	3,220	8 21	21
1,170	2 4	6·5	3,280	8 34	21·5
1,250	2 16	7	3,330	8 47	22
1,330	2 28	7·5	3,390	9 0	22·5
1,410	2 40	8	3,450	9 13	23
1,490	2 52	8·5	3,510	9 26	23·5
1,570	3 4	9	3,560	9 39	24
1,650	3 16	9·5	3,620	9 52	24·5
1,720	3 28	10	3,680	10 5	25
1,800	3 40	10·5	3,730	10 18	25·5
1,870	3 52	11	3,790	10 31	26
1,950	4 5	11·5	3,840	10 44	26·5
2,020	4 18	12	3,890	10 57	27
2,090	4 31	12·5	3,940	11 0	27·5
2,160	4 44	13	3,990	11 14	28
2,230	4 57	13·5	4,040	11 28	28·5
2,300	5 10	14	4,090	11 42	29
2,370	5 24	14·5	4,140	11 56	29·5
2,440	5 38	15	4,190	12 10	30
2,510	5 52	15·5			

RANGE TABLE for 25-PR. R.M.L. GUN of 18 cwt.

Charge, 4 lbs. R.L.G. powder.

Projectile, common and shrapnel shell.

Range.	Elevation.	Fuze Scale.	Range.	Elevation.	Fuze Scale.
Yards.	° ′		Yards.	° ′	
210	0 19	1	2,700	6 33	16
300	0 28	1·5	2,770	6 49	16·5
390	0 37	2	2,840	7 5	17
480	0 46	2·5	2,910	7 21	17·5
570	0 56	3	2,980	7 37	18
660	1 6	3·5	3,050	7 53	18·5
750	1 16	4	3,120	8 9	19
840	1 27	4·5	3,190	8 25	19·5
930	1 39	5	3,260	8 41	20
1,020	1 51	5·5	3,330	8 58	20·5
1,100	2 3	6	3,400	9 15	21
1,180	2 14	6·5	3,470	9 33	21·5
1,260	2 25	7	3,540	9 51	22
1,340	2 37	7·5	3,600	10 9	22·5
1,420	2 49	8	3,600	10 26	23
1,500	3 1	8·5	3,720	10 43	23·5
1,580	3 13	9	3,780	11 0	24
1,660	3 25	9·5	3,840	11 17	24·5
1,740	3 38	10	3,900	11 34	25
1,820	3 51	10·5	3,960	11 51	25·5
1,900	4 4	11	4,020	12 8	26
1,980	4 17	11·5	4,080	12 25	26·5
2,060	4 31	12	4,130	12 42	27
2,140	4 45	12·5	4,180	12 58	27·5
2,220	4 59	13	4,230	13 14	28
2,300	5 13	13·5	4,280	13 30	28·5
2,380	5 29	14	4,330	13 46	29
2,460	5 45	14·5	4,380	14 2	29·5
2,540	6 1	15	4,430	14 18	30
2,620	6 17	15·5			

Range Table for 40-Pr. R.M.L. Gun of 35 cwt.

Charge, 7 lbs. R.L.G. powder.

Projectile, common shell with gas check, 40 lbs.

Range.	Elevation.	Fuze Scale.	Range.	Elevation.	Fuze Scale.
Yards.	° ′		Yards.	°	
220	0 19	1	2,870	6 27	16
340	0 30	1·5	2,940	6 40	16·5
460	0 41	2	3,010	6 53	17
570	0 52	2·5	3,080	7 7	17·5
680	1 3	3	3,150	7 21	18
780	1 14	3·5	3,220	7 35	18·5
880	1 25	4	3,290	7 49	19
980	1 36	4·5	3,350	8 2	19·5
1,080	1 47	5	3,410	8 15	20
1,170	1 58	5·5	3,470	8 28	20·5
1,260	2 10	6	3,530	8 41	21
1,350	2 22	6·5	3,590	8 54	21·5
1,440	2 34	7	3,650	9 7	22
1,530	2 46	7·5	3,710	9 20	22·5
1,620	2 59	8	3,770	9 33	23
1,710	3 12	8·5	3,830	9 46	23·5
1,790	3 25	9	3,890	9 59	24
1,870	3 38	9·5	3,950	10 12	24·5
1,950	3 51	10	4,010	10 25	25
2,030	4 4	10·5	4,070	10 42	25·5
2,110	4 17	11	4,120	10 54	26
2,190	4 30	11·5	4,170	11 4	26·5
2,270	4 43	12	4,220	11 16	27
2,350	4 56	12·5	4,270	11 28	27·5
2,430	5 9	13	4,320	11 40	28
2,510	5 22	13·5	4,370	11 52	28·5
2,590	5 35	14	4,420	12 4	29
2,660	5 48	14·5	4,470	12 16	29·5
2,730	6 1	15	4,510	12 28	30
2,800	6 14	15·5			

Range Table for 64-Pr. R.M.L. Gun, 64 cwt. and 64-Pr. Converted Guns of 71 and 58 cwts.

Charge, 8 lbs. R.L.G powder.

Projectile, common shell.

Range.	Elevation.	Fuze Scale.	Range.	Elevation.	Fuze Scale.
Yards.	° ′		Yards.	° ′	
170	0 17	1	2,350	5 34	16
260	0 26	1·5	2,410	5 46	16·5
350	0 36	2	2,470	5 48	17
430	0 46	2·5	2,530	6 10	17·5
510	0 56	3	2,590	6 22	18
590	1 6	3·5	2,650	6 34	18·5
670	1 16	4	2,710	6 46	19
750	1 26	4·5	2,770	6 58	19·5
830	1 36	5	2,830	7 11	20
910	1 46	5·5	2,890	7 24	20·5
980	1 56	6	2,940	7 37	21
1,060	2 6	6·5	3,000	7 50	21·5
1,140	2 16	7	3,060	8 3	22
1,210	2 26	7·5	3,110	8 16	22·5
1,290	2 37	8	3,170	8 29	23
1,360	2 48	8·5	3,220	8 42	23·5
1,430	2 59	9	3,280	8 55	24
1,500	3 10	9·5	3,330	9 8	24·5
1,570	3 21	10	3,380	9 21	25
1,640	3 32	10·5	3,440	9 34	25·5
1,710	3 43	11	3,490	9 47	26
1,780	3 54	11·5	3,540	10 0	26·5
1,850	4 5	12	3,590	10 13	27
1,910	4 16	12·5	3,640	10 26	27·5
1,980	4 27	13	3,690	10 39	28
2,040	4 38	13·5	3,740	10 52	28·5
2,110	4 49	14	3,790	11 5	29
2,170	5 0	14·5	3,840	11 18	29·5
2,230	5 11	15	3,890	11 30	30
2,290	5 22	15·5			

Range Table for 64-Pr. R.M.L. of 64 cwt.

Charge, 10 lbs. R.L.G powder.

Projectile, common shell.

Range.	Elevation.	Fuze Scale.	Range.	Elevation.	Fuze Scale.
Yards.	° ′		Yards.	° ′	
160	0 12	1	2,580	5 21	16
250	0 19	1·5	2,660	5 33	16·5
330	0 27	2	2,730	5 45	17
420	0 35	2·5	2,800	5 58	17·5
500	0 43	3	2,870	6 11	18
590	0 51	3·5	2,950	6 24	18·5
670	0 59	4	3,020	6 37	19
760	1 7	4·5	3,090	6 50	19·5
840	1 16	5	3,160	7 4	20
930	1 25	5·5	3,230	7 18	20·5
1,010	1 34	6	3,300	7 32	21
1,090	1 43	6·5	3,370	7 46	21·5
1,170	1 52	7	3,440	8 0	22
1,260	2 2	7·5	3,510	8 14	22·5
1,340	2 13	8	3,580	8 28	23
1,420	2 24	8·5	3,650	8 42	23·5
1,500	2 35	9	3,710	8 56	24
1,590	2 46	9·5	3,780	9 10	24·5
1,670	2 57	10	3,850	9 24	25
1,750	3 9	10·5	3,920	9 38	25·5
1,830	3 21	11	3,980	9 52	26
1,910	3 33	11·5	4,050	10 6	26·5
1,990	3 45	12	4,120	10 20	27
2,060	3 57	12·5	4,180	10 34	27·5
2,140	4 9	13	4,250	10 49	28
2,210	4 21	13·5	4,310	11 4	28·5
2,290	4 33	14	4,380	11 19	29
2,360	4 45	14·5	4,440	11 34	29·5
2,440	4 57	15	4,510	11 50	30
2,510	5	15·5			

Range Table for 64-Pr. Rifled M.L. Gun of 64 cwt.

Charge, battering 12 lb. R.L.G.

Projectile, battering with gas-check, 90 lb.

Range.	Elevation.	Range.	Elevation.
Yards.	° ′	Yards.	° ′
100	—	1,300	2 18
200	—	1,400	2 33
300	—	1,500	2 48
400	0 10	1,600	3 3
500	0 23	1,700	3 18
600	0 36	1,800	3 34
700	0 50	1,900	3 50
800	1 4	2,000	4 6
900	1 18	2,100	4 22
1,000	1 33	2,200	4 38
1,100	1 48	2,300	4 54
1,200	2 3	2,400	5 11

RANGE TABLE for 80-PR. CONVERTED R.M.L. GUN of 5 tons.

Charge, 10 lb. R.L.G. powder.

Projectile, common shell.

Range.	Elevation.	Fuze Scale.	Range.	Elevation.	Fuze Scale.
Yards.	° ′		Yards.	° ′	
150	0 15	1	2,440	5 56	16
230	0 26	1·5	2,510	6 9	16·5
310	0 36	2	2,570	6 21	17
390	0 46	2·5	2,640	6 34	17·5
470	0 56	3	2,700	6 46	18
550	1 6	3·5	2,770	7 0	18·5
630	1 16	4	2,830	7 13	19
710	1 27	4·5	2,890	7 26	19·5
790	1 38	5	2,960	7 40	20
870	1 49	5·5	3,020	7 53	20·5
950	2 0	6	3,090	8 8	21
1,030	2 11	6·5	3,150	8 22	21·5
1,110	2 22	7	3,210	8 36	22
1,190	2 33	7·5	3,270	8 50	22·5
1,270	2 44	8	3,330	9 4	23
1,350	2 55	8·5	3,390	9 18	23·5
1,420	3 6	9	3,450	9 32	24
1,500	3 18	9·5	3,500	9 46	24·5
1,580	3 30	10	3,560	10 0	25
1,650	3 42	10·5	3,610	10 14	25·5
1,720	3 54	11	3,670	10 28	26
1,800	4 6	11·5	3,720	10 42	26 5
1,870	4 18	12	3,780	11 0	27
1,950	4 30	12·5	3,830	11 14	27·5
2,020	4 42	13	3,890	11 30	28
2,090	4 54	13·5	3,940	11 44	28·5
2,160	5 6	14	3,990	12 0	29
2,230	5 18	14·5	4,040	12 16	29·5
2,300	5 30	15	4,080	12 32	30
2,370	5 43	15·5			

RANGE TABLE for 7-INCH R.M.L. GUN of 90 cwt.

Charge, 22 lb. P powder.

Projectile—Palliser, Shrapnel, or Common Shell.								Double Shell.		
Distance of Object.	Elevation.	Time of Flight.	Fuze Scale.	Distance of Object.	Elevation.	Time of Flight.	Fuze Scale.	Distance of Object.	Elevation.	Time of Flight.
Yds.	° ′	″		Yds.	° ′	″		Yds.	° ′	″
140	—	0·36	1·0	2,110	4 21	5·93	14·5	100	—	0·27
200	0 1	0·54	1·5	2,190	4 33	6·16	15·0	200	—	0·54
260	0 9	0·72	2·0	2,260	4 45	6·39	15·5	300	0 13	0·82
330	0 15	0·90	2·5	2,330	4 57	6·63	16·0	400	0 26	1·10
400	0 22	1·09	3·0	2,410	5 8	6·87	16·5	500	0 40	1·39
470	0 31	1·28	3·5	2,490	5 20	7·11	17·0	600	0 54	1·69
550	0 40	1·47	4·0	2,570	5 32	7·35	17·5	700	1 9	1·99
630	0 49	1·66	4·5	2,640	5 44	7·60	18·0	800	1 24	2·30
700	0 58	1·85	5·0	2,720	5 56	7·85	18·5	900	1 39	2·61
780	1 8	2·05	5·5	2,800	6 8	8·10	19·0	1,000	1 55	2·92
850	1 18	2·25	6·0	2,870	6 20	8·35	19·5	1,100	2 11	3·23
920	1 28	2·45	6·5	2,940	6 32	8·60	20·0	1,200	2 27	3·54
990	1 38	2·65	7·0	3,020	6 44	8·85	20·5	1,300	2 44	3·85
1,060	1 48	2·85	7·5	3,090	6 56	9·11	21·0	1,400	3 1	4·16
1,130	1 58	3·05	8·0	3,170	7 8	9·38	21·5	1,500	3 18	4·47
1,200	2 8	3·26	8·5	3,250	7 20	9·65	22·0	1,600	3 35	4·78
1,280	2 19	3·47	9·0	3,320	7 32	9·92	22·5	1,700	3 53	5·10
1,350	2 30	3·68	9·5	3,400	7 44	10·20	23·0	1,800	4 11	5·42
1,430	2 41	3·90	10·0	3,480	7 57	10·48	23·5	1,900	6 29	5·74
1,510	2 52	4·12	10·5	3,550	8 10	10·76	24·0	2,000	4 47	6·06
1,590	3 3	4·34	11·0	3,630	8 23	11·05	24·5	2,100	5 5	6·38
1,660	3 14	4·56	11·5	3,710	8 36	11·35	25·0	2,200	5 24	6·70
1,730	3 25	4·78	12·0	3,790	8 49	11·65	25·5	2,300	5 43	7·02
1,810	3 36	5·01	12·5	3,870	9 03	11·95	26·0	2,400	6 02	7·33
1,890	3 47	5·24	13·0	3,950	9 17	12·26	26·5	2,500	6 21	7·63
1,960	3 58	5·47	13·5							
2,040	4 09	5·70	14·0							

Range Table for 7-inch R.M.L. Gun of 6½ tons.

Charge, 14 lbs. R.L.G. powder.
Projectile, common shell.

Range.	Elevation.	Fuze Scale.	Range.	Elevation.	Fuze Scale.
Yards.	° ′		Yards.	° ′	
170	0 16	1	2,330	5 21	16
250	0 24	1·5	2,390	5 32	16·5
330	0 33	2	2,450	5 43	17
410	0 42	2·5	2,510	5 54	17·5
480	0 51	3	2,570	6 5	18
560	1 0	3·5	2,630	6 15	18·5
630	1 10	4	2,680	6 27	19
710	1 20	4·5	2,740	6 39	19·5
790	1 30	5	2,800	6 49	20
870	1 40	5·5	2,850	7 0	20·5
940	1 50	6	2,910	7 11	21
1,010	2 0	6·5	2,960	7 22	21·5
1,090	2 10	7	3,020	7 33	22
1,170	2 20	7.5	3,070	7 43	22·5
1,230	2 30	8	3,120	7 53	23
1,310	2 40	8·5	3,180	8 3	23·5
1,380	2 50	9	3,230	8 13	24
1,460	3 0	9·5	3,280	8 23	24·5
1,530	3 10	10	3,330	8 33	25
1,600	3 20	10·5	3,380	8 43	25·5
1,660	3 31	11	3,420	8 53	26
1,730	3 42	11·5	3,470	9 3	26·5
1,800	3 53	12	3,520	9 12	27
1,870	4 4	12·5	3,560	9 21	27·5
1,930	4 15	13	3,600	9 30	28
2,000	4 26	13·5	3,640	9 39	28·5
2,070	4 37	14	3,690	9 48	29
2,130	4 48	14·5	3,740	9 57	29·5
2,200	4 59	15	3,770	10 5	30
2,270	5 10	15·5			

Range Table for 7-inch R.M.L. Gun of 6½ tons.

Charge, 22 lb. R.L.G. powder.

Projectile, Palliser or common shell.

Range.	Elevation.	Fuze Scale.	Range.	Elevation.	Fuze Scale.
Yards.	° '		Yards.	° '	
200	0 15	1	2,470	4 24	16
280	0 21	1·5	2,540	4 38	16·5
370	0 28	2	2,600	4 42	17
450	0 35	2·5	2,660	4 51	17·5
530	0 42	3	2,720	5 0	18
610	0 49	3·5	2,780	5 9	18·5
690	0 56	4	2,830	5 18	19
770	1 4	4·5	2,890	5 27	19·5
850	1 12	5	2,950	5 36	20
940	1 20	5·5	3,000	5 45	20·5
1,010	1 28	6	3,060	5 54	21
1,090	1 36	6·5	3,110	6 3	21·5
1,170	1 44	7	3,170	6 12	22
1,250	1 52	7·5	3,220	6 21	22·5
1,330	2 0	8	3,270	6 30	23
1,410	2 9	8·5	3,320	6 39	23·5
1,480	2 18	9	3,370	6 46	24
1,560	2 27	9·5	3,410	6 54	24·5
1,640	2 36	10	3,460	7 2	25
1,710	2 45	10·5	3,510	7 10	25·5
1,790	2 54	11	3,560	7 18	26
1,860	3 3	11·5	3,600	7 26	26·5
1,930	3 12	12	3,640	7 34	27
2,010	3 21	12·5	3,690	7 42	27·5
2,080	3 30	13	3,740	7 50	28
2,150	3 89	13·5	3,780	7 58	28·5
2,210	3 49	14	3,820	8 5	29
2,280	3 57	14·5	3,860	8 12	29·5
2,350	4 6	15	3,890	8 18	30
2,410	4 15	15·5			

RANGE TABLE for 7-inch R.M.L. GUN of 6½ tons.

Charge battering, 30 lb. P. powder.

Projectile, Palliser or common shell.

Range.	Elevation.	Fuze Scale.	Range.	Elevation.	Fuze Scale.
Yards.	° ′		Yards.	° ′	
190	0 9	1	2,600	3 58	16
280	0 15	1·5	2,670	4 7	16·5
380	0 22	2	2,740	4 16	17
470	0 29	2·5	2,810	4 25	17·5
560	0 36	3	2,880	4 34	18
650	0 43	3·5	2,950	4 43	18·5
740	0 50	4	3,010	4 52	19
830	0 57	4·5	3,080	5 7	19·5
920	1 4	5	3,150	5 16	20
1,000	1 11	5·5	3,210	5 25	20·5
1,090	1 18	6	3,280	5 34	21
1,170	1 25	6·5	3,340	5 43	21·5
1,260	1 32	7	3,410	5 52	22
1,340	1 39	7·5	3,470	6 1	22·5
1,420	1 46	8	3,530	6 10	23
1,500	1 54	8·5	3,600	6 21	23·5
1,580	2 2	9	3,660	6 30	24
1,660	2 10	9·5	3,720	6 40	24·5
1,730	2 18	10	3,780	6 50	25
1,810	2 26	10·5	3,840	7 0	25·5
1,880	2 34	11	3,900	7 10	26
1,960	2 42	11·5	3,960	7 20	26·5
2,030	2 50	12	4,020	7 30	27
2,110	2 58	12·5	4,080	7 40	27·5
2,180	3 6	13	4,140	7 50	28
2,250	3 14	13·5	4,200	8 0	28·5
2,320	3 22	14	4,260	8 10	29
2,390	3 31	14·5	4,310	8 20	29·5
2,460	3 40	15	4,370	8 30	30
2,530	3 49	15·5			

RANGE TABLE for 7-inch R.M.L. GUN of 7 tons.

Charge, 14 lb. R.L.G powder.
Projectile, common shell.

Range.	Elevation.	Fuze Scale.	Range.	Elevation.	Fuze Scale.
Yards.	° ′		Yards.	° ′	
140	0 15	1	2,410	5 28	16
230	0 24	1·5	2,480	5 39	16·5
310	0 33	2	2,540	5 49	17
400	0 42	2·5	2,600	5 59	17·5
480	0 51	3	2,660	6 9	18
560	1 0	3·5	2,720	6 19	18·5
640	1 10	4	2,780	6 29	19
720	1 20	4·5	2,840	6 39	19·5
800	1 30	5	2,900	6 49	20
890	1 40	5·5	2,950	6 59	20·5
960	1 50	6	3,010	7 9	21
1,040	2 0	6·5	3,070	7 19	21·5
1,120	2 10	7	3,120	7 29	22
1,200	2 21	7·5	3,170	7 39	22·5
1,270	2 32	8	3,220	7 48	23
1,350	2 43	8·5	3,280	7 57	23·5
1,430	2 54	9	3,320	8 6	24
1,500	3 5	9·5	3,370	8 15	24·5
1,580	3 16	10	3,420	8 24	25
1,650	3 27	10·5	3,460	8 33	25·5
1,720	3 38	11	3,510	8 42	26
1,790	3 49	11·5	3,550	8 50	26·5
1,860	4 0	12	3,590	8 58	27
1,930	4 11	12·5	3,630	9 6	27·5
2,010	4 22	13	3,670	9 14	28
2,080	4 33	13·5	3,710	9 21	28·5
2,150	4 44	14	3,750	9 28	29
2,210	4 55	14·5	3,790	9 34	29·5
2,280	5 6	15	3,820	9 40	30
2,350	5 17	15.5			

Range Table for 7-inch R.M.L. Gun of 7 tons.

Charge, 22 lb. R.L.G. powder.
Projectile, common shell.

Range.	Elevation.	Fuze Scale.	Range.	Elevation.	Fuze Scale.
Yards.	° ′		Yards.	° ′	
200	0 10	1	2,590	4 27	16
290	0 16	1·5	2,650	4 36	16·5
370	0 22	2	2,710	4 45	17
470	0 29	2·5	2,780	4 54	17·5
550	0 36	3	2,840	5 3	18
630	0 43	3·5	2,900	5 12	18·5
720	0 50	4	2,950	5 21	19
800	0 57	4·5	3,010	5 30	19·5
890	1 4	5	3,070	5 39	20
930	1 12	5·5	3,130	5 48	20·5
1,050	1 20	6	3,190	5 56	21
1,140	1 28	6·5	3,240	6 4	21·5
1,220	1 36	7	3,290	6 12	22
1,310	1 44	7·5	3,340	6 20	22·5
1,390	1 53	8	3,390	6 28	23
1,470	2 2	8·5	3,440	6 36	23·5
1,550	2 11	9	3,490	6 44	24
1,630	2 22	9·5	3,530	6 52	24·5
1,710	2 31	10	3,580	6 59	25
1,790	2 40	10·5	3,620	7 6	25·5
1,860	2 49	11	3,660	7 13	26
1,940	2 58	11·5	3,710	7 20	26·5
2,020	3 8	12	3,750	7 27	27
2,100	3 18	12·5	3,790	7 33	27·5
2,170	3 28	13	3,820	7 39	28
2,240	3 38	13·5	3,860	7 45	28·5
2,320	3 48	14	3,900	7 51	29
2,380	3 58	14·5	3,930	7 57	29·5
2,450	4 8	15	3,930	8 3	30
2,520	4 18	15·5			

RANGE TABLE for 8-inch R.M L. GUN of 9 tons.

Charge, 20 lb. R.L.G. powder
Projectile, common shell.

Range.	Elevation.	Fuze Scale.	Range.	Elevation.	Fuze Scale.
Yards.	° ′		Yards.	° ′	
170	0 17	1	2,380	5 28	16
250	0 26	1·5	2,450	5 40	16·5
330	0 35	2	2,520	5 52	17
410	0 44	2·5	2,590	6 4	17·5
480	0 53	3	2,650	6 16	18
560	1 2	3·5	2,720	6 28	18·5
640	1 11	4	2,780	6 40	19
720	1 21	4·5	2,840	6 52	19·5
790	1 31	5	2,910	7 4	20
870	1 41	5·5	2,970	7 16	20·5
940	1 51	6	3,030	7 28	21
1,020	2 1	6·5	3,100	7 40	21·5
1,090	2 11	7	3,160	7 51	22
1,170	2 21	7·5	3,220	8 3	22·5
1,240	2 31	8	3,280	8 15	23
1,320	2 42	8·5	3,340	8 26	23·5
1,390	2 53	9	3,390	8 37	24
1,470	3 4	9·5	3,450	8 49	24·5
1,540	3 15	10	3,510	9 1	25
1,620	3 26	10·5	3,580	9 13	25·5
1,690	3 37	11	3,630	9 24	26
1,760	3 48	11·5	3,690	9 36	26·5
1,830	3 59	12	3,740	9 47	27
1,900	4 10	12·5	3,800	9 59	27·5
1,970	4 21	13	3,860	10 11	28
2,040	4 32	13·5	3,920	10 23	28·5
2,110	4 43	14	3,970	10 35	29
2,180	4 54	14·5	4,030	10 47	29·5
2,250	5 5	15	4,080	10 59	30
2,320	5 16	15·5			

RANGE TABLE for 8-inch R.M.L. GUN of 9 tons.

Charge, 30 lb. R.L.G. powder.

Projectile, Palliser or common shell.

Range.	Elevation.	Fuze Scale.	Range.	Elevation.	Fuze Scale.
Yards.	° ′		Yards.	° ′	
190	0 17	1	2,600	4 59	16
270	0 24	1·5	2,670	5 10	16·5
360	0 32	2	2,740	5 21	17
440	0 40	2·5	2,810	5 32	17·5
530	0 48	3	2,880	5 43	18
610	0 56	3·5	2,940	5 54	18·5
700	1 5	4	3,010	6 5	19
790	1 14	4·5	3,080	6 16	19·5
880	1 23	5	3,150	6 27	20
960	1 32	5·5	3,220	6 38	20·5
1,040	1 41	6	3,290	6 49	21
1,120	1 50	6·5	3,360	7 0	21 5
1,210	1 59	7	3,430	7 11	22
1,290	2 8	7·5	3,500	7 22	22·5
1,380	2 17	8	3,560	7 33	23
1,460	2 26	8·5	3,630	7 44	23·5
1,550	2 35	9	3,690	7 55	24
1,630	2 44	9·5	3,760	8 6	24·5
1,710	2 54	10	3,820	8 17	25
1,790	3 4	10·5	3,890	8 28	25·5
1,870	3 14	11	3,950	8 39	26
1,940	3 24	11·5	4,010	8 50	26·5
2,010	3 34	12	4,080	9 2	27
2,090	3 44	12·5	4,140	9 14	27·5
2,170	3 54	13	4,210	9 26	28
2,250	4 6	13·5	4,270	9 38	28·5
2,320	4 16	14	4,330	9 50	29
2,390	4 26	14·5	4,400	10 2	29·5
2,460	4 37	15	4,450	10 14	30
2,530	4 48	15·5			

RANGE TABLE for 8-inch R.M.L. GUN of 9 tons.

Charge, 35 lb. P. powder

Projectile, Palliser or common shell.

Range.	Elevation.	Fuse Scale.	Range.	Elevation.	Fuze Scale.
Yards.	° ′		Yards.	° ′	
180	0 11	1	2,610	4 53	16
270	0 19	1·5	2,680	5 3	16·5
360	0 27	2	2,740	5 13	17
450	0 35	2·5	2,810	5 23	17·5
540	0 43	3	2,870	5 33	18
630	0 51	3·5	2,940	5 44	18·5
710	1 0	4	3,000	5 54	19
800	1 9	4·5	3,070	6 4	19·5
880	1 18	5	3,130	6 14	20
970	1 27	5·5	3,200	6 25	20·5
1,050	1 36	6	3,260	6 36	21
1,140	1 45	6·5	3,300	6 47	21·5
1,220	1 54	7	3,390	6 57	22
1,310	2 3	7·5	3,460	7 7	22·5
1,390	2 12	8	3,520	7 17	23
1,470	2 21	8·5	3,590	7 28	23·5
1,550	2 30	9	3,650	7 38	24
1,630	2 39	9·5	3,720	7 49	24·5
1,710	2 48	10	3,780	7 59	25
1,790	2 57	10·5	3,850	8 9	25·5
1,870	3 7	11	3,910	8 19	26
1,950	3 17	11·5	3,980	8 30	26·5
2,030	3 28	12	4,040	8 41	27
2,100	3 39	12·5	4,110	8 52	27·5
2,180	3 50	13	4,170	9 2	28
2,250	4 1	13·5	4,230	9 13	28·5
2,330	4 12	14	4,290	9 24	29
2,400	4 22	14·5	4,360	9 35	29·5
2,470	4 32	15	4,420	9 45	30
2,540	4 43	15·5			

Range Table for 9-inch R.M.L. Gun of 12 tons.

Charge, 30 lbs. R.L.G. powder.

Projectile, common shell.

Range.	Elevation.	Fuze Scale.	Range.	Elevation.	Fuze Scale.
Yards.	° ′		Yards.	° ′	
160	0 18	1	2,580	5 57	16
250	0 28	1·5	2,650	6 8	16·5
330	0 38	2	2,720	6 19	17
420	0 49	2·5	2,790	6 30	17·5
500	1 0	3	2,860	6 40	18
580	1 11	3·5	2,930	6 51	18·5
670	1 22	4	3,000	7 2	19
750	1 33	4·5	3,070	7 13	19·5
830	1 44	5	3,142	7 24	20
920	1 55	5·5	3,210	7 35	20·5
1,000	2 6	6	3,280	7 47	21
1,080	2 17	6·5	3,340	7 57	21·5
1,170	2 28	7	3,410	8 9	22
1,250	2 40	7·5	3,470	8 20	22·5
1,330	2 52	8	3,540	8 31	23
1,420	3 4	8·5	3,600	8 42	23·5
1,500	3 16	9	3,670	8 54	24
1,580	3 27	9·5	3,740	9 5	24·5
1,660	3 39	10	3,800	9 16	25
1,740	3 51	10·5	3,870	9 28	25·5
1,820	4 3	11	3,930	9 39	26
1,900	4 15	11·5	3,990	9 50	26·5
1,980	4 27	12	4,040	10 1	27
2,060	4 39	12·5	4,100	10 11	27·5
2,130	4 51	13	4,150	10 21	28
2,210	5 2	13·5	4,210	10 30	28·5
2,290	5 13	14	4,260	10 39	29
2,360	5 24	14·5	4,310	10 48	29·5
2,430	5 35	15	4,360	10 57	30
2,510	5 46	15·5			

RANGE TABLE for 9-inch R.M.L. GUN of 12 tons.

Charge, 43 lb. R.L.G. powder.

Projectile, Palliser or common shell.

Range.	Elevation.	Fuze Scale.	Range.	Elevation.	Fuze Scale.
Yards.	° ′		Yards.	° ′	
200	0 18	1	2,840	5 48	16
310	0 29	1·5	2,910	6 0	16·5
410	0 39	2	2,990	6 12	17
510	0 49	2·5	3,060	6 24	17·5
620	0 59	3	3,140	6 36	18
720	1 9	3·5	3,210	6 48	18·5
820	1 19	4	3,290	7 0	19
910	1 29	4·5	3,360	7 12	19·5
1,000	1 39	5	3,430	7 24	20
1,090	1 49	5·5	3,510	7 36	20·5
1,180	1 59	6	3,580	7 48	21
1,270	2 9	6·5	3,650	8 0	21·5
1,360	2 19	7	3,720	8 13	22
1,440	2 29	7·5	3,800	8 26	22·5
1,530	2 39	8	3,870	8 38	23
1,660	2 56	8·5	3,940	8 50	23·5
1,790	3 12	9	4,010	9 2	24
1,830	3 17	9·5	4,080	9 15	24·5
1,870	3 23	10	4,150	9 28	25
1,950	3 34	10·5	4,220	9 41	25·5
2,030	3 45	11	4,290	9 54	26
2,120	3 57	11·5	4,360	10 7	26·5
2,200	4 9	12	4,430	10 20	27
2,280	4 21	12·5	4,500	10 33	27·5
2,370	4 34	13	4,560	10 46	28
2,450	4 47	13·5	4,630	11 0	28·5
2,530	4 59	14	4,700	11 14	29
2,610	5 12	14·5	4,770	11 28	29·5
2,680	5 24	15	4,830	11 42	30
2,760	5 36	15·5			

RANGE TABLE for 9-inch R.M.L. GUN of 12 tons.

Charge, battering, 50 lbs. P. powder.
Projectile, Palliser or common shell.

Range.	Elevation.	Fuze Scale.
Yards.	° ′	
180	0 11	1
280	0 18	1·5
370	0 25	2
460	0 33	2·5
550	0 41	3
640	0 49	3·5
730	0 57	4
820	1 5	4·5
910	1 14	5
1,000	1 23	5·5
1,090	1 32	6
1,180	1 41	6·5
1,270	1 50	7
1,350	1 59	7·5
1,440	2 9	8
1,520	2 19	8·5
1,610	2 29	9
1,690	2 39	9·5
1,780	2 49	10
1,860	2 59	10·5
1,940	3 9	11
2,030	3 20	11·5
2,100	3 30	12
2,180	3 40	12·5
2,260	3 51	13
2,330	4 1	13·5
2,410	4 12	14
2,480	4 22	14·5
2,560	4 33	15
2,630	4 43	15·5
2,700	4 52	16
2,770	5 2	16·5
2,840	5 12	17
2,900	5 22	17·5
2,970	5 32	18
3,030	5 42	18·5
3,100	5 52	19
3,160	6 2	19·5
3,230	6 12	20
3,290	6 22	20·5
3,360	6 32	21
3,420	6 42	21·5
3,490	6 52	22
3,550	7 2	22·5
3,620	7 12	23
3,680	7 23	23·5
3,750	7 34	24
3,810	7 45	24·5
3,880	7 56	25
3,940	8 7	25·5
4,010	8 18	26
4,070	8 29	26·5
4,140	8 40	27
4,200	8 51	27·5
4,260	9 2	28
4,330	9 13	28·5
3,390	9 24	29
4,450	9 35	29·5
4,510	9 46	30

Range Table for 10 R.M.L. Gun of 18 tons.

Charge, 40 lb. R.L.G. or 44 lb. P. powder.
Projectile, common shell.

Range.	Elevation.	Fuze Scale.	Range.	Elevation.	Fuze Scale.
Yards.	° ′		Yards.	° ′	
160	0 19	1	2,300	5 35	16
240	0 28	1·5	2,400	5 47	16·5
320	0 38	2	2,460	5 58	17
400	0 48	2·5	2,530	6 10	17·5
480	0 58	3	2,590	6 21	18
560	1 8	3·5	2,660	6 34	18·5
630	1 18	4	2,720	6 46	19
710	1 28	4·5	2,790	6 59	19·5
780	1 38	5	2,850	7 11	20
860	1 49	5·5	2,910	7 23	20·5
930	1 59	6	2,970	7 35	21
1,000	2 9	6·5	3,030	7 47	21·5
1,080	2 20	7	3,090	7 59	22
1,150	2 30	7·5	3,150	8 11	22·5
1,220	2 40	8	3,210	8 23	23
1,300	2 51	8·5	3,270	8 35	23·5
1,370	3 1	9	3,330	8 48	24
1,440	3 12	9.5	3,390	9 0	24·5
1,510	3 22	10	3,440	9 12	25
1,580	3 33	10·5	3,500	9 25	25·5
1,650	3 43	11	3,560	9 38	26
1,720	3 54	11·5	3,620	9 51	26·5
1,790	4 5	12	3,670	10 2	27
1,860	4 16	12·5	3,730	10 15	27·5
1,930	4 27	13	3,790	10 29	28
2,000	4 39	13·5	3,840	10 42	28·5
2,070	4 50	14	3,900	10 56	29
2,140	5 2	14·5	3,950	11 8	29·5
2,200	5 13	15	4,010	11 23	30
2,270	5 25	15·5			

RANGE TABLE for 10 inch R.M.L. GUN of 18 tons.

Charge, battering, 70 lb. P. powder.

Projectile, common shell.

Range.	Elevation.	Fuze Scale.	Range.	Elevation.	Fuze Scale.
Yards.	° ′		Yards.	° ′	
170	0 15	1	2,570	4 42	16
260	0 23	1·5	2,610	4 52	16·5
340	0 31	2	2,710	5 2	17
430	0 39	2·5	2,780	5 12	17·5
510	0 47	3	2.850	5 22	18
600	0 55	3·5	2,920	5 32	18·5
680	1 3	4	2,990	5 42	19
770	1 12	4·5	3,060	5 52	19·5
850	1 21	5	3,130	6 2	20
940	1 30	5·5	3,200	6 12	20·5
1,020	1 39	6	3,260	6 22	21
1,110	1 48	6·5	3,330	6 32	21·5
1,190	1 57	7	3,400	6 42	22
1,270	2 6	7·5	3,460	6 51	22·5
1,350	2 15	8	3,530	7 1	23
1,430	2 24	8·5	3,590	7 10	23·5
1,510	2 33	9	3,660	7 21	24
1,590	2 42	9·5	3,720	7 31	24·5
1,670	2 52	10	3,780	7 41	25
1,750	3 2	10·5	3,850	7 52	25·5
1,820	3 11	11	3,910	8 1	26
1,900	3 20	11·5	3,970	8 11	26·5
1,970	3 29	12	4,030	8 20	27
2,050	3 38	12·5	4,090	8 30	27·5
2,120	3 47	13	4,150	8 40	28
2,200	3 56	13·5	4,210	8 50	28·5
2,270	4 5	14	4,270	9 0	29
2,350	4 14	14·5	4,330	9 10	29·5
2,420	4 23	15	4,380	9 18	30
2,500	4 32	15·5			

Range Table for 11-inch R.M.L. Gun of 25 tons.

Charge, 60 lb. P. powder.

Projectile, common or shrapnel shell.

Range.	Elevation.	Fuze Scale.	Range.	Elevation.	Fuze Scale.
Yards.	° ′		Yards.	° ′	
150	0 19	1	2,400	5 47	16
230	0 30	1·5	2,470	5 58	16·5
300	0 39	2	2,540	6 9	17
380	0 50	2·5	2,620	6 22	17·5
460	1 1	3	2,690	6 33	18
540	1 12	3·5	2,760	6 44	18·5
610	1 22	4	2,830	6 55	19
690	1 33	4·5	2,910	7 8	19·5
760	1 43	5	2,980	7 19	20
840	1 54	5·5	3,050	7 31	20·5
910	2 4	6	3,120	7 42	21
990	2 15	6·5	3,200	7 56	21·5
1,060	2 25	7	3,270	8 7	22
1,140	2 36	7·5	3,340	8 19	22·5
1,210	2 46	8	3,410	8 31	23
1,290	2 58	8·5	3,490	8 45	23·5
1,360	3 9	9	3,560	8 57	24
1,430	3 19	9·5	3,630	9 9	24·5
1,510	3 31	10	3,700	9 21	25
1,580	3 42	10·5	3,780	9 34	25·5
1,660	3 54	11	3,850	9 47	26
1,730	4 4	11·5	3,920	9 59	26·5
1,810	4 16	12	3,990	10 12	27
1,880	4 27	12·5	4,070	10 26	27·5
1,960	4 39	13	4,140	10 39	28
2,030	4 49	13·5	4,210	10 51	28·5
2,110	5 1	14	4,280	11 4	29
2,180	5 12	14·5	4,350	11 18	29·5
2,250	5 23	15	4,420	11 29	30
2,330	5 35	15·5			

RANGE TABLE for 12-inch R.M.L. GUN of 25 tons

Charge, 50 lb. R.L.G. or 55 lb. P. powder.
Projectile, common shell.

Range.	Elevation.	Fuze Scale.	Range.	Elevation.	Fuze Scale.
Yards.	° ′		Yards.	° ′	
160	0 19	1	2,560	5 54	16
240	0 28	1·5	2,630	6 5	16·5
320	0 38	2	2,710	6 18	17
410	0 49	2·5	2,780	6 30	17·5
490	0 59	3	2,860	6 43	18
570	1 8	3·5	2,930	6 55	18·5
650	1 19	4	3,000	7 7	19
740	1 31	4·5	3,080	7 20	19·5
820	1 41	5	3,150	7 32	20
900	1 52	5·5	3,220	7 44	20·5
980	2 2	6	3,290	7 57	21
1,070	2 14	6·5	3,370	8 11	21·5
1,150	2 25	7	3,440	8 24	22
1,230	2 36	7·5	3,510	8 37	22·5
1,310	2 47	8	3,580	8 50	23
1,390	2 58	8·5	3,650	9 3	23·5
1,470	3 9	9	3,720	9 16	24
1,550	3 21	9·5	3,790	9 30	24·5
1,630	3 32	10	3,860	9 44	25
1,710	3 43	10·5	3,930	9 58	25·5
1,790	3 55	11	4,000	10 12	26
1,870	4 7	11·5	4,070	10 26	26·5
1,940	4 18	12	4,130	10 40	27
2,020	4 30	12·5	4,200	10 54	27·5
2,100	4 42	13	4,270	11 8	28
2,180	4 54	13·5	4,330	11 22	28·5
2,250	5 5	14	4,400	11 36	29
2,330	5 17	14·5	4,460	11 48	29·5
2,410	5 30	15	4,530	12 2	30
2,480	5 51	15·5			

Range Table for 12-inch R.M.L. Gun of 25 tons.

Charge, battering, 85 lb. P. powder.
Projectile, common shell.

Range.	Elevation.	Fuze Scale.	Range.	Elevation.	Fuze Scale.
Yards.	° ′		Yards.	° ′	
180	0 9	1	2,710	4 57	16
280	0 16	1·5	2,790	5 7	16·5
370	0 23	2	2,860	5 17	17
470	0 32	2·5	2,940	5 28	17·5
560	0 41	3	3,010	5 38	18
650	0 50	3·5	3,090	5 49	18·5
740	1 0	4	3,160	5 59	19
830	1 10	4·5	3,230	6 9	19·5
910	1 19	5	3,300	6 19	20
1,000	1 29	5·5	3,370	6 29	20·5
1,080	1 38	6	3,440	6 40	21
1,170	1 47	6·5	3,510	6 50	21·5
1,250	1 56	7	3,580	7 0	22
1,340	2 6	7·5	3,650	7 10	22·5
1,420	2 16	8	3,710	7 20	23
1,510	2 27	8·5	3,790	7 33	23·5
1,590	2 36	9	3,850	7 43	24
1,680	2 47	9·5	3,910	7 53	24·5
1,760	2 57	10	3,980	8 5	25
1,840	3 6	10·5	4,040	8 15	25·5
1,920	3 16	11	4,110	8 27	26
2,000	3 26	11·5	4,170	8 38	26·5
2,080	3 36	12	4,230	8 48	27
2,160	3 46	12·5	4,300	9 0	27·5
2,240	3 56	13	4,360	9 10	28
2,320	4 6	13·5	4,420	9 20	28·5
2,400	4 17	14	4,480	9 30	29
2,480	4 27	14·5	4,540	9 40	29·5
2,560	4 37	15	4,600	9 51	30
2,640	4 47	15·5			

Range Table for 12-inch R.M.L. Gun of 35 tons.

Charge, 85 lb. P. powder.

Projectile, common shell.

Range.	Elevation.	Fuze Scale.	Range.	Elevation.	Fuze Scale.
Yards.	° ′		Yards.	° ′	
200	0 19	1	2,640	5 4	16
300	0 29	1·5	2,710	5 13	16·5
390	0 38	2	2,780	5 23	17
490	0 48	2·5	2,850	5 33	17·5
580	0 57	3	2,910	5 41	18
680	1 7	3·5	2,980	5 51	18·5
770	1 17	4	3,040	6 0	19
860	1 27	4·5	3,110	6 9	19·5
950	1 37	·5	3,170	6 18	20
1,040	1 47	5·5	3,230	6 27	20·5
1,120	1 56	6	3 290	6 36	21
1,210	2 6	6·5	3,350	6 45	21·5
1,290	2 15	7	3,410	6 54	22
1,380	2 24	7·5	3,470	7 3	22·5
1,460	2 34	8	3,530	7 12	23
1,540	2 43	8·5	3,590	7 21	23·5
1,620	2 53	9	3,650	7 30	24
1,700	3 3	9·5	3,710	7 39	24·5
1,780	3 12	10	3,770	7 48	25
1,850	3 21	10·5	3,830	7 57	25·5
1,930	3 31	11	3,880	8 5	26
2,000	3 40	11·5	3,940	8 14	26·5
2,080	3 50	12	4,000	8 24	27
2,150	3 59	12·5	4,060	8 33	27·5
2,220	4 8	13	4,120	8 43	28
2,300	4 19	13·5	4,180	8 52	28·5
2,370	4 28	14	4,230	9 0	29
2,440	4 37	14·5	4,280	9 8	29·5
2,510	4 46	15	4,330	9 16	30
2,580	4 55	15·5			

Range Table for 12·5-inch R.M.L. Gun of 38 tons.

Charge, 100 lb. P^2. powder.

Projectile, common or shrapnel shell with gas check.

Range.	Elevation.	Fuze Scale.	Range.	Elevation.	Fuze Scale.
Yards.	° ′		Yards.	° ′	
180	0 15	1	2,700	4 41	16
280	0 25	1·5	2,770	4 50	16·5
370	0 34	2	2,840	4 58	17
460	0 43	2·5	2,910	5 6	17·5
550	0 52	3	2,990	5 16	18
640	1 1	3·5	3,060	5 24	18·5
730	1 10	4	3,100	5 33	19
820	1 19	4·5	3,200	5 42	19·5
910	1 28	5	3,270	5 50	20
1,000	1 38	5·5	3,340	5 58	20·5
1,080	1 46	6	3,400	6 6	21
1,160	1 54	6·5	3,470	6 15	21·5
1,250	2 4	7	3,540	6 24	22
1,340	2 14	7·5	3,600	6 33	22·5
1,420	2 22	8	3,660	6 41	23
1,510	2 32	8·5	3,730	6 50	23·5
1,590	2 40	9	3,790	6 58	24
1,670	2 48	9·5	3,850	7 6	24·5
1,750	2 57	10	3,920	7 15	25
1,830	3 5	10·5	3,980	7 24	25·5
1,920	3 14	11	4,040	7 33	26
2,000	3 22	11·5	4,100	7 42	26·5
2,080	3 31	12	4,160	7 51	27
2,160	3 40	12·5	4,220	8 0	27·5
2,240	3 48	13	4,290	8 9	28
2,320	3 57	13·5	4,350	8 18	28·5
2,400	4 6	14	4,410	8 27	29
2,470	4 15	14·5	4,460	8 36	29·5
2,550	4 24	15	4,520	8 45	30
2,630	4 33	15·5			

RANGE TABLE for 12·5-inch R.M.L. GUN of 38 tons.

Charge, 160 lb. P². with air space, 30 cubic inches per lb. of powder.

Projectile, Palliser and common shell with gas check.

Range.	Elevation.	Fuse Scale.	Range.	Elevation.	Fuse Scale.
Yards.	° ′		Yards.	° ′	
200	0 18	1	2,690	4 30	16
280	0 25	1·5	2,760	4 39	16·5
370	0 32	2	2,840	4 48	17
460	0 40	2·5	2,910	4 57	17·5
550	0 48	3	2,980	5 6	18
640	0 56	3·5	3,060	5 14	18·5
730	1 4	4	3,130	5 22	19
810	1 12	4·5	3,210	5 30	19·5
890	1 20	5	3,280	5 39	20
980	1 28	5·5	3,360	5 48	20·5
1,060	1 34	6	3,430	5 57	21
1,140	1 42	6·5	3,500	6 6	21·5
1,220	1 50	7	3,580	6 15	22
1,310	1 58	7·5	3,650	6 24	22·5
1,400	2 6	8	3,720	6 34	23
1,480	2 14	8·5	3,800	6 44	23·5
1,560	2 23	9	3,880	6 54	24
1,640	2 32	9·5	3,960	7 4	24·5
1,730	2 41	10	4,020	7 14	25
1,810	2 50	10·5	4,100	7 24	25·5
1,890	2 59	11	4,180	7 34	26
1,980	3 8	11·5	4,250	7 44	26·5
2,060	3 17	12	4,320	7 54	27
2,140	3 26	12·5	4,400	8 4	27·5
2,220	3 35	13	4,480	8 14	28
2,300	3 44	13·5	4,560	8 24	28·5
2,380	3 53	14	4,640	8 34	29
2,450	4 2	14·5	4,710	8 44	29·5
2,530	4 12	15	4,790	8 54	30
2,610	4 21	15·5			

RANGE TABLE for 6·3-inch R.M.L. HOWITZER.

Charge, 1 lb. R.L.G. powder.

Projectile, common shell weighted = 70 lb., including gas check.

Mean elevation due to each 50 yards of range by interpolation.

Range.	Drift. Right.	Elevation.	Deflection. Left.	Value of 5 minutes.		Time of flight.
				Elevation. In range.	Deflection.	
Yards.	Yards.	° ′	° ′	Yards.	Yards.	Seconds.
400	2·6	10 35	0 23	3·1	0·56	3·9
450	3·5	12 00	0 28	3·0	0·63	4·3
500	4·6	13 30	0 33	2·8	0·70	4·8
550	5·7	15 00	0 37	2·8	0·76	5·3
600	7·0	16 45	0 42	2·4	0·83	5·9
650	8·5	18 30	0 47	2·4	0·90	6·5
700	10·3	20 40	0 53	1·9	0·97	7·2
750	12·3	23 00	0 59	1·8	1·04	7·9
800	14·8	25 30	1 08	1·7	1·11	8·7
850	17·9	28 30	1 18	1·4	1·18	9·5
900	22·1	32 15	1 28	1·1	1·25	10·6
950	28·8	37 30	1 49	0·8	1·32	12·1
			Charge 2 lbs.			
400	—	4 30	—	7·1	0·56	2·3
500	0·5	5 45	0 4	7·1	0·70	3·0
600	1·8	7 00	0 11	7·1	0·83	3·7
700	3·5	8 15	0 18	6·7	0·97	4·4
800	5·5	9 35	0 25	6·3	1·11	5·1
900	7·7	11 00	0 31	5·9	1·25	5·8
1,000	10·2	12 30	0 37	5·6	1·39	6·6
1,100	13·0	14 05	0 43	5·3	1·53	7·4
1,200	16·5	15 45	0 49	5·0	1·67	8·2
1,300	20·0	17 30	0 55	4·8	1·81	9·1
1,400	24·0	19 20	1 02	4·5	1·95	10·0
1,500	28·5	21 15	1 09	4·3	2·09	10·9
1,600	33·6	23 15	1 16	4·1	2·13	11·8
1,700	40·0	25 30	1 25	3·7	2·27	12·8
1,800	47·2	28 00	1 34	3·3	2·41	13·9
1,900	57·3	31 00	1 49	2·8	2·55	15·2
2,000	72·0	35 00	2 10	2·1	2·69	16·9
2,100	91 8	40 00	2 37	1·7	2·91	19·0

Range Table for 6·3-inch R.M.L. Howitzer.

Charge, 3 lb. R.L.G. powder.

Projectile, common shell weighted = 70 lb., including gas check.

Mean elevation due to each 50 yards of range, by interpolation.

Range.	Drift. Right.	Elevation.	Deflection. Left.	Value of 5 minutes. Elevation. In range.	Value of 5 minutes. Deflection.	Time of flight.
Yards.	Yards.	° ′	° ′	Yards.	Yards.	Seconds.
400	—	2 20	—	9·1	0·56	1·7
500	0·5	3 15	0 4	9·1	0·70	2·3
600	1·0	4 10	0 6	9·1	0·83	2·9
700	2·1	5 05	0 11	9·1	0·97	3·6
800	3·3	6 00	0 14	9·1	1·11	4·2
900	4·5	6 55	0 17	9·1	1·25	4·8
1,000	5·8	7 50	0 20	9·1	1·39	5·4
1,100	7·1	8 45	0 23	9·1	1·53	6·0
1,200	8·5	9 40	0 26	9·1	1·67	6·6
1,300	10·2	10 35	0 29	9·1	1·81	7·2
1,400	12·4	11 30	0 32	9·1	1·95	7·8
1,500	14·6	12 35	0 35	9·1	2·09	8·3
1,600	16·9	13 20	0 38	9·1	2·13	8·9
1,700	19·2	14 20	0 41	8·3	2·27	9·5
1,800	22·0	15 20	0 44	8·3	2·41	10·1
1,900	25·0	16 25	0 47	7·7	2·55	10·8
2,000	28·2	17 35	0 50	7·1	2·69	11·5
2,100	31·5	18 45	0 54	7·1	2·91	12·2
2,200	35·5	20 00	0 58	6·7	3·10	12·9
2,300	40·0	21 20	1 03	6·3	3·19	13·6
2,400	45·0	22 50	1 08	5·6	3·33	14·4
2,500	51·0	24 30	1 14	5·0	3·47	15·4
2,600	58·2	26 15	1 21	4·8	3·61	16·4
2,700	66·0	28 15	1 28	4·1	3·75	17·5
2,800	75·0	30 30	1 36	3·7	3·89	18·7
2,900	85·6	33 00	1 46	3·3	4·03	20·0
3,000	99·0	36 00	1 59	2·8	4·17	21·6
3,100	126·0	39 40	2 26	2·3	4·30	23·5

Range Table for 6·3-inch R.M.L. Howitzer.

Charge, 4 lb. R.L.G. powder.

Projectile, common shell weighted = 70 lb., including gas check.

Mean elevation due to each 50 yards of range, by interpolation.						
Range.	Drift Right.	Elevation.	Deflection. Left.	Value of 5 minutes.		Time of flight.
				Elevation. In range.	Deflection.	
Yards.	Yards.	° ′	° ′	Yards.	Yards.	Seconds.
400	0·00	1 45	0 0	16·7	0·56	1·7
500	0·3	2 15	0 2	16·7	0·70	2·1
600	0·8	2 50	0 5	14·3	0·83	2·5
700	1·5	3 25	0 8	14·3	0·97	3·0
800	2·3	4 00	0 10	14·3	1·11	3·5
900	3·0	4 35	0 12	14·3	1·25	3·9
1,000	4·0	5 10	0 14	14·3	1·39	4·4
1,100	5·0	5 45	0 16	14·3	1·53	4·9
1,200	6·0	6 20	0 18	14·3	1·67	5·3
1,300	7·3	6 55	0 20	14·3	1·81	5·7
1,400	8·5	7 30	0 22	14·3	1·95	6·1
1,500	10·0	8 05	0 24	14·3	2·09	6·6
1.600	12·0	8 45	0 27	12·5	2·13	7·1
1,700	14·0	9 25	0 30	12·5	2·27	7·6
1,800	16·0	10 05	0 32	12·5	2·41	8·1
1,900	18·3	10 45	0 35	12·5	2·55	8·6
2,000	20·5	11 30	0 37	11·2	2·69	9·3
2,100	23·0	12 15	0 39	11·2	2·91	9·7
2,200	25·8	13 00	0 42	11·2	3·10	10·3
2,300	28·8	13 45	0 45	11·2	3·19	10·8
2,400	31·8	14 30	0 48	11·2	3·33	11·3
2,500	35·0	15 15	0 51	11·2	3·47	11·9
2,600	39·0	16 05	0 54	10·0	3·61	12·5
2,700	43·3	17 00	0 57	9·1	3·75	13·1
2,800	48·0	18 00	1 00	8·4	3·89	13·8
2,900	53·0	19 00	1 05	8·4	4·03	14·5
3,000	58·3	20 00	1 10	8·4	4·17	15·3
3,100	63·8	21 00	1 14	8·4	4·30	15·8
3,200	69·2	22 00	1 18	8·4	4·44	16·5
3,300	75·0	23 05	1 22	7·7	4·58	17·3
3,400	81·5	24 10	1 26	7·7	4·72	18·0
3,500	88·6	25 25	1 31	6·6	4·86	18·8
3,600	96·0	26 45	1 36	6·3	5·00	19·6
3,700	104·0	28 15	1 41	5·5	5·14	20·6
3,800	123·0	29 45	1 56	5·5	5·28	21·6
3,900	135·0	31 45	2 05	4·2	5·42	22·8
4,000	152·0	35 00	2 17	2·6	5·56	24·9
4,100	180·0	39 30	2 38	1·9	5·70	27·6

RANGE TABLE for 8-inch R.M.L. HOWITZER of 46 cwt.

Charge, 2 and 3 lbs. R.L.G. powder.

Projectile, common shell with gas check, 185 lbs.

Mean elevation due to each 100 yards of range, by interpolation.

Range.	Drift.	Elevation.	Deflection.	Value of 5 minutes.		Time of flight.
				Elevation in range.	Deflection.	
Yards.	Yards.	° ′	° ′	Yards.	Yards.	Seconds.
			2 lb. charge.			
400	11·0	18 15	1 42	2·00	0·56	5·4
500	14·5	24 00	1 44	1·44	0·70	6·9
600	22·0	35 00	2 12	0·75	0·83	9·6
			3 lb. charge.			
400	7·5	10 45	1 8	3·8	0·56	4·0
500	9·3	13 50	1 8	3·5	0·70	4·9
600	11·6	16 15	1 10	2·9	0·83	5·9
700	14·5	19 30	1 15	2·6	0·97	7·0
800	18·2	23 45	1 22	2·1	1·11	8·4
900	23·2	29 00	1 33	1·6	1·25	10·2
1,000	33·2	38 00	2 00	0·9	1·39	12·9

RANGE TABLE for 8-inch R.M.L. HOWITZER of 46 cwt.

Charge, 4 and 5 lb. R.L.G. powder.

Projectile, common shell with gas check, 185 lbs.

Mean elevation due to each 100 yards of range, by interpolation.						
Range.	Drift.	Elevation.	Deflection.	Value of 5 minutes.		Time of flight.
				Elevation in range.	Deflection.	
Yards.	Yards.	° ′	° ′	Yards.	Yards.	Seconds.
			4 lbs. charge.			
400	5·0	8 00	0 45	6·3	0·56	3·5
500	6·5	9 30	0 46	5·6	0·70	4·2
600	8·3	11 10	0 49	5·0	0·83	4·9
700	10·2	13 00	0 52	4·5	0·97	5·6
800	12·4	15 00	0 56	4·2	1·11	6·4
900	15·0	17 20	1 00	3·6	1·25	7·4
1,000	18·0	20 00	1 04	3·1	1·39	8·5
1,100	22·0	23 00	1 12	2·8	1·53	9·7
1,200	28·0	27 15	1 24	2·0	1·67	11·3
1,300	35·0	32 00	1 36	1·7	1·80	13·0
1,400	45·0	38 15	1 58	1·3	1·94	15·2
			5 lbs. charge.			
400	3·5	5 40	0 31	7·1	0·56	2·9
500	5·0	7 00	0 36	6·3	0·70	3·5
600	6·5	8 20	0 39	6·3	0·83	4·2
700	8·0	9 45	0 41	6·0	0·97	4·9
800	10·0	11 10	0 45	6·0	1·11	5·6
900	12·0	12 45	0 48	5·3	1·25	6·3
1,000	14·3	14 30	0 51	4·8	1·39	7·1
1,100	16·8	16 15	0 55	4·3	1·53	8·0
1,200	20·0	18 25	1 00	3·8	1·67	9·0
1,300	23·5	20 35	1 05	3·8	1·80	10·0
1,400	27·5	23 00	1 10	3·5	1·94	11·0
1,500	32·5	26 00	1 17	2·8	2·08	12·2
1,600	39·5	29 30	1 28	2·4	2·22	13·7
1,700	48·0	33 30	1 41	2·1	2·36	15·3
1,800	61·0	39 00	2 02	1·5	2·50	17·4

RANGE TAALE for 8-inch R.M.L. HOWITZER of 46 cwt.

Charge, 6 lbs. R.L.G. powder.

Projectile, common shell with gas check, 185 lbs.

Mean elevation due to each 100 yards of range, by interpolation.						
Range.	Drift.	Elevation.	Deflection.	Value of 5 minutes.		Time of flight.
				Elevation in range.	Deflection.	
Yards.	Yards.	°	° ′	Yards.	Yards.	Seconds.
400	1·50	4 30	0 13	7·7	0·56	2·6
500	3·00	5 35	0 22	7·7	0·70	3·2
600	4·75	6 45	0 28	7·1	0·83	3·8
700	6.50	7 55	0 33	7·1	0·97	4·4
800	8·50	9 05	0 38	7·1	1·11	5·0
900	10·50	10 15	0 42	7·1	1·25	5·7
1,000	12·50	11 30	0 45	6·7	1·39	6·4
1,100	14·75	12 50	0 48	6·2	1·53	7·1
1,200	17·25	14 15	0 51	5·9	1·67	7·8
1,300	20·00	15 45	0 55	5·5	1·80	8·6
1,400	22·75	17 15	0 58	5·5	1·94	9·4
1,500	26·00	18 55	1 2	5·0	2·08	10·2
1,600	29·75	20 45	1 7	4·5	2·22	11·1
1,700	33·75	22 45	1 11	4·1	2·36	12·1
1,800	38·75	25 00	1 17	3·7	2·50	13·1
1,900	44·75	27 30	1 25	3·3	2·64	14·3
2,000	52·25	30 25	1 34	2·9	2·78	15·6
2,100	62·00	33 45	1 46	2·5	2·92	17·1
2,200	81·00	39 00	2 12	1·5	3·06	19·3

RANGE TABLE for 8-inch R.M.L. HOWITZER of 46 cwt.

Charge, 7 lbs R.L.G powder.

Projectile, common shell with gas check, 185 lbs.

Mean elevation due to each 100 yards of range, by interpolation.						
Range.	Drift.	Elevation.	Deflection.	Value of 5 minutes.		Time of flight.
				Elevation in range.	Deflection	
Yards.	Yards.	° ′	° ′	Yards.	Yards.	Seconds.
400	0·5	3 25	0 4	9·1	0·56	2·1
500	2·0	4 20	0 14	9·1	0·70	2·7
600	3·6	5 20	0 21	8·3	0·83	3·3
700	5·3	6 20	0 27	8·3	0·97	3·9
800	7·1	7 20	0 31	8·3	1·11	4·5
900	8·9	8 20	0 35	8·3	1·25	5·1
1,000	10·8	9 20	0 38	8·3	1·39	5·7
1,100	12·8	10 20	0 41	8·3	1·53	6·3
1,200	14·8	11 25	0 44	8·0	1·67	6·9
1,300	17·0	12 35	0 47	7·1	1·80	7·6
1,400	19·5	13 45	0 50	7·1	1·94	8·2
1,500	22·0	14 55	0 53	7·1	2·08	8·9
1,600	25·0	16 15	0 56	6·3	2·22	9·6
1,700	28·5	17 35	1 00	6·3	2·36	10·4
1,800	32 0	19 05	1 04	5·5	2·50	11·2
1,900	36·0	20 40	1 08	5·2	2·64	12·0
2,000	40·0	22 25	1 12	4·8	2·78	12·9
2,100	45·5	24 25	1 18	4·2	2·92	13·9
2,200	52·0	26 30	1 25	4·0	3·06	15·0
2,300	60·0	28 50	1 33	3·6	3·20	16·2
2,400	70·0	31 30	1 45	3·1	3·34	17·6
2,500	85·0	35 00	2 02	2·6	3·48	19·2
2,600	106·0	38 40	2 27	2·3	3·61	20·8

Range Table for 8-inch R.M.L. Howitzer of 46 cwt.

Charge, 8 lbs. R.L.G. powder.

Projectile, common shell with gas check, 185 lbs.

Mean elevation due to each 100 yards of range, by interpolation.

Range.	Drift.	Elevation.	Deflection.	Value of 5 minutes.		Time of flight.
				Elevation in range.	Deflection.	
Yards.	Yards.	° ′	° ′	Yards.	Yards.	Seconds.
400	1·3	3 05	0 11	10·0	0·56	2·1
500	2·5	3 55	0 16	10·0	0·70	2·7
600	3·9	4 45	0 23	10·0	0·83	3·3
700	5·5	5 35	0 28	10·0	0·97	3·8
800	7·1	6 25	0 31	10·0	1·11	4·3
900	8·8	7 25	0 35	10·0	1·25	4·9
1,000	10·6	8 10	0 38	10·0	1·39	5·4
1,100	12·5	9 05	0 40	9·1	1·53	6·0
1,200	14·4	10 00	0 43	9·1	1·67	6·6
1,300	16·4	10 55	0 45	9·1	1·80	7·2
1,400	18·6	11 50	0 47	9·1	1·94	7·8
1,500	20·9	12 45	0 50	9·1	2·08	8·3
1,600	23·3	13 45	0 52	8·3	2·22	8·9
1,700	25·8	14 45	0 54	8·3	2·36	9·5
1,800	28·4	15 45	0 56	8·3	2·50	10·1
1,900	31·4	16 50	0 59	8·0	2·64	10·8
2,000	34·8	18 00	1 02	7·1	2·78	11·5
2,100	38·8	19 20	1 06	6·6	2·92	12·2
2,200	43·0	20 40	1 10	6·6	3·06	13·0
2,300	47·5	22 05	1 14	6·3	3·20	13·8
2,400	53·0	23 35	1 19	5·9	3·34	14·7
2,500	59·0	25 20	1 24	4·7	3·48	15·6
2,600	67·0	27 10	1 32	4·5	3·61	16·6
2,700	76·0	29 20	1 41	3·8	3·75	17·7
2,800	89·0	31 45	1 54	5·5	3·89	19·0
2,900	107·0	35 00	2 12	2·6	4·03	20·6
3,000	135·0	39 00	2 42	2·1	4·17	22·5

RANGE TABLE for 8-inch R.M.L. HOWITZER of 46 cwt.

Charge, 9 lbs. R.L.G. powder.

Projectile, common shell with gas check, 185 lbs.

Mean elevation due to each 100 yards of range, by interpolation.						
Range.	Drift.	Elevation.	Deflection.	Value of 5 minutes.		Time of flight.
				Elevation in range.	Deflection.	
Yards.	Yards.	° ′	° ′	Yards.	Yards.	Seconds.
400	0·5	2 10	0 4	12·5	0·56	1·7
500	1·7	2 55	0 12	11·1	0·70	2·2
600	3·0	3 40	0 18	11·1	0·83	2·7
700	4·4	4 25	0 22	11·1	0·97	3·2
800	5·8	5 10	0 26	11·1	1·11	3·7
900	7·3	5 55	0 27	11·1	1·25	4·2
1,000	8·8	6 40	0 31	11·1	1·39	4·7
1,100	10·4	7 25	0 34	11·1	1·53	5·2
1,200	12·1	8 15	0 36	10·0	1·67	5·8
1,300	14·0	9 05	0 38	10·0	1·80	6·4
1,400	16·0	9 55	0 41	10·0	1·94	7·0
1,500	18·1	10 45	0 43	10·0	2·08	7·6
1,600	20·3	11 35	0 45	10·0	2·22	8·1
1,700	22·5	12 25	0 47	10·0	2·36	8·7
1,800	25·0	13 20	0 50	9·1	2·50	9·3
1,900	27·8	14 25	0 52	9·1	2·64	9·9
2,000	30·8	15 20	0 55	9·1	2·78	10·5
2,100	34·0	16 15	0 58	9·1	2·92	11·1
2,200	37·5	17 15	1 01	8·3	3·06	11·7
2,300	41·2	18 15	1 04	8·3	3·20	12·3
2,400	45·0	19 15	1 07	8·3	3·34	12·9
2,500	50·0	20 30	1 12	8·0	3·48	13·6
2,600	55·0	21 40	1 16	7·1	3·61	14·4
2,700	61·5	23 05	1 22	5·9	3·75	15·2
2,800	69·0	24 35	1 28	5·5	3·89	16·1
2,900	77·5	26 15	1 36	5·0	4·03	17·1
3,000	88·5	28 05	1 46	4·5	4·17	18·1
3,100	102·0	30 10	1 58	4·0	4·31	19·3
3,200	119·0	32 30	2 13	3·6	4·44	20·6
3,300	141·0	35 15	2 33	3·0	4·58	22·0
3,400	168·0	38 30	2 57	2·6	4·72	23·7

RANGE TABLE for 8-inch R.M.L. HOWITZER of 46 cwt.

Charge, 10 lbs. R.L.G. powder.

Projectile, common shell with gas check, 185 lbs.

Mean elevation due to each 100 yards of range, by interpolation.

Range.	Drift right.	Elevation.	Deflection left.	Value of 5 minutes.		Time of flight.
				Elevation in range.	Deflection.	
Yards.	Yards.	° ′	° ′	Yards.	Yards.	Seconds.
400	0·9	2 00	0 8	14·3	0·56	1·7
500	1·8	2 35	0 13	14·3	0·70	2·1
600	2·8	3 10	0 17	14·3	0·83	2·6
700	4·0	3 50	0 21	12·5	0·97	3·1
800	5·3	4 30	0 24	12·5	1·11	3·6
900	6·6	5 10	0 26	12·5	1·25	4·1
1,000	8·0	5 50	0 29	12·5	1·39	4·6
1,100	9·5	6 30	0 31	12·5	1·53	5·1
1,200	11·0	7 10	0 33	12·5	1·67	5·6
1,300	12·7	7 50	0 35	12·5	1·80	6·1
1,400	14·5	8 35	0 37	11·1	1·94	6·6
1,500	16·4	9 20	0 39	11·1	2·08	7·1
1,600	18·5	10 05	0 42	11·1	2·22	7·6
1,700	20·7	10 55	0 44	11·1	2·36	8·1
1,800	22·9	11 35	0 46	11·1	2·50	8·6
1,900	25·2	12 20	0 48	11·1	2·64	9·1
2,000	27·6	13 05	0 50	11·1	2·78	9·6
2,100	30·6	13 55	0 52	10·0	2·92	10·2
2,200	33·7	14 45	0 55	10·0	3·06	10·8
2,300	37·3	15 35	0 58	10·0	3·20	11·3
2,400	41·5	16 30	1 02	9·1	3·34	11·9
2,500	45·5	17 25	1 06	9·1	3·48	12·5
2,600	51·0	18 25	1 11	8·3	3·61	13·1
2,700	57·0	19 25	1 16	8·3	3·75	13·7
2,800	64·0	20 30	1 23	8·0	3·89	14·4
2,900	72·0	21 40	1 29	7·1	4·03	15·2
3,000	81·0	22 50	1 37	7·1	4·17	16·0
3,100	90·0	24 00	1 45	7·1	4·31	16·7
3,200	100·0	25 15	1 53	6·6	4·44	17·4
3,300	111·0	26 45	2 01	5·5	4·58	18·3
3,400	123·0	28 25	2 10	5·0	4·72	19·3
3,500	137·0	30 15	2 21	4·5	4·86	20·4
3,600	154·0	32 25	2 33	3·8	5·00	21·6
3,700	176·0	35 00	2 51	3·2	5·13	23·1
3,800	204·0	38 15	3 11	2·6	5·27	24·9

ST OF SERVICE O

Cla O na	ges and Fuzes.	Ba	Energy at Muzzle.	Piercing Power. Iron Plates.† In Inches.		Proof.				Nature of Ordnance.
						Number of Rounds.	Charge.		Projectile.	
	†† Fuze S.S.	Weigh Empty		At 500 Yards.	At 1,000 Yards.		P² or P.	R.L.G.		
		lbs. oz	f. t.				lbs. ozs.	lbs. ozs.	lbs. ozs.	R. MUZZLE-LOADERS.
	—	—	—	—	—	—	—	—	—	17·72-inch, 100 tons.
	—	1700	27,213	27·5	26·4		‡			16-inch, 80 tons.
	—	790	11,727	18·7	18	1 2	160 0 180 0	30 c. in. to lb.	818 0	12·5-inch, 38 tons.
Armour Piercing.	—	688	8,340	15·4	14·6	1 2	110 0 115 0	—	700 0	12-inch, 35 tons.
Armour Piercing.	—	586	7,200	13·9	13·1	2	—	87 8	600 0	12-inch, 25 tons.
Armour Piercing.	—	529	6,590	13·8	·13·1	1 2	85 0 95 0	—	530 0	11-inch, 25 tons.
Armour Piercing.	—	393	5,318	12·7	12·0	1 2	70 0 75 0	—	400 0	10-inch, 18 tons.
Armour Piercing.	—	244	3,635	10·4	9·6	1 2	50 0 58 0	—	250 0	9-inch, 12 tons.
Armour Piercing.	—	174 11	2,492	9·0	8·25	2	—	37 8	180 0	8-inch, 9 tons.
Armour Piercing.	—	112 1	1,891	8·0	7·2	2	—	27 8	115 0	7-inch, 7 tons.
	Pettman's G.S. 9, or 15 sec.	112 1	1,855	8·0	7·1	2	—	37 8	115 0	7-inch, 6½ tons.
				Energy of Common Shell.						
				At 1,000 Yards.	At 2,000 Yards.					
	Do. ...	—	1,246	901	739	2	—	17 8	115 0	7-inch, 90 cwt.
Medium	—	—	853	591	484	2	—	12 8	80 0	80-pr., 5 tons converted.
Medium	—	—	607	422	336	2	—	10 0	64 0	64 cwt. Mark I.,

INDEX.

A

	Page.
Air, resistance of, to projectiles	339
Air, spacing, and its object	38
American employment of cast iron for ordnance	8
" guns, Rodman construction	24
Annual returns	371, 372
" " S.B. ordnance	375
" " rifled "	376
Aprons, lead, for S.B. ordnance	64
Arcs, elevating R.M.L. guns, pivots for	206
" graduated or racers	184
Armaments, Foreign Powers	86
Armstrong guns, for horse and field artillery	71
" " heavy natures introduced	71
" " superior to S.B. guns	71
" " liability to accidents	71
" R.B.L. construction, adoption of	71
" " guns, manufacture of	71
" Sir William, appointed to R.G.F. as Superintendent	71
Artillery, its ascendancy	69
Austrian experiments with bronze and bronze steel	6
" Uchatius guns	24

B

Ballistic power of guns	333
Bar copying	115
" iron, its manufacture	103, 104
" " for heaviest guns, forged	107
" " scarfing and welding	104
" rifling, its movements of rotation and translation	114, 115
Barlow, Professor, theory as to circumferential tensions	21
Barrels, solid forged, used for R.B.L. and some R.M.L. guns	123
Bashforth's chronoscope, and experiments with	340
" formula for resistance of air, and examples as to	340, 341
" tables	354, 355
Bearers, shot R.B.L. guns	184
" " R.M.L. "	202
" " $\frac{80}{64}$-pr.	241
Bell crank lever as used in machinery	98
Bell, mouthing of 10 guns and upwards, also the 6"·3 Howitzer	176, 177
Belts or bands	96
Bevil wheels	96
Binding slabs in a forging	107
Bit, vent, Armstrong B.L. guns	134
" " S.B. guns	64
Blocks, wood, taking impressions	296
Bloomfield, C.I. S.B. guns	56

Page.
Blooms, puddled and wrought iron 102, 103
Blueing, operation of 377
Bolt, iron, elevating eye complete for use with 25, 16 and 9-pr. guns 203
„ steel, for pieces, pivots, M.K. II. for overbank carriage 203
Bore, powder pressure in, how measured 346, 347
„ R.B.L., different diameters of 124
„ R.M.L., end of, marked on upper surface of gun 181
„ „ length to be considered 87
„ „ length, how determined 87
„ „ powder pressure, attempts at reduction in 88
Boring, fine and finish, 7-inch steel A. tube 174
„ operation of 112, 113
„ rough 7-inch steel A. tube 163
Bracket, foresight, for converted guns 203, 241
Brass Foundry at Woolwich Arsenal, 1717 56
Breech coils for heavy guns, manufacture of 110
„ coil, 7-inch R.M.L., manufacture of 166
„ fittings, preparation of R.B.L. guns for 124
„ „ for R.B.L. guns 125
„ „ table of, and weights 144
„ loading, its defects and advantages 31, 38
„ piece R.B.L. guns, forged 123
„ „ abolished in present construction 29
„ screw, manufacture, material, thread and pitch of 127, 128
„ „ marks on 127
Britten's and Jeffrey's rifling 54
Brittleness, definition of, as applied to metals 2
Broaching, operation of 114
Bronze, properties of, as a gun metal 5
„ S. B. ordnance, and manufacture of 56, 59
„ for foreign rifled ordnance.. 6
„ manufacture of R.M.L. 9-pr. of 5, 78
„ Austrian, Italian, French, and Russian experiments with 6
„ improvements in, for gun construction 5
„ phosphor 6
Bronzing, operation of 377
Browning 377, 378
Building up, systems of gun construction by 24
Bushes, breech, iron 7-inch R.B.L. gun 135
„ „ copper, R.B.L. 135
„ vent 50, 60, 177, 236
„ „ removable 51
„ „ piece copper, R.B.L. guns 135
Bushing R.B.L. guns .. .. 125, 306, to 309
„ „ 7-inch guns, double bushed 135

C

Calibre of rifled guns, considerations which determine the 87
„ Service R.B.L. and R.M.L pieces 180, 291, 292
Cams, use of, in machinery 97
Carronades, when introduced, &c. 57, 59
Cartridge, length of, abnormal pressures due to 87
Cascable screw, with what guns used 161
„ „ its manufacture 167
„ „ fitting, screwing in, and securing 167
Cast iron guns, manufacture of 56
„ ordnance, classification of 57
„ „ natures in service 65
„ „ venting of 60
Centring of projectile 42
„ operations in manufacture 111
Chambers, enlarged, object of 37, 38
„ R.B.L. guns 124
„ R.M.L. „ conical, why required 174, 175
„ S.B. pieces 57

Page.

Change wheels, as used in machinery 97
Chase sights 184, 193, 194
Chronograph' Le Boulengé, employment of 339
" " detailed description of 334, 339
Chronoscope, Bashforth's 340
" Noble's 349, 353
Clamps, metal, for wood scales, R.M.L. 203
" moveable, tangent sights 133, 135, 203, 241
Clamping arrangements for sights 132
Clinometers and their use 184, 198
Coils and coiling, manufacture of 104, 106
" breech 7-inch R.M.L. manufacture of 166
" heavier used for Fraser construction.. 27
" prepared for uniting 108
Coins, graduated for S.S. 63
Collar, cast iron, for converted guns 235
" leather, for 12-pr. R.B.L. breech screws 135
Compression, testing amount of 100
" due to skrinking 26
Committee, Armstrong and Whitworth 77, 82
" 1866 M.L. v. B.L.. 73
" 1868. Guns for India.. 73
" 1869. High angle fire.. 75
" 1869. On explosives 85
" 1870. Special, B.L. or M.L. field guns 73
" 1871. On mitrailleurs.. 75
Condie's hammer 98
Cone pulleys used in machinery 97
Condemnation, provisional, of gun.. 301
Construction, original or Armstrong 28, 82
" Fraser.. .. ..28, 84
Converted guns, comparison with S.B. pieces 94
" " manufacture of 233
Conversion of S.B. mortars to rifled pieces attempted 79
Coppering R.B.L. guns 306, 309
Copying bar for rifling machine 115, 116
Cranks used in machinery 98
Cross head for vent-pieces 126
Crusher and cutting plugs, use of 351, 352
Crutch iron 136
Cups, tin, for R.B.L. guns 125

D

Deflection, permanent, angle of, ascertained 52
" angle of, for R.M.L. guns 212, 213
" theories as to 53
Defects at proof, impressions preserved 178
" exterior, on guns, little importance of 304
" how noted in returns 297
" in S.B. guns.. 295
" large exterior, noted on examination memo. 304
" in Rifled guns 298, 299
" in steel A. tubes 299
" in vents, and sentence 301
" in wrought iron A. tubes 298
Denmark, employment of C.I. for Ordnance 9
Densities, table of 359
Derrick, muzzle 203, 204
" " how fixed 397
Designation of S.B. ordnance 57
" Rifled 160, 291, 292
Dickson's S.B. guns 50
Double bush in 7-inch B.L. guns 125
Drawings, working 119
Drilling, operation of 118

Page.

Drilling for sights R.M.L. guns .. 180
Drip hole, R.B.L. guns .. 124
Driving side of groove .. 42
Drumming scrap iron to remove rust .. 103
Ductility, definition of, as applied to metals .. 2
" of wrought iron .. 10
Dundas, S.B. guns .. 56

E

Effects produced by a gun, how measured .. 333
" required from field, siege, and heavy guns .. 344, 345
Elastic limit, definition of .. 2
" " governs amount of shrinkage .. 27
" " of wrought-iron .. 10
" " of steel, how tested .. 99
" " of metals used in R.G.F. .. 18
Elasticity, definition and measure of .. 24
Elasticities, varying, so called principle of .. 23
Elevating patch, R.M.L. .. 189
" plates " (metal) .. 205, 206
Energy of projectiles from service guns .. 356, 357, 360
English and Foreign ordnance, comparison of .. 91, 93
Engraving Royal monogram .. 174
Examination and proof before issue .. 293
" of breech fittings R.B.L. guns .. 303
" of exterior of guns S.B. or Rifled .. 304
" " necessary should a shell burst in the bore .. 304
" of guns before and after proof .. 177
" " at out stations .. 293, 372
" " rules, home service .. 372
" " " foreign " .. 373
" memo. of .. 294
" of work during manufacture .. 119
" vents of S.B. and Rifled guns .. 300
Examining and cleaning ordnance S.B. and Rifled, with tools .. 317, 319
Examples, Bashforth's for finding ranges and velocities .. 354, 355
" to determine length of bore .. 37
Expansion, heat required for, prior to shrinking .. 109
Experiments, as to accuracy of S.B. v. R.B.L. guns .. 70
" Armstrong and Whitworth guns .. 71
" Okehampton, 16 and 9-prs. .. 77
" 68-pr. carronades rifled .. 79
" Bashforth's Chronoscope .. 340
" cast-iron 9-inch howitzer rifled .. 79
" comparative value of German and British 9-prs. .. 75
" French, at Bourges .. 76
" Gatling guns .. 381
" guns v. armour plates .. 81
" R.M.L. howitzers .. 85
" strengthening and rifling S.B. pieces .. 233
" to test efficiency of British ordnance .. 89
" " " Fraser system .. 28
" 16-inch gun .. 86, 287
" 38-ton guns, 12-inch and 12·5-inch .. 85
" 16-inch gun 80 tons .. 86
Experimental field pieces .. 77
Extractors, tin cup, 7-inch B.L. guns .. 136
Exterior form of heavy service guns .. 246
Eyes, elevating for B.L. guns .. 136

F

Factory, name of, upon left trunnion .. 181
Facing implements, R.B.L. .. 137, 321

Page.
Faucet, meaning of the term 108
Field R.M.L. ordnance, manufacture of, when commenced 88
Fine and finish boring 174
Fired gunpowder, work realised (Capt. A. Noble) 361
" " tables, showing work from 362, 363
Fissures or hair lines in bore, S.B. bushed guns 302
" " " R.M.L. guns 303
Fittings, breech, R.B.L., preparing gun for 124
" " " proof of 129
" " " weight of (table).. 144
" for R.M.L. guns 201
" preservation of 305
" R.M.L., table of 231
" screw holes for, fixing screws 182
Foreign ordnance compared with English.. .. ..91, 93
Forgings, solid, how made 107
" " for A. tubes.. 123
Form, exterior of heavy guns 246
Formula, Hélie's 340
" piercing iron plates 345
Friction tube pin 64, 139
France, employment of C.I. for heavy ordnance 9
" present field guns 91
Fraser construction 23
" " experiments as to 29
" " employed since 1868, its advantages 161
" " proposed in 1866 88
" " adopted in 1866–67 29
French experiments with phosphor bronze 6
" modified rifling introduced 73
" " " with what guns used 45
" " " competitive trial of 54
Furnace, Price's Retort 103
" reverberatory for puddling 102
" for heating coils 104

G

Gas checks, used to prevent erosion 36
" escapes, their object 245
" " how formed 167, 245
" " not required for 25-prs. and under 161
Gatling guns, introduced, employment, &c. 381, 384
" " committee on 75
Gauges, crusher, description of 347
" " table of pressures shown by 348
" pressure, Rodman's 346
" R.G.F. cylinder for bores and grooves 121
Gauging interior of tubes for shrinking 119
Graduations on tangent sights as wood scales 195
Gravity, centre of, line denoting 181
Grip in R.B.L. guns.. 124
Guard, metal, vent, 10-inch guns 205
Guide plates 137, 205
Grooves, commencement of 117
" form of 43
" French 46
" " modified 46
" loading and driving side of 42
" of rifling, how cut.. 114
" plain 43
" pressure on 42, 346
" R.B.L. guns 124
" size, shape, and number of 43, 176
" shunt 44
" " objection to 45
" Woolwich 45

Page.
Grooves, Woolwich, splay of heavy guns .. 176
" wearing of .. 298
Guns, ammunition and stores, simplicity not requisite .. 36
" Armstrong principle of construction .. 28
" Austrian Uchatius .. 24
" Ball mouthing for 10″ guns and upwards, also the 6·3 inch howitzer .. 177
" British and foreign, comparative value .. 91, 93
" casting or building up of .. 20
" chambers enlarged in R.B.L. and R.M.L. .. 37, 86
" circumferential strength .. 20
" construction, general consideration as to .. 20
" field, history of .. 74, 76
" foreign, examples of failures of .. 17
" Fraser, construction since 1868 .. 161
" Gatling, introduction of .. 75
" " construction of .. 381
" " for S.S. .. 75
" heavy, early principles of construction .. 28
" " history of .. 80, 84
" Krupp's, and their breech mechanism .. 32
" Lancaster, failure of .. 81
" limit of thickness of metal of, when homogeneous .. 21
" "original" construction to 1867 .. 161
" Palliser, principle of construction of .. 25, 30
" resumé, as to service, heavy .. 87
" Rodman .. 24
" service R.B.L. introduced in 1859 .. 69, 74
" " " construction, faults in .. 71, 74
" " " " of .. 122
" " " manufacture of, ceased in 1868 .. 74
" " " proof of .. 129
" siege, history of .. 78
" Woolwich, principles of construction .. 25, 30, 83
" " " " adopted in 1868 .. 87
" rifled B.L., ammunition and stores complicated .. 36
" " " Armstrong adopted .. 71
" " " " manufacture commenced 1859 .. 71
" " " for horse and field batteries .. 76
" " " heavy, manufacture commenced .. 71
" " " historical précis as to each nature of .. 147
" " " liability to damage on service .. 71
" " " manufacture suspended .. 72
" " " numbers of, in the service .. 143
" " " objections to, for naval service .. 82
" " " steel tubes for .. 30
" " " siege and position discarded .. 78
" " " superiority over S.B. .. 70
" " " table of dimensions of, &c. &c. .. 160
" " " (wedge) introduced 1864 .. 81
" " " " manufacture stopped .. 81
" " " v. R.M.L., accuracy of fire .. 35
" " " v. " generally .. 31
" " " 40-pr., detail of manufacture .. 122, 125
" " " 7-inch, construction and short description of .. 147, 148
" " " 40-pr. of 35 cwts. .. 151
" " " 40-pr. " 32 " .. 149
" " " 20-pr. " 16 " .. 153
" " " 20-pr. " 15 " .. 153
" " " 20-pr. " 13 " .. 154
" " " 12-pr. " 8 " .. 155
" " " 9-pr. " 6 " .. 157
" " " 6-pr. " 3 " .. 158
" " " 64-pr. and 40-pr. wedge .. 159
" " M.L. built on Fraser principle .. 88
" " " heavy, when introduced for our service .. 80, 88
" " " field .. 69
" " " siege .. 78, 80

Page.

Guns, rifled M.L., mountain 77
" " " advantages over B.B.L. 36
" " " conversion, how carried out 233
" " " " S.B. guns proposed for 233
" " " " Palliser's system of 233
" " " converted, historical précis as to 80
" " " " 64-pr., 71 cwt., detail of conversion .. 234, 238
" " " " proof and examination 238
" " " " $\frac{64}{32}$-pr., 58 cwt. 238, 265
" " " " $\frac{80}{68}$-pr., 5 tons 239, 266
" " " " comparison of power of 94
" " " conversion from cast iron S.B. an economical measure 88, 233
" " " examination before and after proof 238
" " " " of generally 293
" " " first employed in the field by France.. 69
" " " field, foreign powers 71, 91, 93
" " " " bronze, 9-prs., adopted and discontinued for India .. 5, 73
" " " " wrought iron with steel tube recommended in 1866 .. 74
" " " field manufacture of commenced 76
" " " Fraser modification 84
" " " heavy foreign powers 86, 95
" " " " necessity for 85
" " " " their importance 80
" " " " accuracy and rapidity of fire of 85
" " " manufacture of latest patterns 161
" " " "marks," of present construction 162
" " " " " all guns in service 245
" " " original construction, its expense 83
" " " progress made since 18th century 81
" " " proof of 177
" " " shunt rifling approved of 82
" " " steel tubes of 30, 163
" " " v. R.B.L., summary 74
" " " 17"·72 100 tons, its introduction 88
" " " 16" 80 tons, its introduction 88
" " " " " manufacture 170
" " " " " as a 14·5. and 15" 86
" " " " " not scored after 94 rounds 36
" " " " " trials at Shoeburyness 87
" " " 12"·5 38 tons, its introduction.. 85
" " " " " accuracy of 35
" " " " " calibre decided 85
" " " " " manufacture of 169
" " " 12" 35 tons approved for S.S... 85
" " " " " manufacture of 169
" " " " " description of 284
" " " " 25 tons introduced 85
" " " " " manufacture of 169
" " " " " description of different marks 283
" " " 11" 25 tons introduced 85
" " " " " manufacture of 169
" " " " " description of different marks 282
" " " 10" 18 tons introduced 84
" " " " " description of different marks 279
" " " " " and upwards, manufacture 169
" " " 9" 12 tons introduced 84
" " " " " Marks IV. and V., manufacture of 168
" " " " " description of different marks 278
" " " " " trials of Fraser construction 29
" " " " " introduction of Fraser construction 88
" " " " " final trials 84
" " " 8" 9 tons introduced 83
" " " " " manufacture of 168
" " " " " description of different marks 274
" " " 7" 7 tons introduced 83
" " " " " Mark III, detail of manufacture 162, 168
" " " " " " description of different marks .. 271

Page.
Guns, rifled M.L., 7″ 6½ tons introduced 269
„ „ „ „ „ Mark III., manufacture 272
„ „ „ „ „ „ description of different marks .. 270
„ „ „ „ 90 cwt. I., reduced from 7″ 6½ tons 169
„ „ „ „ „ „ description of 268
„ „ „ 64-pr. 64 cwt. introduced 78, 81
„ „ „ „ „ Fraser construction 29
„ „ „ „ „ III., manufacture of 172
„ „ „ „ „ „ description of different marks 262
„ „ „ 40-pr., 35 cwt., II., introduced 78, 258
„ „ „ „ „ „ manufacture of 172
„ „ „ „ „ „ description of different marks 258
„ „ „ 25 pr., 18 cwt., I., introduced.. .. .. .. 78, 251
„ „ „ „ „ „ manufacture of 172
„ „ „ „ „ „ description of 257
„ „ „ 16-pr., 12 cwt., I., introduced.. 75
„ „ „ „ „ „ manufacture of 172
„ „ „ „ „ „ description of 255
„ „ „ 13-prs., experimental 77, 290
„ „ „ 9-pr., 8 cwt., I., introduced 252
„ „ „ „ „ „ manufacture of 173
„ „ „ „ „ „ description of 253
„ „ „ „ „ „ heavy, for Horse Artillery 76
„ „ „ „ „ II., introduced.. 255
„ „ „ „ „ „ manufacture of 173
„ „ „ „ „ „ description of 255
„ „ „ 6 cwt., I., introduced 76
„ „ „ „ „ „ manufacture of 173
„ „ „ „ „ „ description of 255
„ „ „ „ „ II., introduced.. 76
„ „ „ „ „ „ manufacture of 173
„ „ „ „ „ „ description of 255
„ „ „ „ bronze, cast at Woolwich 73
„ „ „ 7-pr. (steel), 200 lbs., Mark IV., introduced 78
„ „ „ „ „ „ „ manufacture of 173
„ „ „ „ „ „ „ its employment 78
„ „ „ „ „ 150 lbs., for India 78
„ „ „ „ description of different marks 247, 252
„ „ „ „ converted from 8-pr. S.B., bronze 78
„ „ „ „ „ „ comparison in a rifled state .. 79
„ „ „ „ bronze, historical précis as to 247, 248
„ S.B., and R.B.L., power of, compared 90
„ „ bronze, manufacture of 56
„ „ „ natures in the service 59
„ „ „ to be retained or abolished 59
„ „ cast iron, manufacture of 56
„ „ „ to be retained or abolished.. 58
„ „ „ 10-inch 58
„ „ „ 8-inch 58
„ „ „ 68-pr.. 58
„ „ „ 42-pr.. 58
„ „ „ 32-pr.. 58
„ „ „ 24-pr.. 59
„ „ „ 18-, 12-, 9-, and 6-prs. 59
„ „ „ strengthening of, for rifled pieces 79
„ „ classification of 57
„ „ stores belonging to 63
„ „ proof of 59
„ „ marks on.. 59
„ „ wrought iron, 150-prs... .. .. 56, 82
„ „ „ 100-prs... .. .. 56, 82
Gunpowder, fired, work performed by, in gun 362
Gutta percha for impressions 297

H

Hammers, Nasmyth's and Condie's 98
„ table showing power of 99

Page

Hanging scales, description and use of 200
Hart, Dr., theory as to circumferential tensions of a cylinder 21
Hélie's formula for finding velocity of shot 340
Horizontal axis, lines of, on R.B.L. guns, mode of obtaining 129
" " " R.M.L. " 179
" " " S.B. 62
Howitzers, bronze and iron S.B. 59
" rifled, proposed 79
" " 9″ cast iron, experiments with 79
" " 8″, 9″, and 10″, designs for, submitted 79
" " 10″, 6 tons, manufacture of 171
" " 8″, 46 cwt. 171
" " 6·3″, 18 cwt., design for, submitted 79
" " 8″, 70 " manufacture of, proposed 289
" " 6·6″, 36 " " " " 290
" " comparison of power with that of S.B. mortars 290
" " further experiments with 85
" " sights, exceptional 183
" " " not inclined 52
" " 8″, sights for 186
" smooth bore 63
" " Gomer chambers of 57
Hydraulic pressure, forcing on outer portion by 26

I

Implements, facing R.B.L. guns 137, 306, 321
Impressions arranged for recording defects 297
" before and after proof 178
" labelling of 298
" of vents 300
" when and how taken on service 294, 374
Increasing twist of rifling 40, 116
Index plates 184, 198, 208
Indicator rings 128, 140
Ingots, steel, for our guns, manufacture and tests of 99, 163
Instruments for taking impressions of bore 137, 317
" " " vents 300, 317
" sighting 138, 314, 320
" " wood, temporary, and how used.. 301
Inventions in ordnance 1615 to 1851 68
Iron, cast, America, its employment for ordnance 9
" " as a material for ordnance 7
" " characteristics of 8
" " how obtained 8
" " melting point of 8
" " puddled 102
" " suitable for S.B., but not for rifled ordnance 8
" " varieties of and qualities 8
Irons, priming 63
Iron, wrought 9, 10
" " advantages 10
" " bars, how made for coils 103
" " barrels for converted guns 10
" " cold short, definition of 10
" " departmental tests applied to 10
" " ductility of 10
" " elastic limit of 10, 18
" " fibrous construction and exemplification 9
" " for ordnance 9
" " how obtained.. 9
" " impurities, difficulty of removal 9
Iron, wrought, malleability 10
" " manufacture of 102
" " not thoroughly homogeneous 10

10

Page.
Iron, wrought, properties of, summarized 15
" " red short, meaning of 9
" " scarfing and welding of bars 104
" " tenacity or tensile strength 10
" " tensile strength of 18
" " tests applied to 10, 100
" " why used for exterior of our guns 10, 15
Italy, experiments with phosphor bronze 6

J

Jacket, component parts of, and manufacture 110
" ready for welding (diagram) 110
" 7″ R.M.L. gun, manufacture of 166
Jeffrey's and Britten's rifling, competitive trial of 54

K

Keep pins for R.B.L. guns 129
" " their use 128
" " pivot elevating R.M.L.. 206
Krupp guns 17, 32, 76

L

Lacquering and painting 304, 317, 330
" R.M.L. guns 182, 305
Lancaster guns 87
" " failure of 81
" " rifling, competitive trial of 54
Lands, definition of, in rifling 134
Lapping, operation of 114, 175, 179
Lathe used for turning coils, tubes, &c. 112
Le Boulengé chronograph, mode of employment 334
" " detail description of 335
Length of S.B. ordnance, how measured 51
" R.B.L. " " " 160
" R.M.L. " " " 291
Levers, for R.B.L., guns 128
" " " " their use 128
" releasing vent piece, 7″ guns 138
Limit of elasticity defined 2
Line of metal, S.B. guns 62
" horizontal and vertical axis 62
" quarter sight, S.B. guns 62
Lines, centre of gravity, and half weight 181
" defining end of bore, and rifling 181
" horizontal and vertical, on B.L. guns 129
" " " ascertained, R.M.L. guns, and their use .. 179
" of metal ascertained, R.M.L. guns.. 179
Loading side of groove 43

M

Machinery, generally as used in R.G.F. 96, 99
" operations 111
Machines, rifling 114
" " accuracy of 69
" " hand 138
" screw cutting 118
" for testing tenacity, &c., &c. 100

Page.

Malleability, definition of, as applied to metals .. 29
Manufacture, work examined during .. 119
Manufacturing operations in the case of rifled ordnance .. 99
Manufacture of R.B.L. guns .. 122
" R.M.L. 7" guns, detail of .. 162
Memo. of examination .. 294
Metal, line of .. 62
Metallurgy, rapid progress in .. 1
Metals, absolute strength .. 4
" considerations necessary as to use, and properties of .. 1
" definition of brittleness, ductility, softness, elasticity, elastic limit, tenacity, tensile strength .. 2
" physical treatment of, to be considered .. 1
" summary of, for gun purposes .. 15
" used for construction of ordnance .. 4
" used in R.G.F., table of elastic limit and tenacity .. 18
Mandrels, of what made .. 106
Marks of R.M.L. guns now manufactured .. 245
" on S.B. guns .. 59
" " " as to vents .. 61
" vent-pieces .. 126
" breech-screws .. 127
" R.B.L. guns .. 129
" R.M L. guns now manufactured .. 181
Material of inner barrel marked on muzzle .. 181
Materials used for gun construction .. 99
Millar's sights, S.B. guns .. 62
" cast iron guns .. 62
Mitre wheels as employed in machines .. 96
Moncrieff sights .. 184
" " their use .. 191
Monk's cast iron S.B. guns .. 56
Monogram, Royal, engraved .. 129
Mortars, rifled, experimental, why tried and failure .. 78
" S.B. .. 57
Mountain guns, R.M.L., manufacture commenced .. 77
Muzzle-loading, advantages and disadvantages of .. 31

N

Nasmyth's hammer .. 98
" S.B. " .. 58

O

Operations, manufacturing, as to rifled ordnance .. 99
" machine .. 111
Ordnance (*see* Carronades, Guns, Howitzers, and Mortars)
" for high angle fire .. 78
" preparation of for transport .. 306
" R.M.L., classification of .. 243
" " effects produced by .. 344
" rifled, service, table of .. 429
" S.B., and stores .. 56
" S.B. to be retained .. 65
" " abolished .. 66
Ordnance, R.M.L., dimensions, &c., table of .. 291, 292
" " proportion for siege train .. 257
Original construction R.B.L, guns .. 28, 122
" " R.M.L. " .. 161

P

Page.
Painting and lacquering 304, 317, 330
Palliser sytem of converting S.B. cast iron into rifled ordnance 233
" guns, wrought iron tubes used 25, 30
Patch metal, elevating B.L. guns 139
Patterns, standard, sealed 120
Pawl, its employment in machines 98
Pendulum, ship's 64
Penetration of 10-inch R.M.L. guns 84
" iron plates, and formula for 345
Phosphor bronze 6
Pin, iron, friction tube 64, 139, 205, 242
" " for converted guns 235
" keep, for R.B.L. guns 128, 139
" " pivot, elevating R.M.L. 206
Pinion used in machines 97
Pitch of thread, how measured 97
Pivots and pivot pieces, R.M.L. guns 206
" countersunk for 10″ and upwards 205
" the various marks of.. 207
" steel, for R.B.L. guns 140
Plain groove, and by what pieces used 45
Planing, operation of 119
Plates, elevating R.M.L. guns 205
" formulæ as to piercing 305
" guide 64
" index, and readers 208
" " " how used 197
Plate, preserving muzzle sight 9-pr. S.S. gun of 6 cwt. 6″ 3 & 8″ howitzer 205, 231
Plug, crusher and cutter 347
" vent 64
Pointer, moveable, wood scale for R.B.L. guns 141
" " " R.M.L. " 195
Porter bar 107
Powder, loss of effect beyond a certain length of bore 37
Power, mechanical, how communicated in machinery 96
" of guns, how ascertained 332
Practical considerations as to construction of ordnance 1, 19
Preponderance, S.B. ordnance 51
" rifled " and how ascertained.. 179
Preservation of ordnance on service 304
" of fittings of sights 305
Pressure on studs, grooves, &c. 42, 346
Pricker or priming iron 64, 208, 242
Proof of bursting of a 9-inch R.M.L. gun 29
" powder, how determined 345
" " attempts to reduce 39
" water test, steel tubes 164
" " S.B. guns 59
" S.B. guns 59
" R.B.L. guns and fittings 129
" R.M.L. guns 177
" " 80-ton guns 87
Projectiles, centring of 42
" form of for rifled pieces 39
" stored up work in 333, 343
Pulleys as used in machinery 98
Puddling 103
Puddled blooms 103

Page.

Q

Quadrants .. 184
Quadrants, their use .. 198
Quadrant planes on howitzers .. 182, 260
Quarter-sight line and scales, S.B. gun .. 62

R

Racks and pinions as used in machinery .. 97, 115
„ elevating .. 207
„ „ wrench for fixing .. 142
Range tables, how made out .. 385
„ „ 12″·5, 38 tons, R.M.L. .. 417, 418
„ „ 12″ 35 „ „ .. 416
„ „ 12″ 25 „ „ .. 414, 415
„ „ 11″ 25 „ „ .. 413
„ 10″ 18 „ „ .. 411, 412
„ „ 9″ 12 „ „ .. 408, 410
„ „ 8″ 9 „ „ .. 405, 407
„ „ 7″ 7 „ „ .. 403, 404
„ „ 7″ 6½ „ „ .. 400, 402
„ „ 7″ 90 cwt. „ .. 399
„ „ $\frac{12}{8}$-pr., converted, 5 cwt., R.M.L. .. 398
„ „ 8″ howitzer, 46 cwt. .. 422, 423
„ „ 64-pr., 64 cwt. (12 lbs.) .. 397
„ „ 64-pr., 64 cwt. (10 lbs. charge) .. 396
„ „ 64-pr., 64 cwt. (8 lbs. charge) .. 395
„ „ 64-pr., 58 and 71 cwts., converted (8 lbs. charge) .. 395
„ „ 40-pr., 35 cwt., Mark II. .. 394
„ „ 25-pr.. 18 „ .. 393
„ „ 16-pr., 12 „ .. 392
„ „ 9-pr., 8 „ .. 390
„ „ 7-pr., 150 lbs. (steel) .. 386
„ „ „ 200 lbs. (steel) .. 388, 389
Ratchet wheel, as used in machinery .. 98
„ brace .. 322
Readers, for index plates .. 184, 198, 208
Repairs at certain stations .. 314
„ general, as to guns in the field .. 306
„ carried out in an arsenal .. 316
Replacing broken clamping screw for sight .. 312
Re-tubing a gun .. 316
Returns, annual, ordnance, to be furnished .. 371
„ „ rifled ordnance „ .. 376
„ „ S.B. „ .. 375
Re-venting converted guns .. 316
„ R.B.L. „ .. 309
„ S.B. or R.M.L. guns .. 314
„ with cone vent .. 315
„ with through vent .. 316
Rifling, end of, marked on gun surface .. 181
„ French, competitive trial of .. 54
„ „ modification introduced .. 75
„ „ modified .. 45
„ Jeffrey's and Britten's, competitive trial of .. 54
„ Lancaster's, „ „ .. 54
„ machine in detail, and its action .. 114
„ modern polygroove .. 46
„ operation of .. 114
„ plain groove .. 235
„ R.B.L. and twist .. 124
Rifling, 7″ R.M.L., detail .. 175

Page.
Rifling, Scott's, competitive trial .. 54
" shunt, " " .. 55
" " superseded .. 83
" Woolwich, introduced .. 83
Ring, copper, for vent-piece, R.B.L. .. 124, 126, 140
" indicator " " .. 128, 140
" tappet " " .. 128, 140
" trunnion, manufacture of .. 107, 113
" " R.B.L. guns .. 123
" " welded to breech coil .. 29
Robin's, ballistic pendulum .. 334
Rodman guns.. .. 24
" pressure gauge .. 346
Rolling bar, iron .. 103
Rotation necessary, how ascertained .. 343
" of projectiles, pressure due to, on studs .. 42
Russia, experiments with phosphor bronze .. 6
" present field pieces .. 93

S

Saddles, metal, for 7-inch R.B.L. guns .. 140
Sand, why used when welding .. 104
Scale, hanging, for siege ordnance .. 184
" " how used .. 200
" heel S.S. R.M.L. guns .. 181
" quarter sight, S.B. guns .. 62
" tangent, bronze S.B. ordnance .. 63
" " R.B.L. ordnance .. 130
" " wood, L.S. S.B. cast iron .. 63
" " generally (*see* Sights).
" wood, R.M.L. guns .. 184, 195, 241
" " " " adjustment of .. 195
" " " " table showing marks of .. 228
" " side R.B.L. guns .. 134, 140
" " S.B. " .. 63
Scrap, shavings .. 107
Screw, breech, R B.L. guns, manufacture, material, &c. .. 127
" cascable, with what guns used .. 161
" " testing as to accuracy of fitting .. 167
" copper, set .. 132, 141, 208
" cutting in machine .. 118
" fixing .. 142, 208, 242
" " brackets and sockets .. 241
" holes, for fittings, R.M.L. .. 182
" pitch defined, right and left hand, double thread, &c. .. 97
" preserving .. 142, 208, 242, 305
" thread .. 97
Scoring, its cause and prevention by gas checks .. 299
Scott's rifling, competitive trial of .. 53
Sentence as to bore of rifled ordnance .. 299
" " vents " " .. 301
Shackles, vent-piece, R.B.L. guns .. 126
Shafting used for machinery .. 96
Shoulders on A. tubes .. 174
Shot bearers, 7-inch R.B.L. guns .. 134
Shrinkage, defined .. 26
Shrinking, operation of heat required for, &c. .. 108
rough jacket together .. 110
" R.B.L. gun together .. 124
" R.M.L. 7-inch gun together .. 167
Shunt, rifling, competitive trial of .. 55

Page.

Shunt, rifling, introduced .. 81
„ „ objections to .. 45
Siege R.M.L. ordnance .. 78, 243, 257
Sighting converted guns .. 239
„ generally .. 51
„ instrument for R.B.L. guns .. 137
„ plates, wood .. 310
„ R.B.L. guns, deflection allowed .. 130
„ R.M.L. „ .. 180
„ S.B. ordnance tools for .. 320
Sights, angle of permanent deflection of, in rifled ordnance, how determined .. 52
„ barrel-headed, for R.B.L. guns .. 131
„ „ „ „ graduations .. 132
„ blueing and bronzing .. 377
„ centre, hind, fore and muzzle, general arrangement of, for 64-pr. .. 240
„ „ „ „ R.M.L. guns 187
„ „ „ table of Marks I. and upwards .. 225
„ „ „ lengthened .. 187
„ „ „ socket deepened for .. 378
„ chase, and how used .. 184, 193
„ clamping arrangement for .. 182
„ clamps, moveable .. 183, 135
„ drilling of sockets, R.M.L. guns .. 180
„ drop .. 133, 188
„ fore, adjusting 7-pr., 9-pr., and 16-pr. .. 312
„ front or fore, generally .. 51
„ „ R.M.L. guns .. 183
„ graduations, above or below 40-inch radius .. 196
„ hexagonal, gun metal, for 7-inch 72 cwt. and 12-pr. R.B.L. guns .. 130
„ hind .. 184
„ L.S. and S.S. for R.B.L. guns .. 130, 134
„ „ „ sliding leaf, R.B.L. guns .. 132
„ preservation on service .. 305
„ rectangular steel bars, R.B.L. guns .. 130
„ screw .. 133, 189
„ „ adjusting new leaf on service .. 310
„ „ adjustment .. 133
„ tangent, 80-pr., 5 tons .. 239
„ „ 64-pr., 71 cwt. .. 239
„ „ L.S. or S.S. .. 184
„ „ "set" of, marked on gun .. 181, 184
„ „ table of latest patterns R.B.L. .. 143
„ „ „ Mark I. and upwards R.B.L. .. 146
„ „ „ „ „ R.M.L. .. 212
„ „ „ „ „ „ .. 212, 218
„ trunnion 80-pr., 5 tons, and 64 pr., 71 cwts, converted guns .. 239
„ „ 16-pr. R.M.L. .. 183, 189
„ S.B ordnance .. 62
„ „ „ Millar's .. 62
„ „ „ „ adjustment of .. 314
„ R.M.L. guns, Moncrieff .. 184
„ „ „ „ how used and adjusted .. 191
„ R.B.L. „ .. 142
„ R.M.L. „ .. 183, 208
„ „ „ 40-pr. and below .. 183
„ „ „ 64-pr. and upwards .. 183
„ „ Howitzer, 8-inch, and 6·3 inch .. 183, 186, 189
„ „ „ „ exceptional .. 183
„ rifled ordnance generally .. 51
„ R.M.L., special .. 183
„ „ telescope how used at present .. 195
„ „ turret .. 184
„ „ „ how used .. 190
Slabs, manufacture of .. 107
Slides, tangent, wood or brass, S.S. S.B. ordnance .. 63
Slotting, operation of .. 119
Slots, vent .. 124

Page.
Sockets, metal, centre hind sight, converted guns 242
„ „ R.B.L. guns.. 142
„ „ sight 206
„ „ „ fitting R.M.L. guns 180
Softness, definition of, as applied to metals 2
Solid forgings, how made 107
„ „ barrels made of 123
Space, unrifled in bore of R.M.L. guns, line denoting 181
Spanner box, foresight, use of 209
„ trunnion sight 209
Spikes used, description of 64
Splay of grooves in heavy guns 176
Steel, advantages of, for gun construction 13
„ Bessemer 12
„ cast, used in R.G.F., and definition of 12
„ crucible.. 12
„ definition of 11
„ elastic limit, and tests applied 100
„ fibrous structure of 13
„ guns, foreign, their failure 17
„ hardening properties 13
„ puddled or cement 12
„ Siemen's or Siemen-Martin's.. 12
„ soft, hard, and hardened, defined 2
„ summary of qualities required for our guns 13
„ tests applied to, in R.G.F. 15, 100
„ tenacity and elastic limit, variation of 13
„ tube for R.M.L. guns, manufacture of 168
„ uncertainty and remedies tried 13
„ used by us for gun barrels 30
„ „ for ordnance 99
Stops in bore for projectile 46, 260
Stores for R.B.L. guns 130
„ „ „ „ table of 145
„ „ R.M.L. „ 201, 231
„ „ „ „ table of 231
„ „ „ „ converted 241
„ „ S.B. „ 63
Straight-edge for R.B.L. guns 142, 303
Studs, pressure, on, with uniform and increasing twist 42
„ trunnion 209
Sweden, employment of cast iron for ordnance 8

T

Tables (see p. v.).
Taps for screw cutting 118
Tappet rings, R.B.L. guns 128, 140
Tenacity, cast iron 8
„ definition of 2
„ how measured 3
„ of metals used in R.G.F. 18
„ of steel as supplied to R.G.F 100
„ of wrought iron 9
Tensile strength, definition of 2, 4
Tension, circumferential 20
„ due to shrinkage 28
„ initial 2
Tests of departmental iron 104
„ steel tube by water 164
Testing machine for metals 100
Theoretical construction of ordnance 1
Thread, pitch of, how mechanically obtained 97
„ used with breech-screw guns 127

Page.

Thread used with cascables .. 167
Tooth wheels, used in machinery .. 96
Toughness, definition of .. 2
Toughening steel tube of 7" R.M.L. in oil .. 163
Transport, preparing ordnance for .. 306
Trunnion, marks on .. 181
„ rings, boring of .. 113
„ „ from solid forgings .. 107
„ „ manufacture of .. 107
„ „ R.B.L. guns .. 123
„ studs, and their use .. 209
Tubes, steel, variation in thickness .. 247
„ A., 7" R.M.L. manufacture .. 163
„ „ materials of, stamped on muzzle .. 181
„ „ shoulders on .. 174
„ B., 7" R.M.L. manufacture .. 165
Turning, operation of .. 112
„ R.B.L. guns .. 124
Turret sights .. 184
„ how used .. 190
Twist of rifling R.B.L. guns .. 124
„ increasing .. 116
„ uniform .. 116

U

Uchatius, bronze, steel, and experiments .. 6
„ guns, Austrian .. 6
Uniting two coils end to end .. 108
Uniform twist of rifling .. 40, 116

V

Values, tables of, $\frac{d^2}{w}.s$, Professor Bashforth .. 366, 367
„ „ $\frac{d^2}{w}.t$, „ .. 368, 369
Velocity, angular, imparted by rifling .. 41
„ deduced from M.V. .. 339
„ linear, of rotation .. 41
„ muzzle, mode of arriving at .. 334
„ obtained at any distance .. 339
„ rotation .. 343
„ taken by Le Boulengé Chronograph .. 334
„ table of, with various charges .. 354, 355
„ „ heavy guns .. 362
„ translation .. 334
„ 80-ton gun .. 87
Vent bits, Armstrong, for B.L. guns .. 141
„ bush, Frazer's removable .. 51
„ bushes, general question of .. 48, 50
„ channel of vent pieces .. 175
Vents, regulations as to sentencing .. 301
„ marks as to S.B. ordnance .. 61
Venting converted guns exceptional .. 286
„ „ „ tools for .. 287, 315
„ generally .. 48
„ rifled .. 48
„ R.M.L. before proof .. 177
„ R.B.L. 7" guns, through the body .. 125
„ S.B. ordnance .. 48, 60
„ tools, list A .. 322
„ „ list B distribution .. 326

	Page.
Venting 9″ R.M.L. and upwards	180
„ 8″ „ and under, similar to S.B. ordnance	180
Vent-piece, R.B.L. guns	125, 132
„ component parts	126
„ slot R.B.L. guns	124
Vertical line of axis, R.B.L. guns	130
„ „ „ R.M.L. „	179
„ „ „ S.B „	62
Viewing	119

W

Washer, iron trunnion for 12″·5 guns	210
Water escape, R.B.L. guns	124
„ proof, S.B. guns	59
„ „ R. guns	177
„ shrinking, why used	109
„ test for toughened steel tubes	80, 164
„ used in certain cases during hammering	106
Wedge guns, R.B.L..	81, 159
Weight, actual, of rifled guns, how ascertained	186
Welding of iron, sand used, scarfing	104
„ jacket, heavy guns	110
„ properties of wrought iron	9
„ „ steel	12
Wheels, used in machinery, bevil, change, mitre, spur, tooth, worm	96
„ change, for altering direction of motion	97
Windage, S.B. ordnance	58
Woolwich groove, and by what pieces used	45
Work done upon gun	346
„ stored up in projectiles	333, 344
Wrench, fixing elevating racks	142, 210
„ pin, friction tube	211, 243
„ removing elevating plates M. IV.	211

ORDNANCE BRONZE R.M.L. 7-PR. 200 LBS.

Scale 3/4 inch - 1 foot.

MARK III.

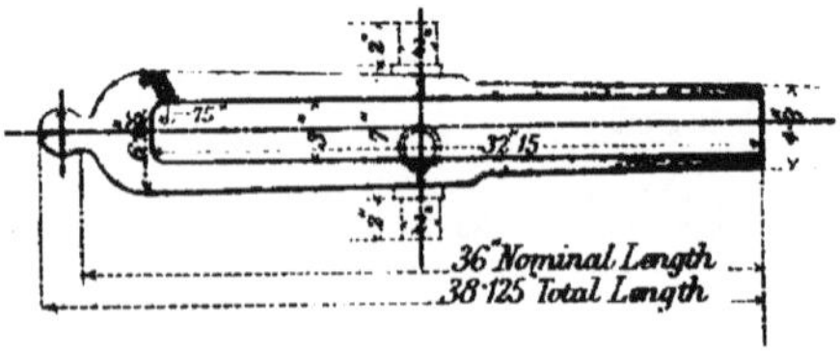

ORDNANCE STEEL R.M.L. 7-PR. 150 LBS.

MARK III.

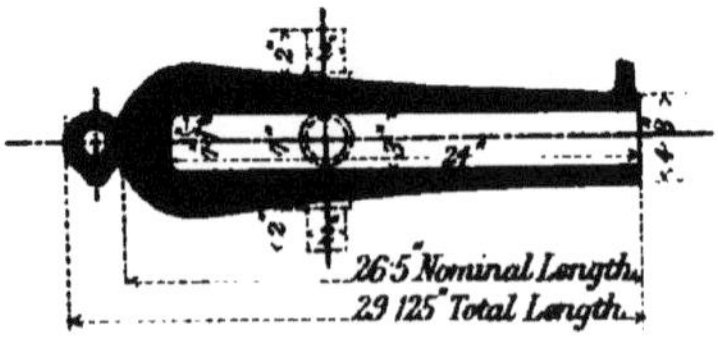

ORDNANCE STEEL R.M.L. 7-PR. 200 LBS.

MARK IV.

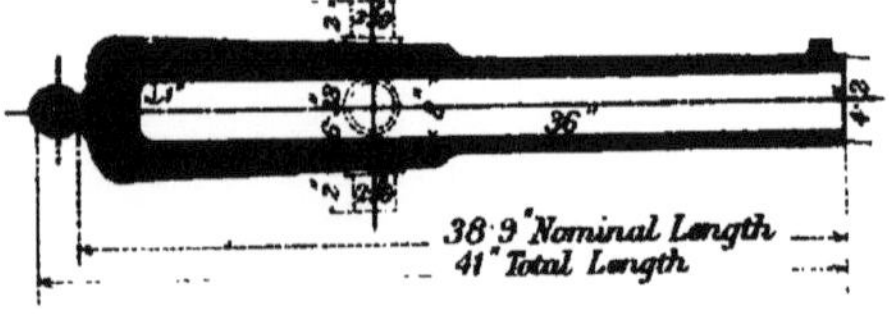

Dangerfield Lith 24 Bedford St Covent Garden

ORDNANCE WROUGHT IRON R.M.L 9-Pr. 6 CWT.

Scale ¾ inch–1 foot

MARK I S.S.

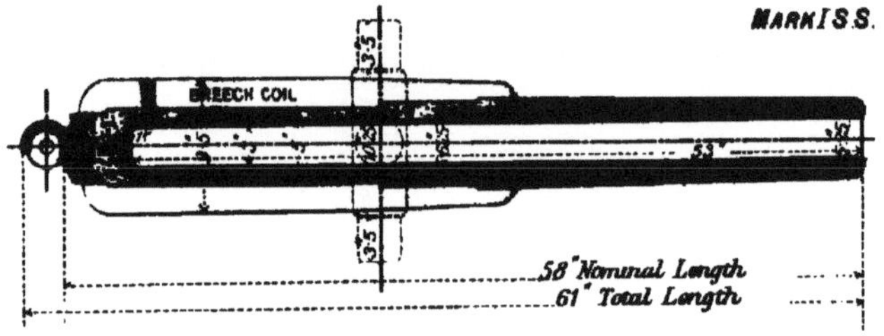

ORDNANCE WROUGHT IRON R.M.L 9-Pr. 8 CWT.

MARK I L.S.

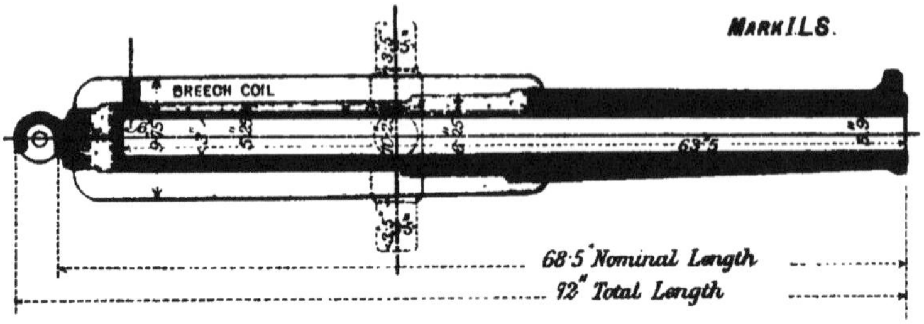

9 Pr. 8 Cwt. Mark II S.S.

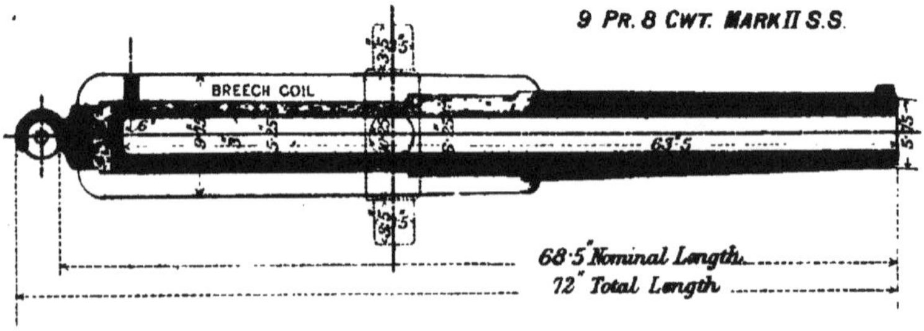

9 Pr. 6 Cwt. Mark III L.S.

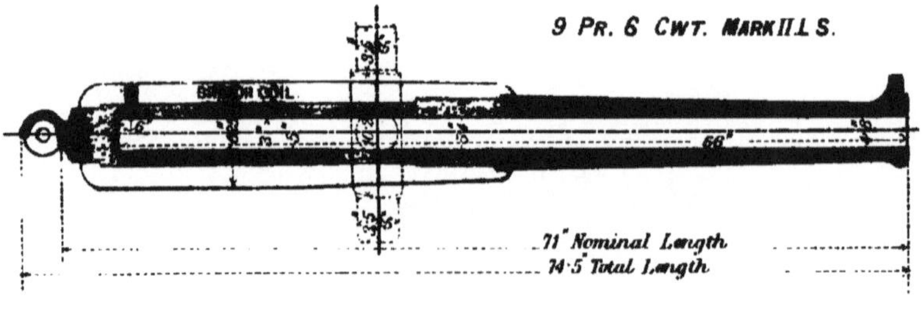

ORDNANCE WROUGHT IRON R.M.L. 16-PR. 12 CWT. L.S.

Scale ¾ inch–1 foot.

MARK I.

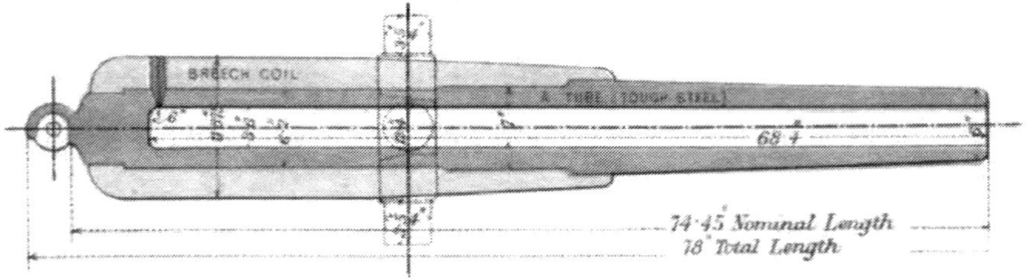

ORDNANCE WROUGHT IRON R.M.L. 25-PR. 18 CWT.

MARK II.

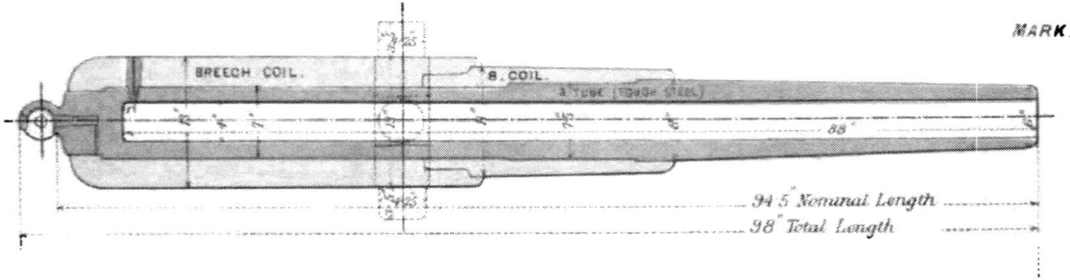

40 P[R] WROT IRON GUN 34 CWT.

MARK I.

Scale ¾ inch - 1 Foot.

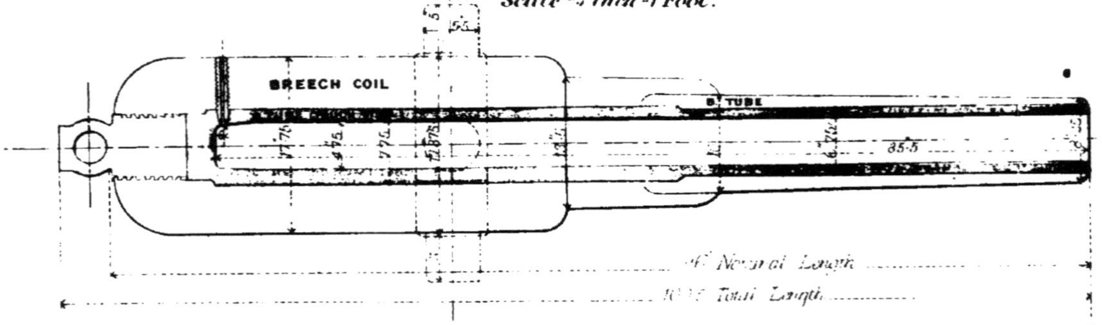

40 P[R] WROT IRON GUN 35 CWT.

MARK II.

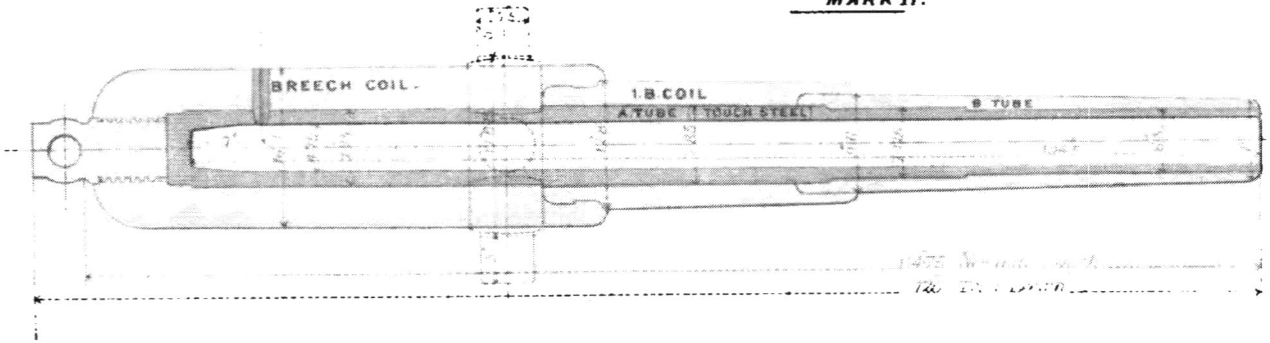

PL. IV.

Pl. V.

WROT IRON RIFLED MUZZLE LOADING 64 PR. GUNS OF 64 CWT.

Scale 3/8 Inch – 1 Foot.

MARK I.

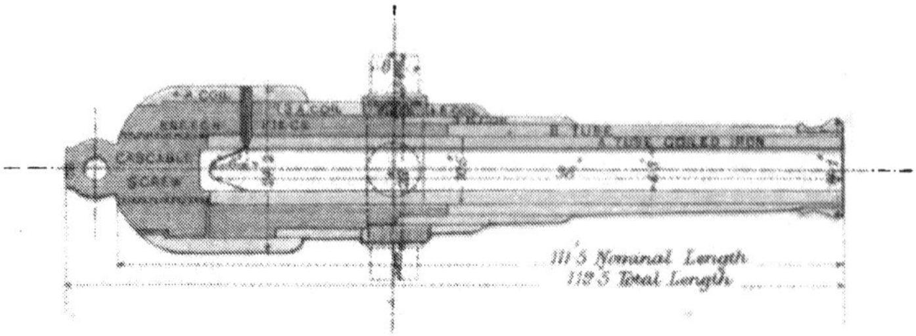

MARK II.

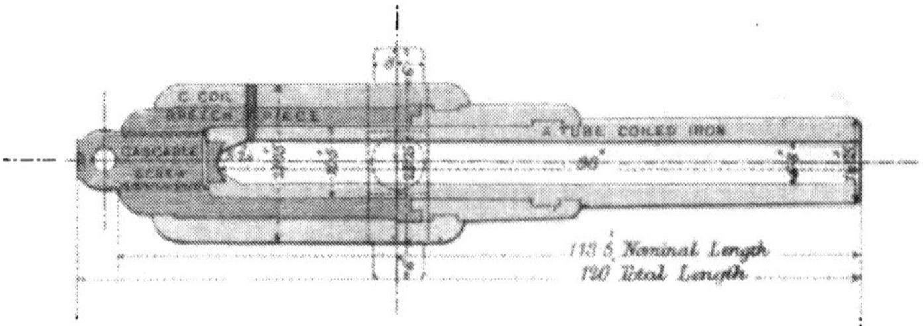

MARK III.

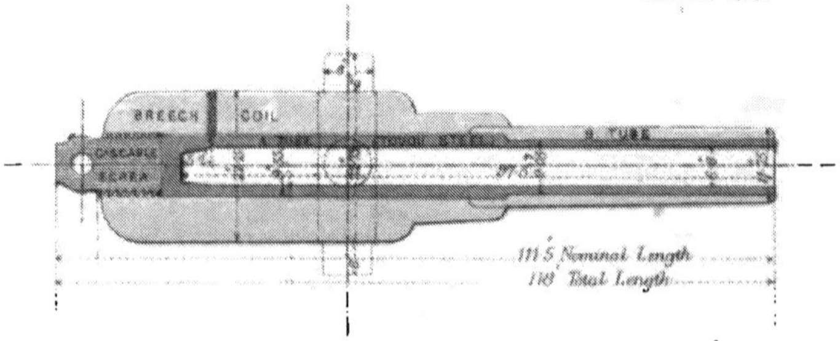

Dangerfield 22 Bedford St Covent Garden

Pl. VI.

CONVERTED RIFLED M. L. GUNS.

Scale ⅜ Inch – 1 Foot.

64 Pr 58 CWT. FROM 32 Pr 58 CWT.

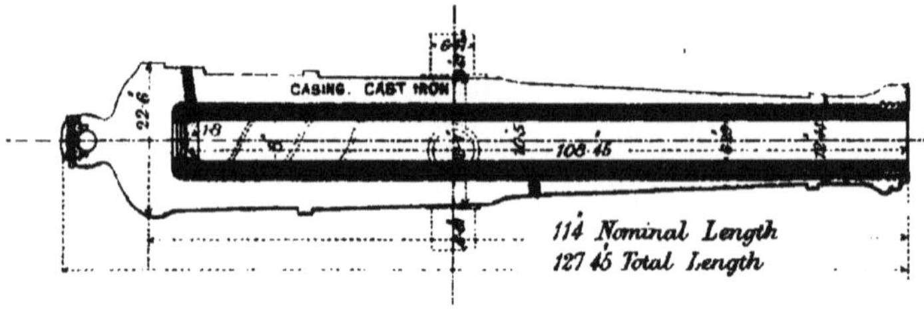

64 Pr 71 CWT. FROM 8 In. 65 CWT.

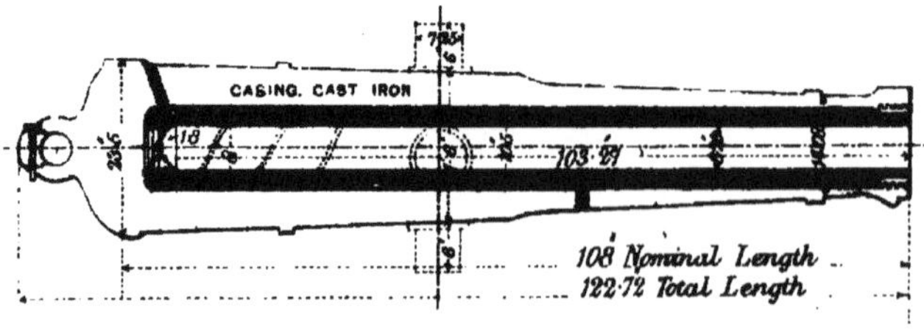

80 Pr 5 TONS FROM 68 Pr 95 CWT.

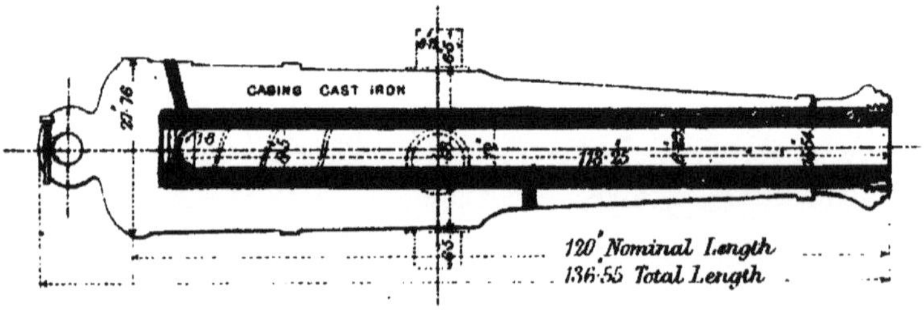

Dangerfield. 22 Bedford St Covent Garden.

Royal Gun Factories.

ORDNANCE WROUGHT IRON RIFLED M. L. HOWITZER 6·3 INCH 18 CWT.

WEIGHT 17 CWT. 2 QRS. 21 LBS.

PREPONDERANCE ·8 .

Scale 1 inch = 1 foot.

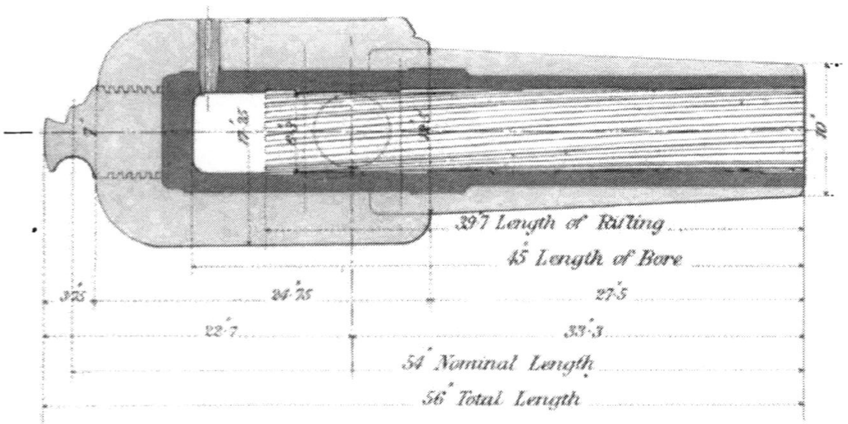

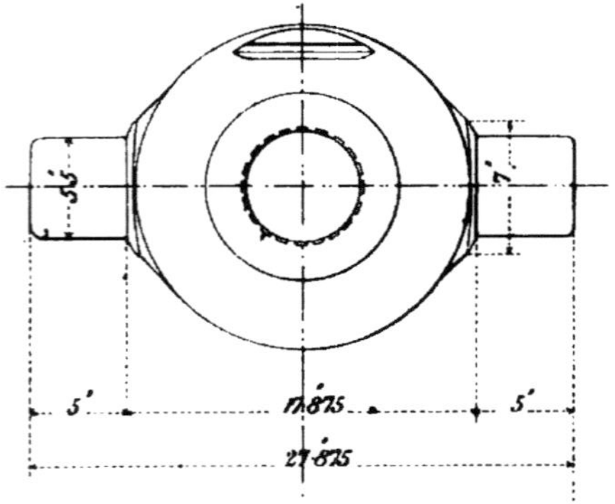

SECTION OF GROOVE.

FULL SIZE.

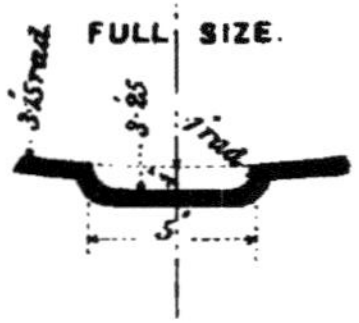

Number of Grooves 20.

Rifling, an increasing twist, from 1 turn in 100 Calibres at Breech, to 1 turn in 35 Calibres at Muzzle.

Pl. VII.

ORDNANCE WROT IRON MUZZLE LOADING HOWITZER 8 INCH 46 CWT. R. MARK I.

WEIGHT.................46 CWT. 0 QR. 12 LB.
PREPONDERANCE......2 . 0 . 12 .

Scale 1 inch – 1 foot.

9 . 12 . 72 73/3/412

SECTION OF GROOVE.

Full size.

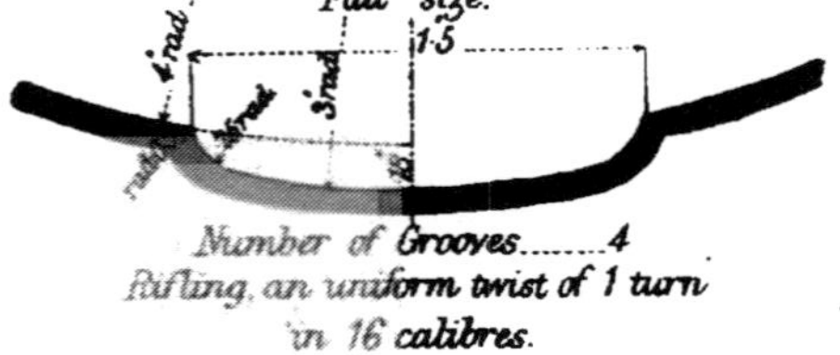

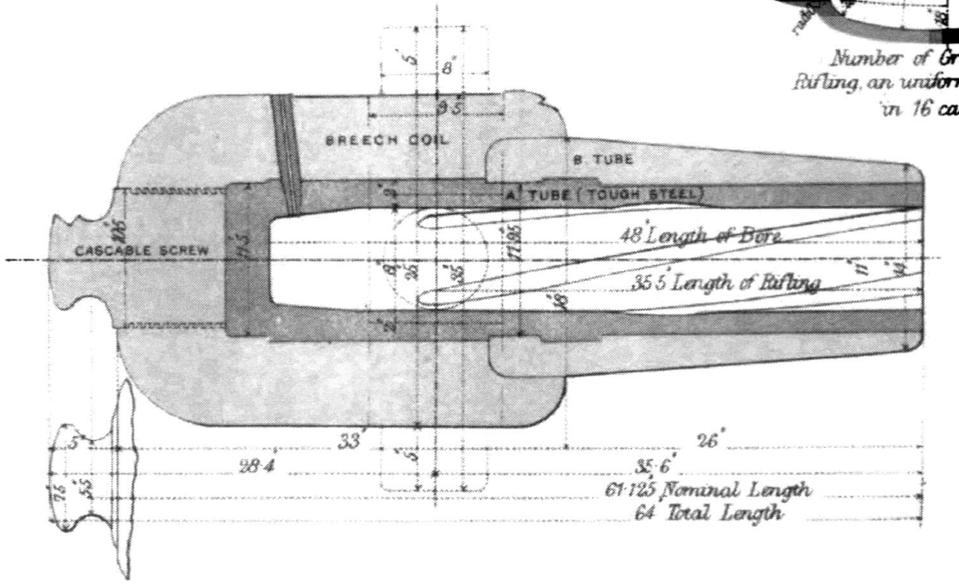

Pl. VIII.

Pl. IX.

WROT IRON RIFLED MUZZLE LOADING 7 INCH GUNS OF 7 TONS.

Scale, ⅜ Inch – 1 Foot.

MARK I.

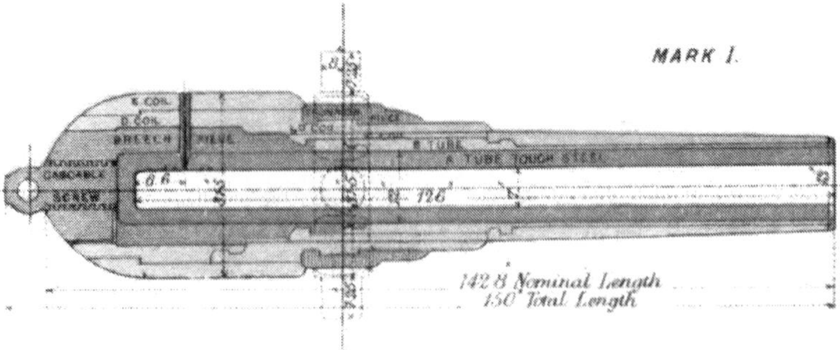

MARK II.

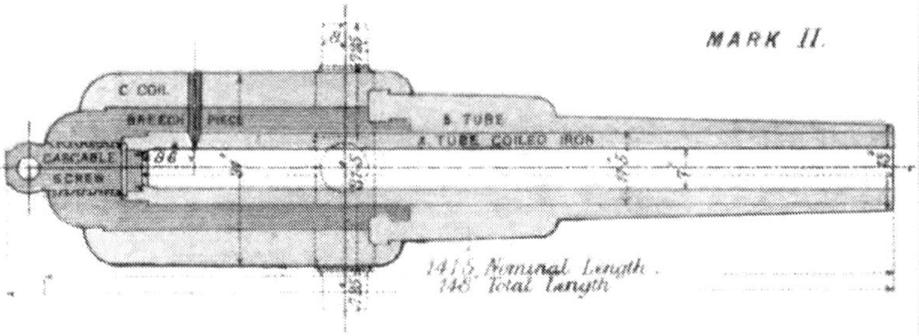

MARK III.

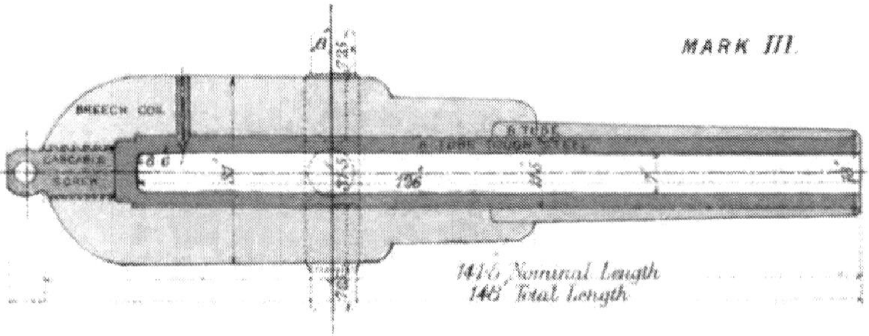

Pl. X.

WROT IRON RIFLED MUZZLE LOADING 7 INCH GUNS OF 6½ TONS.

Scale ⅜ Inch – 1 Foot.

MARK I.

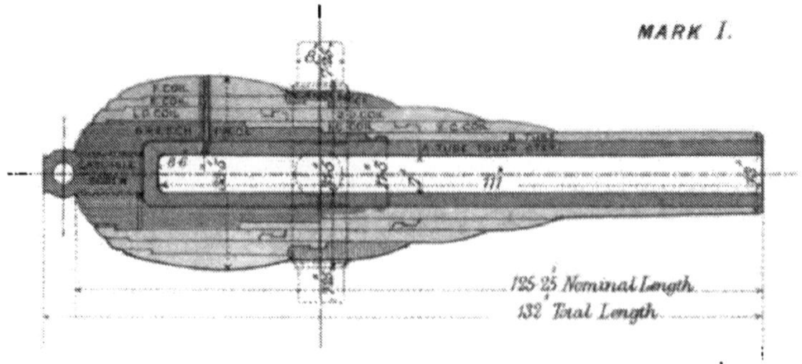

MARK II.

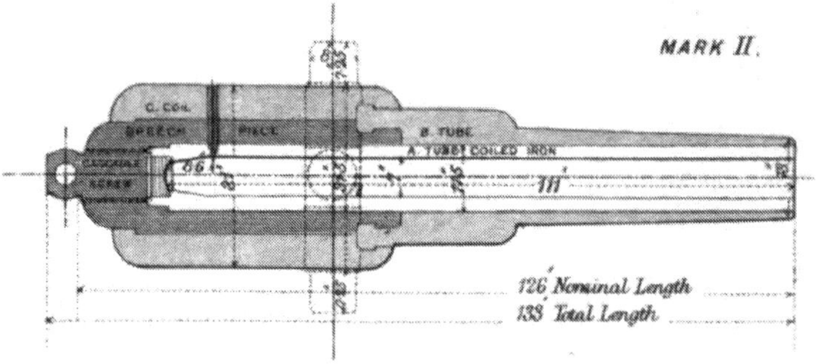

MARK III.

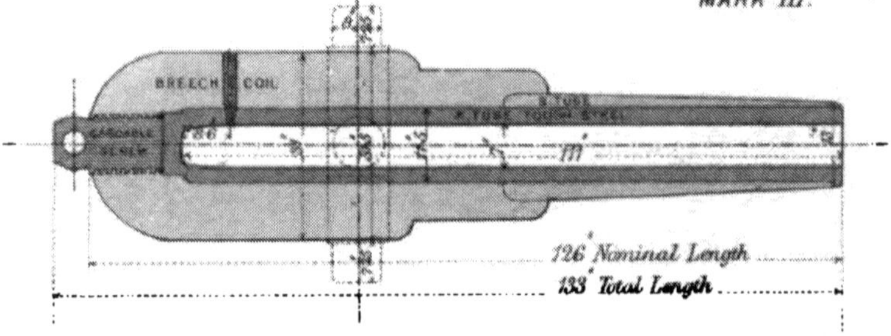

Day & Son 22 Bedford St Covent Garden

7 INCH RIFIED MUZZLE LOADING GUNS.

Scale ½ Inch - 1 Foot.

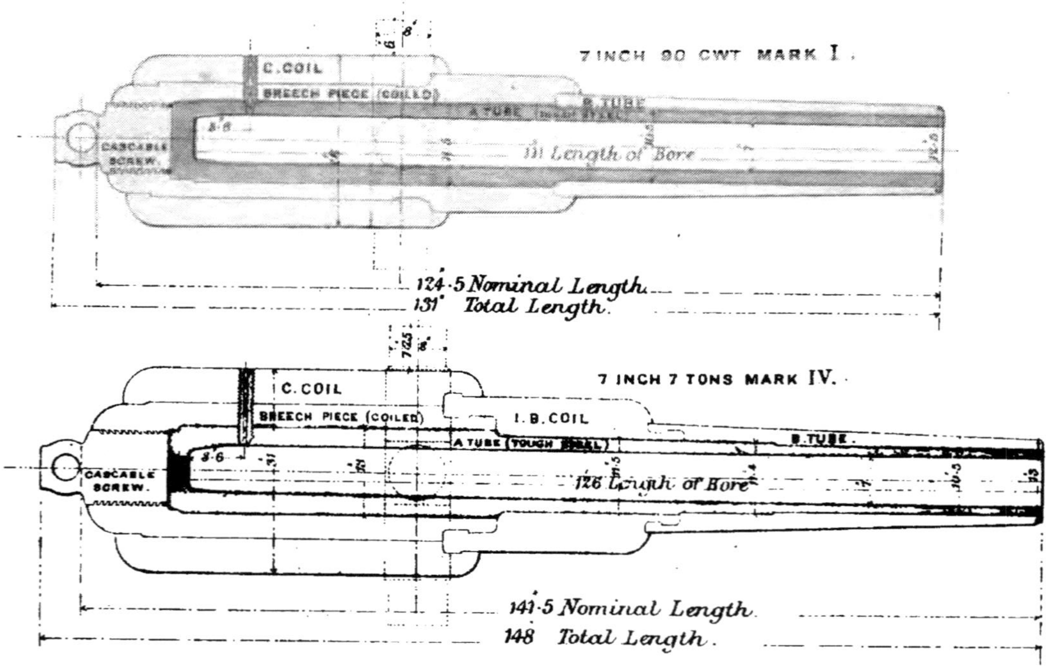

Dangerfield Lith 22, Bedford St. Covent Gard.

WROT IRON RIFLED MUZZLE LOADING 8 INCH GUNS OF 9 TONS.

Scale ⅜ Inch – 1 Foot.

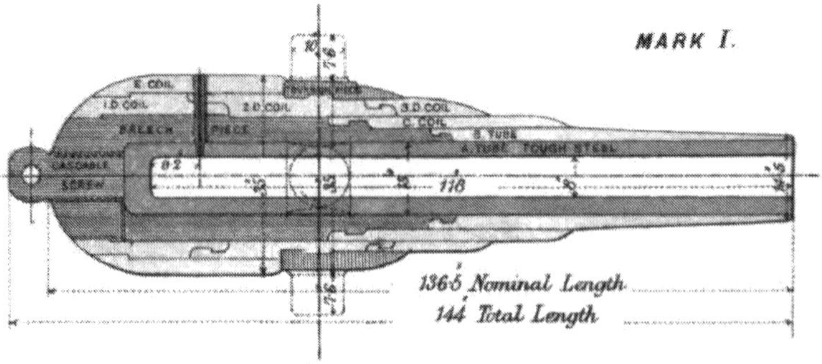

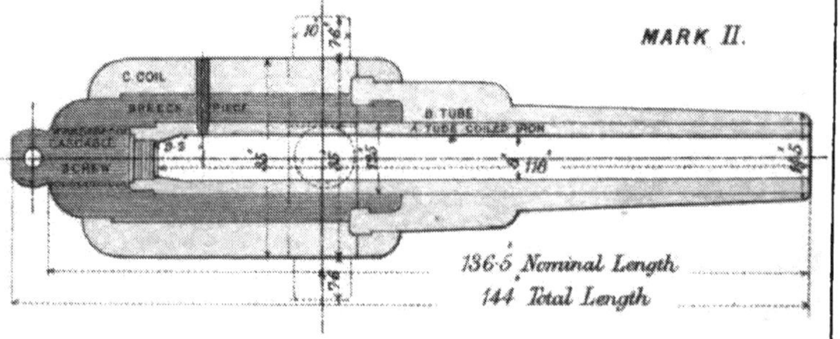

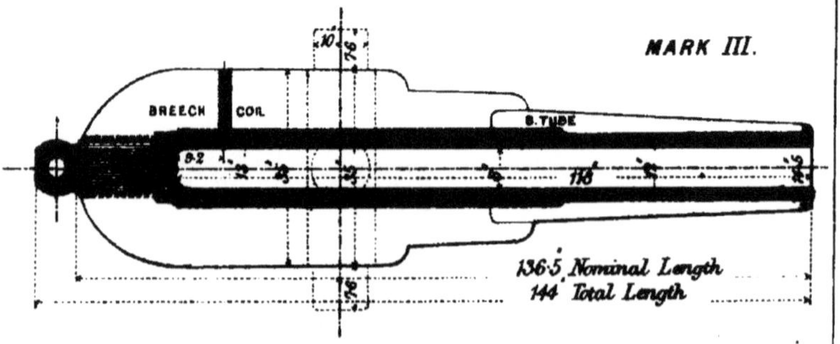

Dangerfield. 22 Bedford St Covent Garden

Pl. XIII.

WROT IRON RIFLED MUZZLE LOADING 9 INCH GUNS OF 12 TONS.

Scale ⅝ Inch – 1 Foot.

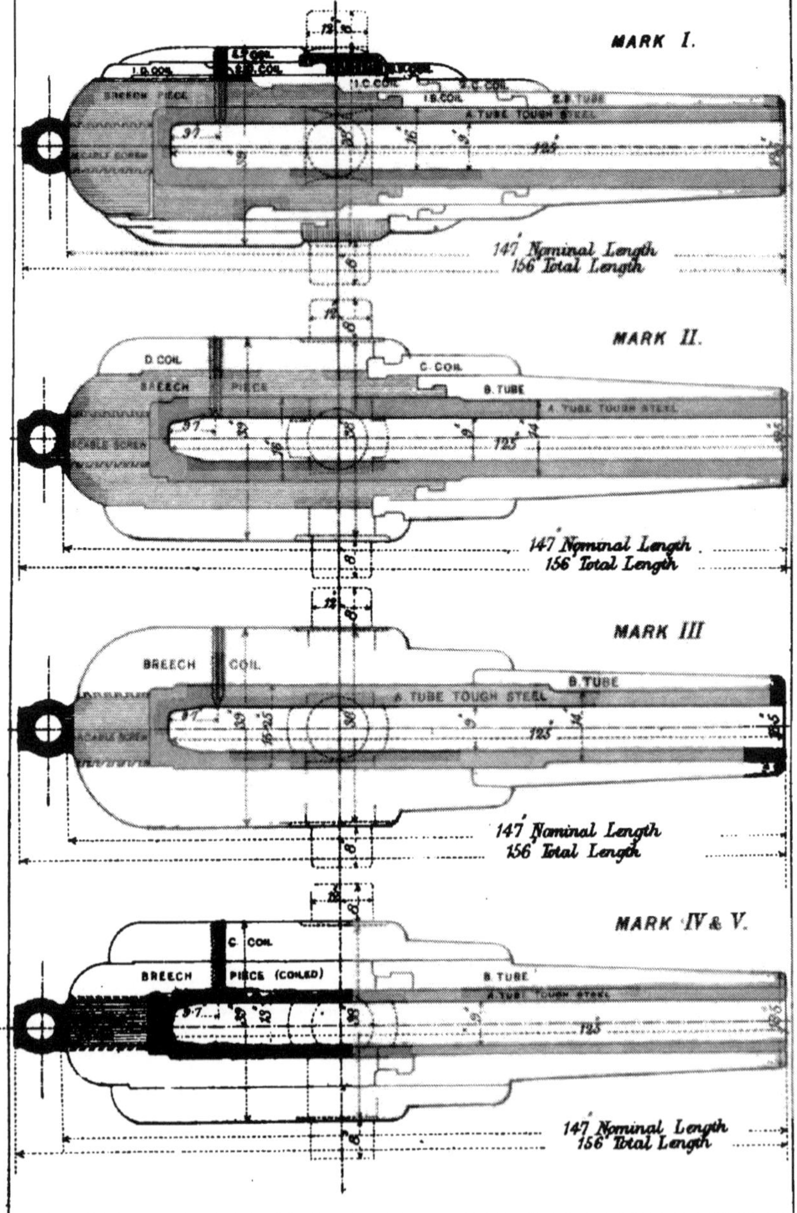

Dangerfield 22 Bedford St Covent Garden

Pl. XIV.

WROUGHT IRON RIFLED MUZZLE-LOADING 10 INCH GUN 18 TONS.

Scale ⅜ inch - 1 Foot.

Note. These Guns when intended for Broadside use in the Navy and for Land Service are vented upon the right-hand side at an angle of 45° with the vertical axis of gun. But when employed in double gun turrets are vented right and left i.e. upon the outsides of their respective positions.

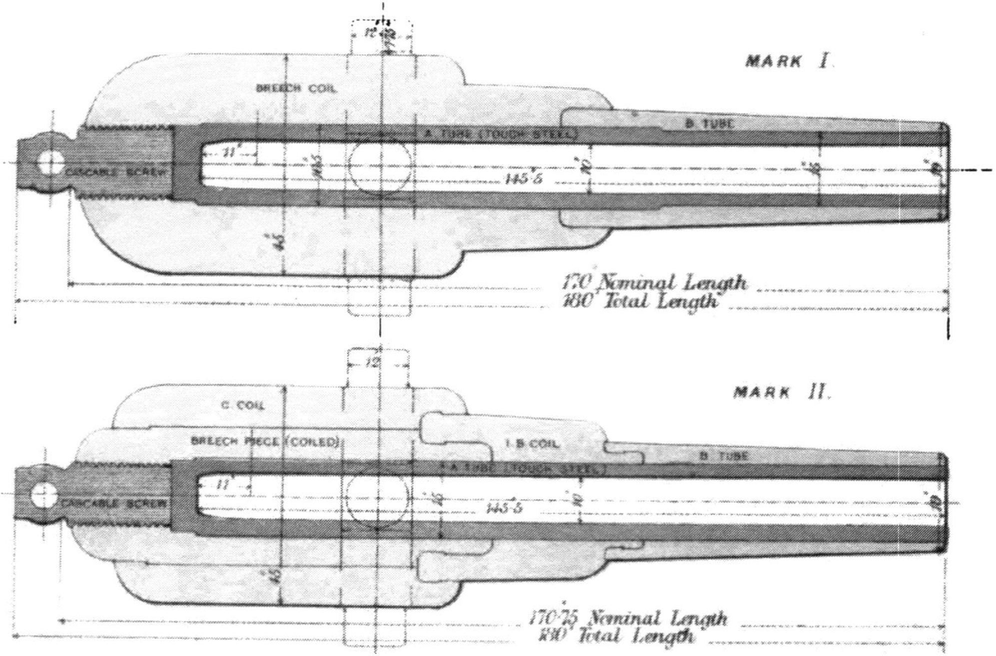

Pl. XV.

WROUGHT IRON RIFLED MUZZLE-LOADING 11 INCH GUN 25 TONS.

Scale ⅝ inch - 1 Foot.

Note. These Guns when intended for Broadside use in the Navy and for Land Service are vented upon the right-hand side at an angle of 45° with the vertical axis of gun. But when employed in double gun turrets are vented right and left i.e. upon the outsides of their respective positions.

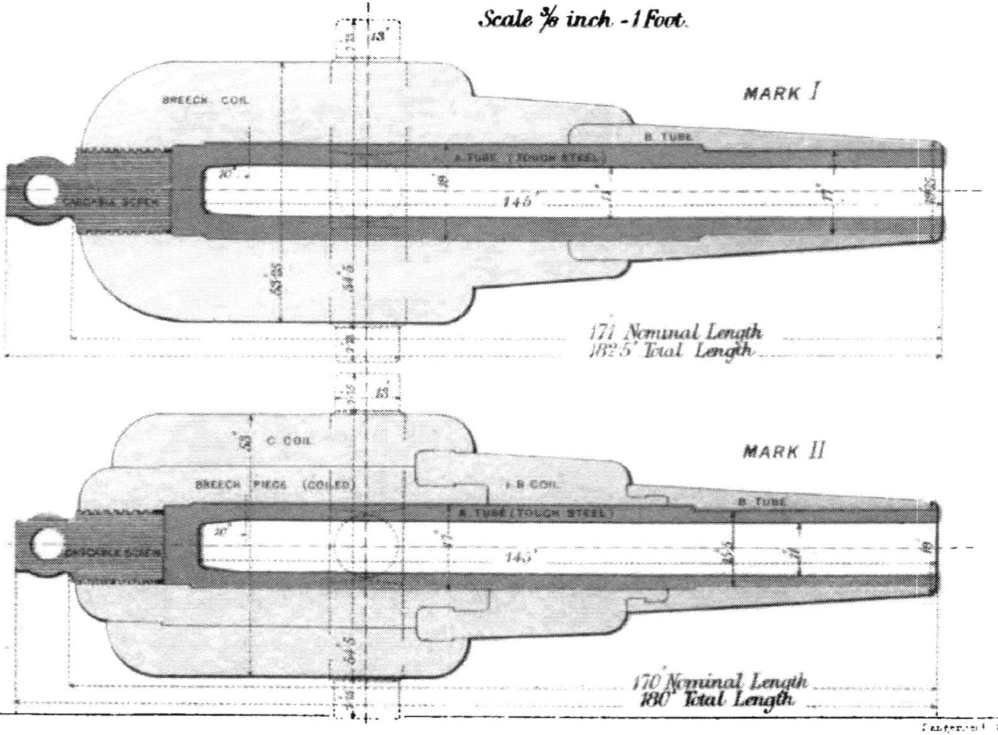

Pl. XVI.

WROUGHT IRON RIFLED MUZZLE-LOADING 12 INCH GUN 25 TONS.

Scale ⅝ inch - 1 Foot.

Note. These Guns when intended for Broadside use in the Navy and for Land Service are vented upon the right-hand side at an angle of 45° with the vertical axis of gun. But when employed in double gun turrets are vented right and left i.e. upon the outsides of their respective positions.

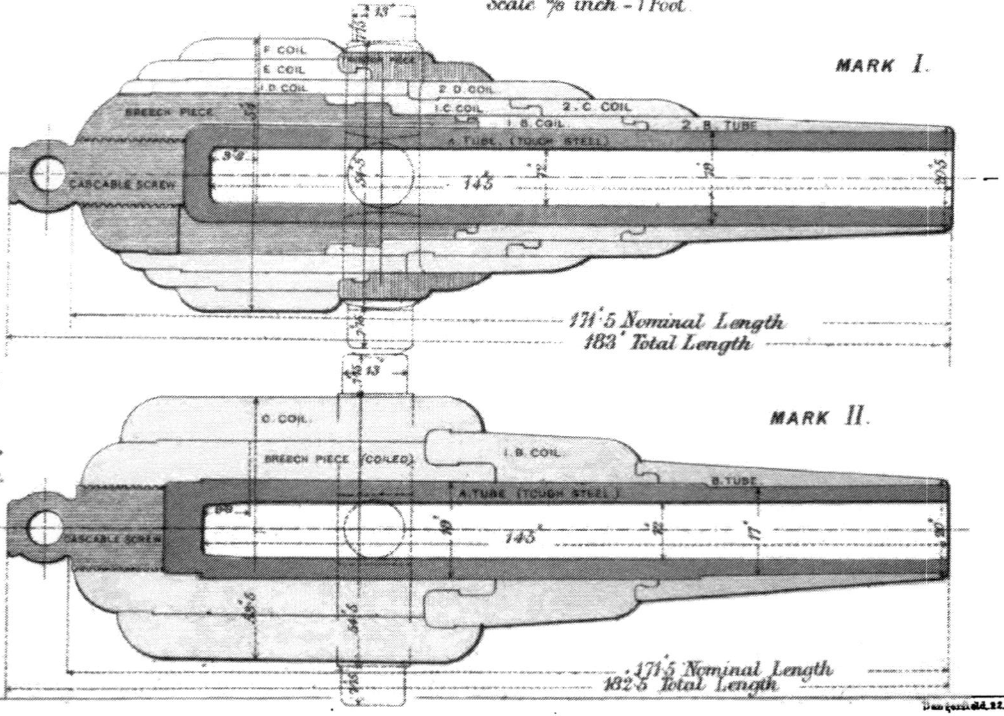

Dangerfield, 22 Bedford St Covent Garden.

WROUGHT IRON RIFLED MUZZLE-LOADING 12 INCH GUN 35 TONS.

Scale 3/8 inch - 1 Foot.

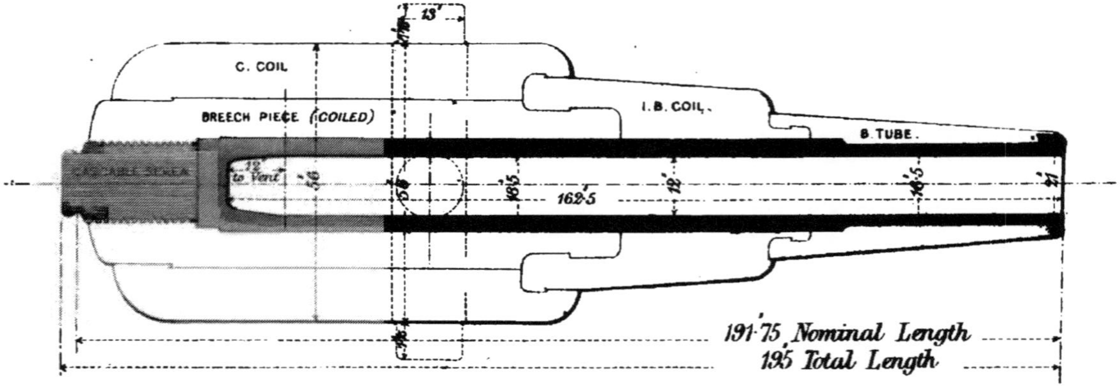

Pl. XVII.

Dangerfield. 22 Bedford St Covent Garden

ORDNANCE WROUGHT IRON RIFLED M. L. 12·5 INCH 38 TONS MARK 1.

§ 2792.

7, 11, 74 $\frac{74}{\frac{2}{3259}}$

	TONS	CWT	QR	LB
WEIGHT	38	0	0	0
PREPONDERANCE	NIL			

Scale ⅜ inch – 1 Foot.

SECTION OF GROOVE

Full size.

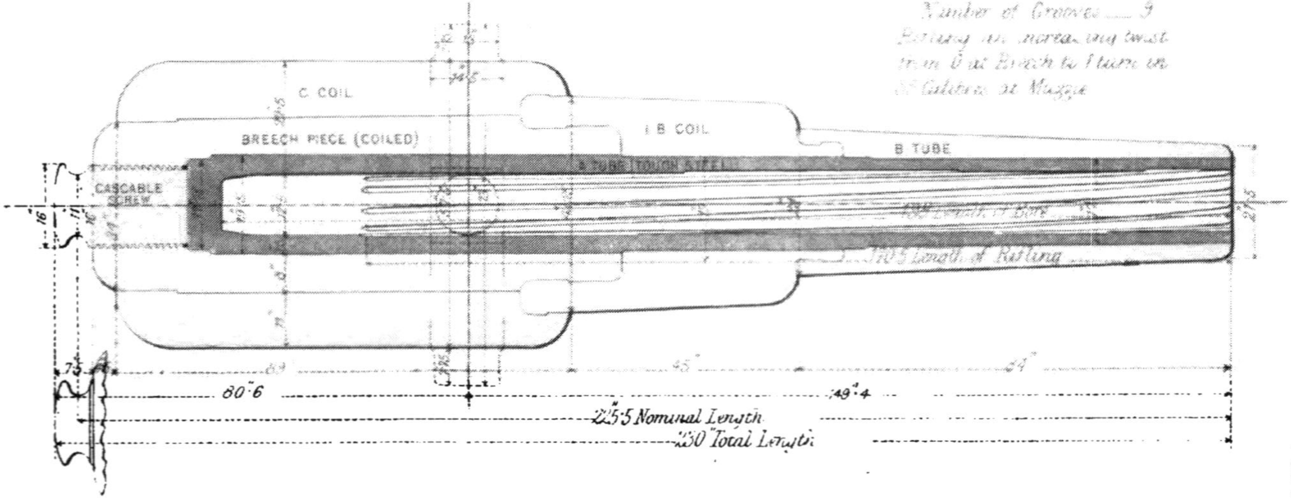

Pl. XVIII.

ORDNANCE WROT IRON RIFLED MUZZLE-LOADING 16 IN 80 TONS.

Scale 12/18 inch = 1 foot.

ESTIMATED	WEIGHT	81 TONS.
D°	PREPONDERANCE	NIL

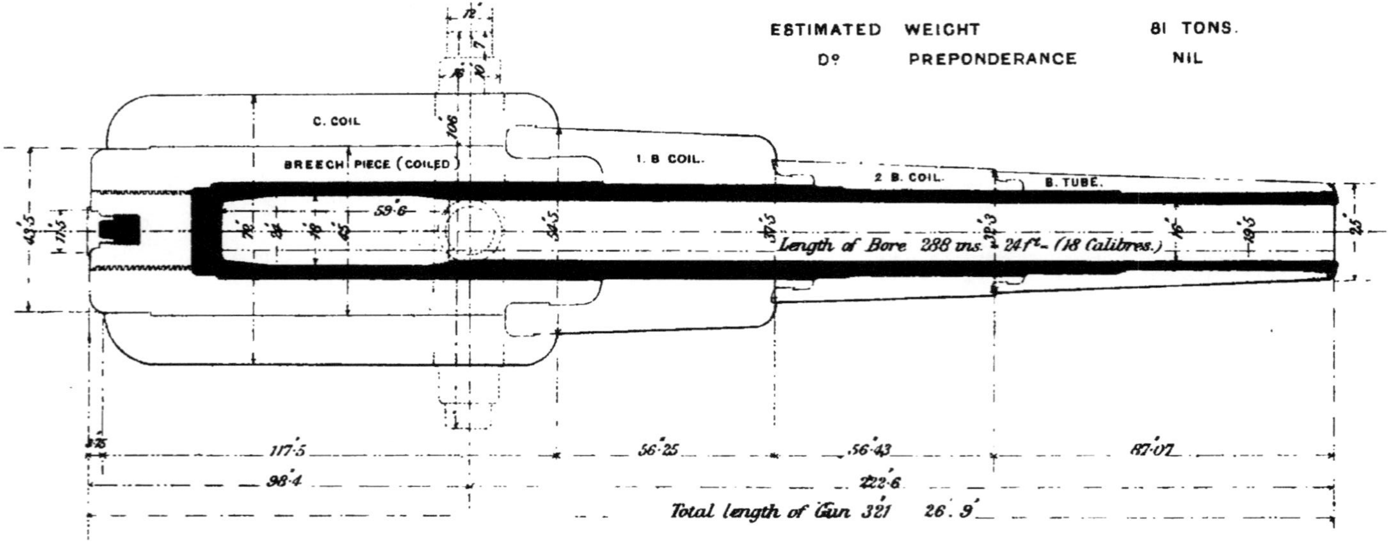

Pl. XIX

ORDNANCE WROT IRON RIFLED MUZZLE-LOADING 17·72 INCH 100 TONS.

Scale ¼ inch - 1 foot.

	TONS.	CWT	QRS.	LB.
WEIGHT	102	4	3	24
PREPONDERANCE	4	10	2	0

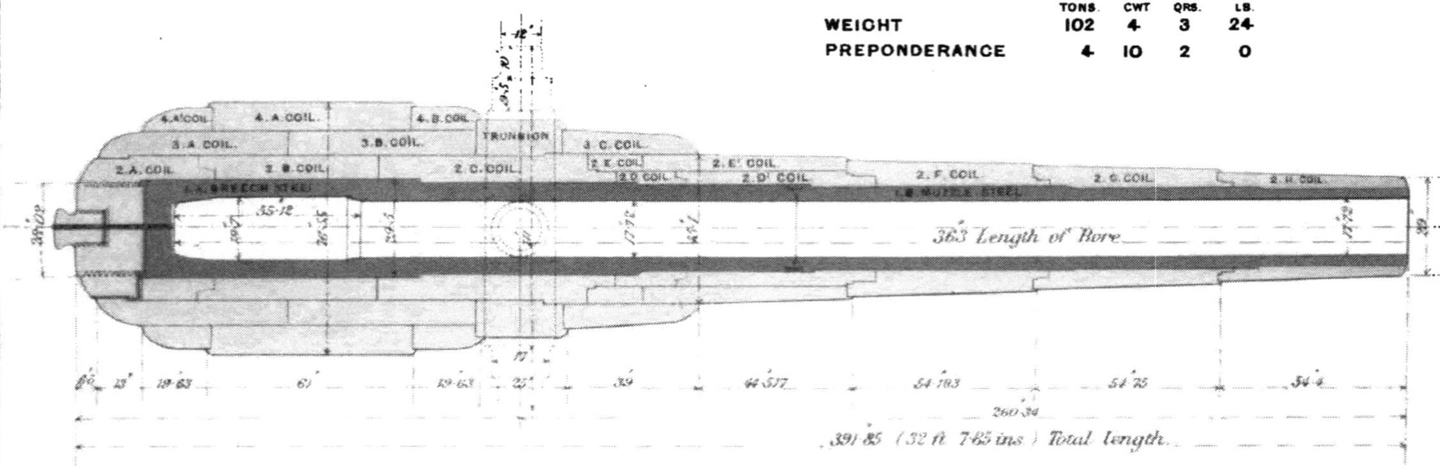

Number of Grooves, 28.

Rifling, an Increasing twist of 1 turn in 150 Calibres at Breech, to 1 turn in 50 Calibres at 2·88 inches from Muzzle. To 2·88 inches a uniform twist of 1 turn in 50 Calibres.

SECTION OF GROOVE.

FULL SIZE.

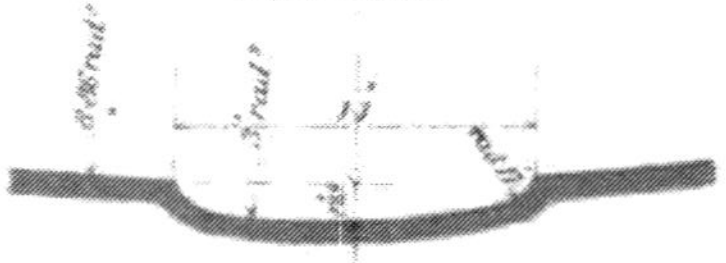

PL XX.

Disturnell & Co. Lith 73 Bedford St Covent Garden

Referred to at p.

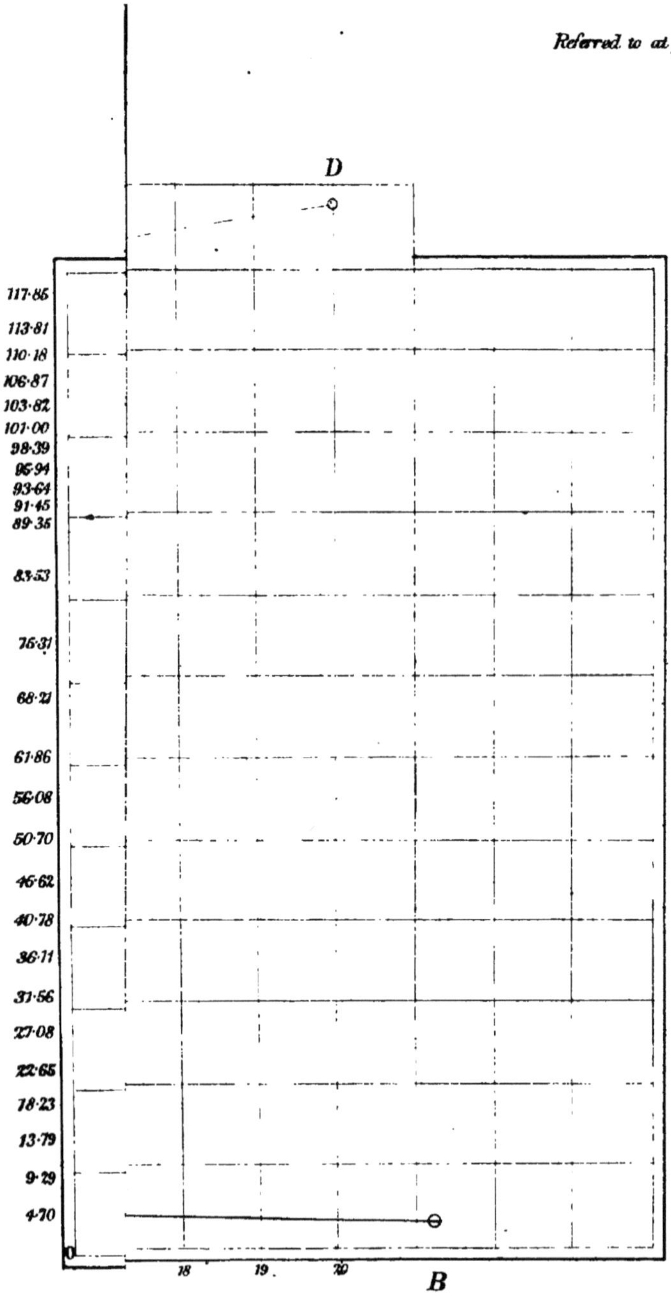

-tridge.
corres-
that

ding.

Sangerfield Lith 2, Bedford St. Covent Garden.

Referred to at p.

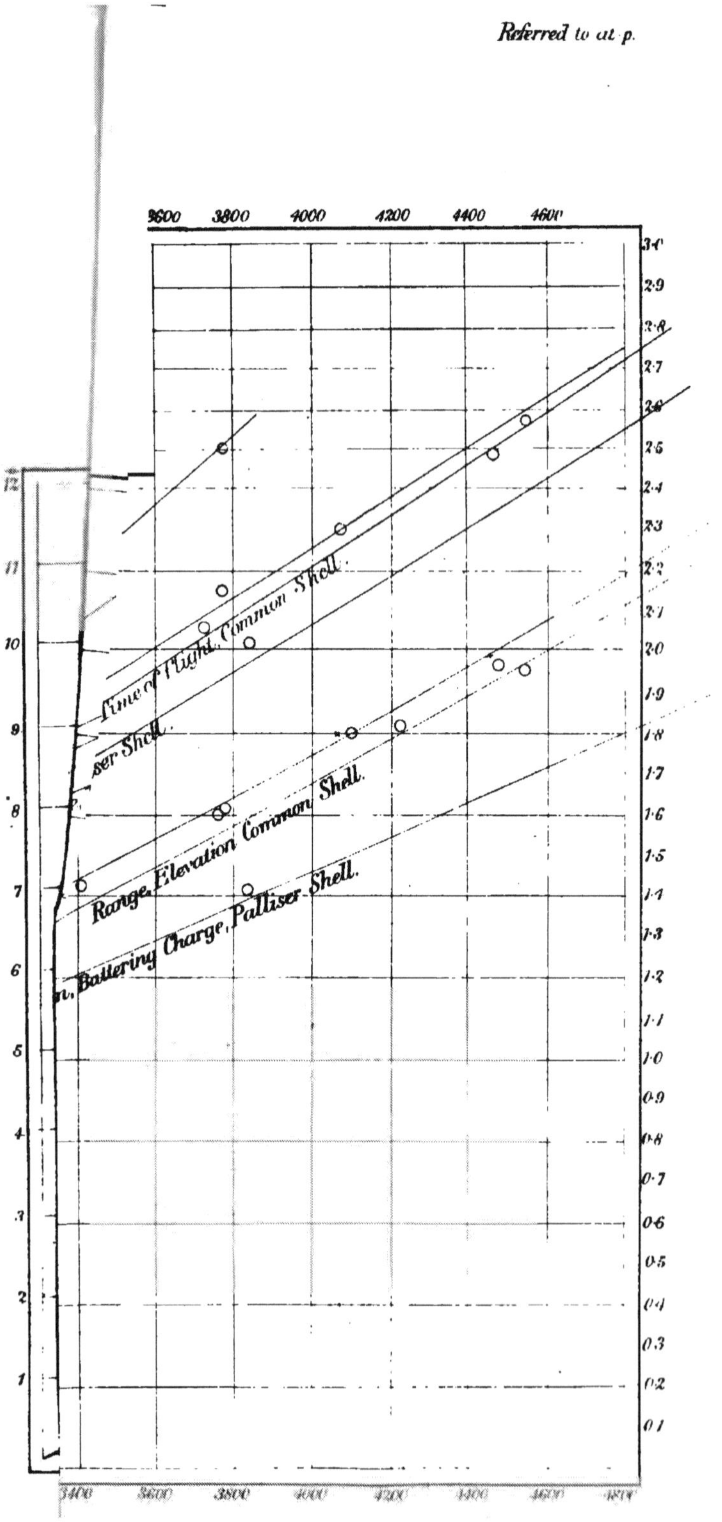

3600
3800
4000
4200
4400
4600
Time of Flight, Common Shell.
ser Shell.
Range, Elevation Common Shell.
n, Battering Charge, Palliser Shell.
12
11
10
9
8
7
6
5
4
3
2
1
3·0
2·9
2·8
2·7
2·6
2·5
2·4
2·3
2·2
2·1
2·0
1·9
1·8
1·7
1·6
1·5
1·4
1·3
1·2
1·1
1·0
0·9
0·8
0·7
0·6
0·5
0·4
0·3
0·2
0·1
3400
3600
3800
4000
4200
4400
4600
4800

98

LONDON:
PRINTED FOR HER MAJESTY'S STATIONERY OFFICE,
BY HARRISON AND SONS,
PRINTERS IN ORDINARY TO HER MAJESTY.
(1,000 10 | 78. 665. Wt. 897.)

Milton Keynes UK
Ingram Content Group UK Ltd.
UKHW051918140823
426877UK00005B/138

9 781016 029612